Lecture Notes in Computer Scie|

Commenced Publication in 1973
Founding and Former Series Editors:
Gerhard Goos, Juris Hartmanis, and Jan van Leeuwen

Hans-Jörg Kreowski   Ugo Montanari
Fernando Orejas   Grzegorz Rozenberg
Gabriele Taentzer (Eds.)

Formal Methods in Software and Systems Modeling

Essays Dedicated to Hartmut Ehrig
on the Occasion of His 60th Birthday

 Springer

Volume Editors

Hans-Jörg Kreowski
University of Bremen, Department of Computer Science
28334 Bremen, Germany
E-mail: kreo@informatik.uni-bremen.de

Ugo Montanari
Università di Pisa, Dipartimento di Informatica
Via F. Buonarroti 2, 56127 Pisa, Italy
E-mail: ugo@di.unipi.it

Fernando Orejas
Universitat Politècnica de Catalunya, Department LSI
Jordi Girona 1-3, 08034 Barcelona, Spain
E-mail: orejas@lsi.upc.es

Grzegorz Rozenberg
Universiteit Leiden, Leiden Institute of Advanced Computer Science
2300 RA Leiden, The Netherlands
E-mail: rozenberg@liacs.nl

Gabriele Taentzer
Technische Universität Berlin, Fakultät IV
Franklinstr. 28/29, 10587 Berlin, Germany
E-mail: gabi@cs.tu-berlin.de

Library of Congress Control Number: 2005920315

CR Subject Classification (1998): F.4.2-3, F.3, D.2, G.2.2, D.3

ISSN 0302-9743
ISBN 3-540-24936-2 Springer Berlin Heidelberg New York

Springer is a part of Springer Science+Business Media

springeronline.com

© Springer-Verlag Berlin Heidelberg 2005
Printed in Germany

Typesetting: Camera-ready by author, data conversion by Olgun Computergrafik
Printed on acid-free paper   SPIN: 11392910   06/3142   5 4 3 2 1 0

Hartmut Ehrig

VI

Preface

This Festschrift is dedicated to Hartmut Ehrig on the occasion of his 60th birthday on December 6, 2004. The contributions discuss various aspects of the formal and visual modeling of software and systems. The authors are some of Hartmut Ehrig's former students and collaborators who are established researchers in their fields. All essays were invited, but they nevertheless went through a reviewing process.

Hartmut Ehrig is a leading, very enthusiastic and highly inspiring scientist who has made lasting contributions to the theoretical foundation of software and system modeling and in particular to graph transformation, algebraic specification and net theory. For more than 30 years, his name is associated with the double-pushout approach, which is the most frequently used and most successful framework in graph transformation. For nearly as long, his work on structuring, parameterization, refinement, and modularization of algebraic specifications has helped to develop this area in an important and sustainable way. Also net theory owes him a very powerful notion and fundamental study of high-level nets. While Hartmut Ehrig is a category theorist and has advocated the use of category theory in most of his research, he has also undertaken many successful efforts to cooperate with researchers in applied areas such as data base systems, software engineering, and even mechanical engineering.

The essays in this book are divided in three parts, each consisting of eight papers: graph transformation, algebraic specification and logic, and formal and visual modeling. Five papers from the first part concern syntactic and semantic aspects of graph transformation (concurrent semantics, interconnection of graph transformation modules, graph processes, graph transformation with variables, and changing labels in the double-pushout approach). The other three papers relate graph transformation with net theory, software engineering, and molecular biology. The papers from the second part address a wide spectrum of topics ranging from data types, coalgebras and interfaces, through functorial semantics of rewrite theories and interactive formal reasoning, to the integration of logics and schema theory. Moreover, one paper relates conditional specifications and interaction charts. The third part contains all further contributions concerning formal and visual modeling including four papers on statechart models, link graphs, architectural connectors for UML, and concurrent object-based systems. Two papers deal with Petri nets considering them as a foundation for a system theory for transportation on one hand and providing them with a loose semantics on the other hand. And the other two papers in this part discuss nested constraints for high-level systems and transformation units with interlinking semantics.

We felt privileged to be able to edit this volume for Hartmut, expressing in this way our admiration for his scientific work and our thanks for his friendship and collaboration. We would like to express our gratitude to all contributors to this volume. We are also indebted to the referees and in particular to Roberto Bruni and Horst Reichel, who served as reviewers without being authors. We are grateful to Peter Knirsch for his support in editing the book and careful unification of all the print files. Very special thanks go to DADARA, who provided the beautiful cover illustration. Finally, we would like to acknowledge the excellent cooperation with Springer, the publisher of this Festschrift.

Dezember 2004

Hans-Jörg Kreowski
Ugo Montanari
Fernando Orejas
Grzegorz Rozenberg
Gabriele Taentzer

Table of Contents

Formal and Visual Modeling

Part I

Graph Transformation

On the Concurrent Semantics
of Algebraic Graph Grammars[*]

Paolo Baldan[1] and Andrea Corradini[2]

[1] Dipartimento di Informatica, Università Ca' Foscari di Venezia, Italy
`baldan@dsi.unive.it`
[2] Dipartimento di Informatica, Università di Pisa, Italy
`andrea@di.unipi.it`

Abstract. Graph grammars are a powerful model of concurrent and distributed systems which can be seen as a proper extension of Petri nets. Inspired by this correspondence, a truly concurrent semantics has been developed along the years for the algebraic approaches to graph grammars, based on Winskel's style unfolding constructions as well as on suitable notions of processes. A basic role is played in this framework by the study of contextual and inhibitor nets, two extensions of ordinary nets which can be seen as intermediate models between ordinary Petri nets and algebraic graph grammars.

This paper presents a survey of these results, discussing in a precise, even if informal way, some of the main technical contributions that made possible the development of such a theory.

Introduction

Petri nets [40, 42] are one of the most widely used models of concurrency. Since their introduction they have attracted the interest of both theoreticians and practitioners. Along the years Petri nets have been equipped with satisfactory semantics, doing justice to their intrinsically concurrent nature. These semantics have served as basis for the development of a variety of modelling and verification techniques. However, the simplicity of Petri nets, which is one of the reasons of their success, represents also a limit in their expressiveness. If one is interested in giving a more structured description of the state, or if the kind of dependencies between steps of computation cannot be reduced simply to causality and conflict, Petri nets are likely to be inadequate.

This paper summarizes the work presented by the authors in a series of papers [2–4, 7–10, 12], most of which written jointly with Ugo Montanari, and some with Nadia Busi, Michele Pinna and Leila Ribeiro. Such papers are the outcome of a project aimed at proposing graph transformation systems as an alternative model of concurrency, extending Petri nets. The basic intuition underlying the use of graph transformation systems for formal specifications is to represent the

[*] Research partially supported by the EU FET-GC Project IST-2001-32747 AGILE, and by the EC RTN 2-2001-00346 SEGRAVIS.

H.-J. Kreowski et al. (Eds.): Formal Methods (Ehrig Festschrift), LNCS 3393, pp. 3–23, 2005.

states of a system as graphs (possibly attributed with data-values) and state transformations by means of rule-based graph transformations. Needless to say, the idea of representing system states by means of graphs is pervasive in computer science. Whenever one is interested in giving an explicit representation of the interconnections, or more generally of the relationships among the various components of a system, a natural solution is to use (possibly hierarchical and attributed) graphs. The possibility of giving a suggestive pictorial representation of graphical states makes them adequate for the description of the meaning of a system specification, even to a non-technical audience. A popular example of graph-based specification language is given by the Unified Modelling Language (UML), but we recall also the more classical Entity/Relationship (ER) approach, or Statecharts, a specification language suited for reactive systems. Moreover, graphs provide a privileged representation of systems consisting of a set of processes communicating through ports.

When one is interested in modelling the *dynamic aspects* of systems whose states have a graphical nature, graph transformation systems are clearly one of the most natural choices. Since a graph rewriting rule has only a local effect on the state, it is natural to allow for the parallel application of rules acting on independent parts of the state, so that a notion of concurrent computation naturally emerges in this context. The research in the field, mainly that dealing with the so-called *algebraic approaches* to graph transformation [25, 22, 27], has led to the attempt of equipping graph grammars with a satisfactory semantical framework, where their truly concurrent behaviour can be suitably described and analyzed. After the seminal work [31], which introduced the notion of *shift equivalence*, many original contributions to the theory of concurrency for algebraic graph transformation systems have been proposed during the last ten years, most of them inspired by their relation with Petri nets. In particular, for the *double-pushout* (DPO) approach to graph transformation, building on some ideas of [31], a *trace semantics* has been proposed in [18, 22]. Resorting to a construction in the style of Mazurkiewicz, the trace semantics has been used to derive an *event structure semantics* [20, 19] for DPO graph grammars. Graph grammars have been endowed also with a process semantics with the introduction of *graph processes* [21], further refined with the notion of concatenable (deterministic) processes [8]. A Winskel's style unfolding construction [51] has been defined both for the *single pushout* (SPO) and the DPO approaches [43, 9, 10], and has been exploited for providing, through suitable chains of functors, such grammars with more abstract semantics based on event structures and domains.

In this survey paper, after recalling the basics of the algebraic approaches to graph transformation and their relationship with Petri nets, we will summarize the functorial, unfolding semantics of Petri nets and the elegant way in which it can be reconciled with the event structure semantics based on deterministic processes. Next we will discuss how this approach has been generalized to algebraic graph grammars. This required the definition of new structures and constructions that will be briefly outlined in the following sections. We shall focus mainly on the definition of two generalizations of prime event structures, called *asymmetric* and *inhibitor* event structures, explaining why they were necessary, and to which extent they made possible to generalize to graph grammars the constructions and results originally developed for Petri nets. It is worth stressing here that this research activity contributed to the theory of Petri nets as well, by generalizing the functorial semantics to the classes of *contextual* and *inhibitor nets*, already introduced in the literature. Such nets can be considered as intermediate models between Place/Transition (P/T) nets and graph grammars.

The rest of this paper is organized as follows. In Section 1 we introduce the DPO and SPO approaches to graph transformation, discussing their relation with Petri nets, and we stress the role of contextual and inhibitor nets as intermediate models between Petri nets and graph grammars. This allows us to organize the mentioned models in an ideal partial order where each model generalizes its predecessors. Then Section 2 outlines the approach to the truly concurrent semantics of ordinary Petri nets which is proposed as a paradigm. Section 3 describes the semantical framework that has been generalized from Petri nets to graph grammars, and Section 4 gives an overview of the results, by explaining how and to what extent such semantical framework has been lifted along the chains of models, first to contextual and inhibitor nets and then to DPO and SPO graph grammars. Finally, Section 5 discusses some open problems and directions of future research.

1   Graph Grammars and Their Relation with Petri Nets

In this section we present the algebraic approaches to graph transformation and we discuss how ordinary Petri nets can be seen as special algebraic graph grammars. The new features with which graph grammars extend ordinary Petri nets establish a close relationship between graph grammars and two generalizations of Petri nets in the literature, i.e., contextual and inhibitor nets.

1.1   The Algebraic Approaches to Graph Transformation

Generally speaking, a graph grammar consists of a start graph together with a set of *graph productions*, i.e., rules of the kind $p : L \rightsquigarrow R$, specifying that, under certain conditions, once an occurrence (a *match*) of the left-hand side L in a graph G has been detected, it can be replaced by the right-hand side R. The form of graph productions, the notion of match, and the mechanisms establishing how a production can be applied to a graph, and what the resulting graph is, depend on the specific graph rewriting formalism.

Here we consider the *algebraic approaches* to graph rewriting [25, 18, 27], where the basic notions of production and direct derivation are defined in terms of constructions and diagrams in a suitable category. Consequently, the resulting theory is very general and flexible, easily adaptable to a very wide range of structures, simply by changing the underlying category.

In the *double-pushout* approach, a graph production consists of a left-hand side graph L, a right-hand side graph R and a (common) interface graph K embedded both in R and in L, as depicted in the top part of Fig. 1. Informally, to apply such a rule to a graph G we must find a *match*, namely an occurrence

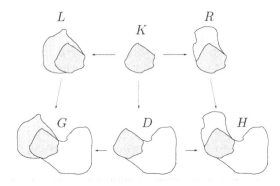

Fig. 1. A (double-pushout) graph rewriting step.

of its left-hand side L in G. The rewriting mechanism first removes the part of the left-hand side L which is not in the interface K producing the graph D, and then adds the part of the right-hand side R which is not in the interface K, thus obtaining the graph H. Formally, this is obtained by requiring the two squares in Fig. 1 to be pushouts in the category of graphs and *total* graph morphisms, hence the name of the approach. The interface graph K is "preserved": it is necessary to perform the rewriting step, but it is not affected by the step itself. Notice that the interface K plays a fundamental role in specifying how the right-hand side has to be glued with the graph D. Working with productions having an empty interface graph K, the expressive power would drastically decrease: only disconnected subgraphs could be added.

In the *single-pushout* approach, a production consists instead of a partial, injective graph morphism $p : L \rightarrowtail R$ from the left- to the right-hand side graph. By looking at the partial morphism p as a span of total morphisms

$$L \hookleftarrow dom(p) \rightarrow R$$

one sees that the domain of p plays here the role of the interface K of DPO rules. To apply such a production to a given match of L in a graph G, i.e., to a *total* morphism $L \rightarrow G$, we have to compute the pushout of $p : L \rightarrowtail R$ and $L \rightarrow G$ in the category of graphs and *partial* graph morphisms.

The most relevant difference between the DPO and SPO approaches (see [27]) is the fact that while the construction of the double pushout diagram may fail if the match $L \rightarrow G$ does not satisfy the so-called *gluing conditions* with respect to the given rule, the construction of the pushout of an SPO rule $L \rightarrowtail R$ and a match $L \rightarrow R$ is always possible. We shall come back on this when relevant in the rest of the paper.

1.2   Relation with Petri Nets

A basic observation belonging to the folklore (see, e.g., [17] and references therein) regards the close relationship existing between graph grammars and

Petri nets. Basically a Petri net can be viewed as a graph transformation system that acts on a restricted kind of graphs, namely discrete, labelled graphs (that can be considered as sets of tokens labelled by places), the productions being the transitions of the net. For instance, Fig. 2 presents a Petri net transition t and the corresponding DPO and SPO graph productions which consume nodes corresponding to two tokens in s_0 and one token in s_1 and produce new nodes corresponding to one token in s_2 and one token in s_3. The interface is empty in the DPO rule and the domain of the morphism is empty in the SPO rule, since nothing is explicitly preserved by a net transition. It is easy to check that both representations satisfy the properties one would expect: the production can be applied to a given marking if and only if the corresponding transition is enabled, and the double or single pushout construction produces the same marking as the firing of the transition.

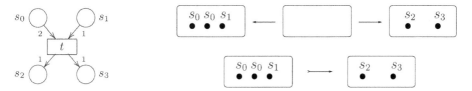

Fig. 2. A Petri net transition and the corresponding DPO and SPO productions.

In this view, general graph transformation systems can be seen as a *proper* extension of ordinary Petri nets in two dimensions:

1. they allow for general productions, possibly with non-empty interface, specifying rewriting steps where a part of the state is *preserved*, i.e., required, but not affected by the rewriting step;
2. they allow for a *more structured description of the state*, that is an arbitrary, possibly non-discrete, graph.

The first capability is essential to give a faithful representation of concurrent accesses to shared resources. In fact, the part of the state preserved in a rewriting step, i.e., the (image of the) interface graph in the DPO or the domain of the production in the SPO approach, can be naturally interpreted as a part of the state which is accessed in a read-only manner by the rewriting step. Coherently with such interpretation, several productions can be applied in parallel sharing (part of) the interface. It is worth remarking that the naïve technique of representing a read operation as a consume/produce cycle may cause a loss of concurrency since it imposes an undesired serialization of the read-only accesses to the shared resource.

As for the second capability, even if multisets may be sufficient in many situations, as already mentioned in the introduction, graphs are more appropriate when one is interested in giving an explicit representation of the interconnections among the various components of the systems, e.g., if one wants to describe the topology of a distributed system and the way it evolves.

These distinctive features of graph grammars establish a link with two extensions of ordinary Petri nets in the literature, introduced to overcome some deficiencies of the basic model: *contextual nets* and *inhibitor nets*.

1.3   Contextual Nets

Contextual nets [37], also called nets with test arcs in [16], activator arcs in [30] or read arcs in [49], extend ordinary nets with the possibility of checking for the presence of tokens which are not consumed. Concretely, besides the usual preconditions and postconditions, a transition of a contextual net has also some *context* conditions, which specify that the presence of some tokens in certain places is necessary to enable the transition, but such tokens are not affected by the firing of the transition. Following [37], non-directed (usually horizontal) arcs are used to represent context conditions: for instance, transition t in the left part of Fig. 3 has place s as context.

Clearly the context of a transition in a contextual nets closely corresponds to the interface graph of a DPO production and to the domain of an SPO production, seen as a partial morphism. As suggested by Fig. 3, a contextual net corresponds to a graph grammar still acting on discrete graphs, but where productions may have a non-empty interface/domain.

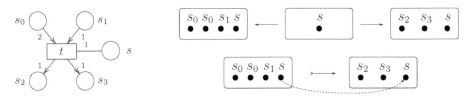

Fig. 3. A contextual Petri net transition and the corresponding DPO and SPO productions.

For their ability of faithfully representing concurrent read-only accesses to shared resources, contextual nets have been used to model the concurrent access to shared data (e.g., for serializability problems for concurrent transactions in a database) [23, 44], to give a concurrent semantics to concurrent constraint programs [14] where several agents access a common store, to model priorities [29] and to compare temporal efficiency in asynchronous systems [49].

1.4   Inhibitor Nets

Inhibitor nets (or nets with *inhibitor arcs*) [1] further generalize contextual nets with the possibility of checking not only for the presence, but also for the *absence* of tokens in a place. For each transition an *inhibitor set* is defined and the transition is enabled only if no token is present in the places of its inhibitor set. When a place s is in the inhibitor set of a transition t we say that s *inhibits* (the firing of) t. The fact that a place s inhibits a transition t is graphically represented by drawing a dotted line from s to t, ending with an empty circle, as shown in the left part of Fig. 4.

While, at a first glance, this could seem a minor extension, it definitely increases the expressive power of the model. In fact, many other extensions of ordinary nets, like nets with reset arcs or prioritized nets, can be simulated in a direct way by using nets with inhibitor arcs (see, e.g., [39]). Indeed the crucial observation is that ordinary nets can easily simulate all the operations of RAM machines, with the exception of the *zero-testing*. Enriching nets with inhibitor arcs is the simplest extension which allows to overcome this limit, thus giving the model the computational power of Turing machines.

Fig. 4. Correspondence between inhibitor Petri nets and DPO graph grammars.

In this case the relation with algebraic graph grammars is less straightforward, and it only concerns the DPO approach. We must recall that in a graph transformation system each rewriting step is required to preserve the consistency of the graphical structure of the state, namely each step must produce a well-defined graph. Hence, as required by a part of the application condition of the DPO approach, the so-called *dangling condition*, a production q which removes a node n cannot be applied if there are edges having n as source or target, which are not removed by q: in fact, such edges would remain *dangling* in the resulting graph. In other words the presence of such edges *inhibits* the application of q. This is informally illustrated by Fig. 4, where place s which inhibits transition t in the left part, becomes an edge which would remain dangling after the execution of t, in the right part. As in the case of contextual nets, this intuitive relation can be made formal, but here, for lack of space, we cannot give the details of the correspondence.

It is worth stressing, again informally, that in the SPO approach the dangling condition is not necessary. By the nature of pushouts in the category of graphs and partial morphisms, a rule which deletes a node can be applied to any match of its left-hand side: any edge attached to that node is automatically erased by the construction, as a kind of side-effect.

2   Truly Concurrent Semantics of Petri Nets

Along the years Petri nets have been equipped with several semantics, aimed at describing, at the right degree of abstraction, the truly concurrent nature of their computations. The approach that we propose as a paradigm, comprises the semantics based on *deterministic processes*, whose origin dates back to an early proposal by Petri himself [41] and the semantics based on the *nondeterministic unfolding*, introduced in a seminal paper by Nielsen, Plotkin and Winskel [38], and shows how the two may be reconciled in a satisfactory framework.

2.1 Deterministic Process Semantics

The notion of *deterministic process* naturally arises when trying to give a truly concurrent description of net computations, taking explicitly into account the *causal* dependencies ruling the occurrences of events in *single* computations.

The prototypical example of Petri net process is given by the *Goltz-Reisig processes* [28]. A Goltz-Reisig process of a net N is a (deterministic) occurrence net O, i.e., a finite net enjoying suitable acyclicity and conflict freeness properties, plus a mapping to the original net $\varphi : O \rightarrow N$. The flow relation induces a partial order on the elements of the net O, which can be naturally interpreted as causality. The mapping essentially labels places and transitions of O with places and transitions of N, in such a way that places in O can be thought of as tokens in a computation of N and transitions of O as occurrences of transition firings in such computation. For instance, Fig. 5 depicts a Petri net and a deterministic process of such a net, representing the sequential execution of two occurrences of t_1 followed by t_2, in parallel with t_3.

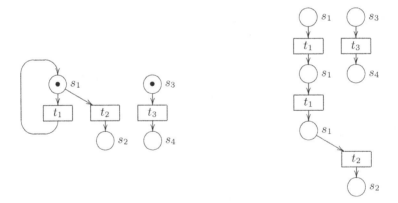

Fig. 5. A Petri net and a deterministic process for the net.

A refinement of Goltz-Reisig processes, the so-called *concatenable processes* [24], form the arrows of a category $\mathbf{CP}[N]$, where objects are markings (states of the net) and arrow composition models the sequential composition of computations. It turns out that such category is a *symmetric monoidal category*, in which the tensor product represents faithfully the parallel composition of processes.

2.2 Unfolding Semantics

A deterministic process represents only a single, deterministic computation of a net. Nondeterminism is captured implicitly by the existence of several different "non confluent" processes having the same source. An alternative classical approach to the semantics of Petri nets is based on an *unfolding construction*, which maps each net into a single branching structure, representing all the possible events that can occur in all the possible computations of the net and the relations existing between them. This structure expresses not only the causal ordering between the events, but also gives an explicit representation of the branching (choice) points of the computations.

In the seminal work of Nielsen, Plotkin and Winskel [38], the denotation of a *safe net* is a *coherent finitary prime algebraic Scott domain* [47] (briefly *domain*), obtained via a construction which first unfolds the net into a (nondeterministic) occurrence net which is then abstracted to a prime event structure, which, finally, gives rise to a domain. Building on such result, Winskel [51] proves the existence of a chain of categorical coreflections (a particularly nice kind of adjunction), leading from the category **S-N** of safe (marked) P/T nets to the category **Dom** of finitary prime algebraic domains, through the categories **O-N** of occurrence nets and **PES** of prime event structures.

$$\textbf{S-N} \xrightleftharpoons[\mathcal{U}]{\overset{\mathcal{I}_{Occ}}{\underset{\bot}{\longleftarrow}}} \textbf{O-N} \xrightleftharpoons[\mathcal{E}]{\overset{\mathcal{N}}{\underset{\bot}{\longleftarrow}}} \textbf{PES} \xrightleftharpoons[\mathcal{L}]{\overset{\mathcal{P}}{\underset{\sim}{\longleftarrow}}} \textbf{Dom}$$

The first step unwinds a safe net N into a *nondeterministic occurrence* net $\mathcal{U}(N)$, which can be seen as a "complete" nondeterministic process of the net N, representing in its branching structure *all* the possible computations of the original net N. The construction exploits the fact that in a safe Petri net, a specific occurrence of a transition t can be identified uniquely by its *history*, namely by the finite set of transition occurrences starting from the initial marking which are strictly necessary to enable to considered occurrence of t. Any two distinct transition occurrences t_1 and t_2 can be related in four possible, mutually exclusive, ways:

1. t_2 is *causally dependent* on t_1 (denoted $t_1 < t_2$) if any computation including t_2 includes also t_1;
2. t_1 is causally dependent on t_2 ($t_2 < t_1$) in the symmetric case;
3. t_1 and t_2 are in *conflict* ($t_1 \# t_2$) if they do not appear together in any computation;
4. t_1 and t_2 are *concurrent* if none of the previous conditions holds.

The relations of causality and conflict are easily shown to be generated by the *direct causality*, which relates a transition occurrence which produces a token with all those which consume it, and by the *direct conflict*, which relates two transition occurrences which would consume the same token. The occurrence net $\mathcal{U}(N)$ obtained as the unfolding of a safe net N records exactly all this information.

The subsequent step abstracts such occurrence net to a *prime event structure* (PES). The PES is obtained from the unfolding simply by forgetting the places, and remembering only the transition occurrences and the causality and conflict

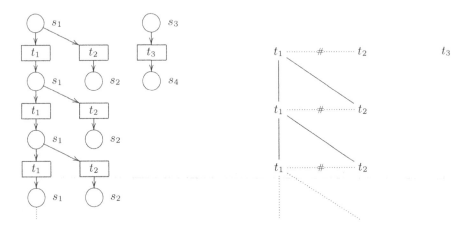

Fig. 6. Unfolding and event structure semantics of Petri nets.

relations among them. From a prime event structure E it is possible to generate freely an occurrence net $\mathcal{N}(E)$ which is the "most general" among those having E as underlying PES. Such a net is obtained by considering the events of E as transition occurrences, and introducing, among others, one fresh place for every pair of events related by causality or conflict in E, in order to enforce the same relationships in $\mathcal{N}(E)$.

The last step (which establishes an equivalence between the category of prime event structures and the category of domains) maps any event structure to its domain of configurations. Fig. 6 presents the unfolding and event structure corresponding to the net in Fig. 5.

In [36] it has been shown that essentially the same construction applies to the category of *semi-weighted* nets, i.e., P/T nets in which the initial marking is a set and transitions can generate at most one token in each post-condition. Besides strictly including safe nets, semi-weighted nets also offer the advantage of being characterized by a "static condition", not involving the behaviour but just the structure of the net.

2.3 Reconciling Deterministic Processes and Unfolding

Since the unfolding is essentially a "maximal" nondeterministic process of a net, one would expect the existence of a clear relation between the unfolding and the deterministic process semantics. Indeed, as shown in [35], the domain associated to a net N through the unfolding construction can be equivalently characterized as the set of deterministic processes of the net starting from the initial marking, endowed with a kind of prefix ordering. This result is stated in an elegant categorical way by resorting to concatenable processes. Given a (semi-weighted) net N with initial marking m, the comma category $\langle m \downarrow \mathbf{CP}[N] \rangle$ is shown to be a preorder, whose elements are intuitively finite computations starting from the initial state, and if φ_1 and φ_2 are elements of the preorder, $\varphi_1 \preceq$

φ_2 when φ_1 can evolve to φ_2 by performing appropriate steps of computation. Then the ideal completion of such preorder, which includes also the infinite computations of the net, is shown to be isomorphic to the domain generated from the unfolding.

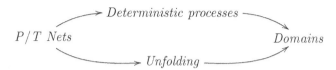

3 Concurrent Semantics: From Nets to Graph Grammars

In this section, guided by the relationship between graph grammars and Petri nets, we describe the way the semantical framework described in the previous section has been generalized to graph grammars.

The main complications which arise in the treatment of graph grammars are related to the possibility of expressing rewritings where part of the state is preserved and, just for the DPO approach, to the need of preserving the consistency of the graphical structure of the state, a constraint which leads to the mentioned "inhibiting effects" between production applications. Therefore, not surprisingly, contextual and inhibitor nets play an essential role in the extension in that they offer a technically simple framework, where problems which are conceptually relevant to graph grammars can be studied in isolation.

Intuitively, we can organize the considered formalisms in an ideal partial ordering leading from Petri nets to graph transformation systems

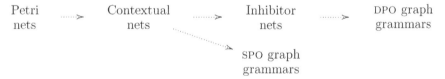

and for each one of such formalisms we develop a similar theory by following a common schema which can be summarized as follows:

1. We define a *category of systems* **Sys**, where morphisms, which basically origin from an algebraic view of the systems, can be interpreted as simulations.
2. We develop an *unfolding semantics*, expressed as a coreflection between **Sys** and a subcategory **O-Sys**, where objects, called "occurrence" systems, are suitable systems exhibiting an acyclic behaviour. From the unfolding we extract an (appropriate kind of) *event structure*, the transformation being expressed as a functor from **O-Sys** to the considered category of event structures **ES**. In the case of contextual nets and of SPO grammars this functor establishes a coreflection between **O-Sys** and **ES**. Finally, a connection is established with domains and PES by showing that the category **ES** of generalized event structures coreflects into the category **Dom** of domains.

Summing up, we obtain the following chain of functors, leading from systems to event structures and domains

$$\mathbf{Sys} \xrightleftharpoons[\perp]{} \mathbf{O\text{-}Sys} \longrightarrow \mathbf{ES} \xrightleftharpoons[\perp]{} \mathbf{Dom} \xrightleftharpoons[\sim]{} \mathbf{PES}$$

The last step in the chain is the equivalence between the categories **Dom** of domains and **PES** of prime event structures, due to Winskel.

3. We introduce a notion of *deterministic process* for systems in **Sys**. Relying on the work in point (2), a general (possibly nondeterministic) *process* of a system $\mathcal{S}$ is defined as an "occurrence system" in **O-Sys**, plus a (suitable kind) of morphism back to the original system $\mathcal{S}$ (the prototypical example of nondeterministic process being the unfolding). Then, roughly speaking, a process is *deterministic* if it contains no conflict, or, in other words, if the corresponding event structure has a configuration including all the events. The deterministic processes of a system $\mathcal{S}$ are turned into a category $\mathbf{CP}[\mathcal{S}]$, by endowing them with a notion of concatenation, modelling the sequential composition of computations.

4. We show that the deterministic process and the unfolding semantics can be reconciled by proving that, as for ordinary nets, the comma category $\langle Initial\ State \downarrow \mathbf{CP}[\mathcal{S}]\rangle$, is a preorder whose ideal completion is isomorphic to the domain obtained from the unfolding, as defined at point (2).

It is fair to point here that the steps (3) and (4) above have not been completely worked out for SPO grammars.

Observe that, differently from what happens for ordinary nets, the unfolding semantics (essentially based on nondeterministic processes) is defined before developing a theory of deterministic processes. To understand why, note that for ordinary nets the only source of nondeterminism is the the presence of pairs of different transitions with a common precondition, and therefore there is an obvious notion of "deterministic net". When considering contextual nets, inhibitor nets or graph grammars the situation becomes less clear: the dependencies between event occurrences cannot be described only in terms of causality and conflict, and the deterministic systems cannot be given a purely syntactical characterization. Consequently, a clear understanding of the structure of nondeterministic computations becomes essential to be able to single out which are the good representatives of deterministic computations.

4   Some Insights into the Technical Problems

For each one of the considered models the core of the developed theory is point (2) and, more specifically, the formalization of the kind of dependencies among events which can occur in their computations. As mentioned above, such dependencies cannot be faithfully reduced to causality and conflict and thus appropriate generalizations of Winskel's event structures must be defined. Next we give some more details on the specific problems that we found for each formalism and on the way we decided to face them.

Grammar $\mathcal{G}_1$

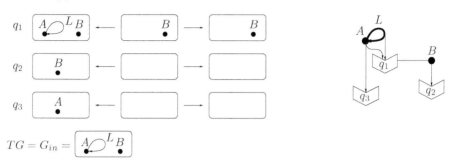

$$TG = G_{in} = \boxed{A \, \overset{L}{\circlearrowright} \, B}$$

Fig. 7. The safe graph grammar $\mathcal{G}_1$ and its net-like representation.

4.1   From Prime to Asymmetric Event Structures

In the case of algebraic graph grammars, both in the DPO and in the SPO approaches, the presence of a context in a production, i.e., of items that are needed for the application of a production but which are not consumed, introduces a new kind of dependency among production occurrences, making prime event structures not completely satisfactory as a semantic domain.

As an example, consider the (typed) DPO graph grammar $\mathcal{G}_1$ of Figure 7. On the left-hand side, the grammar is represented as usual in the DPO approach, consisting of a set of productions (spans of injective graph morphisms) and a start graph, all of them typed over the type graph TG (i.e., equipped with a homomorphism to TG), which, in this case, coincides with the start graph G_{in}. On the right-hand side, a net-like pictorial representation of the same grammar is shown, where the productions and the items of the type graph play the rôle of transitions and of places of a Petri net, respectively. This net-like representation can be given for *strongly safe* graph grammars, which, intuitively, are the graph grammar counterpart of safe nets (see [10] for more details).

Let us focus on productions q_1 and q_2. Both are applicable to the start graph G_{in}. But notice that if we apply q_2 first, then q_1 cannot be applied anymore because q_2 deletes node B; on the other hand, if we apply q_1 first, then q_2 can still be applied. This phenomenon has been extensively studied for *contextual nets*. For example, in the net N_1 of Fig. 8(a), transitions t_1 and t_2 play the same rôle as productions q_1 and q_2 of the above grammar, and place s, which is a context of q_1 and a precondition of q_2, is like node B above.

The possible firing sequences in net N_1 are given by the firing of t_1, the firing of t_2, and the firing of t_1 followed by t_2, denoted $t_1; t_2$, while $t_2; t_1$ is not allowed. This situation cannot be modelled in a direct way within a prime event structure: t_1 and t_2 are neither in conflict nor concurrent nor causally dependent. Simply, as for an ordinary conflict, the firing of t_2 prevents t_1 to be executed, so that t_1 can never follow t_2 in a computation, but the converse is not true, since t_2 *can* fire after t_1. This situation can be interpreted naturally as an *asymmetric conflict* between the two transitions. Equivalently, since t_1 precedes

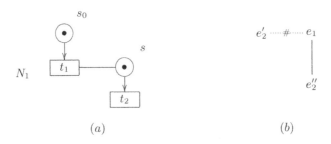

Fig. 8. A simple contextual net and a prime event structure representing its behaviour.

t_2 in any computation where both transitions fire, t_1 acts as a cause of t_2 in such computations. However, differently from a true cause, t_1 is not necessary for t_2 to be fired. Therefore we can also think of the relation between the two transitions as a *weak* form of *causality*.

A possible way to encode this situation in a PES is to represent the firing of t_1 with an event e_1 and the firing of t_2 with two distinct mutually exclusive events: e_2', representing the execution of t_2 that prevents t_1, thus in conflict with e_1, and e_2'', representing the execution of t_2 after t_1, thus caused by e_1. Such PES is depicted in Fig. 8.(b), where causality is represented by a plain arrow and conflict is represented by a dotted line, labelled by #. However, this solution is not completely satisfactory with respect to the interpretation of contexts as "read-only resources": since t_1 just reads the token in s without changing it, one would expect the firing of t_2, preceded or not by t_1, to be represented by a single event.

In order to provide a more direct, event based representation of contextual net computations, *asymmetric event structure* (AES) were introduced in [7]. An AES, besides of the usual causality relation $\leq$ of a prime event structure, has a relation $\nearrow$, called the *asymmetric conflict relation*, that allows one to specify the new kind of dependency described above simply as $t_1 \nearrow t_2$. Informally, in an AES each event has a set of "strong" causes (given by the causality relation) and a set of weak causes (due to the presence of the asymmetric conflict relation). To be fired, each event must be preceded by all strong causes and by a (suitable) subset of the weak causes. Therefore, differently from PES's, an event of an AES can have more than one history. Moreover the usual symmetric binary conflict $e\#e'$ can be represented easily by using cycles of asymmetric conflicts: if $e \nearrow e'$ and $e' \nearrow e$ then clearly e and e' can never occur in the same computation, since each one should precede the other.

The main result of [7] shows that Winskel's functorial semantics for safe nets can be generalized to the following, similar chain of adjunctions for contextual nets, where asymmetric event structures play a central rôle.

| Semi-weighted Contextual Nets | $\xleftarrow{\quad}$ $\underset{\mathcal{U}_a}{\overset{\bot}{\rightleftarrows}}$ | Occurrence Contextual Nets | $\underset{\mathcal{E}_a}{\overset{\mathcal{N}_a}{\rightleftarrows}}$ $\bot$ | Asymmetric Event Structures | $\underset{\mathcal{L}_a}{\overset{\mathcal{P}_a}{\rightleftarrows}}$ $\bot$ | Domains |

Grammar $\mathcal{G}_2$

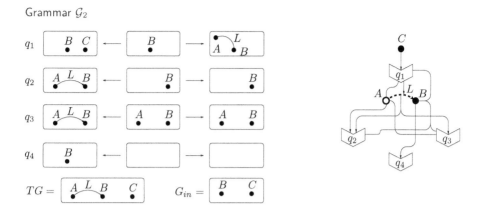

Fig. 9. Graph grammar $\mathcal{G}_2$ and its net-like representation.

4.2   From Asymmetric to Inhibitor Event Structures

Unfortunately, AES's are not yet sufficient to capture all relationships among the production occurrences in a DPO graph grammar. Consider grammar $\mathcal{G}_2$ of Figure 9 (in the net-like representation, on the right-hand side, nodes with empty interior and dashed edges can be seen as empty places). More specifically, let us focus on the relationships among the various productions. Notice that q_4 can be applied to the start graph G_{in} consisting of nodes B and C, but if we first apply q_1, then the application of q_4 is prevented by the dangling condition: removing the node B would leave edge L without its target node, so, basically, q_4 cannot be applied for ensuring a structural property of the state. Production q_4 could be applied again if we first delete edge L, by applying production q_2 or q_3.

Such complex relationships have been analyzed in depth for *inhibitor Petri nets*. Consider the inhibitor net N_2 in Fig. 10 where the place s, which inhibits transition t, is in the post-set of transition t' and in the pre-set of t_0. The execution of t' inhibits the firing of t, which can be enabled again by the firing of t_0. Thus t can fire before or after the "sequence" $t'; t_0$, but not in between the two transitions. Roughly speaking there is a sort of atomicity of the sequence $t'; t_0$ with respect to t. The situation can be more involved since many transitions t_0, ..., t_n may have the place s in their pre-set (see the net N_3 in Fig. 10). Therefore, after the firing of t', the transition t can be re-enabled by any of the conflicting transitions t_0, ..., t_n. This leads to a sort of *or*-causality, but only when t fires after t'. With a logical terminology we can say that t causally depends on the implication $t' \Rightarrow t_0 \vee t_1 \vee \ldots \vee t_n$.

In order to model these complex relationships in a direct way, a generalization of PES's and AES's has been introduced, called *inhibitor event structures* (IES's). A IES is equipped with a ternary relation, called *DE-relation (disabling-enabling relation)* and denoted by $\vdash\!\!\circ\ (\cdot, \cdot, \cdot)$, which allows one to model the dependencies between transitions in N_3 as $\vdash\!\!\circ\ (\{t'\}, t, \{t_0, \ldots, t_n\})$. It is possible to show that the DE-relation is sufficient to represent both causality and asymmetric conflict and thus, concretely, it is the only relation of an IES.

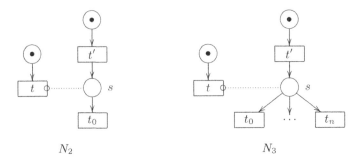

Fig. 10. Two inhibitor nets.

Using inhibitor event structures and the DE-relation as basic tools, a functorial semantics in Winskel's style has been proposed for (semi-weighted) inhibitor nets in [4,3] as summarised by the following diagram:

$$\text{Semi-weighted} \xleftarrow{\quad} \text{Occurrence} \qquad \text{Inhibitor Event} \xleftarrow{\quad \mathcal{P}_i \quad} \text{Domains}$$

Besides of the fact that inibitor event structures replace asymmetric ones, even at this level of abstraction, it is possible to see another relevant difference between the functorial semantics of semi-weighted inhibitor nets and the simpler case of contextual nets. In fact, the functor from the category of inhibitor occurrence nets to the category of IES's *does not have* a left adjoint and thus the whole semantic transformation is not expressed as a coreflection. Indeed, by making only very mild assumptions, it has been shown in [3] that such a left adjoint does not exist, essentially because of the presence of a restricted kind of or-causality in inhibitor occurrence nets.

4.3   Lifting the Results to DPO Graph Grammars

When we finally turn our attention to DPO graph grammars we are rewarded of the effort spent on generalized Petri nets, since basically nothing new has to be invented. Inhibitor event structures are expressive enough to model the structure of DPO graph grammar computations and the theory developed for inhibitor nets smoothly lifts, at the price of some technical complications, to DPO grammars. Furthermore, not only the process and the unfolding semantics developed for DPO graph grammars are shown to agree, but also they have been shown to be consistent with the classical theory of concurrency for DPO grammar in the literature, basically relying on *shift-equivalence*. More specifically:

1. A Winskel's style semantics for DPO graph grammars is presented in [2, 9, 10], as summarized by the following diagram:

$$\text{DPO Graph} \xleftarrow{\quad} \text{DPO Occurrence} \qquad \text{Inhibitor Event} \xleftarrow{\quad \mathcal{P}_i \quad} \text{Domains}$$

The unfolding construction associates to each graph grammar a nondeterministic occurrence grammar describing its behaviour. Such a construction establishes a coreflection between suitable categories of DPO grammars and the category of occurrence grammars. The unfolding is then abstracted to an inhibitor event structure and finally to a prime algebraic domain (or equivalently to a prime event structure).

2. *Nondeterministic graph processes* are introduced in [2,10], generalizing the deterministic processes of [21]. The notion fits nicely in the theory since a graph process of a DPO grammar $\mathcal{G}$ is defined simply as a (special kind of) grammar morphism from an occurrence grammar to $\mathcal{G}$, while in [21] an *ad hoc* mapping was used.

3. *Concatenable graph processes* are introduced in [8], as a variation of deterministic, finite processes, endowed with an operation of concatenation which models sequential composition of computations. The appropriateness of this notion is confirmed by the fact that the category $\mathbf{CP}[\mathcal{G}]$ of concatenable processes of a DPO grammar $\mathcal{G}$ turns out to be isomorphic to the truly concurrent model of computation of $\mathcal{G}$, as defined in [22] using the classical notions of shift-equivalence and of traces.

4. The event structure obtained via the unfolding is shown in [9] to coincide both with the one defined in [20] via a comma category construction on the category of concatenable derivation traces, and with the one proposed in [46], based on a deterministic set-theoretical variant of the DPO approach. These results, besides confirming the appropriateness of the proposed unfolding construction, give an unified view of the various event structure semantics for the DPO approach to graph transformation.

4.4   Unfolding Semantics of SPO Graph Grammars

Recently, a Winskel's style unfolding semantics has been developed for the SPO approach to graph transformation as well, as reported in [12]. Apart from the technical differences in the way rules are defined, the main difference with respect to the DPO approach lies in the fact that there are no conditions on rule applications, i.e., whenever a match is found the corresponding rule can always be applied.

It turned out that a coreflective unfolding semantics for SPO graph grammars can be defined, leading to the following chain of adjunctions:

$$\textbf{SPO Graph Grammars} \xleftarrow[\;\mathcal{U}_s\;]{} \textbf{SPO Occurrence Grammars} \xleftarrow[\;\mathcal{E}_s\;]{\mathcal{N}} \textbf{Asymmetric Event Structures} \xleftarrow[\;\mathcal{L}_a\;]{\mathcal{P}_a} \textbf{Domains}$$

The first step of the above diagram is obtained as a slight variation of the unfolding construction for SPO grammars proposed in [43]. The rest of the construction differs from and improves that for DPO graph grammars recalled above, for the following facts:

– Due to the absence of the dangling condition, asymmetric event structures are sufficient to represent adequately the dependencies among production occurrences: inhibitor event structures are not necessary.

– A novel construction, inspired by the work on contextual nets [7], allows us to build a canonical occurrence SPO graph grammar $\mathcal{N}(A)$ from any given asymmetric event structure A. This provides the left-adjoint functor (indeed a coreflection) which is missing in the corresponding chain for inhibitor nets and DPO grammars. Given an asymmetric event structure A, the corresponding grammar has the events of A as productions. The graph items are freely generated in order to induce the right kind of dependencies between events. More specifically, first the nodes of the graph are freely generated according to the dependencies in A. Then for any pair of nodes, edges connecting the two nodes are freely generated according to the dependencies in A and the specific restrictions of the SPO rewriting mechanism.

5   Conclusions

In this paper we surveyed several results proposed by our coauthors and ourselves in a series of papers, contributing to the development of a systematic theory of concurrency for algebraic graph grammars, aimed at closing the existing gap between graph transformation systems and Petri nets. A second achievement of this research activity is the development of an analogous unifying theory for two widely diffused generalizations of Petri nets, namely contextual and inhibitor nets. In fact, while a theory of deterministic processes for these kind of nets was already available in the literature (see, e.g., [37, 15]), the Winskel-style semantics, comprising the unfolding construction, its abstraction to a prime algebraic domain semantics, as well as its relation with the deterministic process semantics were missing.

The truly concurrent semantics for graph grammars (and generalized nets) is intended to represent the basis for defining more abstract observational semantics to be used for the analysis and verification of the modelled systems. For instance, the notions of process and of event structure associated to a process naturally lead to the definition of a behavioural equivalence, called *history preserving bisimulation* (HP-bisimulation) [48], which, differently from ordinary bisimulation, takes into account the properties of concurrency of the system. A generalization of this approach to graph grammars has been proposed in [11].

The unfolding semantics of Petri nets has been used successfully for the analysis of finite-state systems: as shown in [34], a finite fragment can be extracted from the (possibly infinite) unfolding, which is still useful to study some relevant properties of the system, like reachability, deadlock freeness, and liveness and concurrency of transitions. Such an approach has been extended to contextual nets in [50]. Inspired by this line of research, recently we started to develop, in joint works with Barbara and Bernhard König, a methodology for the verification of algebraic graph grammars using finite approximations of the unfolding, and a suitable graph logic for expressing relevant properties. Papers [5, 13] address the verification of possibly infinite-state systems, while [6] is more closely related to [34] as it consider finite-state systems only.

Finally, although we considered only graph rewriting acting on directed (typed) graphs, it would be interesting to understand if the presented constructions and results can be extended to more general structures. While the generalization to hypergraphs looks trivial, developing a similar theory for more general structures and for abstract categories (e.g., High Level Replacement Systems [26], or the recently introduced Adhesive Categories [32]) is not immediate and represents an interesting topic of further investigation.

References

1. T. Agerwala and M. Flynn. Comments on capabilities, limitations and "correctness" of Petri nets. *Computer Architecture News*, 4(2):81–86, 1973.

2. P. Baldan. *Modelling concurrent computations: from contextual Petri nets to graph grammars*. PhD thesis, Department of Computer Science, University of Pisa, 2000. Available as technical report n. TD-1/00.

3. P. Baldan, N. Busi, A. Corradini, and G.M. Pinna. Domain and event structure semantics for Petri nets with read and inhibitor arcs. *Theoretical Computer Science*, to appear, 2004.

4. P. Baldan, N. Busi, A. Corradini, and G.M. Pinna. Functorial concurrent semantics for Petri nets with read and inhibitor arcs. In C. Palamidessi, editor, *CONCUR'00 Conference Proceedings*, volume 1877 of *LNCS*, pages 442–457. Springer Verlag, 2000.

5. P. Baldan, A. Corradini, and B. König. A static analysis technique for graph transformation systems. In *Proc. of CONCUR 2001*, pages 381–395. Springer, 2001. LNCS 2154.

6. P. Baldan, A. Corradini, and B. König. Veryfing Finite-State Graph Grammars: an Unfolding-Based Approach. In *Proc. of CONCUR 2004*, pages 83–98. Springer, 2004. LNCS 3170.

7. P. Baldan, A. Corradini, and U. Montanari. Contextual Petri nets, asymmetric event structures and processes. *Information and Computation*, 171(1):1–49, 2001.

8. P. Baldan, A. Corradini, and U. Montanari. Concatenable graph processes: relating processes and derivation traces. In *Proceedings of ICALP'98*, volume 1443 of *LNCS*, pages 283–295. Springer Verlag, 1998.

9. P. Baldan, A. Corradini, and U. Montanari. Unfolding and Event Structure Semantics for Graph Grammars. In W. Thomas, editor, *Proceedings of FoSSaCS '99*, volume 1578 of *LNCS*, pages 73–89. Springer Verlag, 1999.

10. P. Baldan, A. Corradini, and U. Montanari. Unfolding of double-pushout graph grammars is a coreflection. In G. Ehrig, G. Engels, H.J. Kreowsky, and G. Rozemberg, editors, *TAGT'98 Conference Proceedings*, volume 1764 of *LNCS*, pages 145–163. Springer Verlag, 1999.

11. P. Baldan, A. Corradini, and U. Montanari. Bisimulation Equivalences for Graph Grammars. In *Formal and Natural Computing*, W. Brauer, H. Ehrig, J. Karhumäki, A. Salomaa eds., number 2300 in LNCS, pages 158–190. Springer Verlag, 2002.

12. P. Baldan, A. Corradini, U. Montanari, and L. Ribeiro. Coreflective Concurrent Semantics for Single-Pushout Graph Grammars. *Proceedings WADT 2002*, volume 2755 of *LNCS*, pages 165–184. Springer Verlag, 2002.

13. P. Baldan and B. König. Approximating the behaviour of graph transformation systems. In A. Corradini, H. Ehrig, H.-J. Kreowski, and G. Rozenberg, editors, *Proceedings of the First International Conference on Graph Transformation (ICGT 2002)*, volume 2505 of *LNCS*, pages 14–30. Springer, 2002.

14. F. Bueno, M. Hermenegildo, U. Montanari, and F. Rossi. Partial order and contextual net semantics for atomic and locally atomic CC programs. *Science of Computer Programming*, 30:51–82, 1998.
15. N. Busi. *Petri Nets with Inhibitor and Read Arcs: Semantics, Analysis and Application to Process Calculi*. PhD thesis, University of Siena, Department of Computer Science, 1998.
16. S. Christensen and N. D. Hansen. Coloured Petri nets extended with place capacities, test arcs and inhibitor arcs. In M. Ajmone-Marsan, editor, *Applications and Theory of Petri Nets*, volume 691 of *LNCS*, pages 186–205. Springer Verlag, 1993.
17. A. Corradini. Concurrent graph and term graph rewriting. In U. Montanari and V. Sassone, editors, *Proceedings of CONCUR'96*, volume 1119 of *LNCS*, pages 438–464. Springer Verlag, 1996.
18. A. Corradini, H. Ehrig, M. Löwe, U. Montanari, and F. Rossi. Abstract graph derivations in the double-pushout approach. In H.-J. Schneider and H. Ehrig, editors, *Proceedings of the Dagstuhl Seminar 9301 on Graph Transformations in Computer Science*, volume 776 of *LNCS*, pages 86–103. Springer Verlag, 1994.
19. A. Corradini, H. Ehrig, M. Löwe, U. Montanari, and F. Rossi. An event structure semantics for safe graph grammars. In E.-R. Olderog, editor, *Programming Concepts, Methods and Calculi*, IFIP Transactions A-56, pages 423–444. North-Holland, 1994.
20. A. Corradini, H. Ehrig, M. Löwe, U. Montanari, and F. Rossi. An event structure semantics for graph grammars with parallel productions. In J. Cuny, H. Ehrig, G. Engels, and G. Rozenberg, editors, *Proceedings of the 5th International Workshop on Graph Grammars and their Application to Computer Science*, volume 1073 of *LNCS*. Springer Verlag, 1996.
21. A. Corradini, U. Montanari, and F. Rossi. Graph processes. *Fundamenta Informaticae*, 26:241–265, 1996.
22. A. Corradini, U. Montanari, F. Rossi, H. Ehrig, R. Heckel, and M. Löwe. Algebraic Approaches to Graph Transformation I: Basic Concepts and Double Pushout Approach. In Rozenberg [45], chapter 3.
23. N. De Francesco, U. Montanari, and G. Ristori. Modeling Concurrent Accesses to Shared Data via Petri Nets. In *Programming Concepts, Methods and Calculi, IFIP Transactions A-56*, pages 403–422. North Holland, 1994.
24. P. Degano, J. Meseguer, and U. Montanari. Axiomatizing the algebra of net computations and processes. *Acta Informatica*, 33:641–647, 1996.
25. H. Ehrig. Tutorial introduction to the algebraic approach of graph-grammars. In H. Ehrig, M. Nagl, G. Rozenberg, and A. Rosenfeld, editors, *Proceedings of the 3rd International Workshop on Graph-Grammars and Their Application to Computer Science*, volume 291 of *LNCS*, pages 3–14. Springer Verlag, 1987.
26. H. Ehrig, A. Habel, H.-J. Kreowski, and F. Parisi-Presicce. Parallelism and concurrency in High-Level Replacement Systems. *Mathematical Structures in Computer Science*, 1:361–404, 1991.
27. H. Ehrig, R. Heckel, M. Korff, M. Löwe, L. Ribeiro, A. Wagner, and A. Corradini. Algebraic approaches to graph transformation II: Single pushout approach and comparison with double pushout approach. In Rozenberg [45], chapter 4.
28. U. Golz and W. Reisig. The non-sequential behaviour of Petri nets. *Information and Control*, 57:125–147, 1983.
29. R. Janicki and M. Koutny. Invariant semantics of nets with inhibitor arcs. In *Proceedings of CONCUR '91*, volume 527 of *LNCS*. Springer Verlag, 1991.
30. R. Janicki and M. Koutny. Semantics of inhibitor nets. *Information and Computation*, 123:1–16, 1995.

31. H.-J. Kreowski. *Manipulation von Graphmanipulationen.* PhD thesis, Technische Universität Berlin, 1977.
32. S. Lack and P. Sobociński. Adhesive categories. In I. Walukiewicz, editor, *Foundations of Software Science and Computation Structures*, volume 2987 of *LNCS*, pages 273–288. Springer, 2004.
33. M. Löwe. Algebraic approach to single-pushout graph transformation. *Theoretical Computer Science*, 109:181–224, 1993.
34. K.L. McMillan. *Symbolic Model Checking.* Kluwer, 1993.
35. J. Meseguer, U. Montanari, and V. Sassone. Process versus unfolding semantics for Place/Transition Petri nets. *Theoretical Computer Science*, 153(1-2):171–210, 1996.
36. J. Meseguer, U. Montanari, and V. Sassone. On the semantics of Place/Transition Petri nets. *Mathematical Structures in Computer Science*, 7:359–397, 1997.
37. U. Montanari and F. Rossi. Contextual nets. *Acta Informatica*, 32(6), 1995.
38. M. Nielsen, G. Plotkin, and G. Winskel. Petri Nets, Event Structures and Domains, Part 1. *Theoretical Computer Science*, 13:85–108, 1981.
39. J.L. Peterson. *Petri Net Theory and the Modelling of Systems.* Prentice-Hall, 1981.
40. C.A. Petri. *Kommunikation mit Automaten.* PhD thesis, Schriften des Institutes für Instrumentelle Matematik, Bonn, 1962.
41. C.A. Petri. Non-sequential processes. Technical Report GMD-ISF-77-5, Gesellshaft für Mathematik und Datenverarbeitung, Bonn, 1977.
42. W. Reisig. *Petri Nets: An Introduction.* EACTS Monographs on Theoretical Computer Science. Springer Verlag, 1985.
43. L. Ribeiro. *Parallel Composition and Unfolding Semantics of Graph Grammars.* PhD thesis, Technische Universität Berlin, 1996.
44. G. Ristori. *Modelling Systems with Shared Resources via Petri Nets.* PhD thesis, Department of Computer Science - University of Pisa, 1994.
45. G. Rozenberg, editor. *Handbook of Graph Grammars and Computing by Graph Transformation. Vol. 1: Foundations.* World Scientific, 1997.
46. G. Schied. On relating rewriting systems and graph grammars to event structures. In H.-J. Schneider and H. Ehrig, editors, *Proceedings of the Dagstuhl Seminar 9301 on Graph Transformations in Computer Science*, volume 776 of *LNCS*, pages 326–340. Springer Verlag, 1994.
47. D. S. Scott. Outline of a mathematical theory of computation. In *Proceedings of the Fourth Annual Princeton Conference on Information Sciences and Systems*, pages 169–176, 1970.
48. R. van Glabbeek and U. Goltz. Equivalence notions for concurrent systems and refinement of actions. In A. Kreczmar and G. Mirkowska, editors, *Proceedings of MFCS'89*, volume 39 of *LNCS*, pages 237–248. Springer Verlag, 1989.
49. W. Vogler. Efficiency of asynchronous systems and read arcs in Petri nets. In *Proceedings of ICALP'97*, volume 1256 of *LNCS*, pages 538–548. Springer Verlag, 1997.
50. W. Vogler, A. Semenov, and A. Yakovlev. Unfolding and finite prefix for nets with read arcs. In *Proceedings of CONCUR'98*, volume 1466 of *LNCS*, pages 501–516. Springer-Verlag, 1998.
51. G. Winskel. Event Structures. In *Petri Nets: Applications and Relationships to Other Models of Concurrency*, volume 255 of *LNCS*, pages 325–392. Springer Verlag, 1987.

From Graph Transformation
to Software Engineering and Back

Luciano Baresi[1] and Mauro Pezzè[2]

[1] Politecnico di Milano,
Dipartimento di Elettronica e Informazione,
Milano, Italy
baresi@elet.polimi.it
[2] Università degli Studi di Milano-Bicocca,
Dipartimento di Informatica Sistemistica e Comunicazione,
Milano, Italy
pezze@disco.unimib.it

Abstract. Software engineers usually represent problems and solutions using graph-based notations at different levels of abstractions. These notations are often semi-formal, but the use of graph transformation techniques can support reasoning about graphs in many ways, and thus can largely enhance them.

Recent work indicates many applications of graph transformation to software engineering and opens new research directions. This paper aims primarily at illustrating how graph transformation can help software engineers, but it also discusses how software engineering can ameliorate the practical application of graph transformation technology and its supporting tools.

1 Introduction

Software engineering aims at developing large software systems that meet quality and cost requirements. The development process that moves from the initial problem to the software solution is based on models that describe and support the development during the different phases. Models are key elements of many software engineering methodologies for capturing structural, functional and non-functional aspects. Popular methodologies prescribe various models – with different degrees of formality, flexibility, and analyzability – to solve the different problems. For example, UML proposes some semi-formal diagrammatic languages: class, object, component, and deployment diagrams for modeling structural aspects, and use case, sequence, activity, collaboration, and statechart diagrams for modeling behavioral aspects [24].

The syntax and semantics of these models are defined informally with different degrees of precision. Although users and tools agree on the main syntactic and semantic aspects, important details are often given different – and frequently incompatible – interpretations.

The scientific community agrees on the need to improve current practice by increasing the degree of formality of these notations. Unfortunately, formal

H.-J. Kreowski et al. (Eds.): Formal Methods (Ehrig Festschrift), LNCS 3393, pp. 24–37, 2005.

methods have not found wide application so far, and cannot be easily used in conjunction with popular modeling notations [6]. This is where graph transformation (*GT*) provides unique features to strengthen diagrammatic models by adding formality. Similarly to grammars for textual languages, GT can formally describe the concrete and abstract syntaxes of modeling languages, but it can also formalize the semantic aspects, and thus provides a strong basis for reasoning on diagrammatic models at all levels.

Formalizing the concrete and abstract syntaxes eliminates ambiguities and contradictions and supports automatic checks for consistency and correctness. GT allows the designer to describe the semantics both operationally and denotationally. Operational semantics can be given by describing the legal evolutions of models in terms of GT rules. Denotational semantics can be expressed by mapping models onto semantic domains by means of rules that mimic syntactical changes onto the chosen semantic domain [28].

GT is well supported by tools, which are useful for solving many problems and validating new ideas and applications, but often they do not scale well. When the size of the application grows, and requires significant mediation between theory and performance, software engineering principles can provide useful ideas. They can contribute significantly in this direction and help GT experts move towards complex problems and applications.

The main goal of this paper is to frame the opportunities offered by GT to software engineering by illustrating sample cases and proposing additional applications not fully explored yet. The paper also suggests ways for improving GT technology and tools, inspired by well-known software engineering principles, to address practical problems.

The paper is organized as follows. Section 2 discusses the opportunities offered by GT as modeling language by touching the use of GT for modeling and reasoning on particular aspects of software systems. Section 3 illustrates the potentiality of GT for modeling and verifying notations, i.e., for formalizing the concrete and abstract syntaxes as well as the semantics of notations, thus providing a uniform framework for modeling heterogeneous notations, and fostering analysis tools. Section 4 indicates how software engineering can help experiment and introduce graph transformation in current modeling practice. Section 5 proposes some future directions and concludes the paper.

2   Graph Transformation for Models

GT provides a formal and intuitive way for describing systems and their evolution. For example, GT has been proposed to model software architectures. Architectural styles constrain the connectivity among components to guarantee the regularity of architectural models, and thus improve the maintainability and evolvability of systems [10].

GT provides a natural way for formalizing architectural styles by means of rules that support the creation of models that comply with the style by construction. Le Metayér [19] first and then Hirsch et al. [12] investigated different approaches to model software architectural styles by using GT; Baresi et al. [3]

extend these approaches for service-oriented applications and emphasize the analyzability of such systems.

Type graphs, or metamodels according to the OMG terminology, define the elements that belong to a style. GT uses these definitions to create new models and make existing ones evolve. For example, Fig. 1, taken from [3], shows a UML class diagram – used as type graph – that models a fragment of the service-oriented architectural style. This means that all service-oriented applications must share this type graph and be suitable instantiation of these classes (types). In other words, the node types of the graph behind these architectures must belong to the set of nodes defined in Fig. 1. The type graph says that a *Component* dynamically searches for *Services* that meet given *Requirements* through a *Connect Request*. A *Service* is a special *Component* that satisfies a *Service Specification*. A *Component knows* the *Service Specification* of available *Services* and can *require services for* given *Requirements*, which *could satisfy* the given specification. The *Connect Request* establishes the link.

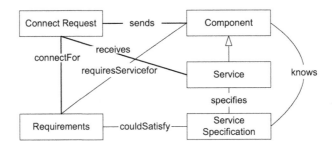

Fig. 1. An excerpt of the type graph for service-oriented architectures [3].

Fig. 2 presents an example GT rule that describes how to establish a new connection. A GT rule comprises two parts: a left- and a right-hand side. The left-hand side describes the conditions for the application of the rule. In this example, the *Component* must *know* a *Service Specification* that *could satisfy* its *Requirements*. The right-hand side describes the effect of the modification. In this case, the *Component* can *send* a *Connect Request*, newly added, for the required services to the *Service*.

The rule applies to all instances of the SoA style to enable generic components to request services according to given specifications. It creates a new *Connect Request*, called *req* to connect the *Requirements*, *Service*, and *Component* accordingly. The rule formally describes one of the steps to build and configure architectural models, without changing either the syntax or semantics. A set of rules can formally describe all possible building steps, thus formalizing completely the semantics of an architectural style, i.e., the semantics of an entire family of models.

Architectures often evolve with systems. Architecture reconfiguration refers to control the evolution of architectures according to given guidelines. The for-

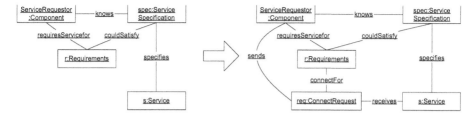

Fig. 2. A GT rule that models a reconfiguration of a service oriented architecture, where a component plays the role of a service requester and sends a request to the service it would like to connect to (The figure is taken from [3]).

malization of reconfiguration guidelines and constraints with GT provides a means to automatically check for consistency of architecture reconfigurations, and in perspective also a means for predicting and comparing the effects of different reconfiguration approaches.

Security is another interesting problem that presents variegate aspects that can be modeled with diagrammatic languages and GT. For example, Fig. 3 shows an Alloy model of the type system of a role based access control (RBAC) policy taken from [17]. The figure formalizes the relations among the entities in the system. *Roles* can be organized hierarchically (a role can be a *super-* or a *sub-role*, but each role has a unique super-role); roles' duty are separated (*sod*); a *session belongs* only to *users* whose *roles are activated* for the considered sessions.

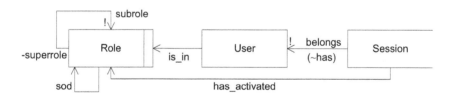

Fig. 3. An Alloy model of a role based access control (RBAC) policy (The figure is taken from [17]).

Policies can be expressed with rules that describe the evolutions of the relations among roles. A rule may for example allow users to activate only a subset of authorized roles when in a session. Such rule can be formally expressed with the GT rule of Fig. 4 taken from [17]. Koch and Parisi-Presicce show that GT rules allow to model not only types and policy rules, as UML or Alloy, but also policy constraints, and thus support a larger set of analyses that include constraint checking and conflict solving.

Also the behavior of mobile systems depends on the current configuration that changes dynamically. Mobile systems can be modeled with graphs that

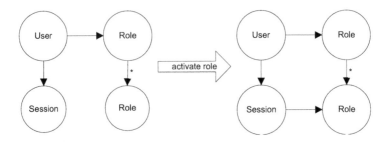

Fig. 4. A GT rule that allows a *User* within a *Session* to activate a *Role* only if it belongs to a specific subset of *Roles* that represent the roles the user is allowed to activate (The figure is taken from [17]).

evolve when agents move from node to node, activate or deactivate, and leave the network. Typed graphs can model different configurations, while graph transformation can formalize the rules that describe how configurations evolution, as observed in an early work by Corradini et al. [7].

So far, we introduced graph transformation mainly as a modeling means, but recently we have seen proposals that go a step beyond and address the validation of modeled GT systems. The pairing of graph transformation and model checking techniques is proposed by Varró [29]. His approach transforms a graph transformation system, along with an initial configuration of the system under analysis, into a Promela specification, which is the representation required by SPIN. The model checker starts from the initial configuration and searches all the possible sequences of rules. This way, we can check if a given sequence of rules is feasible or try to identify the right sequence to obtain a target configuration of the system under analysis.

3   Graph Transformation for Notations

In the previous section, we have illustrated how GT can be used as modeling language. In this section, we show how GT can be used to define specification languages. Figure 5 uses the well-known MVC (Model-View-Control) pattern to identify the different parts that belong to a diagram notation ([1]). The *graphical elements* are the *view*, i.e., the set of lines, boxes, bubbles, and labels perceived by the user. The *concrete syntax*, the *control*, captures the structure of diagrams in terms of their graphical elements and the relationships among them. The *abstract syntax* stands for the *model* and defines the structure of the diagram in terms of the diagram notation itself. Even if we could move directly from the graphical elements to their abstract syntax, it is important that we distinguish between *visualization* and *interpretation*. The concrete syntax deals with visualization and identifies the spatial relationships among the graphical elements. The abstract syntax deals with interpretation and specifies the tokens of the language and how they are composed to obtain a meaningful sentence (i.e., a model). The interpretation may require a *semantic domain*, that is, an external

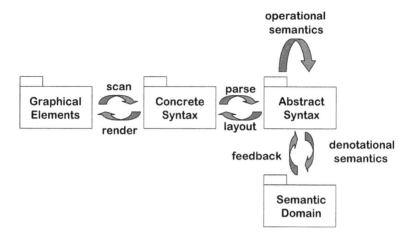

Fig. 5. Layered view of a *diagram language*.

formal model. The meaning of a diagram becomes the result of its translation into the semantic domain.

GT systems can be employed to define both the different views and the relationships among them. This means, for example, that we can use a GT system to formalize the abstract syntax of a diagram language, but we can also imagine a pair of GT systems to specify the relationships between the concrete and abstract syntaxes of a language. More precisely, the pairs of rules define how to *parse* models if we move from concrete to abstract, while they specify the *layout* of models if we move from abstract to concrete.

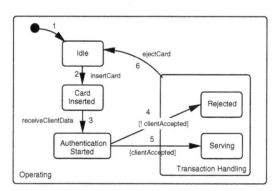

Fig. 6. A simple Statechart model of a CashBox (The figure is taken from [5]).

Let us consider for example the simple Statechart diagram of Fig. 6, taken from [5], as running example. It models the behavior of a *CashBox*. After switching it on, the *CashBox* enters the super-state *Operating* and moves to its default sub-state *Idle*. As soon as a card is inserted (i.e., event *insertCard* occurs), the component enters state *Card Inserted*. When it receives the client data from the card (event *receiveClientData*), it moves to state *Authentication Started*. Here, the choice of the next state depends on the data received. If the client is accepted, the component enters state *Serving*, otherwise it moves to state *Rejected*. In both cases and after processing the transaction, the component returns to state *Idle* as soon as the card is ejected (event *cardEjected*).

The first aspect that characterizes a diagram like this is the set of *graphical elements* that define its structure. These elements are specific to the chosen format and concur in defining the *concrete syntax* of the notation. If we chose an XML representation, like Scalable Vector Graphic (SVG) [30], the schema associated with the notation specifies how to identify a line, a bubble, a rectangle, or a label. These representation-specific elements are the starting point to construct and understand the graphical sentences behind a set of pictorial symbols.

The *concrete syntax* abstracts away from the particular representation and identifies the relationships among the graphical shapes. No matter of XML or an object-oriented language, states in Fig. 6 are represented through rectangles with rounded corners, initial states with black bubbles, and state transitions with directed edges. Relationships among these elements can be: a line *connects* two rectangles, a rectangle *contains* other rectangles, or an element *is on the left/right* of another element. At this level, a GT system defines the concrete syntax of the language in terms of the steps necessary to build a correct model. These rules can be conceived with the idea of scanning an existing graphical representation to produce the concrete syntax model, but it could also be defined with respect to a user that uses a syntax-directed editor for the supported notation. In this latter case, the transformation system defines all the correct user actions on the editor.

For example, Minas in [14] proposes a complete hyperedge grammar for *editing* well-formed Statechart diagrams, to feed the DiaGen tool for automatically generating a graphical editor for Statechart diagrams. Formally, the grammar defines all correct Spatial Relationship Hypergraphs (SRHG), that is, hypergraphs with edges like *label, rectangle, edge*, etc. and nodes representing the points where the hyperedges are connected. SRHGs through a further set of transformation rules become Reduced Hypergraph Model (HGM). These graphs represent the abstract syntax of the example Statecharts.

The *abstract syntax* defines the modeling elements supplied by the notation, without the concrete "sugar", and the relationships among them. Tokens at this level are related to the semantic interpretation, that is, we think of the example of Fig. 6 in terms of states (initial, AND-decomposed, and OR-decomposed) transitions, events, and so on. Figure 7 shows a simplified abstract syntax graph for the Statechart diagram of Fig. 6. Nodes are instances of a simple type graph that comprises: *startStates, ORStates, States*, and *Transitions* (further details are omitted for the sake of clarity). Edges connect the nodes to render the connections between states and transitions in the Statechart diagram.

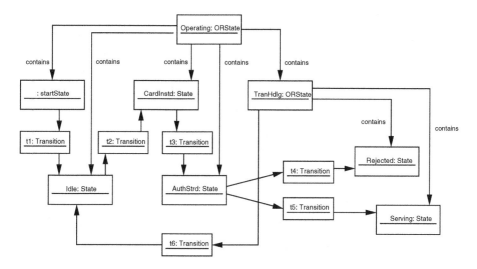

Fig. 7. Abstract syntax graph for the Statechart diagram of Fig. 6 (The figure is taken from [5]).

If we wanted to map abstract to concrete syntaxes, this implies the definition of the concrete *layout* of models. The grammar in this case defines how abstract concepts should be rendered at the concrete level, but also the correct positioning of each element on the canvas. Special-purpose algorithms for defining the layout of user models can be implemented using GT rules and textual attributes to compute the coordinates of each graphical symbol.

GT systems can be used to specify the rules that govern the elements at a given level, but also to define the mappings and transformations between levels. The choice is between two parallel systems, with paired rules, or a single complex system. In the first case each pair would comprise a rule that specifies a step in the first domain and the corresponding step in the other domain. The selection and application of the first rule would also trigger the application of the second rule. In the second case, we could foresee complex rules that start from symbols of the first domain and rewrite them into their equivalent elements of the second domain.

Moving to the semantics of the model of Fig. 6, we can specify it both *operationally* and *denotationally*.

An operational semantics can be given directly on the abstract syntax of the language through yet another GT system that specifies an interpreter for the language ([8]). Each model can also be "compiled" into a set of dedicated rules ([18]) to specify the behavior of each single model separately. For example, transitions for *syntactically correct* Statechart diagrams can be generated by applying the rules of the transformation unit *term(S)* presented in [18]. Fig. 8 shows the result for some of the transitions of Fig. 6. Rule *t1* moves the current state from the start state to *Idle*; Rule *t5* moves the current state from *Authentication Started* to the hierarchy *Transaction Handling / Serving*. Rule *t6* moves

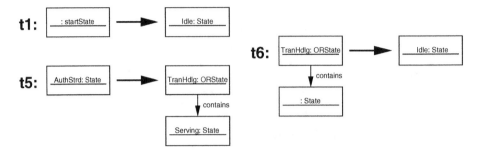

Fig. 8. Some transitions of Fig. 6 as GT rules (The figure is taken from [5]).

the current state from the hierarchy *Transaction Handling / any contained state* back to *Idle*.

Denotational semantics is given by mapping the abstract syntax to a *semantic domain*. The role played by GT depends on the chosen semantic domain. If it is a textual one, the productions of the grammar that defines the abstract syntax can be augmented with textual annotations to build the semantic representation. More generally, the productions can be paired with those of the textual grammar that specify the semantic models, and the application of a production of the abstract syntax grammar automatically triggers the application of the paired textual production [27].

For example, Engels et al. define the dynamic semantics of Statecharts through CSP (Communicating Sequential Processes [13]). The left-hand side of each rule defines how UML-like metamodel instances can be built for Statecharts (GT rules); the right-hand side codes how the corresponding CSP specification must be modified accordingly (textual grammar productions). Baresi uses high-level timed Petri nets as semantic domain. The mapping is defined with pairs of GT production: the first production defines the evolution of the abstract syntax representation, while the second production states the corresponding changes on the semantic Petri nets [2, 4]. Figure 9 taken from [2] shows a simple transformation rule that formalizes the connection of two Statechart states with a Statechart transition. A transformation rule comprises two productions. The left-hand side production applies to pairs of Statechart states and connects them through a Statechart transition, as indicated by the right-hand side of the rule. The right-hand side production applies to the Petri net places corresponding to the selected Statechart states and adds a Petri net transition and two arcs to connect them. The figure omits the textual annotations that can be found in [2].

More sophisticated mappings could be implemented using triple graph grammars [28], that is, besides the two grammars that define the abstract syntax and the corresponding modifications of the semantic domain, a third grammar would state the mapping between the two paired productions explicitly.

The different models defined so far pave the ground to additional analysis techniques, e.g., modern refactoring approaches. For example, Mens et al. start from the abstract syntax representation of UML class diagrams to reason on

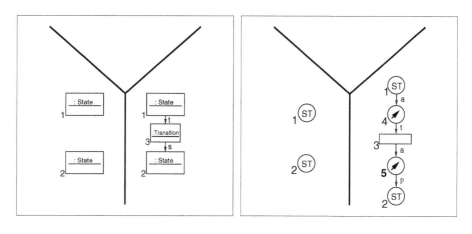

(a) Abstract Syntax Model   (b) Semantic Model

Fig. 9. A simple transformation rule taken from [2].

them and improve the quality of designed models [21]. Transformation rules work at this level to modify and improve user-defined class diagrams.

Similarly, with the help of pair grammars, transformation rules can be used to maintain, enforce or access the consistency of different views on the same problem. The fact that a modification of a given model can trigger modifications in other models, that is, the application of a rule on a model triggers the application of other rules on other models, is a way to enforce the consistency among views in complex models. Rules can be used both as a generative means, i.e., consistency is enforced by construction, or they can be a validation means and thus access the consistency of views defined independently. For example Hausmann et al. show how to use GT to enforce and access the consistency between requirements and high-level design of software systems [11].

4   Software Engineering for Graph Transformation

After discussing how GT can be proficiently used in software engineering, we briefly discuss how graph transformation can benefit from software engineering principles and techniques.

Software engineers can help graph transformation experts improve tools. Currently, the most popular tools for GT are excellent supports for early experimentation and validation of new ideas, but do not always meet the needs of software developers. They work well with research-size models, but become slow or crash when the size of models increases. They support well the direct use of graph transformation, but provide less help to software experts that would like to exploit GT without being proficient in the formalism. They need open tools that interoperate with sophisticated development environments [22].

Software engineers look for "problem-oriented", rather than "solution-oriented" tools. They are not ready to use graph transformation systems per-se, but they want means to solve development problems. This means that many details and technicalities should be hidden, while the adoption of standard notations and formats should be emphasized. This approach was adopted by Fujaba with its use of UML, as standard modeling notation (class diagrams to define concepts, object diagrams to define rules, and activity diagrams to compose them), and its hiding of many details related to graph transformation. The GT community is using standard notations to renders type graphs and rules. UML is the de-facto standard also in this community and the adoption of this notation is a first step towards the readability and usability of GT.

Software engineers need techniques that interface well with other CASE tools. Software engineering can contribute to GT with well-known standards for information exchange like XMI or MOF [23]. Besides special-purpose XML-based standards to encode graphs and graph transformation (GXL and GTXL, namely), GT tools should also support common "generic" standards to import and export artifacts and improve their interfaceability with other tools in the development process.

When designing real(istic) artifacts, modularization becomes an important issue. Nowadays, many graph transformation systems adopt a flat organization and the number of rules is constrained. Some proposals were presented to add modules to graph transformation; none of them has been widely accepted, but modules are a key element to handle real problems through graph transformation. They help understand the specification, by allowing the designer to reason at different levels, and foster the idea of reusing organized and coherent sets of rules. Something like a *package* in UML, to group the rules, and a *component* to show its interfaces would clearly improve the organization and management of graph transformation systems.

Software engineering can also help validate designed graph transformation systems. Flow-based analysis techniques, but also unit and integration testing techniques may help understand how a set of rules (i.e., a system) works. Other more formal approaches could also exploit process calculi and model checking techniques to reason on the dynamic behavior of such systems and prove interesting properties.

The last aspect concerns the methodologies and heuristics usually adopted by software engineers to model the different aspects of a software system, reason on them, and solve possible problems. This help is nothing concrete, but the right mix between capabilities in abstracting and modeling and the knowledge of graph transformation systems should pave the ground to better graph-based models of addressed aspects.

5   Conclusions

This overview was motivated by the idea that graphs provide a direct and intuitive way to represent, visualize and analyze a variety of structures that are typical to software engineering, and graph transformation provides general means for both describing and analyzing changes on graphs and specifying their governing rules and constraints. Sections 2 and 3 sketch some approaches, which are intended as example applications and do not aim at providing a complete survey of existing experiments with graph transformation and software engineering.

Moving to future directions, modern software development paradigms and applications, such as mobility, pervasive computing and component-based development raise new critical challenges. They come from the impossibility of predicting all possible uses of the software at development time. Traditional quality assurance techniques are mostly based on pre-deployment test and analysis, and provide limited support to these new challenges. Unpredicted interactions with new components or agents may lead to unexpected behaviors not previously verified. Current research investigates techniques such as in-field testing, run-time monitoring and self-healing mechanisms. They make use of run-time information to capture and analyze the in-field behavior of complex systems ([20, 25, 26, 9, 16]).

A main challenge of "post-deployment verification" is modeling the expected and experienced run-time behavior of the system, to identify and record potentially harmful runs. An excellent example of application of post-deployment modeling is the STAT intrusion detection framework proposed by Kemmerer and Vigna [15]. STAT models suspect security threats with finite state machines. The run-time behavior of the monitored systems is compared to the STAT model that detects and signals possible security threats. Finite state machines are effective in this context because the behaviors of interest are fairly simple and can be captured with a simple model. In general, the behavior of interest may be very complicated and may not be easily captured with a simple finite state machine. Graph transformation systems provide an excellent support for modeling complex evolutions, as illustrated in the previous sections, and thus are an obvious candidate to extend post-deployment modeling for monitoring complex system evolutions.

We think that these examples are only the first fruits of a cooperation that has all the chances to become wider. The marriage will be complete as soon as the knowledge also addresses the other way around. Software engineers could contribute to improve the development of graph transformation tools and standard exchange formats for graphs and graph transformation. They can lend their experience on modularity and components, to add these features to graph transformation specifications, and on analysis and testing to validate produced GT systems.

References

1. M. Andries, G. Engels, and J. Rekers. How to represent a visual specification. In K. Marriott and B. Meyer, editors, *Visual Language Theory*, pages 241–255. Springer-Verlag, 1997.
2. L. Baresi. *Formal customization of graphical notations.* PhD thesis, Dipartimento di Elettronica e Informazione – Politecnico di Milano, 1997. In Italian.

3. L. Baresi, R. Heckel, S. Thone, and D. Varro. Modeling and validation of service-oriented architectures: application vs. style. In *Proceedings of the 9th European software engineering conference held jointly with 10th ACM SIGSOFT international symposium on Foundations of software engineering*, pages 68–77. ACM Press, 2003.

4. L. Baresi, A. Orso, and M. Pezzè. Introducing formal methods in industrial practice. In *Proceedings of the 20th International Conference on Software Engineering*, pages 56–66. ACM Press, 1997.

5. L. Baresi and R. Heckel. Tutorial introduction to graph transformation: A software engineering perspective. In *Proceedings of the First International Conference on Graph Transformation (ICGT 2002)*, volume 2505 of *Lecture Notes in Computer Science*, pages 402–429. Springer-Verlag, 2002.

6. E. Clarke, J. Wing, R. Alur, R. Cleaveland, D. Dill, A. Emerson, S. Garland, S. German, J. Guttag, A. Hall, T. Henzinger, G. Holzmann, C. Jones, R. Kurshan, N. Leveson, K. McMillan, J. Moore, D. Peled, A. Pnueli, J. Rushby, N. Shankar, J. Sifakis, P. Sistla, B. Steffen, P. Wolper, J. Woodcock, and P. Zave. Formal methods: State of the art and future directions. *ACM Computing Surveys*, 28(4):626–643, 1996.

7. A. Corradini, F. Dotti, and L. Ribeiro. A graph transformation view on the specification of applications using mobile code. In *Proceedings of the International Symposium on Graph Transformation and Visual Modeling Techniques (GT-VMT)*, volume 50 (3). Electronic Notes in Computer Science, 2001.

8. G. Engels, J.H. Hausmann, R. Heckel, and S. Sauer. Dynamic meta modeling: A graphical approach to the operational semantics of behavioral diagrams in UML. In A. Evans, S. Kent, and B. Selic, editors, *Proc. UML 2000, York, UK*, volume 1939 of *Lecture Notes in Computer Science*, pages 323–337. Springer-Verlag, 2000.

9. D. Garlan and B. Schmerl. Model-based adaptation for self-healing systems. In *Proceedings of the first workshop on Self-healing systems*, pages 27–32. ACM Press, 2002.

10. D. Garland. Software architecture: a roadmap. In *The Future of Software Engineering*, pages 91–101. ACM Press, 2000.

11. J. Hausmann, R. Heckel, and G. Taentzer. Detecting conflicting functional requirements in a use case driven approach: A static analysis technique based on graph transformation. In *Proceedings of the International Conference on Software Engineering (ICSE'2002)*, pages 105–155. ACM Press, May 2002.

12. D. Hirsch, P. Inverardi, and U. Montanari. Graph grammars and constraint solving for software architecture styles. In *ISAW '98: Proceedings of the Third International Workshop on Software Architecture*, pages 69–72, 1998.

13. C. Hoare. Communicating sequential processes. *Communicat. Associat. Comput. Mach.*, 21(8):666–677, 1978.

14. B. Hoffmann and M. Minas. A generic model for diagram syntax and semantics. In *Proc. ICALP2000 Workshop on Graph Transformation and Visual Modelling Techniques, Geneva, Switzerland*. Carleton Scientific, 2000.

15. R.A. Kemmerer and G. Vigna. Intrusion Detection. *IEEE Computer*, 2002. Special publication on Security and Privacy.

16. J.O. Kephart and D.M. Chess. The vision of autonomic computing. *IEEE Computer*, 36(1):41–50, January 2003.

17. M. Koch, L. V. Mancini, and F. Parisi-Presicce. A graph based formalism for RBAC. *ACM Transactions on Information and System Security (TISSEC)*, 5(3):332–365, August 2002.

18. S. Kuske. A formal semantics of UML state machines based on structured graph transformation. In M. Gogolla and C. Kobryn, editors, *Proc. UML 2001*, volume 2185 of *Lecture Notes in Computer Science*. Springer-Verlag, 2001.
19. D. Le Métayer. Software architecture styles as graph grammars. In *Sigsoft*, pages 15–23. ACM Pres, 1996.
20. S. McCamant and M. Ernst. Predicting problems caused by component upgrades. In *Proceedings of ESEC/FSE 2003*, pages 287–296. ACM Press, 2003.
21. T. Mens, N. Van Eetvelde, D. Janssen, and S. Demeyer. Formalising refactorings with graph transformations. *Journal of Software Mainetnance and Evolution*, pages 1001–1025, 2004.
22. T. Mens, A. Scürr, and G. Taenzer. *Proceedings of the Workshop on Graph-Based Tools*. ENTCS, 2002.
23. OMG. Meta object facility (MOF) specification, September 1999.
24. OMG. *Unified Modeling Language (UML), version 1.5*. OMG Standard, 2003.
25. A. Orso, D. Liang, M. Harrold, and R. Lipton. Gamma system: continuous evolution of software after deployment. In *Proceedings of the international symposium on Software testing and analysis*, pages 65–69, Roma, Italy, 2002. ACM Press.
26. C. Pavlopoulou and M. Young. Residual test coverage monitoring. In *International Conference on Software Engineering*, pages 277–284, 1999.
27. T. W. Pratt. Pair grammars, graph languages and string-to-graph translations. *Journal of Computer and System Sciences*, 5:560–595, 1971.
28. A. Schürr. Specification of graph translators with triple graph grammars. In *Proceedings of the 20th International Workshop on Graph-Theoretic Concepts in Computer Science*, volume 904 of *LNCS*, pages 228–253. Springer Verlag, 1994.
29. D. Varró. Towards symbolic analysis of visual modelling languages. In Paolo Bottoni and Mark Minas, editors, *Proc. GT-VMT 2002: International Workshop on Graph Transformation and Visual Modelling Techniques*, volume 72 of *ENTCS*, pages 57–70, Barcelona, Spain, October 11-12 2002. Elsevier.
30. W3C. *SVG: Scalable Vector Graphics (SVG) version 1.2*. W3C, May 2004. http://www.w3.org/TR/2004/WD-SVG12-20040510/.

Flexible Interconnection
of Graph Transformation Modules
A Systematic Approach

Gregor Engels, Reiko Heckel, and Alexey Cherchago

University of Paderborn, Germany
{engels,reiko,cherchago}@upb.de

Abstract. Modularization is a well-known concept to structure software systems as well as their specifications. Modules are equipped with export and import interfaces and thus can be connected with other modules requesting or providing certain features.

In this paper, we study modules the interfaces of which consist of behavioral specifications given by typed graph transformation systems. We introduce a framework for classifying and systematically defining relations between typed graph transformation systems. The framework comprises a number of standard ingredients, like homomorphisms between type graphs and mappings between sets of graph transformation rules.

The framework is applied to develop a novel concept of substitution morphism by separating preconditions and effects in the specification of rules. This substitution morphism is suited to define the semantic relation between export and import interfaces of requesting and providing modules.

1 Introduction

One of the most successful principles of software engineering is *encapsulation*, i.e., the containment of implementations in classes, modules, or components accessible through well-defined interfaces only. This reduces possible dependencies of clients to those functions provided in the interface and allows to replace implementations without affecting the client.

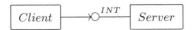

Fig. 1. Server component implementing interface INT that is used by Client component.

As it is obvious from Fig. 1, the developer of the Client component requires knowledge about the interface of the Server. That means, the development of the two components can not easily be decoupled and the architectural dependencies have to be known at design time.

In the service-oriented paradigm, but also in more advanced component models, this picture is extended by distinguishing between *provided* and *required* interfaces. While provided interfaces describe existing implementations, required

H.-J. Kreowski et al. (Eds.): Formal Methods (Ehrig Festschrift), LNCS 3393, pp. 38–63, 2005.

interfaces are specifications of *virtual components* whose existence is assumed at design time to capture the context dependencies of the components under development.

Fig. 2. Requestor and provider components.

As shown in Fig. 2, composing two components (or services) now means to connect their required and provided interfaces. This is possible if the operations asked for in the required interface are guaranteed by the provided interface. In programming languages like Java and component models like Corba such a relation between interfaces is verified by the compiler, matching the signatures (names and parameter types) of these operations.

However, in a truly open scenario, as it is typical for Web services or, more generally, service-oriented architectures, we cannot assume that, e.g., the name of an operation has any global meaning or that the types of the parameters convey enough information about the purpose and usage of the operation. In such case, it is inevitable that both required and provided interfaces contain *behavioral specifications* which are taken into account when interfaces are matched.

1.1   Module and Component Models with Behavioral Interfaces

The first steps in this direction have been made in the context of algebraic and logic specifications. An algebraic specification module MOD (see, e.g., [7]) consists of a body BOD providing the implementation and of interfaces IMP for import and EXP for export describing, respectively, required and provided functionality. (In addition, a parameter PAR is provided to allow for generic modules, but this feature will not be relevant for our purposes.) All specifications are connected through algebraic specification morphisms.

The composition of modules MOD and MOD' is based on morphisms, too, connecting the import (required) interface of IMP of MOD with the export (provided) interface of EXP' of MOD'. Hence, algebraic specification modules realize the idea illustrated in Fig. 2 that components are connected indirectly through the matching of required and provided interfaces.

In [7] the relation between IMP and EXP' is described by standard morphisms of algebraic specifications. That means, for example, that matching operations are required to have the same number of parameters of corresponding types. A more flexible approach to the connection of required and provided interfaces is presented by Zaremski and Wing in [25] and [26], who have developed sophisticated matching procedures at the level of both signatures and specifications.

In object-oriented programming, the extension of interfaces with behavioral information became known under the name of *Design by Contract* in the context

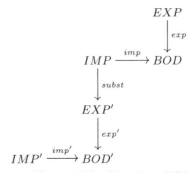

Fig. 3. Conformation of the TGTS-modules.

of the Eiffel language [22]. Here, preconditions and effects of operations are specified by means of logic predicates, and object-oriented subtyping rules are extended to capture the compatibility of, for example, the contracts of a newly introduced subclass with that of a superclass.

1.2   Modules of Graph Transformation Systems

Since the mid nineties [5], there is an increasing interest in the transfer of modularity concepts from algebraic specifications and programming languages to graph transformation systems [21, 24, 10] (see also the survey [19]). Modules of typed graph transformation systems (TGTS modules) [10] follow the structure of algebraic specification modules, replacing the specifications BOD, IMP, and EXP by graph transformation systems related by different kinds of morphisms. In particular, IMP and BOD are related by a simple inclusion morphism ($IMP \subseteq BOD$), whereas EXP and BOD are connected by a refinement morphism, allowing a sequential or parallel decomposition of rules.

In [1, 15] it has been observed that graph transformation rules could provide a more abstract, visual representation of contracts, specifying preconditions and effects of operations. In order to check the desired behavioral compatibility of contracts between required and provided interfaces, a matching relation has been defined which can be syntactically verified. Roughly speaking, the semantic idea of this compatibility was the *substitution principle*, i.e., it should be safe to replace the required rule by a matching provided rule: The applicability of the first should imply the applicability of the latter, and the effect of applying the latter should satisfy the expectations of the first.

However, it has been assumed that the two rules to be compared are defined over the same types, based on the assumption that the matching is performed by a central discovery agency which represents both provided and required contracts over a common ontology. This assumption, which is satisfied in service-oriented architectures, is not in general true for modules or components.

It is the purpose of this paper to define a flexible matching relation enabling retyping as a morphism of typed graph transformation systems representing the required and provided (import and export) interfaces of modules.

1.3   Morphisms of Graph Transformation Systems

A survey of the literature reveals at least five fundamentally different proposals for morphisms of graph transformation system [2, 23, 13, 17, 20]. They represent different objectives, like inclusions, refinements, or views and enjoy different semantic properties.

So far there has been no general and systematic approach for comparing and relating different notions. Hence, before going on to extend the list by a new proposal, so-called substitution morphisms, we will survey possible definitions and provide a four-step recipe for deriving the appropriate definition from given semantic requirements.

We will apply this recipe to derive a notion of morphism between graph transformation systems with application conditions which, as it turns out, are essential for a flexible and yet semantically meaningful relation between required and provided interfaces.

1.4   Outline of the Paper

The rest of this paper is organized as follows. After recalling in the next section the basic concepts of the DPO and DPB approaches to graph transformation, in Section 3 we introduce a framework enabling to classify and systematically define morphisms of typed graph transformation systems. In particular, two examples of morphisms existing in the literature will be discussed informally in Section 3.1. After that, the constituents of the framework will be identified in Section 3.2, and aggregated in the definitions of the sample morphisms in Section 3.3. Then, a new concept of a substitution morphism playing a role of an inter-connector between two modules requiring and offering a specific service will be considered in Section 3.4.

The flexibility of the substitution morphism depends on the rule structure which is refined in Section 4 via separating preconditions and effects. Here, we will revise all the necessary definitions on the graph transformation rules with application conditions in Section 4.1, and formally define the substitution morphism in Section 4.2. In Section 5, the substitution morphism as well as the other sample morphisms are illustrated via their application as intra- and inter-connectors of the modules. We conclude with the summary of our work in Section 6.

2   Basic Definitions

In this section we review some of the basic notions of the *double-pushout* (DPO) [8] and *double-pullback* (DPB) [18] approaches to graph transformation. The DPB approach represents a loose version of the classical DPO, assuming that rules may be incomplete specifications of the transformations to be performed and thus allowing additional, unspecified effects. Both approaches are presented using typed graphs [3].

By *graphs* we mean directed unlabeled graphs $G = \langle G_V, G_E, src^G, tar^G \rangle$ with set of vertices G_V, set of edges G_E, and functions $src^G : G_E \to G_V$ and $tar^G : G_E \to G_V$ associating with each edge its source and target vertex. A graph homomorphism $f : G \to H$ is a pair of functions $\langle f_V : G_V \to H_V, f_E : G_E \to H_E \rangle$ preserving source and target, that is, $src^H \circ f_E = f_V \circ src^G$ and $tar^H \circ f_E = f_V \circ tar^G$. With componentwise identities and composition this defines the category **Graph**.

Given a graph TG, called *type graph*, a TG-*typed (instance) graph* consists of a graph G together with a typing homomorphism $g : G \to TG$ (cf. Fig. 4 on the left) associating with each vertex and edge x of G its type $g(x) = t$ in TG. In this case, we also write $x : t \in G$. A TG-typed graph morphism between two TG-typed instance graphs $\langle G, g \rangle$ and $\langle H, h \rangle$ is a graph morphism $f : G \to H$ which preserves types, that is, $h \circ f = g$. With composition and identities this defines the category **Graph**$_{TG}$, which is the comma category **Graph** over TG.

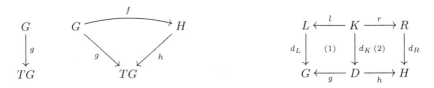

Fig. 4. Typed graph and graph morphism (left) and double-pushout (or -pullback) diagram (right).

Definition 1 (typed graph transformation system). *A TG-typed graph transformation rule is specified by a span $(L \xleftarrow{l} K \xrightarrow{r} R)$ of injective TG-typed graph morphisms (cf. Fig. 4 on the right).*

*Given TG-typed graph transformation rules $p = (L \xleftarrow{l} K \xrightarrow{r} R)$ and $q = (L' \xleftarrow{l'} K' \xrightarrow{r'} R')$, a typed rule morphism $f : p \to q$ is a tuple (f_L, f_K, f_R) of TG-typed graph morphisms commuting with the span morphisms l, l', r and r' (cf. Fig. 5). With componentwise identities and composition this defines the category **Rule**$_{TG}$, which is the comma category **Rule** over TG.*

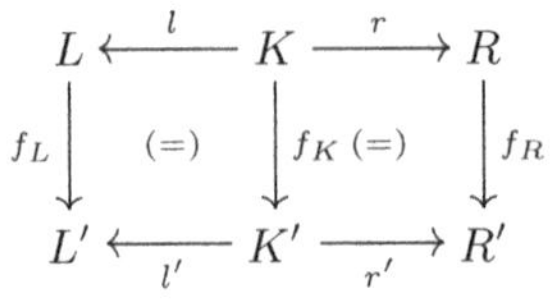

$$L \xleftarrow{l} K \xrightarrow{r} R$$
$$f_L \downarrow \quad (=) \quad \downarrow f_K \, (=) \quad \downarrow f_R$$
$$L' \xleftarrow{l'} K' \xrightarrow{r'} R'$$

Fig. 5. Typed rule morphism.

A *typed graph transformation system* $GTS = (TG, P, \pi)$ *consists of a type graph TG, a set of rule names P, and a mapping $\pi : P \to |\mathbf{Rule}_{TG}|$ associating with each rule name p a TG-typed rule $\pi(p)$.*

The *left-hand side L* of a rule contains the items that must be present for an application of the rule, the *right-hand side R* those that are present afterwards, and the *interface graph K* specifies the "gluing items", i.e., the objects which are read during application, but are not consumed.

As running example, a specification of a mutual exclusion algorithm with deadlock detection [16] is developed throughout the paper.

Example 1 (MUTEX). The typed graph transformation system in Fig. 6 models a distributed algorithm for mutual exclusion (MUTEX). This example is derived from a small case study [16] and tailored for our presentation. Two basic types, processes P (drawn as black nodes) and resources R (drawn as light boxes), constitute the type graph shown in the upper-left corner. A *request* is modeled by an edge going from a process to a resource. The fact that the resource is currently *held_by* the process is shown by an edge in the opposite direction. A token ring algorithm implements the mutual exclusion. The processes in the token ring are arranged in a cycle. Two neighbor processes are connected by an edge running from the antecedent to the *next* process. This edge is given by a loop in the type graph. A default position for introducing new processes and resources is marked by a pointer *head*.

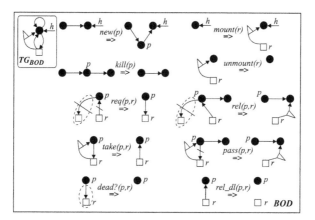

Fig. 6. Graph transformation system modeling MUTEX algorithm.

An edge with a white flag denotes a token which is passed from process to process along the ring. In order to get an access to a resource a process waits for the corresponding token. Mutual exclusion is achieved by uniqueness of the token for each resource in the system.

Now we discuss the rules of the graph transformation system. The first four rules are used for creating and killing processes (*new* and *kill*), and for mounting and unmounting resources (*mount* and *unmount*). The rules *req*, *take*, and *rel* allow processes to issue requests, take resources, and release them upon *regular* completion of their task. The negative application conditions [14] for *req* ensure that a process can not issue more then one request at a time. The negative application condition for *rel* prevents the release of a resource *r* while the process requests another resource, since *r* may still be required to complete the given task.

The last two rules are intended for application in possibly deadlock situations resulting from competition of processes for non-sharable resources. The MUTEX algorithm does not know how to detect deadlocks, therefore the rule *dead?* represents external features, to be imported from another module. The dotted part of this rule is a positive application condition (cf. Section 4.1) representing items that must be present for the rule application, but are not consumed. This condition restricts the applicability of rule *dead?* to situations where the process has a pending request for a resource.

In general, positive application conditions can be encoded by extending both the left- and the right-hand side of a rule by the required elements: they become part of the context. That means rules with positive application conditions can easily be transformed into ordinary rules. The use of positive application conditions makes a difference, however, when we consider relations between different systems, as shall be demonstrated in Section 4.2.

Rule *rel_dl* finally implements the resolution of detected deadlocks by forcing the release of the resource held by the involved process.

In the DPO approach, transformation of graphs is defined by a pair of pushout diagrams, a so-called double-pushout construction. Operationally speaking that means: the elements of G matched by $L \setminus l(K)$ are removed, and a copy of $R \setminus r(K)$ is added to D.

A *double-pushout (DPO) diagram d* is a diagram as in Fig. 4 on the right, where (1) and (2) are pushouts. Gluing the graphs L and D over their common part K yields again the given graph G, i.e., D is a so-called *pushout complement* and the left-hand square (1) is a pushout square. Only in this case the application is permitted. Similarly, the derived graph H is the gluing of D and R over K, which forms the right-hand side pushout square (2).

This formalization implies that only vertices that are preserved can be merged or connected to edges in the context. It is reflected in the *identification* and the *dangling conditions* of the DPO approach. The *identification condition* states that objects from the left-hand side may only be identified by the match if they also belong to the interface (and are thus preserved). The *dangling condition* ensures that the structure D obtained by removing from G all objects that are to be deleted is indeed a graph, that is, no edges are left "dangling" without source or target node.

Definition 2 (DPO graph transformation). *Given a typed graph transformation system $GTS = (TG, P, \pi)$, a (DPO) transformation step in GTS from G to H via p is denoted by $G \xRightarrow{p/d} H$, or simply by $G \xRightarrow{p} H$ if the DPO diagram d is understood.*

A transformation sequence $\rho = \rho_1 \ldots \rho_n : G \Rightarrow^* H$ in GTS via $p_1, \ldots, p_n$ is a sequence of transformation steps $\rho_i = (G_i \xRightarrow{p_i/d_i} H_i)$ such that $G_1 = G, H_n = H$ and consecutive steps are composable, that is, $G_{i+1} = H_i$ for all $1 \le i < n$.

The category of transformation sequences over GTS *denoted by* **Trf**(GTS) *has all graphs* $G \in$ **Graph**$_{TG}$ *as objects and all transformation sequences in GTS as arrows.*

A sample transformation step is shown in Fig. 7. It applies the rule *new* inserting a new process in the token ring.

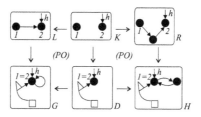

Fig. 7. A sample DPO transformation step.

The DPO approach ensures that the changes to the given graph H are exactly those specified by the rule. A more liberal notion of rule application is provided by the double-pullback (DPB) approach to graph transformation [18], where *at least* the elements of G matched by $L \setminus l(K)$ are removed, and *at least* the elements matched by $R \setminus r(K)$ are added. The DPB approach introduces *graph transitions* and generalizes DPO by allowing additional, unspecified changes. Formally, graph transitions are defined by replacing the double-pushout diagram of a transformation step with a double-*pullback*.

Definition 3 (DPB graph transitions). *Given a typed graph transformation system* $GTS = (TG, P, \pi)$, *a transition in GTS from* G *to* H *via* p, *denoted by* $G \overset{p/d}{\rightsquigarrow} H$, *is a diagram like in the right of Fig. 4, where both (1) and (2) are pullback squares. A transition is called* injective *if both* g *and* h *are injective graph morphisms. It is called* faithful *if it is injective, and the morphisms* d_L *and* d_R *satisfy the following* identification condition [4] *with respect to* l *and* r: *for all* $x, y \in L$, $y \notin l(K)$ *implies* $d_L(x) \neq d_L(y)$, *and analogously for* d_R.
A transition sequence $\rho = \rho_1 \dots \rho_n : G \rightsquigarrow^* H$ *in GTS via* $p_1, \dots, p_n$ *is a sequence of faithful transitions* $\rho_i = G_i \overset{p_i/d_i}{\rightsquigarrow} H_i$ *such that* $G_1 = G, H_n = H$ *and consecutive steps are composable, that is,* $G_{i+1} = H_i$ *for all* $1 \leq i < n$.
The category of transitions over GTS, *denoted by* **Trs**(GTS), *has all graphs* $G \in$ **Graph**$_{TG}$ *as objects and all transition sequences in GTS as arrows.*

A sample transition is shown in Fig. 8. It also demonstrates an application of the rule *new*. Note that during application of the rule a token is deleted that is unspecified by *new*. Here the left-hand square is not a pushout: the graph G obtaining by the gluing of L and D additionally contains the *token* which is "spontaneously deleted".

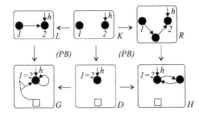

Fig. 8. A sample DPB graph transition.

3   TGTS Morphisms, Systematically

In this section, we provide a framework for classifying and systematically defining morphisms of typed graph transformation systems based on a number of standard "ingredients", like homomorphisms between type graphs and mappings between sets of rules. First, two examples of morphisms will be discussed informally in the context of TGTS modules. Then, in Section 3.2, the constituents of the framework are presented and combined, yielding definitions of the sample morphisms in Section 3.3. In Section 3.4, a novel concept of *substitution morphism* is considered.

3.1   TGTS Morphisms as Intra-connectors of Modules

Each TGTS describes a specific behavior in terms of the transformation or transition sequences obtained via application of its rules. A TGTS morphism $f : GTS \rightarrow GTS'$ defines a relation between the behaviors of GTS and GTS' through an association of their type graphs and rules. Thus, a systematic approach should always start by identifying the kind of semantic relation that shall be expressed.

First, we consider an example of *behavior-preserving* morphisms providing a first attempt at describing the relation between the export interface EXP of a module MOD with its body BOD. The export interface EXP specifies the features offered for import by other modules. The specification of these features should be consistent with their implementation in the body. That means, applicability of EXP rules should imply applicability of the corresponding BOD rules. Behavior-preserving morphisms shall ensure this property.

Example 2 (behavior-preserving morphism). The body of the module MUTEX is given in Fig. 6. One service provided by this module is deadlock resolution described by the rule rel_dl in the export interface EXP (cf. Fig. 9 on the right). (It shall be imported by an external deadlock detection module to break up detected deadlocks.) The embedding of EXP into BOD preserves behavior: Each transformation sequence in EXP implies a corresponding sequence in BOD.

The type graph TG_{EXP} of the export is a subgraph of TG_{BOD} of the body containing all the types relevant for deadlock resolution. More generally, a homomorphism between type graphs ensures that all types of the source (TG_{EXP} in

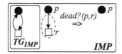 

Fig. 9. TGTSs *IMP* (left) and *EXP* (right) of the module *MOD* modeling the MU-TEX algorithm.

this case) have a correspondence in the target (TG_{BOD}). If the homomorphism is not an inclusion, as in our example, a type in the target may have a different name than its source or two different types in the source may be mapped to the same target type.

Based on the homomorphism, graphs, rules, and also transformations typed over the source can be converted into such typed over the target by a simple renaming of their types. This gives us the opportunity to compare two systems by translating the rules of the source system into ones typed over the target.

Due to the subgraph relation between TG_{EXP} and TG_{BOD} the translation of the *EXP* rule to the *BOD* type graph does not change anything in this rule. The comparison reveals that the rule identical to *rel_dl* is already present in TG_{BOD}, even with the same name. In the general case, we might consider a mapping of rule names as well to use different names for corresponding rules in the two systems.

The behavior-preserving morphisms as discussed above are originally introduced in [9, 11]. In our example, the export interface *EXP* is just a subsystem of *BOD*. More general situations are considered in [11] where the relation between export interface and body may be, e.g., *spatial* or *temporal* refinements. In spatial refinements, a rule of the source system may be associated with an amalgamation of rules of the target system, in temporal refinements with a sequential composition.

The requirements and definitions about behavior-preserving morphisms are presented in Section 3.3. Here, we proceed with an example of *behavior-reflecting* morphisms, determining the relation between the import interface *IMP* and the body *BOD* of a module *MOD*. The idea is that the rules required at *IMP* have at least the effect of the rules specified at *BOD*. Otherwise, the body could not use the imported rules for the internal implementations. This can be expressed as a reflection of the *BOD* transformations by *IMP* transitions.

Example 3 (behavior-reflecting morphism). As mentioned already, deadlock detection represents an external feature abstractly represented in the MUTEX module by the rule *dead?* in the import interface *IMP* (cf. Fig. 9 on the left).

Reflection of *BOD* behavior by *IMP* means that for each transformation in *BOD* we require a corresponding transition in *IMP*. As with behavior-preserving morphisms, we have to specify the relation between the type graphs and rules of the two systems.

For type graphs, a homomorphism from TG_{IMP} to TG_{BOD} ensures that *BOD* has at least the same types as *IMP*. In order to check that transformations in BOD are reflected by transitions in IMP, we have to compare the rules of the two systems. In this case, since we are interested in reflection rather than preservation of steps, we translate the rules of BOD to IMP *against* the direction of the type graph morphism. That means, beside the renaming of types, elements of the rules are removed if their type in BOD does not have a pre-image in IMP under the type graph homomorphism.

Then, the BOD behavior is reflected by IMP if each rule, after the translation, turns out to be a super-rule of the corresponding rule in IMP. In our case, the rule *dead?* of BOD coincides with the one in IMP after the translation.

Morphisms that reflect transformation in the target by transitions in the source have been introduced in [16] to specify the relation between different views of a system model.

Below, we formally introduce the different components of TGTS morphisms.

3.2   Definitions of Ingredients

In this section we define the two main ingredients of TGTS morphisms, i.e., translations between type graphs and subrule relations. We start with forward and backward retyping using the notation of [12].

Definition 4 (retyping). *A graph morphism* $f_{TG} : TG \rightarrow TG'$ *induces a forward retyping functor* $f_{TG}^{>} : \mathbf{Graph}_{TG} \rightarrow \mathbf{Graph}_{TG'}$, $f^{>}(g) = f \circ g$ *and* $f^{>}(k : g \rightarrow h) = k$ *by composition as shown in the diagram below,*

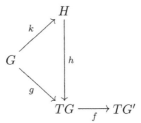

as well as a backward retyping functor $f_{TG}^{<} : \mathbf{Graph}_{TG'} \rightarrow \mathbf{Graph}_{TG}$, $f^{<}(g') = g^{*}$ *and* $f^{<}(k' : g' \rightarrow h') = k^{*} : g^{*} \rightarrow h^{*}$ *by pullbacks and mediating morphisms as shown in the diagram below.*

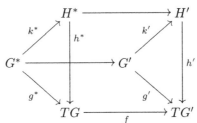

We proceed by listing a number of relations between rules typed over the same type graph.

Definition 5 (subrule relations). *Given TG-typed graph transformation rules $p : (L \xleftarrow{l} K \xrightarrow{r} R)$, $q : (L' \xleftarrow{l'} K' \xrightarrow{r'} R')$, and a typed rule morphism $f : p \to q$ (cf. Fig. 5), we say that*

- *p is* identical to *q, in symbols $p = q$, if f is an identity in $\mathbf{Rule}_{TG}$,*
- *p is a* DPO-subrule *of q, in symbols $p \sqsubseteq_{DPO} q$, if the diagrams (1) and (2) in Fig. 5 are pushouts in $\mathbf{Graph}_{TG}$,*
- *p is a* DPB-subrule *of q, in symbols $p \sqsubseteq_{DPB} q$, if the diagrams (1) and (2) in Fig. 5 are pullbacks in $\mathbf{Graph}_{TG}$ and construct a faithful transition.*

This list could be further extended by relations between a single rule and a collection of rules such as the *spatial* and *temporal* refinements, but this is beyond the scope of this paper.

Having defined retyping and rule relations we are now in a position to combine these ingredients into definitions of TGTS morphisms.

3.3 Recipes for TGTS Morphism

We have already discussed by means of the two examples in Section 3.1 how semantic requirements determine the definition of morphisms of graph transformation systems. In this section, we are going to make this explicit in terms of a four-step recipe. In each step we introduce a number of options and motivate possible choices based on the semantic requirements. First of all we formulate an initial assumption to simplify the presentation.

Assumption: Without loss of generality we assume that the TGTS morphism $f : GTS \to GTS'$ and the type graph morphism $f_{TG} : TG \to TG'$ have the same direction. That means, the target system has at least the types like the ones of the source system, but possibly more.

Step 1. The first variation point is the *relation between the sets of rule names* of GTS and GTS'. Here it is most convenient to use total functions, rather than general relations. For example, a mapping from P to P' designates for each $p \in P$ one corresponding $p' \in P'$: the relation is *left total* and *right unique*. This option should be used for behavior-preserving morphisms, where each transformation of the source system has to be associated with a transformation of the target system. Dually, a mapping in the opposite direction provides for each $p' \in P'$ one $p \in P$: a *left unique* and *right total* relation which is suitable for the behavior-reflecting morphisms.

Step 2. The next alternative is introduced by the *context of comparison*, i.e., where the corresponding rules of the two systems are compared. This can be done either in the context of GTS' using the forward retyping $f_{TG}^{>} : \mathbf{Graph_{TG}} \to \mathbf{Graph_{TG'}}$ of the rules in GTS, or in the context of GTS using the backward retyping $f_{TG}^{<} : \mathbf{Graph_{TG'}} \to \mathbf{Graph_{TG}}$ of the rules in GTS'. The forward retyping is appropriate for behavior-preserving morphism, the objective in this case being the construction of transformations in the target from existing ones in the source system. By analogy backward is used for behavior-reflecting morphisms.

We continue with the specification of the subrule relation required between the rules of the two systems. For pairs of corresponding rules as defined in *Step 1* and modulo the retyping functor selected in *Step 2* this means to decide for the direction of the relation in *Step 3* and its kind in *Step 4*.

Step 3. The *direction of the subrule relation*, i.e., if p is required to be a subrule of p', or vice versa, depends on the desired relation between the sets of transformations or transitions of the two systems. If p is a subrule of p' then each transformation step via p' implies a transition or transformation step via p. Thus, behavior-preserving morphisms generally require that the GTS' rules are subrules of the GTS rule, while behavior-reflecting morphisms specify the dual requirement.

Remark 1. Note that it may be the case that a subrule relation between p and p' holds when considered over the larger type graph GTS' using forward retyping, but not if compared via backward retyping (projection) over the smaller GTS type graph. The converse is also true, i.e. a subrule relation may hold over GTS, but not over GTS'.

This motivates why the comparison of rules is always done over the system where the existence of transformations or transitions should be ensured, i.e., the target system if behavior shall be preserved and the source system if behavior shall be reflected.

Step 4. Finally, we have to select the *kind of subrule relation* that the comparison shall be based upon. The identity of p and p' ensures that all transformations via p are also transformations via p'. If p is a DPO or DPB subrule of p', respectively, then each transformation step via p' implies a transformation ($\sqsubseteq_{DPO}$) or transition ($\sqsubseteq_{DPB}$) via p. (The dual holds if we replace p and p'.)

The relation between the different choices and the implied semantic properties is summarized in Table 1. Combinatorially, we obtain eight different notions. Numbers 4 and 5 represent, respectively, the behavior-reflecting and preserving morphisms discussed above.

Next we introduce formally the semantic requirements of behavior-preservation and -reflection and, subsequently, the actual definitions of the morphisms.

Table 1. Ingredients of TGTS-morphisms.

	forward retyping $f_{TG}^{>}$		backward retyping $f_{TG}^{<}$	
	left-total right-unique relation	left-unique right-total relation	left-total right-unique relation	left-unique right-total relation
$\sqsubseteq$	— $\quad$ 1	— $\quad$ 2	DPO/DPB_ϵ $\quad$ 3	$DPO, DPB, [=]$ $\quad$ 4
$\sqsupseteq$	$=, [DPO/DPB]$ $\quad$ 5	— $\quad$ 6	— $\quad$ 7	— $\quad$ 8

Definition 6 (preservation of behavior). *Given typed graph transformation systems $GTS = (TG, P, \pi)$ called the source system and $GTS' = (TG', P', \pi')$ called the target system. We say that the target system preserves the behavior of the source system if there exists a functor $F : \mathbf{Trf}(GTS) \to \mathbf{Trf}(GTS')$.*

The existence of a functor between two categories of sequences requires that each individual step in GTS is mapped to a sequence in GTS'. By induction, this mapping extends to sequences in GTS. However, we will deal with the simpler case where a step in GTS is actually mapped to a single step in GTS'.

As discussed above, this requires that each rule $p \in P$ has a corresponding rule $p' \in P'$. Hence, a mapping $f : P \to P'$ is chosen in *Step 1*. To ensure the preservation of sequences in GTS', the comparison of rules is done in the context of GTS' and, therefore, forward retyping is applied at *Step 2*.

The mapping in *Step 1* must guarantee the desired relation between the transformations in the two systems. This is achieved if in *Step 3* the rules in GTS' are subrules of those in GTS. The choices in *Step 4* ensuring behavior preservation range from identity to DPB relations. The identity is the most common one because it results in an embedding of GTS into GTS', while any true subrule relations would mean that the rules of GTS are reduced in GTS'.

The behavior-preserving morphism is specified in cell 5 of Table 1 and formally defined below.

Definition 7 (behavior-preserving morphism). *Given typed graph transformation systems $GTS = (TG, P, \pi)$ and $GTS' = (TG', P', \pi')$, a behavior-preserving TGTS morphism $f^{pres} = (f_{TG}, f_P)$ is given by a type graph morphism $f_{TG} : TG \to TG'$ and a mapping $f_P : P \to P'$ between the sets of rule names such that for each $p \in P$, $f^{>}_{TG}(\pi(p)) = \pi'((f_P(p)))$.*

The justifications for the following claim can be found in [9, 11].

Fact 1 *Behavior-preserving morphisms $f^{pres} : GTS \to GTS'$ satisfy the requirements of Def. 6.*

Just to consider another example, the candidate in cell 6 differs from the one above in the direction of the mapping between rule names. That means, to each $p' \in P'$ a $p \in P$ is associated. If we require the existence of the subrule relation for all pairs of rules thus associated, this guarantees a partial preservation of behavior only, i.e., for those transformations in GTS via rules with corresponding rules in GTS'.

To continue on the right-hand side of the table, the semantic requirements for behavior-reflecting morphisms are given below.

Definition 8 (reflection of behavior). *Given typed graph transformation systems $GTS = (TG, P, \pi)$ called the source system and $GTS' = (TG', P', \pi')$ called the target system, we say that the first reflects the behavior of the second if there exists a functor $F : \mathbf{Trf}(GTS') \to \mathbf{Trs}(GTS)$.*

That means, each transformation step in GTS' implies a *transition* in GTS, a liberal requirement compared to reflecting transformations in *transformations*.

By the same arguments as above, in *Step 1* we assume a mapping of rule names from P' to P. The context of comparison is the source system, leading to the use of backward retyping is selected at *Step 2*. To fulfill the semantic requirement, rules in P' are subrules of corresponding rules in P in *Step 3*. Both DPO or DPB subrule relations are reasonable at *Step 4*. The first would, in fact, guarantee the stronger reflection property based on transformations only.

This morphism specified in cell 4 of Table 1 is formally defined below.

Definition 9 (behavior-reflecting morphism). *Given typed graph transformation systems* $GTS = (TG, P, \pi)$ *and* $GTS' = (TG', P', \pi')$, *a behavior-reflecting morphism* $f^{refl} = (f_{TG}, f_P)$ *is given by a type graph morphism* $f_{TG} : TG \to TG'$ *and a mapping* $f_P : P' \to P$ *between rule names such that for each* $p' \in P'$, $\pi(f_P(p')) \sqsubseteq_{DPO/DPB} f_{TG}^{<}(\pi'(p'))$.

The proof of following is obvious.

Fact 2 *Behavior-reflecting morphisms* $f^{refl} : GTS \to GTS'$ *satisfy the requirements of Def. 8.*

In [16] a variant of the above has been used represented by cell 3. The difference from 4 is the direction of the mapping of rule names from P to P', i.e., in the same direction like the mapping of types. Using DPB subrules and assuming in each GTS an empty ϵ-rule, each step in GTS' using a rule without a corresponding rule in GTS is associated with an ϵ-transition. In this way, the behavior is indeed reflected by GTS.

If we consider, instead, DPO subrules we obtain a partial reflection of the target transformations by the source ones.

It turns out that none of the other alternatives in Table 1 preserve or reflect behavior. Variants 1 and 2 are not behavior-preserving, because the subrule relation allows rules in the target system to be larger than in the source. Hence, additional preconditions may be introduced which make rules in GTS' applicable in less situations.

Similarly, variants 7 and 8 are inadequate for the behavior reflection since, due to Remark 1, subrule rule relations are not in general preserved by the retyping.

The preservation properties for subrule relations between the rules of the two systems are detailed in Fig. 10.

In the next section we discuss another kind of semantic relation, called substitutability: Abstract operation specifications in the source system (e.g., the import interface) shall by substituted by their implementations in the target system (e.g., the body of another module).

3.4 Towards Substitution Morphisms

Let us come back to the discussion of the connector between the import interface IMP of the requestor module MOD and the export interface EXP' of the

$$f_{TG}^>(\pi(p)) \sqsubseteq \pi'(p') \quad \overset{\nRightarrow}{\nLeftarrow} \quad \pi(p) \sqsubseteq f_{TG}^<(\pi'(p'))$$

$$f_{TG}^>(\pi(p)) \sqsupseteq \pi'(p') \quad \overset{\Rightarrow}{\nLeftarrow} \quad \pi(p) \sqsupseteq f_{TG}^<(\pi'(p'))$$

Fig. 10. Relation between the different alternatives for the TGTS-morphisms.

provider module MOD'. Since IMP and EXP' are TGTS, the desired relation between them should be described via a TGTS morphism. To construct an appropriate recipe for the morphism between IMP and EXP', it is necessary to understand what are the semantic requirements behind this a relation.

IMP contains the abstract specifications of the required features, so it is natural to interpret its rules as incomplete, with DPB semantics. The provider offers the concrete implementations of the operations, therefore the corresponding rules should be complete, having DPO interpretation.

The concrete rules can be safely substituted for the abstract rules if the following two conditions are true: First, the effect of applying a rule of EXP' should satisfy the expectations described in the rule of IMP for which it was substituted. This is the case if IMP reflects the EXP' behavior. Second, applicability of the IMP rule should imply applicability of the corresponding EXP' rule, i.e., applicability must be preserved from IMP to EXP'. These two requirements are formally given in the following definition.

Definition 10 (substitutability). *Given typed graph transformation systems $GTS = (TG, P, \pi)$, the source system, and $GTS' = (TG', P', \pi')$, the target system, the second is substitutable for the source if there exists a functor $F : \mathbf{Trf}(GTS') \to \mathbf{Trs}(GTS)$ such that for all graphs $G' \in |\mathbf{Trf}(GTS')|$ and for all transition sequences $\rho : F(G') \to \_ \in \mathbf{Trs}(GTS)$ there exists a transformation sequence $\rho' : G' \to \_ \in \mathbf{Trf}(GTS')$ with $F(\rho') = \rho$.*

Let us give an operational interpretation of what happens when the abstract rules $p_i = (L_i \xleftarrow{l_i} K_i \xrightarrow{r_i} R_i)$ are substituted for the concrete rules $p_i' = (L_i' \xleftarrow{l_i'} K_i' \xrightarrow{r_i'} R_i')$. This assumes that requestor and provider are actual components which communicate at runtime.

$$
\begin{array}{ccccccc}
G_0 & \overset{p_1/d_1}{\rightsquigarrow} & H_0 = G_1 & \overset{p_2/d_2}{\rightsquigarrow} & H_1 = G_2 & \cdots \\
{\scriptstyle f_{TG}^<}\Big\uparrow & & {\scriptstyle f_{TG}^<}\Big\uparrow & & {\scriptstyle f_{TG}^<}\Big\uparrow & \\
G_0' & \overset{p_1'/d_1'}{\Longrightarrow} & H_0' = G_1' & \overset{p_2'/d_2'}{\Longrightarrow} & H_1' = G_2' & \cdots
\end{array}
$$

Fig. 11. Substitution in detail.

The starting point is a graph $G_0' \in \mathbf{Graph_{TG'}}$, representing the state of the provider component (cf. Fig. 11).

The substitution consists of the following steps:

- G'_0 is projected to $G_0 \in \mathbf{Graph_{TG}}$ via backward retyping, modeling the requestors incomplete knowledge about the provider.
- If a rule p_1 is applicable to G_0 on the requestor side, the same holds for the corresponding provider rule $p'_1 = f_P(p_1)$.
- A transformation step $G' \overset{p'_1/d'_1}{\Longrightarrow} H'$ is performed by the provider which projects to a transition $G_0 \overset{p_1/d_1}{\rightsquigarrow} H$ via the corresponding rule in the requestor view.

Thus, the requestor receives an update to its local view of the state of the provider, and the cycle can start anew.

After this operational motivation, let us understand the consequences of the semantic requirements of Def. 10, i.e., reflection of behavior and preservation of rule applicability. The first requires that rules are compared over the GTS (backward retyping for *Step 2*) with $\pi(p) \sqsubseteq_{DPB} f^<_{TG}(\pi'(p'))$ (DPB subrule from p to p' for *Steps 3 and 4*). The second is guaranteed if the left-hand sides of the rules p' is contained in that of p (*Steps 3 and 4*), compared over GTS' (forward retyping for *Step 2*). Thus, modulo retyping, p is contained in p', but L' is contained in L, i.e., the rules must be essentially identical.

It is clear that this is not a satisfactory result because it means that, again, requestor and provider components have to be developed in close coordination. We will see in the next section that the solution consists in a separation of the *preconditions* for the application of the rule from the description of the *effects* of the transformation. Indeed, the problem occurs because the left-hand side of a rule mixes up items restricting the applicability with items needed to specify the actual transformations.

4 Separating Preconditions and Effects

As discussed in the previous section, the separate specifications of application conditions and transformations allows for a more flexible notion of substitution morphisms. The desired separation is achieved by extending rules with positive and negative application conditions as introduced below. In Section 4.2, substitution morphisms will be introduced formally. It will be illustrated by an example in Section 5, as well as the other morphisms discussed so far.

4.1 Application Conditions

Negative conditions are well-known to increase the expressive power of rules [14]. This is not the case for positive application conditions which are easily encoded in the left-hand side of a rule (more precisely: in both the left- and the right-hand side if the elements are to be preserved).

However, this encoding, while leading to an identical operational behavior, is not compatible with the semantic requirements for substitution morphisms.

For example, by strengthening the precondition of an operation in the import we should preserve legal substitution relations because the overall requirements towards existing implementations are weakened. Yet, due to the encoding we are enlarging the rule itself, which reduces the collection of legal substitution morphisms outgoing from the import interface.

Therefore, we consider in the following definition negative as well as positive application conditions.

Definition 11 (rules with application conditions). *An application condition $A(p) = (AP(p), AN(p))$ for a graph transformation rule $p : (L \xleftarrow{l} K \xrightarrow{r} R)$ consists of two sets of typed graph morphisms $AP(p), AN(p)$ outgoing from L which contain positive and negative constraints, respectively. $A(p)$ is called positive (negative) if $AN(p)$ $(AP(p))$ is empty.*

Let $L \xrightarrow{\hat{\imath}} \hat{L}$ be a positive or negative constraint and $L \xrightarrow{d_L} G$ be a typed graph morphism (cf. Fig. 12). Then d_L P-satisfies $\hat{\imath}$, if there exists a typed graph morphism $\hat{L} \xrightarrow{d_{\hat{L}}} G$ such that $d_{\hat{L}} \circ \hat{\imath} = d_L$. d_L N-satisfies $\hat{\imath}$, if it does not P-satisfy $\hat{\imath}$.

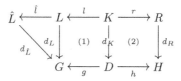

Fig. 12. DPB graph transition and rule with application condition.

Let $A(p) = (AP(p), AN(p))$ be an application condition and $L \xrightarrow{d_L} G$ be a typed graph morphism. Then d_L satisfies $A(p)$, if it P–satisfies at least one positive constraint and N-satisfies all negative constraints from $A(p)$.

A graph transformation rule with application condition is a pair $\hat{p} = (p, A(p))$ consisting of a graph transformation rule $p : s = (L \xleftarrow{l} K \xrightarrow{r} R)$ and an application condition $A(p)$ for p. It is applicable to a graph G via $L \xrightarrow{d_L} G$ if d_L satisfies $A(p)$.

Let $\hat{p} = (p : (L \xleftarrow{l} K \xrightarrow{r} R), A(p))$ be a graph transformation rule with application condition. A graph transition from G to H via the rule $\hat{p}$, denoted by $G \overset{\hat{p}/d}{\rightsquigarrow} H$, is a graph transition via a rule p, such that $d_L \in d$ satisfies the application condition of $\hat{p}$.

Note that positive application conditions consist of a disjunction of positive constraints, in contrast with the conjunction in [14]. That means, $L \xrightarrow{d_L} G$ satisfies $AP(p)$ if it satisfies *at least one* positive constraints. So, positive and negative conditions are, in fact, dual to each other.

As an example of a rule with positive and negative constraints let us consider the rule *req* in Fig. 6. Constraints are represented in the left-hand side of the rule where they are distinguished by dotted borders. If a positive constraint coincides with L, we omit this border. All nodes and edges outside these borders form the left-hand side L while $\hat{L}$ is given by the left-hand side plus one of the bordered parts and $\hat{l}$ or $\hat{k}$ by the corresponding embedding. Two negative constraints and one positive, being identical to the left-hand side, constitute the application condition of the rule *req*.

4.2   Substitution Morphism

We proceed with the definition of substitution morphisms, consisting of two parts. The first one ensures that the applicability of the requestor rule implies the applicability of the associated provider rule. This is similar to behavior-preserving morphisms (cf. cell 5 in Table 1) except that the application condition is considered instead of the actual rule.

The second part of the definition ensures the reflection of effects. Thus, behavior-reflecting morphisms are appropriate here, but only for those rules of EXP' which are associated to rules of IMP, cf. cell 3 of Table 1.

Below, we first deal with reflection of effects and then with preservation of applicability.

Definition 12 (substitution morphism). *Given typed graph transformation systems* $GTS = (TG, P, \pi)$ *and* $GTS' = (TG', P', \pi')$ *containing graph transformation rules with application conditions. A* substitution morphism $f^{sub} = (f_{TG}, f_P)$ *is given by a type graph morphism* $f_{TG} : TG \to TG'$ *and a mapping* $f_P : P \to P'$ *between the sets of rule names, such that for each* $p \in P$ *we have*

1. $\pi(p) \sqsubseteq_{DPB} f_{TG}^<(\pi'(p'))$ *(cf. Fig. 13 on the right)*
2. *applicability of* p *implies that of* $f_P(p) = p'$, *i.e.*
   (a) *for each* $f_{TG}^>(\hat{l} : L \to \hat{L}) \in f_{TG}^>(AP(p))$ *there exist* $\hat{l}' : L' \to \hat{L}' \in AP(p')$ *and a graph homomorphisms* $h_{\hat{L}'_P} : \hat{L}' \to f_{TG}^>(\hat{L})$ *such that the corresponding square in Fig. 13 on the left commutes;*
   (b) *for each* $\hat{k}' : L' \to \hat{L}' \in AN(p')$ *there exist* $f_{TG}^>(\hat{k} : L \to \hat{L}) \in f_{TG}^>(AN(p))$ *and a graph homomorphism* $h_{\hat{L}'_N} : f_{TG}^>(\hat{L}) \to \hat{L}'$ *such that the corresponding square in Fig. 13 on the left commutes.*

The justification for the definition of the substitution morphism is presented in the following theorem.

Theorem 1. *The substitution morphism* $f^{sub} = (f_{TG}, f_P)$ *satisfies the semantic requirements of Def. 10.*

Proof Sketch. It is necessary to show that Def. 12 implies Def. 10, i.e. (1) transformation steps via a GTS' rule can be considered as transitions via the corresponding GTS rule, and (2) the applicability of this GTS rule implies the

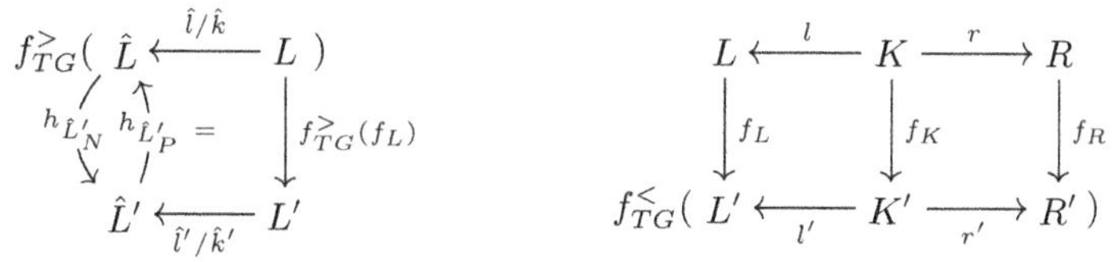

$$f^>_{TG}(\ \hat{L} \xleftarrow{\hat{l}/\hat{k}} L\ )$$

$$\begin{matrix} & \nearrow & \uparrow & & \\ h_{\hat{L}'_N} & h_{\hat{L}'_P} & = & \Big\downarrow f^>_{TG}(f_L) \\ & \searrow & \big\downarrow & & \\ & \hat{L}' & \xleftarrow{\ \hat{l}'/\hat{k}'\ } & L' & \end{matrix}$$

$$\begin{matrix} L & \xleftarrow{\ l\ } & K & \xrightarrow{\ r\ } & R \\ \big\downarrow f_L & & \big\downarrow f_K & & \big\downarrow f_R \\ f^<_{TG}(\ L' & \xleftarrow{\ l'\ } & K' & \xrightarrow{\ r'\ } & R'\ ) \end{matrix}$$

Fig. 13. Substitution morphism of graph transformation rules (the functors $f^>_{TG}$ and $f^<_{TG}$ are applied to the entire constraint of p in the left part of figure and to the entire bottom span in the right part of figure correspondingly).

applicability of the GTS' rule. Assume two graph transformation rules with application conditions $\hat{p} = (\pi(p), A(p))$ in GTS and $\hat{p}' = (\pi'(p'), A(p'))$ in GTS' such that $p' = f_P(p)$.

1. It is necessary to demonstrate that each transformation step via the GTS' rule can be reflected by a transition via the GTS rule. By assumption, for each backward retyped rule $f^<_{TG}(\hat{p}')$ there is a DPO-/DPB-subrule $\hat{p}$ in GTS, i.e. there exist graph homomorphisms between the first and the second rule (f_L, f_K, f_R), forming a faithful transition (cf. Fig. 14 on the right). Now, both transitions can be vertically composed using the composition of the underlying pushout/pullback squares. The faithfulness of the composed transition follows from the preservation of the identification condition under the composition of pushout/pullback squares. Obviously, both transitions have the same underlying span $G \xleftarrow{g} D \xrightarrow{h} H$.

2. We have to show that if $f^>_{TG}(d_L)$ satisfies the application condition of $f^>_{TG}(\hat{p})$, then $d_{L'}$ satisfies the application condition of $\hat{p}'$. This induces two problems:
   (a) $d_{L'}$ (cf. Fig. 14 on the left) must N-satisfy all negative constraints of $\hat{p}'$, i.e., there must not exist $d_{\hat{l}'} : \hat{L}' \rightarrow G'$. This can be proved by assuming existence of $d_{\hat{L}'}$ and showing a contradiction. The full proof of this can be found in [1].
   (b) $d_{L'}$ (cf. Fig. 14 on the left) must P-satisfy some positive constraint of $\hat{p}'$. Since the satisfiability of the positive constraints is defined dually to the negative case, the proof is analogous.
   Combining (a) and (b), we obtain that $d_{L'}$ satisfies the application condition of $\hat{p}'$.

Further we discuss the application of the introduced TGTS morphisms.

5   Application of TGTS Morphisms

In this section we revise the *intra-connectors* relating the import/export interfaces and the body of a module and introduce a new concept of *inter-connector* employing the substitution morphism defined in the previous section. The interconnectors determine a relation between the import and export interfaces of two modules being requestor and provider of a specific service. To illustrate the

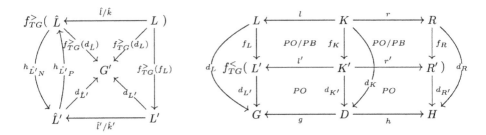

Fig. 14. Substitution morphisms satisfy the semantic requirements (the functors $f^>_{TG}$ and $f^<_{TG}$ are applied to the entire application constraint of p in the left part of figure and to the entire bottom span in the right part of figure correspondingly).

application of the substitution morphism as the inter-connector a module implementing the algorithm for distributed deadlock detection (DDD) is introduced.

5.1   Extended Scenario

The algorithm for distributed deadlock detection is specified by the module MOD' depicted in the lower part of Fig. 15. The upper part of this figure shows the module MOD modeling the algorithm for mutual exclusion discussed in Section 3.1.

The module MOD' offers a deadlock detection service at the export interface EXP' asked for by the module MOD at the import interface IMP (cf. IMP and EXP' in Fig. 15). At the same time, the module MOD' lacks deadlock resolution capabilities provided, in turn, by the module MOD through the export interface EXP (cf. IMP' and EXP in Fig. 15). In general, such a relation between module interfaces, called *cyclic import*, might be problematic for practical realization. However, it properly illustrates different kinds of module connectors.

Example 4 (distributed deadlock detection). The main purpose of MOD' is to observe processes and resources and to detect a deadlock if asked to do so. In a graph representing a system state, a deadlock appears as a cycle of *request* and *held_by* edges, where one process requests a resource held by another process and simultaneously holds a resource requested by it. The distributed deadlock detection uses *blocked* messages, represented by edges with a black flag, in order to detect such cyclic dependencies.

The algorithm is invoked by a process p waiting for a resource r. The process uses rule *dead?* to send a *blocked*-message to r. This feature is offered by MOD' at EXP' for external use, e.g., by MOD. If the resource is held by another process which itself is waiting for a resource, the message is passed on using *waiting*. If this is not the case, which is checked by a negative application condition, the message is deleted by rule *ignore*. Thanks to the mutual exclusion, each resource is held by only one process. Hence, if the message arrives at a resource which is held by the original sender, a cycle has been detected.

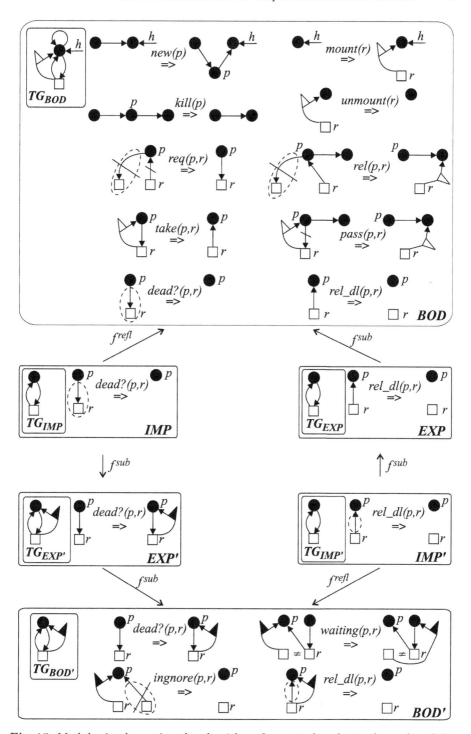

Fig. 15. Modules implementing the algorithms for mutual exclusion (upper) and distributed deadlock detection (lower).

Since MOD' is only destined for deadlock detection, deadlock resolution is described only abstractly by the rule rel_dl, which deletes the *blocked*-message, but does not decide how the deadlock is actually resolved. This rule in the import interface IMP' needs to be replaced by the rule of MOD with the same name. The positive application condition of rel_dl restricts the rule applicability to the system states where a resource is held by the process, i.e. to the situations being meaningful for the deadlock resolution.

5.2 TGTS Morphisms as Intra-connectors of Modules (Revised)

We proceed with the discussion of intra-connectors relating an import interface and a body of a module. In Section 3.1 behavior-reflecting morphisms were proposed for this purpose. We shall verify whether the requirements of Def. 9 are fulfilled for the corresponding constituents of the modules MOD and MOD' in Fig. 15.

First of all, we establish a type graph morphism f_{TG} and a mapping f_P between the sets of rule names. In the module MOD the type graph TG_{IMP} of the import is a subgraph of TG_{BOD} of the body. Similarly, the type graph $TG_{IMP'}$ is a subgraph of $TG_{BOD'}$ in the module MOD'. The type graph morphisms in both cases are given by inclusions. The corresponding rules in the source and target systems are identified by their names, i.e. *dead?* for IMP and BOD, and rel_dl for IMP' and BOD'.

The rule *dead?* in IMP is a subrule of the BOD one, simply because the two rules are identical (cf. Def. 9). The rule rel_dl in IMP' is a subrule of the BOD' rule, because the latter becomes identical to the IMP' rule after the backward retyping. Hence, the specifications at IMP and IMP' conform with BOD and BOD' correspondingly.

In contrast with the import-body connector, the requirements towards the connector between export interface and body shall be strengthened. Behavior-preservation guarantees that the applicability of rules in the export interface implies the applicability of corresponding body rules. However, this property would be satisfied even for empty body rules. In fact, we also require that the effect achieved by the body rules is at least the one promised by the rules in the export interface. Hence, we "upgrade" behavior-preserving morphisms to substitution morphisms. Next we shall demonstrate that the relations between exports and bodies of the modules in Fig 15 are indeed substitution morphism.

To show this one should check preservation of applicability from the export interface to the body and reflection of the effects between the rules in the body and the export interface. The rules rel_dl in the export interface EXP of MOD and *dead?* in EXP' of MOD' are identical to the body rules of the modules, and so the properties required by Def 12 obviously hold.

5.3 TGTS Morphisms as Inter-connectors of Modules

The ultimate aim of matching import and export interfaces of requestor and provider modules is to check whether the corresponding rules in the body of the former can be safely substituted for the rules in the body of the later. That means the obvious choice of morphism for this inter-connector is the substitution morphism.

Let us first discuss the relation between the import interface IMP and the export interface EXP' of the modules in Fig 15. The type graph morphism f_{TG} from TG_{IMP} to $TG_{EXP'}$ is given by an inclusion. The mapping f_P between the sets of rule names is unique, because only one rule *dead?* is contained in each of the interfaces.

After that, one should check the preservation of applicability from IMP to EXP' (cf. Def 12). Each of the rules has one positive application condition being the union of the left-hand side and the dotted part for the IMP rule, and coinciding with the left-hand side for the EXP' rule. The application conditions of the two rules are the same because the forward retyping of the EXP' rule does not introduce any changes. Applicability is thus preserved.

The last step is the reflection of effects. While the backward retyping of the EXP' rule gets rid of the *blocked*-message, it is still bigger in context and effect then the IMP rule. This is allowed by the DPB-subrule relation which can be established between the two rules. Thereby, the import interface IMP is associated with the export interface EXP' by a substitution morphism.

Now we discuss the relation between the import interface IMP' and the export interface EXP. The type graphs $TG_{IMP'}$ and TG_{EXP} of the two systems are the same, consequently the retyping does not change the rules. The positive application conditions of the rules *rel_dl* coincide, that means preservation of applicability from IMP' to EXP. The reflection of effects is ensured by the DPO-subrule relation between the two rules in spite of the bigger context of the EXP' rule additionally containing the *held_by* edge. Hence, the import interface IMP' and the export interface EXP are also connected by a substitution morphism.

The fact that import-export and export-body relations are both described by substitution morphisms allows us, by means of their composition, to consider the body of the provider module as a replacement for the export of the requestor. This is the first prerequisite for a composition of modules, i.e., the actual substitution of the import by the body. The detailed analysis of this construction is, however, beyond the scope of this paper.

The final section summarizes the main results of our work.

6   Conclusion

The contributions of this paper can be summarized in two points: a systematic presentation of morphisms of graph transformation systems along with a recipe of how to define new variants, if needed, in a generic framework; and a novel notion of substitution morphism between graph transformation systems with application conditions being uniformly introduced in the context of this framework.

The latter has been motivated by the need to connect import and export interfaces of modules in a flexible way, i.e., such that they can be developed independently of each other. The first result is a reaction to the multitude of proposals and variants that exist in the literature.

Future work will include the further analysis of modules based on the connectors introduced here, in particular their composition, as well as possible generalizations towards refinements of both the general framework and the notion of substitution morphism.

References

1. A. Cherchago and R. Heckel. Specification matching of web services using conditional graph transformation rules. In H. Ehrig, G. Engels, F. Parisi-Presicce, and G. Rozenberg, editors, *Proc. 2nd Int. Conference on Graph Transformation (ICGT'04), Rome, Italy*, volume 3256 of *LNCS*. Springer-Verlag, 2004.
2. A. Corradini, H. Ehrig, M. Löwe, U. Montanari, and J. Padberg. The category of typed graph grammars and their adjunction with categories of derivations. In *5th Int. Workshop on Graph Grammars and their Application to Computer Science, Williamsburg '94, LNCS 1073*, pages 56–74. Springer-Verlag, 1996.
3. A. Corradini, U. Montanari, and F. Rossi. Graph processes. *Fundamenta Informaticae*, 26(3,4):241–266, 1996.
4. A. Corradini, U. Montanari, F. Rossi, H. Ehrig, R. Heckel, and M. Löwe. Algebraic approaches to graph transformation, Part I: Basic concepts and double pushout approach. In G. Rozenberg, editor, *Handbook of Graph Grammars and Computing by Graph Transformation, Volume 1: Foundations*, pages 163–245. World Scientific, 1997. Preprint available as Tech. Rep. 96/17, Univ. of Pisa, http://www.di.unipi.it/TR/TRengl.html.
5. H. Ehrig and G. Engels. Pragmatic and semantic aspects of a module concept for graph transformation systems. In *5th Int. Workshop on Graph Grammars and their Application to Computer Science, Williamsburg '94, LNCS 1073*, LNCS, pages 137–154. Springer-Verlag, 1996.
6. H. Ehrig, G. Engels, H.-J. Kreowski, and G. Rozenberg, editors. *Handbook of Graph Grammars and Computing by Graph Transformation, Volume 2: Applications, Languages, and Tools*. World Scientific, 1999.
7. H. Ehrig and B. Mahr. *Fundamentals of algebraic specification 2: module specifications and constraints*. Springer-Verlag, 1990.
8. H. Ehrig, M. Pfender, and H.J. Schneider. Graph grammars: an algebraic approach. In *14th Annual IEEE Symposium on Switching and Automata Theory*, pages 167–180. IEEE, 1973.
9. M. Große–Rhode, F. Parisi Presicce, and M. Simeoni. Concrete spatial refinement construction for graph transformation systems. Technical Report SI 97/10, Università di Roma La Sapienza, Dip. Scienze dell'Informazione, 1997.
10. M. Große–Rhode, M. Simeoni, and F. Parisi Presicce. Refinements and modules for typed graph transformation systems. In J.L.Fiadeiro, editor, *Proc. WADT'98 (Workshop on Algebraic Development Techniques), at ETAPS'98, Lisbon, April*, number 1589 in LNCS, pages 138 – 151. Springer, 1999.
11. M. Grosse-Rhode, F. Parisi-Presicce, and M. Simeoni. Spatial and temporal refinement of graph transformation systems. In *Proc. of Mathematical Foundations of Computer Science 1998*, volume 1450 of *LNCS*, pages 553–561. Springer-Verlag, 1998.

12. M. Grosse-Rhode, F. Parisi-Presicce, and M. Simeoni. Refinements and modules for typed graph transformation systems. In J.L. Fiadeiro, editor, *Proc. Workshop on Algebraic Development Techniques (WADT'98), at ETAPS'98, Lisbon, April 1998*, volume 1589 of *LNCS*, pages 138–151. Springer-Verlag, 1999.

13. M. Große-Rhode, F. Parisi-Presicce, and M. Simeoni. Refinement of graph transformation systems via rule expressions. In H. Ehrig, G. Engels, H.-J. Kreowski, and G. Rozenberg, editors, *Proc. 6th Int. Workshop on Theory and Application of Graph Transformation (TAGT'98), Paderborn, November 1998*, volume 1764 of *LNCS*, pages 368–382. Springer-Verlag, 2000.

14. A. Habel, R. Heckel, and G. Taentzer. Graph grammars with negative application conditions. *Fundamenta Informaticae*, 26(3,4):287 – 313, 1996.

15. J.H. Hausmann, R. Heckel, and M. Lohmann. Model-based discovery of web services. In *Proc. International Conference on Web Services*, San Diego, USA, July 2004.

16. R. Heckel. *Open Graph Transformation Systems: A New Approach to the Compositional Modelling of Concurrent and Reactive Systems*. PhD thesis, TU Berlin, 1998.

17. R. Heckel, A. Corradini, H. Ehrig, and M. Löwe. Horizontal and vertical structuring of typed graph transformation systems. *Math. Struc. in Comp. Science*, 6(6):613–648, 1996.

18. R. Heckel, H. Ehrig, U. Wolter, and A. Corradini. Double-pullback transitions and coalgebraic loose semantics for graph transformation systems. *Applied Categorical Structures*, 9(1), January 2001. See also TR 97-07 at http://www.cs.tu-berlin.de/cs/ifb/TechnBerichteListe.html.

19. R. Heckel, G. Engels, H. Ehrig, and G. Taentzer. Classification and comparison of modularity concepts for graph transformation systems. In Ehrig et al. [6], pages 669 – 690.

20. R. Heckel, G. Engels, H. Ehrig, and G. Taentzer. A view-based approach to system modelling based on open graph transformation systems. In Ehrig et al. [6], pages 639 – 667.

21. H.-J. Kreowski and S. Kuske. On the interleaving semantics of transformation units - a step into GRACE. In *5th Int. Workshop on Graph Grammars and their Application to Computer Science, Williamsburg '94, LNCS 1073*, pages 89 – 106. Springer-Verlag, 1996.

22. B. Meyer. *Object-Oriented Software Construction*. Prentice Hall International, 1988.

23. L. Ribeiro. *Parallel Composition and Unfolding Semantics of Graph Grammars*. PhD thesis, TU Berlin, 1996.

24. A. Schürr and A.J. Winter. UML packages for PROgrammed Graph REwrite Systems. In *Selected Papers of 6th International Workshop on Theory and Application of Graph Transformations (TAGT'98), Paderborn, Germany*, volume 1764 of *LNCS*, pages 396–409. Springer-Verlag, 1999.

25. A.M. Zaremski and J.M. Wing. Signature matching: a tool for using software libraries. *ACM Transactions on Software Engineering and Methodology (TOSEM)*, 4(2):146 – 170, April 1995.

26. A.M. Zaremski and J.M. Wing. Specification matching of software components. In *Proc. SIGSOFT'95 Third ACM SIGSOFT Symposium on the Foundations of Software Engineering*, volume 20(4) of *ACM SIGSOFT Software Engineering Notes*, pages 6–17, October 1995. Also CMU-CS-95-127, March, 1995.

Simulating Algebraic High-Level Nets by Parallel Attributed Graph Transformation

Claudia Ermel[1], Gabriele Taentzer[1], and Roswitha Bardohl[2]

[1] Technische Universität Berlin, Germany
{lieske,gabi}@cs.tu-berlin.de

[2] Intern. Center for Computer Science, Schloss Dagstuhl, Germany
rosi@dagstuhl.de

Abstract. The "classical" approach to represent Petri nets by graph transformation systems is to translate each transition of a specific Petri net to a graph rule (behavior rule). This translation depends on a concrete model and may yield large graph transformation systems as the number of rules depends directly on the number of transitions in the net. Hence, the aim of this paper is to define the behavior of Algebraic High-Level nets, a high-level Petri net variant, by a parallel, typed, attributed graph transformation system. Such a general parallel transformation system for AHL nets replaces the translation of transitions of specific AHL nets. After reviewing the formal definitions of AHL nets and parallel attributed graph transformation, we formalize the classical translation from AHL nets to graph transformation systems and prove the correctness of the translation. The translation approach then is contrasted to a definition for AHL net behavior based on parallel graph transformation. We show that the resulting amalgamated rules correspond to the behavior rules from the classical translation approach.

1 Introduction

Visual modeling languages (like the Unified Modeling Language (UML), Petri nets, Statecharts, and many more) play a central role for software and system modeling. Visual models are used for system design, simulation, validation, and code generation. Apart from developing visual models, the simulation of a model on the basis of a formal specification is an important issue for testing and validating the system behavior. The simulation of Petri nets, for example, is realized by playing the token game: a transition can fire if it is enabled, a firing step removes tokens from the transition's predomain places and adds tokens to its postdomain places.

Petri net behavior can be defined as graph transformation system where each transition is translated to a graph rule modeling the corresponding change of the marking (deleting and/or adding tokens) in a firing step [15, 3]. This "classical" way to define Petri net behavior by graph transformation assumes a specific Petri net before compiling its transitions into graph rules (*compiler* approach).

Yet, for related visual behavior modeling languages, it is often possible to define a general graph transformation system which is independent of a specific

H.-J. Kreowski et al. (Eds.): Formal Methods (Ehrig Festschrift), LNCS 3393, pp. 64–83, 2005.

model and can be used to interpret arbitrary models of a visual language (*interpreter* approach). An example is a graph transformation system for describing the behavior of a Statechart variant given in [2]. In general, the interpreter approach is much more flexible and scalable than the compiler approach. As it is independent of a concrete model, the graph transformation system defined for the interpreter approach is fixed once for the complete visual language, i.e. the number of behavior rules is finite and does not grow with the size of the model (scalability). In contrast, using the compiler approach, each specific model must be translated to get the model-specific graph transformation system.

Unfortunately, it is difficult to give a general graph transformation system to simulate Petri nets as there may be arbitrary many places connected to a transition, leading to an arbitrary number of behavior rules. Hence, parallel graph transformation concepts have been used to simulate the behavior of Condition-Event nets in [20] and of *Timed Transition Petri Nets* in [4].

Parallel graph transformation was introduced by Ehrig and Kreowski in [6], later generalized to parallel high-level replacement systems [11] by Ehrig and Taentzer, further elaborated and applied to communication-based systems in [20]. The essence of parallel graph transformation is that (possibly infinite) sets of rules which have a certain regularity, so-called rule schemes, can be described by a finite set of rules modeling the elementary actions. For instance, when modeling the firing of a Petri net transition, the elementary actions would be the removal of a token from a place in the transition's predomain and the addition of a token to a postdomain place. For the description of such rule schemes the concept of amalgamating rules at subrules is used which is based on synchronization mechanisms for rules developed first in [5].

The aim of this paper is to present a formal interpreter approach to define the behavior of high-level Petri nets. A specific, well-defined variant of high-level nets are Algebraic High-Level nets, AHL nets for short, introduced by Ehrig, Padberg and Ribeiro in [18]. We present an interpreter approach for the behavior of AHL nets based on parallel attributed graph transformation. Thus, a general graph transformation system for simulating AHL nets replaces the translation of transitions of specific AHL nets. The resulting parallel behavior specification is formally proven to be semantically equivalent to the corresponding compiler approach translating each specific AHL net to a corresponding attributed graph transformation system. This compiler approach for AHL nets has been presented in [1] and is reviewed in a slightly modified form in this paper.

In Section 2, the formal definitions of AHL nets and their behavior are reviewed, using the well-known *Dining Philosophers* as running example. Section 3 presents the concepts of sequential (classical) and parallel attributed graph transformation. The concepts are the basis in Section 4 to formalize the translation from AHL nets to sequential graph transformation systems according to the compiler approach. We prove the semantical compatibility of an AHL net and its translation to a graph transformation system, i.e. we show that a firing sequence in the net corresponds to a graph transformation sequence in the translated graph transformation system. The compiler approach is contrasted by the interpreter approach based on parallel attributed graph transformation, in Section 5. An interaction scheme is presented specifying the elementary actions when simulating an AHL net. From this scheme, amalgamated rules are defined for AHL nets, and it is proven that these rules correspond semantically to the behavior rules of the sequential graph transformation system given in Section 4. The conclusion (Section 6) gives an outlook on how the simulating graph transformation systems for AHL nets are used in the visual language environment GENGED for simulating and animating the behavior of AHL nets.

2   Algebraic High-Level Nets

An AHL net is a combination of a place/transition net [19] and an algebraic datatype specification $SPEC$ describing operations used as arc inscriptions. Tokens are elements of a corresponding $SPEC$-algebra [8, 7]. In this section, we review the definition of AHL nets and their behavior as given in [18], and present our running example, the well-known *Dining Philosophers*.

In contrast to other variants of AHL nets [14, 16] we do not label places with sorts. The pre- and postdomain of a transition is given by a multiset of pairs of terms and places, i. e. as elements of a commutative monoid.

Definition 1 (Algebraic High-Level Net).
An algebraic high-level net $N = (SPEC, P, T, pre, post, cond, A)$ consists of an algebraic specification $SPEC = (S, OP, E; X)$ with equations E and additional variables X over the signature (S, OP), sets P and T of places and transitions respectively, pre- and postdomain functions $pre, post : T \rightarrow (T_{OP}(X) \times P)^{\oplus}$ assigning to each transition $t \in T$ the pre- and postdomains $pre(t)$ and $post(t)$, respectively, a firing condition function $cond : T \rightarrow \mathcal{P}_{fin}(EQNS(S, OP, X))$ assigning to each transition $t \in T$ a finite set $cond(t)$ of equations over the signature (S, OP) with variables X, and an (S, OP, E)-algebra A.

Remarks

- $T_{OP}(X)$ is the set of terms with variables X over the signature (S, OP), and $M^{\oplus}$ is the free commutative monoid over a set M. Thus, $T_{OP}(X) \times P = \{(term, p) | term \in T_{OP}(X), p \in P\}$.
- The predomain function $pre(t)$ (and similar postdomain function $post(t)$) have the form $pre(t) = \sum_{i=1}^{n}(term_i, p_i)$ with $(n \geq 0)$, $p_i \in P, term_i \in T_{OP}(X)$. This means that $\{p_1, ...p_n\}$ is the predomain of t with arc-inscription $term_i$ for the arc from p_i to t if all $p_1, ..., p_n$ differ (unary case) and arc-inscription $term_{i1} \oplus ... \oplus term_{ik}$ for $p_{i1} = ... = p_{ik}$ (multi case). Note that in our sample AHL net (see Example 1) we have the multi case, but as drawing convention we draw separate arcs, each inscribed by one term only. Hence, we allow to draw more than one arc in one direction between a place and a transition.
- AHL nets together with AHL net morphisms build a category **AHLnet** [18].

Definition 2 (Marking and Firing Behavior of AHL Nets).
Let $N = (SPEC, P, T, pre, post, cond, A)$ be an AHL net according to Def. 1.

- A marking m is an element $m \in M^{\oplus}$ with $M = A \times P = \{(a,p)|a \in \bigcup_{s \in S} A_s, p \in P\}$

- Enabling and firing of transitions is defined as follows: For any $t \in T$ let $Var(t)$ be the set of local variables occurring in $pre(t), post(t)$ and $cond(t)$. An assignment $asg_A : Var(t) \to A$ is called consistent wrt. $t \in T$ if the equations $cond(t)$ are satisfied in A under asg_A. Transition t is enabled under a consistent assignment $asg_A : Var(t) \to A$ and a marking $m \in (A \times P)^{\oplus}$, if $pre_A(t, asg_A) \leq m$. The marking $pre_A(t, asg_A)$ – analogously $post_A(t, asg_A)$ – is defined for $pre(t) = \sum_{i=1}^{n}(term_i, p_i)$ by $pre_A(t, asg_A) = \sum_{i=1}^{n}(\overline{asg}_A(term_i), p_i)$, where $\overline{asg}_A : T_{OP}(Var(t)) \to A$ is the extended evaluation of terms under assignment asg_A. The successor marking m' is defined in the case of t being enabled by $m' = m \ominus pre_A(t, asg_A) \oplus post_A(t, asg_A)$ and gives raise to a firing step $m[t, asg_A\rangle m'$.

Example 1 (The Dining Philosophers as AHL Net).
As example we show the AHL net for *The Dining Philosophers* in Fig. 1 (see [19, 18] for the corresponding place/transition net). We identify the five philosophers as well as their chopsticks by numbers. Fig. 1 (a) shows the initial situation where all philosophers are thinking and all chopsticks are lying on the table. Fig. 1 (b) shows the AHL net with the corresponding initial marking. For this marking, the transition take is enabled as a thinking philosopher and his left and right hand side chopsticks are available. The firing of transition take with the variable binding $p = 2$, for example, removes token 2 from place thinking and adds it to place eating, whereas tokens 2 and 3 are removed from place table, as the chopstick computing operation (p mod 5) +1 is evaluated to 3.

 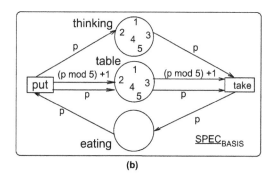

    (a)   (b)

Fig. 1. The *Dining Philosophers* (a) modeled as AHL Net (b)

As datatype specification we take a basic specification for all AHL nets $SPEC_{BASIS}$ consisting of the union of specifications NAT for natural numbers, $BOOL$ for boolean operations, and $STRING$ for strings. The tokens on all places are elements of a corresponding $SPEC_{BASIS}$-algebra, i.e. natural numbers in our example. The arcs are inscribed each by one variable or term from $T_{OP}(X)$ denoting computation operations to be executed on token values.

3   Parallel Attributed Graph Transformation

3.1   Attributed Graph Transformation

In the following, we present attributed graph structures as defined in [9]. For graph transformations in the category of attributed graph structures and homomorphisms with a distinguished class M of morphisms, the Church-Rosser, Parallelism and Concurrency Theorem have been shown in [9].

Definition 3 (Attributed Graph Structure Signatures). *A graph structure signature $GSIG = (S_G, OP_G)$ is an algebraic signature with unary operations $op : s \to s'$ in OP_G only. An attributed graph structure signature $ASSIG = (GSIG, DSIG)$ consists of a graph structure signature $GSIG$ and a data signature $DSIG = (S_D, OP_D)$ with attribute value sorts $S'_D \subseteq S_D$ such that $S'_D = S_D \cap S_G$ and $OP_D \cap OP_G = \varnothing$.*
$ASSIG$ is called well-structured if for each $op : s \to s'$ in OP_G we have $s \notin S_D$.

$ASSIG$-algebras and $ASSIG$-homomorphisms build up a category [9] which is denoted by **ASSIG-Alg**. In the following, we call $ASSIG$-algebras *attributed graphs* and $ASSIG$-homomorphisms *attributed graph morphisms*.

As an example for an attributed graph structure signature we define the signature ASSIG$_{AHL}$ for AHL nets. AHL nets are considered as ASSIG$_{AHL}$-algebras.

Definition 4 (Attributed Graph Structure Signature for AHL Nets).

The attributed graph structure signature for AHL nets (shown visually in Fig. 2) is given by ASSIG$_{AHL}$ = (GSIG$_{AHL}$, DSIG$_{AHL}$). In Fig. 2, the sorts of GSIG$_{AHL}$ are represented as nodes. The operations are the arcs between the sort nodes (the op-links between graph sorts) and from sort nodes to data nodes, (the attr-links between graph sorts and attribute sorts).

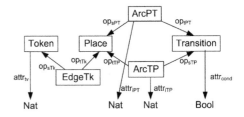

Fig. 2. Abstract Syntax Graph visualizing the ASSIG for AHL Nets

The DSIG part (data signature) consists of the attribute value sorts of the basic specification, i.e. String, Nat and Bool and their usual operations.

The attribute values are used for the arc inscriptions, tokens and transition firing conditions.

Next, we define the double-pushout approach to graph transformation on the basis of category **ASSIG-Alg**.

Proposition 1 (Pushouts of *ASSIG*-Homomorphisms). *Let M be a distinguished class of all homomorphisms f which is defined by $f \in M$ if f_{GSIG} is injective and $f_{DSIG} = id_{DSIG}$ for f in **ASSIG-Alg**. Given $f : A \to B \in M$ and $a : A \to C$ then there exists their pushout in **ASSIG-Alg**.*

Proof: See [9].

Category **ASSIG-Alg** and class M are fixed throughout this section.

Definition 5 (Typed Attributed Graph Transformation System). *A typed attributed graph transformation system $GTS = (S, P)$ based on (**ASSIG-Alg**, M) consists of an ASSIG-algebra S, called start graph and a set P of rules, where*

1. *a rule $p = (L \xleftarrow{l} I \xrightarrow{r} R)$ of ASSIG-algebras L, I and R attributed over the term algebra $T_{DSIG}(X)$ with variable set X of variables $(X_s)_{s \in S_{DSIG}}$, called left-hand side L, interface I and right-hand side R, and homomorphisms $l, r \in M$, i.e. l and r are injective and identities on the data type $T_{DSIG}(X)$,*

2. *a direct transformation $G \xRightarrow{p,m} H$ via a rule p and a homomorphism $m : L \to G$, called match, is given by the diagram to the right, called double-pushout diagram, where (1) and (2) are pushouts in **ASSIG-Alg** (the triple (m, i, m^*) is called rule embedding),*

$$
\begin{array}{ccccc}
L & \xleftarrow{\ l\ } & I & \xrightarrow{\ r\ } & R \\
{\scriptstyle m}\downarrow & (1) & \downarrow{\scriptstyle i} & (2) & \downarrow{\scriptstyle m^*} \\
G & \xleftarrow{\ g\ } & D & \xrightarrow{\ h\ } & H
\end{array}
$$

3. *a typed attributed graph transformation, short transformation, is a sequence $G_0 \Rightarrow G_1 \Rightarrow ... \Rightarrow G_n$ of direct transformations, written $G_0 \xRightarrow{*} G_n$,*
4. *the language $L(GTS)$ is defined by $L(GTS) = \{G \mid S \xRightarrow{*} G\}$.*

Now we add the concept of attribute conditions.

Definition 6 (Attribute Condition). *Given a rule p attributed over the term algebra $T_{DSIG}(X)$, an attribute condition C consists of a set of equations $(a = b)$ over $T_{DSIG}(X)$. An ASSIG-morphism $m : L \to G$ satisfies an attribute condition C, if $m_{DSIG}(a) = m_{DSIG}(b)$ for all $(a = b) \in C$.*

Definition 7 (Conditional Rule and Transformation). *Let $p = (L \xleftarrow{l} I \xrightarrow{r} R)$ be a rule attributed over the term algebra $T_{DSIG}(X)$, and C an attribute condition over $T_{OP}(X)$. Then, $\hat{p} = (p, C, X)$ is a conditional rule. The direct conditional transformation $G \xRightarrow{\hat{p},m} H$ is given by the direct transformation $G \xRightarrow{p,m} H$ if m satisfies C.*

A transformation sequence as well as a graph transformation system and its language based on conditional rules are defined as in Def. 5.

3.2 Parallel Graph Transformation

Parallel graph transformation in the double-pushout approach has been introduced in [20] on the basis of labeled graphs. Here, we extend the concepts to attributed graphs and rules with attribute conditions. The main idea of parallel graph transformation is to apply a number of rules in one parallel step. Their matches are allowed to overlap and can even be conflicting in the general case. Common subactions are described by subrules. Therefore, the notion of subrule embedding is basic to the whole approach.

Definition 8 (Subrule Embedding).

Given a conditional rule $\hat{p} = ((L \xleftarrow{l} I \xrightarrow{r} R), A, Y)$, *a conditional rule* $\hat{s} = ((L_s \xleftarrow{l_s} I \xrightarrow{r_s} R_s), A_s, X)$ *is called* subrule *of* $\hat{p}$ *if* $X \subseteq Y$ *and there are injective morphisms* $e : L_s \to L$, $f : I_s \to I$ *and* $g : R_s \to R$ *in* M *such that* $e \circ l_s = l \circ f$ *and* $g \circ r_s = r \circ f$, *i.e. the diagram to the right commutes.*

$$
\begin{array}{ccccc}
L_s & \xleftarrow{l_s} & I_s & \xrightarrow{r_s} & R_s \\
\downarrow{e} & = & \downarrow{f} & = & \downarrow{g} \\
L & \xleftarrow{l} & I & \xrightarrow{r} & R
\end{array}
$$

   The triple $t = (e, f, g)$ *from* $\hat{s}$ *to* $\hat{p}$ *(short* $t : \hat{s} \to \hat{p}$*) is called* subrule embedding*. In this context,* $\hat{p}$ *is called* extending rule*. Subrule embedding* t *is called* quasi-identical, *if* e, f, *and* g *are isomorphisms. In this case,* $\hat{s}$ *is called* isomorphic *to* $\hat{p}$. *Two subrule embeddings* $t_1 : \hat{s}_1 \to \hat{p}_1$ *and* $t_2 : \hat{s}_2 \to \hat{p}_2$ *are called* isomorphic, *if there are quasi-identical subrule embeddings from* $\hat{s}_1$ *to* $\hat{s}_2$ *and from* $\hat{p}_1$ *to* $\hat{p}_2$ *such that they commute with* t_1 *and* t_2.

All conditional rules and their subrule embeddings build up a category which we call **Rule$_{\mathbf{ASSIG-Alg}}$**. Three rule functors are defined to extract the LHS embeddings, the embeddings of interfaces and the RHS embeddings.

Definition 9 (Rule Functors). *The forgetful functors* V_L, V_I, V_R : **Rule$_{\mathbf{ASSIG-Alg}}$** $\to$ **ASSIG** − **Alg**, *called* rule functors, *are defined in the obvious way, e.g.* $V_L(\hat{p}) = V_L((L \xleftarrow{l} I \xrightarrow{r} R), C, Y) = L$.

To apply a set of rules in parallel in a synchronized way, we have to decide how and how often the rules can be applied to a host graph G. One possibility is to allow a rule to be applied at all different matches it has in G. This would result in a massively parallel application of rules which is not always wanted. To restrict the degree of parallelism, two control features are introduced: the *interaction scheme* and the *covering construction*. The interaction scheme is a set of subrule embeddings and restricts the synchronization possibilities of rule applications. The covering construction restricts the matching possibilities for the rules of the interaction scheme. One special covering construction, called *local*, allows to match a subrule s exactly once to a part $m(s)$ of G, and to match all rules extending s as often as possible to the surroundings of $m(s)$. In this way, a kernel action can be described in a variable context. Another important covering construction, called *fully synchronized* forbids conflicting rule matches, i.e. two rule matches of rules extending the same subrule s have to overlap completely at a match of their common subrule.

   Formally, a covering is described by an instance interaction scheme and a set of matches. The instance interaction scheme contains the concrete number of instances of each rule in the scheme, depending on how many matches into

G have been found for each rule of the interaction scheme. Thus, an interaction scheme can be seen as type information for instance interaction schemes.

Definition 10 (Interaction Scheme). *An* interaction scheme *IS consists of a set of subrule embeddings such that the following conditions hold:*

1. *for each two subrule embeddings $t_1 : \hat{s}_1 \to \hat{p}_1$ and $t_2 : \hat{s}_2 \to \hat{p}_2$ we have $\hat{s}_1 \neq \hat{s}_2$ or $\hat{p}_1 \neq \hat{p}_2$,*
2. *for each two subrule embeddings $t_1 : \hat{s} \to \hat{p}_1$ and $t_2 : \hat{s} \to \hat{p}_2$ in IS with $\hat{s} = (p_s, C_s, X)$, $\hat{p}_1 = (p_1, C_1, Y_1)$ and $\hat{p}_2 = (p_2, C_2, Y_2)$ we have $Y_1 \cap Y_2 = X$.*

IS is called local interaction scheme, *if there is one subrule $\hat{s}$ being the source of at least one subrule embedding to each extending rule.*

Definition 11 (Instance Interaction Scheme). *Given an interaction scheme IS, an interaction scheme IIS is an* instance interaction scheme *of IS, if there is a mapping ins : IIS $\to$ IS such that $\forall t \in$ IIS: if there is an isomorphic subrule embedding $t \xrightarrow{\sim} u$ then ins(t) = u.*

Definition 12 (Covering Construction). *Let IS be an interaction scheme and G an ASSIG-algebra. A* partial covering *COV = (IIS, MA) consists of an instance interaction scheme IIS of IS and a set MA of matches from all rules of all subrule embeddings in IIS to G such that they commute with the subrule embeddings, i.e. for any two subrule embeddings $t_1 : \hat{s} \to \hat{p}_1$ and $t_2 : \hat{s} \to \hat{p}_2$ in IIS there are two matches $m_s : L_s \to G$ and $m_p : L_p \to G$ in MA with $m_p \circ e = m_s$. Let $t_1 : \hat{s} \to \hat{p}_1$ and $t_2 : \hat{s} \to \hat{p}_2$ be any two subrule embeddings in IIS and $m_{p_1} : L_{p_1} \to G$ and $m_{p_2} : L_{p_2} \to G$ corresponding matches in MA.*

1. *COV is called* local, *if IIS is local, and if $\hat{p}_1$ is isomorphic to $\hat{p}_2$, then m_{p_1} has to be non-isomorphic to m_{p_2}.*
2. *COV is called* fully synchronized, *if there are two subrule embeddings $u_1 : \hat{s}' \to \hat{p}_1$ and $u_2 : \hat{s}' \to \hat{p}_2$ such that $m_{p_1}(L_{p_1}) \cap m_{p_2}(L_{p_2}) = m_{s'}(L_{s'})$.*

Since category **ASSIG-Alg** has initial objects being empty graphs attributed over $T_{DSIG}(X)$, and pushouts, it is finitely cocomplete [17], i.e. has all finite colimits. This is the basis to build the amalgamated rule of any partial covering which glues all parallel rules according to their subrule embeddings. Applying the amalgamated rule afterwards according to Def. 5 completes a parallel graph transformation step.

Definition 13 (Amalgamated Rule and Transformation). *Let G be a graph and COV = (IIS, MA) be a covering construction with IIS = $\bigcup_{n \in N} (t_n : \hat{s}_n \to \hat{p}_n)$ being an instance interaction scheme with $\hat{s}_n = ((L_{s_n} \xleftarrow{l_{s_n}} I_{s_n} \xrightarrow{r_{s_n}} R_{s_n}), C_{s_n}, Y_{s_n})$ and $\hat{p}_n = ((L_n \xleftarrow{l_n} I_n \xrightarrow{r_n} R_n), C_n, Y_n)$ and MA = $\bigcup_{n \in N} m_n : L_n \to G$. The* amalgamated rule *$\hat{p}_{COV} = ((L \xleftarrow{l} I \xrightarrow{r} R), C, Y)$ is constructed by the following steps:*

1. *Let L be the colimit object of $\bigcup_{n \in N} V_L(t_n) : V_L(s_n) \to V_L(p_n)$ with $a_n : V_L(p_n) \to L$.*

2. *Let I be the colimit object of $\bigcup_{n \in N} V_I(t_n) : V_I(s_n) \rightarrow V_I(p_n)$ with $b_n : V_I(p_n) \rightarrow I$.*
3. *Let R be the colimit object of $\bigcup_{n \in N} V_R(t_n) : V_R(s_n) \rightarrow V_R(p_n)$ with $c_n : V_R(p_n) \rightarrow R$.*
4. *Morphisms l and r are uniquely determined by the universal property of colimit (I, b_n) such that $a_n \circ l_n = l \circ b_n$ and $c_n \circ r_n = r \circ b_n$.*
5. $C = \bigcup_{n \in N} C_{s_n} \cup \bigcup_{n \in N} C_n.$
6. $Y = \bigcup_{n \in N} Y_{s_n} \cup \bigcup_{n \in N} Y_n.$

Match $m_{COV} : L \rightarrow G$ is uniquely determined by the universal property of colimit (L, a_n), i.e. $m \circ a_n = m_n$. An amalgamated graph transformation is a direct transformation $G \overset{\hat{p}_{COV}, m_{COV}}{\Longrightarrow} H$ applying amalgamated rule $\hat{p}$ at match m.

*A parallel attributed graph transformation system $PAGTS = (S, IScheme)$ based on (**ASSIG-Alg**, M) consists of an ASSIG-algebra S, called start graph and a set $IScheme$ of interaction schemes.*

Parallel transformation sequences and the language of a parallel attributed graph transformation system are defined analogously to Def. 5.

4   Translating AHL Nets to Sequential Graph Transformation Systems

The translation of AHL nets to attributed graph transformation systems generalizes that of P/T nets into graph transformation systems as proposed in the literature [3, 15] and reviews in a slightly modified form the concepts and results in [1]. An initially marked AHL net N together with its behavior is translated to an attributed graph transformation system $AGT = (G, P)$ with start graph G being the translation of the AHL net N with initial marking to an attributed graph typed over the type graph for AHL nets ASSIG_{AHL} (Def. 4), and the set of rules P being behavior rules $\hat{p}_t$, one for each transition $t \in T$ where L and R contain the transition's pre- and postdomain, and the rule application condition corresponds to the firing condition of t.

Definition 14 (Translation of a marked AHL net to an Attributed Graph). *Given an AHL net $N = (SPEC, P, T, pre, post, cond, A)$ with marking $m \in (A \times P)^{\oplus}$. The translation Tr of (N, m) is given by the function $Tr : (AHLnet, (A \times P)^{\oplus}) \rightarrow \mathsf{ASSIG}_{AHL}\text{-Alg}$ from the set of pairs of AHL nets plus markings to the set of algebras wrt. the attributed graph structure signature ASSIG_{AHL} (Def. 4) with*

$$Tr(N, m) = G = (G_{Place}, G_{Trans}, G_{Token}, G_{EdgeTk}, G_{ArcPT}, G_{ArcTP},$$
$$op_{sPT}, op_{tPT}, op_{sTP}, op_{tTP}, op_{sTk}, op_{tTk},$$
$$attr_{tv}, attr_{iPT}, attr_{iTP}, attr_{cond}), \qquad where$$

$G_{DSIG} = T_{OP}(X) \uplus A$ *(disjoint union of the term algebra with variables over ASSIG_{AHL} and A),*

$G_{Place} = P$ *(the place nodes)*, $G_{Trans} = T$ *(the transition nodes)*,
$G_{Token} = \{tk | tk = (a, p, i) \in \widetilde{m}\}$. *The multiset* $m \in (A \times P)^{\oplus}$ *is given by the set* $\widetilde{m} = \{(a, p, i) \in A \times P \times \boldsymbol{N} \,|\, 0 < i \leq m(a, p)\}$, *where multiple occurrences of the same element in* m *are numbered by* i *in* $\widetilde{m}$,
$G_{EdgeTk} = \{e_{tk} | tk \in G_{Token}\}$,
$G_{ArcPT} = \{arcPT | arcPT = (term, p, i) \in PreSet\}$,
$G_{ArcTP} = \{arcTP | arcTP = (term, p, i) \in PostSet\}$, *where the multisets of terms in arc inscriptions are given by the sets* $PreSet = \cup_{t \in T} PreSet_t$ *and* $PostSet = \cup_{t \in T} PostSet_t$ *where* $PreSet_t = \{(term, p, i) | pre(t)(term, p) \geq i > 0\}$ *corresponds to* $pre(t)$ *and, analogously,* $PostSet_t$ *to* $post(t)$.

$op_{sPT} : G_{ArcPT} \rightarrow G_{Place}$ *with* $op_{sPT}(term, p, i) = p \; \forall (term, p, i) \in G_{ArcPT}$,
$op_{tPT} : G_{ArcPT} \rightarrow G_{Trans}$ *with* $op_{tPT}(term, p, i) = t$, *if* $(term, p, i) \in PreSet_t$,
   $\forall (term, p, i) \in G_{ArcPT}$,
$op_{sTP} : G_{ArcTP} \rightarrow G_{Place}, op_{tTP} : G_{ArcTP} \rightarrow G_{Trans}$: *analogously,*
$op_{sTk} : G_{EdgeTk} \rightarrow G_{Token}$ *with* $op_{sTk}(e_{(a,p,i)}) = (a, p, i) \; \forall e_{(a,p,i)} \in G_{EdgeTk}$,
$op_{tTk} : G_{EdgeTk} \rightarrow G_{Place}$ *with* $op_{tTk}(e_{(a,p,i)}) = p \; \forall e_{(a,p,i)} \in G_{EdgeTk}$,

$attr_{tv} : G_{Token} \rightarrow \boldsymbol{N}$ *with* $attr_{tv}((a, p, i)) = a \; \forall (a, p, i) \in G_{TV}$,
$attr_{iPT} : G_{ArcPT} \rightarrow T_{OP}(X)$ *with* $attr_{iPT}((term, p, i)) = term \; \forall (term, p, i) \in G_{ArcPT}$, $attr_{iTP} : G_{ArcTP} \rightarrow T_{OP}(X)$: *analogously,*
$attr_{cond} : G_{Trans} \rightarrow \mathcal{P}_{fin}(EQNS(X))$ *with* $attr_{cond}(t) = cond(t) \; \forall t \in G_{Trans}$

Example 2 (AHL net Dining Philosophers *translated to an attributed graph).*

Fig. 3 shows the attributed graph resulting from the translation of the initially marked AHL net presented in Fig. 1 (b). We visualize Place nodes as ellipses, Transition nodes as rectangles, and Token nodes as coloured circles containing the token value attributes. Token nodes are connected to their places by EdgeTk arcs. ArcPT and ArcTP symbols are drawn as edges which are attributed by the arc inscription terms. In this example we have no firing conditions.

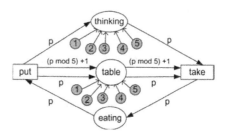

Fig. 3. Translation of AHL net *Dining Philosophers* with initial marking

In addition to the statical structure of the AHL net and the net marking, we now define the translation of a net's firing behavior into a set of graph rules P_{Tr}, the so-called behavior rules. Each behavior rule encorporates the firing behavior of one transition: the left-hand side contains its predomain, the right-hand side its postdomain. The firing condition $cond(t)$ is translated to the attribute condition of the behavior rule for transition t. The firing rules P_{Tr} together with the translated initially marked AHL net $Tr(N, m)$ form an attributed graph transformation system, the translation of the AHL net N including its behavior.

Definition 15 (Translation of AHL net firing behavior to graph rules).
*Let $N = (SPEC, P, T, pre, post, cond)$ be an AHL net. We translate the firing
behavior of N to a set of behavior rules $P_{Tr} = \{p_t = (L_t \xleftarrow{l_t} I_t \xrightarrow{r_t} R_t) | t \in T\}$
where for each transition $t \in T$ the rule components L_t, I_t and R_t are attributed
graphs over ASSIG_{AHL} (Def. 4), defined as follows:*
*The interface I_t contains only nodes of sort Place (the environment of transition
t) and no operations. All sorts and operations in L_t and R_t are empty, except
$\mathsf{Place}, \mathsf{Token}, \mathsf{EdgeTk}$ and the adjacent operations:*

- $L_{Place} = I_{Place} = R_{Place} = \{p | p \in pre(t) \cup post(t)\}$
- $L_{Token}[R_{Token}] = \{tk | tk = (term, p, i) \in PreSet_t[PostSet_t]\}$
- $L_{EdgeTk} = \{e_{tk} | tk \in L_{Token}\}$,

- $op^L_{sTK} : L_{EdgeTk} \rightarrow L_{Token}$ with $op^L_{sTk}(e_{(term,p,i)}) = (term, p, i)$,
- $op^L_{tTK} : L_{EdgeTk} \rightarrow L_{Place}$ with $op^L_{tTk}(e_{(term,p,i)}) = p$,
- $attr^L_{tv} : L_{Token} \rightarrow T_{OP}(X)$ with $attr_{tv}((term, p, i)) = term$
  (analogously for $R_{EdgeTk}, op^R_{sTK}, op^R_{tTK}$ and $attr^R_{tv}$)

*The rule morphisms $L_t \xleftarrow{l_t} I_t$ and $I_t \xrightarrow{r_t} R_t$ are given by $(p_{Place}, p_{Trans}, p_{Token},$
$p_{ArcPT}, p_{ArcTP}, p_{EdgeTk}) = (id_{Place}, \emptyset, \emptyset, \emptyset, \emptyset, \emptyset)$. Let $C = cond(t)$ be a set of
attribute conditions over $T_{OP}(X)$ as defined in Def. 6. Then, $\hat{p}_t = (p_t, C, X)$ is
the conditional rule corresponding to the firing behavior of transition t.*

Remarks. Both L_t and R_t contain only the places of the transition's environment and tokens connected to these places, where the tokens are attributed by
terms of $T_{OP}(X)$. The difference between L_t and R_t is that L_t corresponds to
$pre(t)$ whereas R_t corresponds to $post(t)$. The Token symbols are not in the interface I_t as the rule models the deletion of tokens from the predomain (L_t) and
the addition of tokens to the postdomain (R_t).

Combining the translations of a marked AHL net and of its firing behavior,
we obtain a complete translation of a marked AHL net including its behavior to
an attributed graph transformation system:

Definition 16 (Translation of a marked AHL net and its Firing Behavior to an Attributed Graph Transformation System). *Let N be
an AHL net and m its initial marking. Then the translation $Tr^{AGT}(N, m)$:
$(AHLnet, M^{\oplus}) \rightarrow \mathsf{AGT}$ from the set of pairs of AHL nets plus markings to the
set AGT of attributed graph transformation systems over graph structure signature ASSIG_{AHL} (Def. 4) is defined by $Tr^{AGT}(N, m) = (S_{Tr}, P_{Tr})$ where start
graph $S_{Tr} = Tr(N, m)$ is the translated AHL net marked by m according to
Def. 14, and the set of conditional behavior rules $P_{Tr} = \{(\hat{p}_t, C, X) | t \in T\}$ is the
translation of the firing behavior of all transitions $t \in T$ as defined in Def. 15.*

Example 3 (Attributed graph transformation system for the Dining Philosophers*). Let N be our AHL net as shown in Fig.1 (b), and $S_{Tr} = Tr(N, m)$*

be its translation to an attributed graph as shown in Fig. 3. Then, the behavior transformation system for our AHL net is given by $Tr^{AGT}(N) = (S_{Tr}, P_{Tr})$ with P_{Tr} being the set of two behavior rules constructed according to Def. 15. These behavior rules are shown in Fig. 4. Note that place nodes are preserved by the rule mapping (equal numbers for an object in L and R means that this object is contained in the interface I), and token nodes are deleted (predomain tokens) or generated (postdomain tokens).

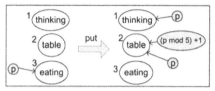 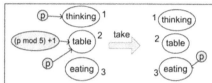

Fig. 4. Translated firing behavior of the AHL net *Dining Philosophers*

In $Tr^{AGT}(N)$ the model behavior is simulated by applying the behavior rules from P_{Tr} to the start graph S_{Tr} and to the sequentially derived graphs which correspond to different markings of N.

Proposition 2 (Semantical Compatibility of AHL net N and its translation to an Attributed Graph Transformation System). *The semantics of an AHL net N with initial marking m_{init} and the semantics of the translation $Tr^{AGT}(N, m_{init})$ are compatible, denoted by $Sem_{AHL}(N, m_{init}) \cong Sem_{AGT}(Tr^{AGT}(N, m_{init}))$, where the semantics of an AHL net is given by a set of firing sequences (firing steps), and the semantics of an attributed graph transformation system by a set of transformation sequences.*

Proof Sketch (For the complete proof see [13]): We show that

1. For each firing step $m[t, asg\rangle m'$ and for $G = Tr(N, m)$ there is a transformation step $d : G \overset{p_t}{\Longrightarrow} H$ where p_t is the behavior rule corresponding to transition t, such that the marking m' is the same as the marking of the backward translated AHL net $Back(H)$. (The backward translation $Back$ of H to the marked AHL net $Back(H)$ is defined formally in [13]).
2. Each firing sequence $\sigma \in Sem_{AHL}(N, m_{init})$ corresponds to a transformation sequence $\sigma' \in Sem_{AGT}(Tr^{AGT}(N, m_{init}))$. This means, for all firing sequences $\sigma_i = (m_i[t_i, asg_i\rangle m_i') \wedge 1 \leq i \leq n$ such that $m_{i-1}' = m_i$, we have $Tr(N, m_{i-1}') = Tr(N, m_i)$ in the corresponding transformation sequence $Tr(N, m_i) \overset{r_{t_i}}{\Longrightarrow} Tr(N, m_i')$, for $1 \leq i \leq n$.

Up to now we discussed an approach of simulating AHL net behavior by graph transformation which is based on compiling the behavior of AHL nets into graph rules. The disadvantage of this approach is that for Petri nets in contrast to other visual languages there is no general behavior transformation system which can be applied to all language elements, but that for each different model (i.e. for each AHL net), the compilation or translation to its corresponding graph transformation system has to be performed according to Def. 16.

In order to have a more general approach for modeling Petri net behavior by graph transformation, we propose to use parallel graph transformation and thus avoid the model-specific translation of transitions to behavior rules.

5   AHL Net Simulation by Parallel Graph Transformation

In this section, we define AHL net behavior by parallel graph transformation (interpreter approach) and compare this approach to the compiler approach presented in Section 4.

For the construction of the covering construction for behavior rules we need a graph to define the set of matches MA from all subrules and rules in the interaction scheme (see Def. 12). This graph needs to supply all the information we need for the behavior rule construction. It contains the predomains of all transitions in form of virtual tokens, i.e. tokens being the terms in $PreSet$ corresponding to the ArcPT inscriptions, and the information about the postdomains in form of ArcTP inscriptions. As we use only "virtual" tokens, we call this graph V *virtually marked AHL net graph*. The amalgamation construction over V then yields amalgamated rules containing the transitions and the adjacent arcs. Thus we apply a restriction functor after the amalgamation and show that the result is equivalent to the sequential behavior rules. Note that so far we do not consider firing conditions in the amalgamation, i.e. the correspondence result (Prop. 4) holds only for AHL nets without firing conditions like the Dining Philosophers.

Definition 17 (Virtually marked AHL net graph). *Let N be an AHL net, and $Tr(N, m)$ the corresponding attributed graph (acc. to Def. 14). The virtually marked AHL net graph V corresponds to $Tr(N, m)$, but is marked by terms $(term, p, i) \in PreSet$ (which virtually enables all transitions):*

$V = Tr(N, m)$ for all sorts except Token, EdgeTk *and the adjacent arc operations:*

$V_{Token} = \{tk | tk = (term, p, i) \in PreSet\}$, $V_{EdgeTk} = \{e_{tk} | tk \in V_{Token}\}$, and the operations op_{sTK}, op_{tTK}, and $attr_{tv}$ are defined as the corresponding operations for the behavior rule sides in Def. 15.

Example 4 (Virtually marked AHL net graph for the Dining Philosophers*). The bottom graph in Fig. 6 shows the virtually marked AHL net graph V_{DIPHI} for our sample AHL net modeling the Dining Philosophers. Note that the marking of the virtually marked AHL net graph denotes the union of predomains of all transitions and has nothing to do with a specific marking as e.g. shown in Fig. 3.*

Next, we define an interaction scheme for AHL nets according to Def. 10.

Definition 18 (Interaction Scheme for AHL Nets).
The interaction scheme IS_{AHL} consists of two subrules glueTrans *and* gluePlace, *two extending rules* get *and* put, *and four subrule embeddings t_1 :* glueTrans $\rightarrow$ get, t_2 : glueTrans $\rightarrow$ put, t_3 : gluePlace $\rightarrow$ get *and t_4 :* gluePlace $\rightarrow$ put.

Fig. 5 shows the interaction scheme IS_{AHL}, i.e. the definitions of the sub-rules, the extending rules, and the four embeddings. For each rule, the algebra is the term algebra $T_{OP}(Y)$ where Y is the set of variables depicted at graph objects in Fig. 5. The interaction scheme IS_{AHL} is local, as e.g. subrule glueTrans *is source of embeddings to both extending rules* get *and* put.

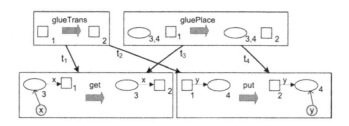

Fig. 5. Interaction Scheme for AHL nets

Example 5 (Partial Covering for the AHL net Dining Philosophers*).*
Given interaction scheme IS_{AHL} as defined in Def. 18. An instance interaction scheme IIS_{take} is shown in the upper part of Fig. 6. (Note that the detailed presentation on the left does not include all gluePlace copies.) Then, $COV_{take} = (IIS_{take}, MA_{take})$ is a partial covering with MA_{take} being a set of matches from IS_{AHL} into V_{DIPHI} as shown at the bottom of Fig. 6. The matches in MA_{take} are indicated in Fig.6 by a fat arc inscribed by MA_{take} and given precisely by node numbers. All matches from all extending rules of all subrule embeddings in IIS_{take} commute with the matches of the subrules.

COV_{take} is local as IIS_{take} is local (the subrule *glueTrans* is embedded in all extending rules) and because the matches from the left-hand sides of all three extending rule instances of get into $G_{AHL_{DIPHI}}$ are non-isomorphic. COV_{take} is additionally fully synchronized, because for each pair of extending rules we find a subrule s.t. the matches of their left-hand sides into $G_{AHL_{DIPHI}}$ overlap only in the match of this common subrule.

Note that, if a graph G and an interaction scheme are given and the covering is characterized (as e.g. for AHL nets the covering must be local, and fully synchronized), then the set of all partial coverings, i.e. the instance interaction schemes and the set of matches MA from all rules and subrules from the instance interaction scheme into G can be computed automatically.

For the covering construction for the AHL net *Dining Philosophers* this means that we can find two basic partial coverings – one for transition take in V_{DIPHI} (as shown in Fig. 6), and the other one for transition put. In the second case, a different instance interaction scheme is computed with three instances of rule put and one instance of rule get. From one instance of the subrule gluePlace there are embeddings into two of the put instances, and from one instance of the subrule glueTrans there are embeddings into all get and put instances.

A desired property of our AHL net covering construction is that it can be computed deterministically in the sense that the rules resulting from the amalgamation are unique. This property will be shown in Proposition 3.

Example 6 (Amalgamated Rule for the AHL net Dining Philosophers).
Let $COV_{take} = (IIS_{take}, MA_{take})$ be the partial covering construction as defined in Def. 5. The LHS (RHS) of the amalgamated rule p_{take} for this partial covering is constructed according to Def. 13 by gluing the instances of the LHS (RHS) of get and put along the objects of the LHS (RHS) of their common subrules.

In the center of Fig. 6, the construction of the amalgamated rule $p_{amalg_{take}}$ from COV_{take} is shown. The embeddings of rules and subrules into the amalgamated rule are indicated by dashed arrows and given precisely by numbers.

The result of the amalgamation, $p_{amalg_{take}}$, is a rule corresponding to the behavior rule for transition take with two slight differences. The variables $x_1, .., x_3$ and y_1 used in the amalgamated rule have to be replaced by the right terms from $T_{OP}(X)$, and the transition and arcs must not appear in the behavior rule. The rewriting step for the variables is given by the matches in MA_{take}, where x_1 is

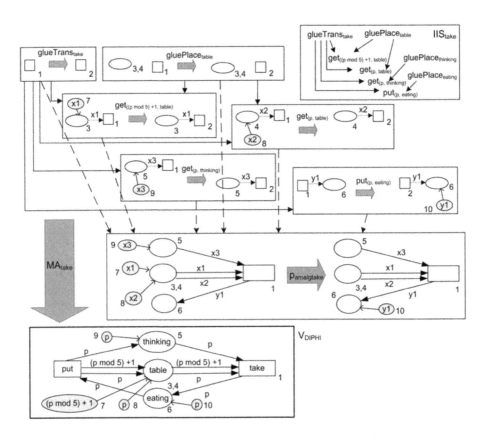

Fig. 6. Covering Construction COV_{take} and Amalgamated Rule $p_{amalg_{take}}$

matched to (p mod 5) + 1, and x_2, x_3 and y_1 are matched to p. The transition and arcs disappear by applying a functor restricting an ASSIG algebra such that the sorts $Trans, ArcPT$ and $ArcTP$ and the adjacent arc operations are empty.

The general construction of a partial covering for a transition $t \in T$ is the basis for the correspondence proof in Proposition 4.

Construction 1 (Partial Coverings for Amalgamated Rules modeling the Firing Behavior of AHL Net Transitions).

Let V be the virtually marked AHL net graph for net N defined in Def. 17. Let $COV_t = (IIS_t, MA_t)$ be the partial covering for a transition $t \in V_{Trans}$ with IIS_t being an instance interaction scheme of IS_{AHL} as defined in Def. 18 and MA_t the set of matches from IIS_t into V. IIS_t and MA_t are defined as follows:

- *Extending rule instances:* For each edge $arcPT \in V_{ArcPT}$ there is one instance get$_{arcPT}$ of the extending rule get. For each edge $arcTP \in V_{ArcTP}$ there is one instance put$_{arcTP}$ of the extending rule put.
- *Subrule instances:* There is one instance of subrule glueTrans for transition t which is embedded into all get and put instances as defined in Def. 18. For each place $p \in N_{Env_t}$ there is one gluePlace instance, called gluePlace$_p$, which is embedded into all those extending rule instances get$_{arcPT}$ with $op_{sPT}(arcPT) = p$ similar as in Def. 18. Analogously, gluePlace$_p$ is embedded into all those extending rule instances put$_{arcTP}$ with $op_{tTP}(arcTP) = p$.
- *Matches in MA_t:* The transitions of all rules and subrules in IIS_t are mapped to $t \in V_{Trans}$. The place nodes from get instances are mapped to place nodes in $pre(t)$ such that the arc inscription and the token value are mapped to the same term and the mappings overlap only in the matches of their subrules in IIS_t. Place nodes from put instances are mapped to place nodes in $post(t)$ such that the mappings overlap only in the matches of their subrules.

Proposition 3 (Existence and Uniqueness of Partial Covering COV_t).
Let V be the virtually marked AHL net graph for net N defined in Def. 17. For each transition $t \in V_{Trans}$ a local, fully synchronized partial covering $COV_t = (IIS_t, MA_t)$ constructed as in Construction 1, exists and is unique.

Proof Sketch (For the complete proof see [13]): We show that

1. there is at least one partial covering COV_t which is local and fully synchronized (due to the instance of glueTrans in IIS_t).
2. COV_t is unique by assuming that there are two different partial coverings $COV1_t$ and $COV2_t$ and by showing that they are equal.

On the basis of the unique construction of the amalgamated rule p_{amalg} : $L_{amalg_t} \to R_{amalg_t}$ using the virtually marked AHL net V as host graph (step (1) in Fig. 7), we get the match $m_{cov} : L_{amalg_t} \to V$ by gluing the matches in MA_t along the matches of the subrules (step (2) in Fig. 7). Then we apply the amalgamated rule p_{amalg} at match m_{cov} to V (step (3) in Fig. 7). The resulting span $V \leftarrow V_I \to V'$ can be interpreted as rule again. This rule still contains all

AHL net places, arcs and the transitions due to V being constructed once for the complete AHL net N. So we now restrict $V \leftarrow V_I \rightarrow V'$ to the elements of the environment of transition t. This transformation step is depicted as step (4) in Fig. 7. The result is the span $V|_{codom(m_{cov})} \leftarrow V_I|_{codom(i)} \rightarrow V'|_{codom(m^*_{cov})}$ which looks similar to our sequential behavior rule p_t with the difference that it still contains the transition and the adjacent arcs. Thus, in a last step (step (5) in Fig. 7) we apply a functor which forgets the transition and its adjacent arcs.

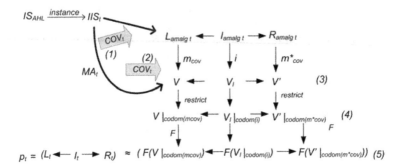

Fig. 7. Correspondence of Amalgamated Rules and Behavior Rules

Proposition 4 now formally states that the rule resulting from this functor application is isomorphic to the sequential behavior rule p_t as defined in Def. 15.

Proposition 4 (Correspondence of Amalgamated Rules and Behavior Rules for AHL Net Simulation).

Let N be an AHL net and $m \in M^{\oplus}$ its initial marking. Let $GTS_N = Tr^{AGT}(N, m) = (S_{Tr}, P_{Tr})$ be the translation of the AHL net marked by m according to Def. 16, with the set of behavior rules $P_{Tr} = \{\hat{p}_t : L_t \rightarrow R_t | t \in T\}$. Let V be the virtually marked AHL net graph for N acc. to Def. 17.

Then for each transition $t \in T$ the following holds: Given $COV_t = (IIS_t, MA_t)$, the partial covering for transition t constructed as in Constr. 1, and $p_{amalg_t} : L_{amalg_t} \rightarrow R_{amalg_t}$, the amalgamated rule for COV_t. Let m_{cov} be the match from L_{amalg_t} to V, with m_{cov} being the gluing of MA_t, and let $V \xrightarrow{P_{amalg_t}, m_{cov}} V'$ be the transformation step. Performing an epi-mono-factorization of the corresponding rule embedding (m_{cov}, i, m^*_{cov}) leads to a new rule $p_{codom} = (codom(m_{cov}) \leftarrow codom(i) \rightarrow codom(m^*_{cov}))$. Let F be a functor that forgets transition and arcs, i.e. the sorts $Trans, ArcPT, ArcTP$ and all adjacent operations are empty. Then, $F(p_{codom}) \widetilde{=} p_t$.

Proof Sketch (For the complete proof see [13]): We construct p_{amalg_t} and $m_{cov} : L_{amalg_t} \rightarrow V$ and show that

1. the transformation step $V \xrightarrow{P_{amalg_t}, m_{cov}} V'$ restricted to the codomain of rule embedding (m_{cov}, i, m^*_{cov}) corresponds to p_t except that it still contains the transition and adjacent arcs.
2. $F((p_{codom})$ is isomorphic to p_t.

6   Conclusion

In this paper we have shown how to define the behavior of AHL nets by parallel, typed and attributed graph transformation systems. This yields the advantage of an interpreter approach for simulating AHL nets. Using parallel graph transformation, possibly infinite rule sets can be described by a finite set of rules (rule schemes) modeling the elementary actions like the firing of a transition in a Petri net. The description of a rule scheme and hence of an infinite rule set is given in a purely categorical way. For AHL nets we defined an interaction scheme and constructed partial coverings. We proved the semantical compatibility between the resulting amalgamated productions and the behavior rules from the sequential graph transformation systems. The categorical definition of sequential AHL net behavior using parallel graph transformation can be extended to define *parallel* firing behavior of AHL nets. Here, the interaction scheme IS_{AHL} needs to be extended by an empty subrule to allow the construction of amalgamated rules containing more than one transition. The instance interaction schemes are still fully synchronized, but do not have to be local anymore. The amalgamation for parallel firing then yields behavior rules for combinations of different transitions and model their parallel firing.

Some restrictions had to be made when defining the behavior of AHL nets by parallel graph transformation. As we provide a general rule scheme (which is not specific to a certain net), we decided to use a fixed data signature, $ASSIG_{AHL}$ for all AHL nets, to make use of the attribute evaluation in attributed graph transformation. Thus we use directly $ASSIG_{AHL}$ terms and avoid to define higher-order functions operating on terms. In our running example all tokens are attributed by natural numbers. Extending the interaction scheme by variants of rules get and put allowing tokens with further kinds of attributes would increase the flexibility of the AHL net simulator. Another restriction concerns the firing conditions for transitions. In this paper, the construction of amalgamated rules and the correspondence result (Prop. 4) are defined for AHL nets without firing conditions, only. As firing conditions are translated to rule conditions in sequential graph transformation systems, this should be reflected also in the amalgamated rule construction, an extension which is planned as future work.

Tool support for AHL net simulation has been realized using GENGED [2], a tool for generating visual modeling environments. In GENGED, an alphabet editor supports the definition of the language vocabulary (alphabet) as graph structure signature and the layout of alphabet symbols by graphical constraints. A visual grammar editor allows to define different kinds of grammars based on the alphabet, e.g. for syntax-directed editing, parsing and/or simulation. Alphabet and grammars configure a specific VL environment, including an editor for the specified language (e.g. an AHL net editor). The behavior rules are used for simulation, where the underlying graph transformations are performed by AGG [21]. Moreover, they are the basis for the definition of an application-specific *animation view* [12]: the original alphabet for AHL nets is merged with a so-called *view alphabet* containing e.g. graphical symbols and layout definitions for the Dining-Philosophers example, i.e. icons for philosophers, a table and chopsticks.

In addition, a so-called *view transformation grammar* is used to extend the AHL net and its behavior rules by corresponding view-specific icons. Applying the view transformation grammar to the AHL net in Fig. 1 (b) yields the animation view shown in Fig. 1 (a), and the application to the behavior rules yields the corresponding *animation rules*. These rules can be enhanced by specific *animation operations* defining e.g. the smooth movement of the philosophers' chopsticks.

In GENGED, the generated environment supports simulation/animation by applying the corresponding simulation/animation rules. Animation scenarios can be exported to the SVG format [22] and viewed by an external SVG viewer which shows continuous state changes according to the defined animation operations. Due to the generic and modular definition of syntax, behavior and animation for behavior models, the GENGED approach reduces considerably the amount of work to realize a domain-specific animation of a system's behavior. Yet, it would be even more desirable to have an interconnection between GENGED and other tools supporting the definition of visual models, e.g. the world of Petri net or UML tools. The motivations for such a tool interconnection are obvious: Petri net tools which are focused on formal analysis could profit from the animation view support offered by GENGED, whereas GENGED might export a Petri net to a Petri net tool for formal analysis. In the DFG researcher group "Petri Net Technology" [10], guided by Ehrig, Reisig and Weber, the Petri net tool infrastructure Petri Net Kernel (PNK) has been developed. As a first step towards tool interchange, an XML-based file exchange between GENGED and the PNK has been realized for place/transition nets.

Last but not least, work is in progress to implement parallel graph transformation in AGG. This extension can serve in future to simulate behavior models such as AHL nets using the interpreter approach as described in this paper.

References

1. R. Bardohl, C. Ermel, and J. Padberg. Formal Relationship between Petri Nets and Graph Grammars as Basis for Animation Views in GenGED. In *Proc. IDPT 2002: Sixth World Conference on Integrated Design and Process Technology.* Society for Design and Process Science (SDPS), 2002.
2. R. Bardohl, C. Ermel, and I. Weinhold. GenGED - A Visual Definition Tool for Visual Modeling Environments. In J. Pfaltz and M. Nagl, eds. *Proc. Application of Graph Transformations with Industrial Relevance (AGTIVE'03)*, Charlottesville/Virgina, USA, September 2003.
3. A. Corradini and U. Montanari. Specification of Concurrent Systems: From Petri Nets to Graph Grammars. In G. Hommel, ed. *Proc. Quality of Communication-Based Systems, Berlin, Germany.* Kluwer Academic Publishers, 1995.
4. J. de Lara, C. Ermel, G. Taentzer, and K. Ehrig. Parallel Graph Transformation for Model Simulation applied to Timed Transition Petri Nets. In *Proc. Graph Transformation and Visual Modelling Techniques (GTVMT) 2004*, 2004.
5. P. Degano and U. Montanari. A model of distributed systems based on graph rewriting. *Journal of the ACM*, 34(2):411–449, 1987.
6. H. Ehrig and H.-J. Kreowski. Parallel graph grammars. In A. Lindenmayer and G. Rozenberg, eds. *Automata, Languages, Development*, pp. 425–447. Amsterdam: North Holland, 1976.

7. H. Ehrig and B. Mahr. *Fundamentals of Algebraic Specification 1: Equations and Initial Semantics*, Vol. 6 of *EATCS Monographs on Theoretical Computer Science.* Springer, Berlin, 1985.

8. H. Ehrig, B. Mahr, F. Cornelius, M. Grosse-Rhode, and P. Zeitz. *Mathematisch Strukturelle Grundlagen der Informatik.* Springer Verlag, 1998.

9. H. Ehrig, U. Prange, and G. Taentzer. Fundamental Theory for Typed Attributed Graph Transformation. In F. Parisi-Presicce, P. Bottoni, and G. Engels, eds. *Proc. 2nd Int. Conf. on Graph Transformation*, Springer LNCS 3256, pp. 161–177, 2004.

10. H. Ehrig, W. Reisig, G. Rozenberg, H. Weber, eds. *Advances in Petri Nets: Petri Net Technology for Communication Based Systems.* Springer LNCS 2472, 2003.

11. H. Ehrig and G. Taentzer. From parallel graph grammars to parallel high- level replacement systems. In *Lindenmayer Systems*, pp. 283–303. Springer, 1992.

12. C. Ermel and R. Bardohl. Scenario Animation for Visual Behavior Models: A Generic Approach. *Software and System Modeling: Special Section on Graph Transformations and Visual Modeling Techniques*, 5, 2004.

13. C. Ermel, G. Taentzer, and R. Bardohl. Simulating Algebraic High-Level Nets by Parallel Attributed Graph Transformation. Long Version. Technical Report 2004–21, TU Berlin, 2004. ISSN 1436-9915.

14. U. Hummert. *Algebraische High-Level Netze.* PhD thesis, TU Berlin, 1989.

15. H.-J. Kreowski. A comparison between Petri-nets and graph grammars. In *Lecture Notes in Computer Science 100*, pp. 1–19. Springer Verlag, 1981.

16. J. Lilius. *On the Structure of High-Level Nets.* PhD thesis, Helsinki University of Technology, 1995. Digital Systems Laoratory, Research Report 33.

17. S. MacLane. *Categories for the Working Mathematician*, Vol. 5 of *Graduate Texts in Mathematics.* Springer, New York, 1971.

18. J. Padberg, H. Ehrig, and L. Ribeiro. Algebraic high-level net transformation systems. *Mathematical Structures in Computer Science*, 5:217–256, 1995.

19. W. Reisig. *Petri Nets*, Vol. 4 of *EATCS Monographs on Theoretical Computer Science.* Springer Verlag, 1985.

20. G. Taentzer. *Parallel and Distributed Graph Transformation: Formal Description and Application to Communication-Based Systems.* PhD thesis, TU Berlin, 1996. Shaker Verlag.

21. G. Taentzer. AGG: A Graph Transformation Environment for System Modeling and Validation. In T. Margaria, ed. *Proc. Tool Exhibition at 'Formal Methods 2003'*, Pisa, Italy, September 2003.

22. WWW Consortium (W3C). *Scalable Vector Graphics (SVG) 1.0 Specification.* http://www.w3.org/TR/svg, 2000.

Graph Processes with Fusions: Concurrency by Colimits, Again*

Fabio Gadducci and Ugo Montanari

Dipartimento di Informatica, Università di Pisa,
Pisa, Italy
{gadducci,ugo}@di.unipi.it

Abstract. Classical concurrency in the DPO approach to graph rewriting, as defined by the shift equivalence construction [7], can also be represented by a graph process, a structure where concurrency and causal dependency are synthetically represented by a partial ordering of rewrites [1]. Interestingly, all shift equivalent derivations, considered as diagrams in the category of graphs, have the same colimit, which moreover exactly corresponds to the graph process. This construction, due to Corradini, Montanari and Rossi, was originally defined for rules with injective right-hand morphisms [6]. This condition turns out to be restrictive when graphs are used for modeling process calculi like ambients [4] or fusion [21], where the coalescing of read-only items is essential [11, 13]. Recently, a paper by Habel, Müller and Plump [16] considered again shift equivalence, extending classical results to non-injective rules. In this paper we look at the graph-process-via-colimit approach: We propose and motivate its extension to non-injective rules in terms of existing computational models, and compare it with the aforementioned results.

Keywords: DPO rewriting, concurrent semantics, process calculi.

1 Introduction

Historically, graph rewriting lies its roots on the late Sixties, as the conceptual extension of the theory of formal languages: The extension was motivated by a wide range of interests, from pattern recognition to data type specification. Nowadays, the emphasis has shifted from the generative aspects of the formalism, moving toward what could be called the "state transformation" view: A graph is considered as a data structure, on which a set of rewriting rules may implement local changes; the transformation mechanism itself is considered as expressing a basic computational paradigm, where graphs describe the states of an abstract machine and rewrites express its possible evolutions. An interest confirmed by the large diffusion of visual specification languages, such as the standard UML, and the use of graphical tools for their manipulation.

To some extent, this is also the intuition behind the introduction of process algebras, such as Milner's CCS [19]: They represent specification languages for

---

* Research partially supported by the EU within the FET – Global Computing Initiative, project AGILE IST-2001-32747 (*Architectures for Mobility*).

H.-J. Kreowski et al. (Eds.): Formal Methods (Ehrig Festschrift), LNCS 3393, pp. 84–100, 2005.

concurrent systems, considered as structured entities interacting *via* some synchronization mechanism. A (possibly distributed) system is just a term over a signature, under the hypothesis that each operator represents a basic feature of the system. The rewriting mechanism (accounting for the interaction between distinct components of a system) is usually described operationally, according to the SOS-style [22], where the rewriting steps are inductively defined by inference rules, driven by the structure of terms. Novel extensions of the process algebra paradigm involve calculi with higher-order features such as name mobility (hence, *nominal calculi*). Here systems are terms, carrying a set of associated *names*, and usually provided with a *structural* congruence, expressing basic observational properties; the reduction mechanism may also change the topology of a system, which formally amounts to changing the associated set of names.

Recent years have seen many proposals concerning the use of graph rewriting techniques for simulating reduction in process algebras, in particular for their mobile extensions. Typically, the use of graphs allows for getting rid of the problems concerning the implementation of reduction over the structural congruence, such as e.g. the α-conversion of (bound) names, since equivalent processes turn out to be mapped into isomorphic graphs. Most of these proposals follow the same pattern: At first, a suitable graphical syntax is introduced, and its operators used for implementing processes. After that, usually ad-hoc graph rewriting techniques allows for simulating the reduction semantics. Most often, the resulting graphical structures are eminently hierarchical (that is, roughly, each node/edge is a structured entity, and possibly a graph). From a practical point of view, this is unfortunate, since the restriction to standard graphs would allows for the reuse of already existing theoretical techniques and practical tools.

Building on our work on the syntactical presentation of rule-based graphical formalisms [3, 5, 12] (using techniques adopted in the algebraic specification community for modelling flow graphs [8]), in recent years we proposed graphical encodings of (possibly recursive) processes of π-calculus [20] and mobile ambients [4] into unstructured graphs, proving suitable soundness and completeness results with respect to the original reduction semantics (see [11] and [13], respectively, also for a comparison with other proposals for the graphical encoding of calculi with name mobility). The use of non hierarchical graphs allows for the reuse of standard graph rewriting theory and tools for simulating the reduction semantics of these calculi, such as the double-pushout (DPO) approach. (A more specific discussion on the advantages offered by our graphical encodings appears in the last paragraph of Section 5.3 of the present paper.)

Having asserted the benefits concerning the encoding of nominal calculi into unstructured graphs and the use of the DPO approach, we face some unresolved problems with respect to the concurrent semantics. The correspondence between graph transformation and process reduction highlights the relevance of concurrency in this setting, since allowing for the simultaneous execution of independent rewrites implicitly defines a concurrent semantics for process reduction.

Unfortunately, the graphical encodings we proposed so far (including that for the simple calculus in Section 5 of the present paper) lies outside the canon of DPO concurrent semantics: More precisely, the matching morphisms (that is, the morphisms identifying the occurrence of the left-hand side of a rule into the graph to be rewritten) have to be restricted, and they are often forced to be injective; more importantly, the right-hand side of the rules resulting from the encoding are usually specified by non-injective morphisms (operationally, they force some node and edge coalescing in the graph to be rewritten).

We recall that concurrency in the DPO approach was originally defined by the *shift equivalence* construction [7], equating those derivations that could be combinatorially related via the repeated application of an interchange operator, i.e., by swapping two consecutive rewriting steps that were *sequentially independent* (roughly, such that they acted on disjoint parts of the graph). As originally proposed in [6], *graph processes* allowed for representing concurrency and causal dependency in a synthetic manner by a partial ordering on the rewrites occurring in a derivation. Furthermore, two derivations are shift equivalent iff the corresponding partial orders are isomorphic, thus stating the substantial uniqueness of the notion of concurrency. Besides its computationally neat presentation, graph processes allowed for lifting the notion of non-sequential process from the Petri nets mold, via the *complete concurrency* property: Each total order on rule instances, compatible with the partial order of the graph process, uniquely characterizes a derivation which is shift equivalent to the original one [1].

As well as shift equivalence, also the graph process semantics was originally defined for rules with injective right-hand morphisms, and thus turns out to be restrictive when modeling nominal calculi, where fusion of read-only items is essential [11, 13]. Building on recent work [16] that extends some of the results concerning sequential independence valid for the classical definition, the main technical achievement of this paper is the extension of the graph process approach to this new setting, its motivation in terms of existing computational models, and its comparison with the aforementioned results.

The paper has the following structure. In Section 2 we recall some basic tools of the DPO approach to (hyper-)graph rewriting, as presented in [7, 9], and we discuss some results on (strong) sequential independence for injective derivations, adapted from [16]. In Section 3 we provide an (obvious) extension of the graph process semantics, as originally proposed in [6], in order to deal with rules sporting non-injective right-hand sides. More importantly, in Section 4 we prove that the correspondence between shift and (graph) process equivalence still holds in the new setting. Finally, Section 5 presents the main example, and motivations, for our work. In Section 5.1 we introduce the simple, yet expressive *solo calculus* [18], first showing how processes are modeled by graphs (Section 5.2), and later providing the rules for simulating also the reduction semantics of the calculus (Section 5.3), arguing on the benefits of the graphical encoding. Finally, in Section 5.4 we discuss the concurrent features of the encoding we propose, proving that it enhances the analysis of the causal dependencies among the possible reductions performed by a process of the solo calculus.

2   A Recollection on Graphs and Sequential Independence

We open the section recalling the definition of (labeled hyper-)graphs, as well as some basic tools of the double-pushout (DPO) approach to (hyper-)graph rewriting, as presented in [7,9]. In particular, we assume in the following a chosen signature (Σ, S), for Σ a set of operators, and S a set of sorts, such that the *arity* of an operator in Σ is a pair (ω_s, ω_t), for ω_s, ω_t strings in S^*.

Definition 1 (graphs). *A graph d (over (Σ, S)) is a five tuple $d = \langle N, E, l, s, t \rangle$, where N, E are the sets of nodes and edges; l is the pair of labeling functions $l_e : E \to \Sigma$, $l_n : N \to S$; $s, t : E \to N^*$ are the source and target functions; and such that for each edge $e \in dom(l)$, the arity of $l_e(e)$ is $(l_n^*(s(e)), l_n^*(t(e)))$, i.e., each edge preserves the arity of its label.*

With an abuse of notation, in the definition above we let l_n^* denote the extension of the function l_n from nodes to strings of nodes. In the following, we denote the components of a graph d by N_d, E_d, l_d, s_d and t_d.

Definition 2 (graph morphisms). *Let d, d' be graphs. A (graph) morphism $f : d \to d'$ is a pair of functions $f_n : N_d \to N_{d'}$, $f_e : E_d \to E_{d'}$ that preserves the labeling, source and target functions.*
  Graphs and graph morphisms form a category, denoted by $\mathbf{G}_{\Sigma,S}$.

Definition 3 (graph production and derivation). *A graph production $p : \sigma$ is composed of a production name p and of a span of graph morphisms $\sigma = (d_L \xleftarrow{l} d_K \xrightarrow{r} d_R)$, with l injective. A graph transformation system (GTS) $\mathcal{G}$ is a set of productions, all with different names. Thus, when appropriate, we denote a production $p : \sigma$ using only its name p.*
  A double-pushout diagram is like the diagram depicted in Figure 1, where top and bottom are spans and (1) and (2) are pushout squares in the category $\mathbf{G}_{\Sigma,S}$. A direct derivation from d_G to d_H via production p and triple $m = \langle m_L, m_K, m_R \rangle$ is denoted by $d_G \xRightarrow{p/m} d_H$.
  We let $d_G \Longrightarrow d_H$ denote the existence of a direct derivation between d_G and d_H, leaving unspecified the applied production and the chosen triple.

Operationally, applying a production p to a graph d_G consists of three steps. First, the *match* $m_L : d_L \to d_G$ is chosen, providing an occurrence of d_L in d_G. Then, all the items of d_G matched by $d_L - l(d_K)$ are removed, leading to the *context graph* d_D. If d_D is well-defined, and the resulting square is indeed a pushout, the items of $d_R - r(d_K)$ are finally added to d_D, further coalescing those nodes and edges identified by r, obtaining the derived graph d_H.

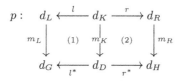

Fig. 1. A direct derivation.

In the following, we will further require *any matching m_L to be injective*, thus denoting a derivation injective if all its matches are so. Besides allowing for an easier check of the existence of the context graph, this condition will be pivotal in our proposed encoding of the solo calculus. The restriction however partially affects the usual properties concerning concurrent execution and shift equivalence, and their correspondence: In order to recover these properties, we first recall the recent results by Habel, Müller and Plump on the characterization of *strong* sequential independence between rewriting steps [16].

Definition 4 (strong sequential independence). *Let $d_G \stackrel{p_1/m_1}{\Longrightarrow} d_H \stackrel{p_2/m_2}{\Longrightarrow} d_M$ be an injective derivation such as in Figure 2. Then, its components are strongly sequentially independent if there exists an independence pair among them, i.e., two graph morphisms $i_1 : d_{R_1} \to d_{D_2}$ and $i_2 : d_{L_2} \to d_{D_1}$ such that*

- $l_2^* \circ i_1 = m_{L_2}$ *and* $r_1^* \circ i_2 = m_{R_1}$;
- $r_2^* \circ i_1$ *is injective.*

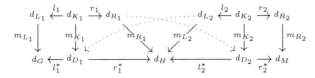

Fig. 2. Strong sequential independence for derivation $\rho = d_G \stackrel{p_1/m_1}{\Longrightarrow} d_H \stackrel{p_2/m_2}{\Longrightarrow} d_M$.

The additional constraint on $r_2^* \circ i_1$ arises for the interplay between the possible fusions operated by r_2^* and the matches being injective. Otherwise, the classical property of sequential independence would fail, namely, the application of the two productions could not be inverted, as shown in Example 6.5 of [16].

Proposition 1 (interchange operator). *Let $\rho = d_G \stackrel{p_1/m_1}{\Longrightarrow} d_H \stackrel{p_2/m_2}{\Longrightarrow} d_M$ be an injective derivation as in Figure 2, and let its components be strongly sequentially independent via an independence pair π. Then, an injective derivation $IC_\pi(\rho) = d_G \stackrel{p_2/m_2^*}{\Longrightarrow} d_{H^*} \stackrel{p_1/m_1^*}{\Longrightarrow} d_M$ can be uniquely chosen, such that its components are strongly sequentially independent via a canonical independence pair π^*.*

In particular, $m_2^* = l_1^* \circ i_1$, and m_1^* is then uniquely induced. The "canonical" bit implies that both the derivation, and its independence pair, can be constructively given: We refer to Theorem 6.4 and Theorem 6.7 of [16].

3   Remarks on the Graph Process Construction

The role of the *interface graph* d_K in a rule is to characterize the elements of the graph to be rewritten that are read but not consumed by a direct derivation. Such a distinction is important when considering *concurrent* derivations, possibly defined as an equivalence class of concrete derivations up-to so-called *shift equivalence* [7], identifying (as for the analogous, better-known *permutation equivalence* of λ-calculus) those derivations which differ only for the scheduling of independent steps. Roughly, the equivalence states the possibility to interchange two direct derivations $d_1 \implies d_2 \implies d_3$ if they act either on disjoint parts of d_1, or on parts that are in the image of the interface graphs.

A more concrete, yet equivalent notion of abstract derivation for a GTS is obtained by means of the so-called *(graph) process semantics*. As for the similar notion on Petri nets [15], a *graph process* describes a derivation abstracting away from the ordering of causally unrelated steps, and thus it offers at the same time a concrete representative for a class of equivalent derivations. The definition below straightforwardly generalizes [1]: The original proposal considered only GTS's where both left- and right-hand sides were injective morphisms.

Definition 5 (graph processes). *Let $\mathcal{G}$ be a GTS and $\rho = d_{G_0} \overset{p_1/m_1}{\implies} \ldots \overset{p_n/m_n}{\implies} d_{G_n}$ a derivation of length n (upper part of Figure 3). The graph process $\Pi(\rho)$ associated to the derivation ρ is the $n+1$-tuple $\langle t_{G_0}, \langle p_1, \pi_1 \rangle, \ldots, \langle p_n, \pi_n \rangle \rangle$: Each π_i is a triple $\langle t_{L_i}, t_{K_i}, t_{R_i} \rangle$, and the graph morphisms $t_{x_i} : d_{x_i} \to d_\rho$, for $x_i \in \{L_i, K_i, R_i\}$ and $i = 1, \ldots, n$, are those uniquely induced by the colimit construction shown in Figure 3.*

Let ρ, ρ' be two derivations of length n, both originating from graph d_{G_0}. They are process equivalent *if the associated graph processes are isomorphic, i.e., if there exists a graph isomorphism $\gamma_\pi : d_\rho \to d_{\rho'}$ and a bijective function $\gamma_p : \{1, \ldots, n\} \to \{1, \ldots, n\}$ such that productions p_i and $p'_{\gamma_p(i)}$ coincide for all $i = 1, \ldots, n$, and all the involved diagrams commute[1].*

A graph process associated to a derivation ρ thus includes, by means of the colimit construction and of the morphisms t_{x_i}, the action of each single production p_i on the graph d_ρ. From the image of each d_{x_i} is then possible to recover a suitable partial order among the direct derivations in ρ, which faithfully mirrors the causal relationship among them. For example, let (Σ_{ex}, S_{ex}) be the one-sorted signature containing just four constants, namely $\{a, b, c, d\}$; and let $\mathcal{G}_{ex}$ be the GTS containing two productions, roughly rewriting a into c and b into d. The derivation ρ_{ex} is represented in Figure 4, where, for the sake of readability, graph morphisms are simply depicted as thick arrows.

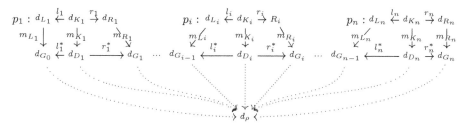

Fig. 3. Colimit construction for derivation $\rho = d_{G_0} \overset{p_1/m_1}{\implies} \ldots \overset{p_n/m_n}{\implies} d_{G_n}$.

---

[1] Explicitly, $\gamma_\pi \circ t_{G_0} = t'_{G_0}$, and $\gamma_\pi \circ t_{x_i} = t'_{x_{\gamma_p(i)}}$ for $x_i \in \{L_i, K_i, R_i\}$ and $i = 1, \ldots, n$.

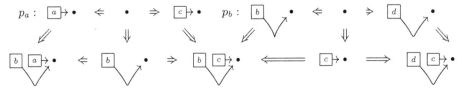

Fig. 4. The derivation $\rho_{ex} = d_{G_0} \overset{p_a/m_a}{\Longrightarrow} d_{G_a} \overset{p_b/m_b}{\Longrightarrow} d_{G_b}$.

Fig. 5. Compact representation for the process $\Pi(\rho_{ex})$.

The process $\Pi(\rho_{ex})$ can be described as in Figure 5, extending the graph $d_{\rho_{ex}}$ with two shaded boxes: They are labeled p_a and p_b, in order to make explicit the mappings t_{x_i} (hence, the action of the productions on the initial graph). Thus, (the application of) the production p_a consumes the a edge (it is in the image of t_{L_a}, but not in the image of t_{K_a}), and this is denoted by the dotted arrow from a into p_a; it then reads the only node (which is indeed in the image of t_{K_a}), denoted by the dotted arrow with no head; and finally, it creates the c edge, denoted by the dotted arrow into c. Similarly, (the application of) the production p_b consumes the b edge, reads the node and creates the d edge.

We feel confident that our example underlines the connection between the process semantics for graphs, and the standard process semantics for Petri nets. This compact representation is further argued upon in the following sections.

4   Some Properties of Graph Processes

The aim of this section is twofold: First of all, we establish a connection between strong sequential independence and the graph process construction. Second, we analyze the difficulties in recovering one of the properties usually associated with process semantics, namely, *complete concurrency*, stating that each possible interleaving of the rules could be related to a concrete derivation.

4.1   Interchanges vs. Colimits

We remember that shift equivalence equates those derivations that are obtained via the repeated application of the interchange operator. Thus, we now prove the correspondence between the equivalences induced on 2-steps derivations by the interchange operator and by the colimit construction. This result states the substantial correspondence between shift equivalence and process semantics, hence, it makes clear that the two different frameworks actually characterize the same notion of *concurrent* computation.

Theorem 1 (from interchange to process). *Let ρ be an injective derivation as in Figure 2, and let its components be strongly sequentially independent via the independence pair π. Then, $\Pi(\rho)$ and $\Pi(IC_\pi(\rho))$ are isomorphic graph processes, and the two derivations are process equivalent.*

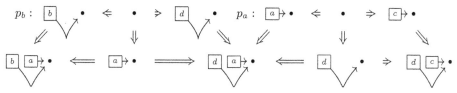

Fig. 6. The derivation $IC_{\langle id,id \rangle}(\rho_{ex})$.

Let us now consider the derivation ρ_{ex} in Figure 4. Its components are clearly strongly sequentially independent, with independence pair given by two isomorphisms. The associated derivation $IC_{\langle id,id \rangle}(\rho_{ex})$ is depicted in Figure 6. The graph processes $\Pi(\rho_{ex})$ and $\Pi(IC_{\langle id,id \rangle}(\rho_{ex}))$ clearly coincide.

Theorem 2 (from process to interchange). *Let $\rho = d_G \overset{p_1/m_1}{\Longrightarrow} d_H \overset{p_2/m_2}{\Longrightarrow} d_M$ and $\rho' = d_G \overset{p_2/m'_1}{\Longrightarrow} d_{H'} \overset{p_1/m'_2}{\Longrightarrow} d_M$ be two injective derivations, and let them be process equivalent via the graph isomorphism $\gamma_\pi : d_\rho \to d_{\rho'}$ (and the obvious swap on productions) between the graphs underlying the processes $\Pi(\rho)$ and $\Pi(\rho')$. Then, γ induces an independence pair π on ρ such that $IC_\pi(\rho) = \rho'$.*

4.2 Derivations out of a Graph Process

When originally proposed in [6], a process associated to a derivation was used to induce a partial order on the family of direct derivations occurring in it. Basically, the intuition tells that, if there is no ordering between the application of two productions, then a different derivation exists, shift equivalent to the former, where the occurrence of those productions is swapped.

Definition 6 (derivation order). *Let $\rho = d_{G_0} \overset{p_1/m_1}{\Longrightarrow} \ldots \overset{p_n/m_n}{\Longrightarrow} d_{G_n}$ be a derivation of length n, and let us consider the associated graph process $\Pi(\rho)$, as in Figure 3. Let us then consider the set given by the (disjoint) union of the sets of nodes N_{d_ρ} and edges E_{d_ρ} of the graph underlying $\Pi(\rho)$, and of the set of production applications p_i's (a distinct instance of a production name p for each occurrence of the production in a direct derivation). Then, the derivation order is the reflexive and transitive closure of the relation induced by*

- *if p_i consumes x then $x \leq p_i$;*
- *if p_i creates x then $p_i \leq x$;*
- *if p_i creates x and p_j preserves x then $p_i \leq p_j$;*
- *if p_i preserves x and p_j consumes x then $p_i \leq p_j$;*

The definition coincides with Definition 17 in [6] (see also Definition 3.4.4 of [1]), and two derivations are said to be *concurrent* if they are not related by the partial order. In fact, it is proved in Theorem 23 of the same paper that, for any total order compatible with the derivation order, there is a derivation where the application of the productions exactly reflects that total order, and whose graph process is isomorphic to the former.

In fact, this is the situation occurring for derivation ρ_{ex}: The partial order does not force any relation between the two productions, and they can be swapped, as it is in fact the case for $IC_{\langle id,id \rangle}(\rho_{ex})$.

As nice as it might be, the property does not hold for those GTS's such that the right morphism may coalesce nodes. This is shown by the derivation ρ_{ex2}, depicted in Figure 7 (inspired by Example 6.5 in [16]): The two components of the derivation are not strongly sequentially independent, so they can not be swapped. This fact goes unnoticed in the associated graph process, where the two productions are unrelated by the partial order, as shown in Figure 8.

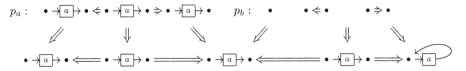

Fig. 7. A derivation with causally related components.

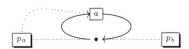

Fig. 8. Compact representation for the process $\Pi(\rho_{ex2})$.

5   Encoding a Simple Process Calculus

We now exploits the results presented in the previous section to discuss about concurrent reductions for a simple (the simplest available, in fact) process calculus, namely, the monadic *solo calculus* [18], one of the dialects of those nominal calculi whose distinctive feature is name fusion [10, 14, 21].

5.1   The Monadic Fragment of the Solo Calculus

This section shortly introduces the monadic variant of the solo calculus, its structural equivalence and the associated reduction semantics.

Definition 7 (processes). *Let $\mathcal{N}$ be a set of names, ranged over by $x, y, w, \ldots$. A process P is a term generated by the syntax*

$$P ::= 0, \; \sigma, \; (\nu x)P, \; P_1 \mid P_2 \qquad \text{for } \sigma \in \{x(y), \overline{x}y\}$$

We let $P, Q, R, \ldots$ range over the set Proc of processes.

The operators $x(y)$ and $\overline{x}y$ are denoted as *input* and *output*, respectively, even if their symmetric behavior makes the distinction (typical instead of other calculi) immaterial; collectively, each instance of them is called a *solo*, to emphasize its lack of connections, except for some possible name sharing, with the other operators. Finally, the first argument of the two operators, indicated by x, is usually called the *channel* where the communication of information take place.

We assume the standard definitions for the set of free names of a process P, denoted by $fn(P)$. Similarly for α-convertibility, with respect to the *restriction* operators $(\nu y)P$: The name y is bound in P, and it can be freely α-converted. Using these definitions, the behavior of a process P is described as a relation obtained by closing a set of basic rules under a suitable congruence.

Definition 8 (reduction semantics). *The* reduction relation *for processes is the relation* $R_\sigma \subseteq Proc \times Proc$, *closed under the structural congruence* $\equiv$ *induced by the equations in Figure 9, generated by the following inference rules*

$$\frac{y \neq w}{(\nu w)(x(y) \mid \overline{x}w \mid P) \to P\{^y/_w\}} \qquad \frac{y \neq w}{(\nu y)(x(y) \mid \overline{x}w \mid P) \to P\{^w/_y\}}$$

$$\frac{}{x(y) \mid \overline{x}y \to 0} \qquad \frac{P \to Q}{(\nu x)P \to (\nu x)Q} \qquad \frac{P \to Q}{P \mid R \to Q \mid R}$$

where $P \to Q$ *means that* $\langle P, Q \rangle \in R_\sigma$.

The two top rules characterize the communication between restricted processes. Consider the second: The process $\overline{x}w$ is ready to communicate the (possibly global) name w along the channel x; it then synchronizes with the process $x(y)$, and the bound name y is thus substituted by w on *all the occurrences* inside the residual process P. Hence, the communication has a global effect, affecting the process as a whole. Note that one of the names among $\{y, w\}$ *has to be bound*, so that, in principle, the rule does not to alter the number of free names floating around: And the possible choice requires the presence of two different rules.

The third rule simply states that there is no reason to bind a name during a reduction, if no substitution has actually to occur.

The two latter rules simply state the closure of the reduction relation with respect to the operators of restriction and parallel composition.

Basically, the axioms state that a process is a collection of solos floating around, much in the tradition of the CHAM paradigm [2], and interacting by forcing some name fusion (in our view, as we will see, this will be equivalent to ask for applying some node coalescing).

The only difference with respect to the syntax and the operational semantics for the monadic fragment of the calculus proposed in Section 2 of [18] is the lack

$$P \mid Q = Q \mid P \qquad P \mid 0 = P \qquad P \mid (Q \mid R) = (P \mid Q) \mid R$$

$$(\nu x)(\nu y)P = (\nu y)(\nu x)P \qquad (\nu x)0 = 0 \qquad (\nu x)(P \mid Q) = P \mid (\nu x)Q \text{ for } x \notin fn(P)$$

Fig. 9. The set of structural axioms.

of a *match* operator $[x = y]$ and the explicit presentation of the three reduction rules, which in [18] are given instead in a more compact way as a unique rule equipped with constraints on the substitution induced by the name fusion. The match would pose no real problem, and it is simply avoided for being as straight as possible in our graphical encoding, as presented in the later section; while the explicit presentation of the reduction rules will make clearer their correspondence with the productions of the GTS associated to the calculus by the encoding.

We conclude the section with a remark on the expressiveness of the calculus. Despite their simple syntax and operational semantics, both the monadic variant with match and the dyadic variant (operators come with two names, besides the channel) are as expressive as the whole fusion calculus (as proved in [18]), which in turn is a symmetric version of the foremost nominal calculi, the π-calculus [20].

5.2   The Graphical Encoding of Solos

This section informally presents a graphical encoding of the solo calculus. Its formal definition could be obtained by easily adapting the proposals for mobile ambients and π-calculus presented by the authors in [11, 13].

In order to help intuition, we begin with an easy description of the normal form for equivalent processes, induced by structural congruence.

Proposition 2 (normal forms). *Let P be a process. Then, P is equivalent to a process of the shape $(\nu x_1)\ldots(\nu x_n)(\sigma_1 \mid \ldots \mid \sigma_m)$ where all x_i's are different, all σ_j' are solos, and the set $X = \{x_1 \ldots x_n\}$ contains only names occurring in $S = \sigma_1 \mid \ldots \mid \sigma_m$, that is, $X \subseteq fn(S)$.*

We could then denote a process in normal form as $(\nu X)\mathcal{P}$, for $\mathcal{P}$ a set of solos, since the order of the restriction operators and of the solos is immaterial. Exploiting that characterization, it is quite easy to think of its graphical correspondence. You just need three hyper-edges, corresponding to the operators of the calculus: They are neatly represented by the *type graph* in Figure 10.

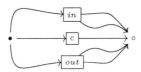

Fig. 10. The type graph.

For graphical convenience, the nodes are represented either by an hollow or as a full circle, in order to distinguish those nodes used for names (the former) from the possible node occurring as a root in the encoding (the latter); similarly for the labels *in* and *out* inside the edges. For example, the encoding of the process (already in normal form) $(\nu w)(x(y) \mid \bar{x}w \mid w(z) \mid \bar{y}z)$ is represented in Figure 11, where nodes are additionally equipped with the name they represent, in order to make the encoding clearer.

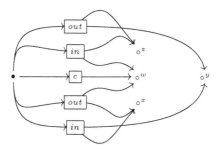

Fig. 11. The encoding of a process.

A final remark on the presence of the root node is now in order. An even simpler graphical encoding could have been obtained by dropping it, since the normal form basically states that solos just float around as in a chemical solution. Indeed, this is the proposal put forward by a former graphical encoding for the calculus, the *solo diagrams* [17]. Our proposal is more general, since it would allow for recovering also the standard interleaving semantics of the calculus (by linking to the root node a flag operator that is deleted and recreated by each of the productions). More importantly, the chosen solution emphasizes the connection between the present encoding and those we previously proposed for mobile ambients and π-calculus: In those calculi the operators may occur nested, thus the root node is necessary for taking care of the syntactical constraints. The similarities between all these encodings allow us for claiming that all the present considerations on concurrent semantics, and our analysis of the causal dependencies due to name fusion among the possible reductions, can be lifted to the syntactically more demanding calculi.

5.3   Encoding Reductions

We can now introduce the GTS $\mathcal{G}_\sigma$, showing how it simulates the reduction semantics for processes. It is a simple system, since it basically contains just three productions (i.e., one for each axiom of the reduction system), plus some instances of them. The first production p_1^σ is depicted in Figure 12: The graph on the left-hand side (center, right-hand side) is $d_L^{\sigma 1}$ ($d_K^{\sigma 1}$ and $d_R^{\sigma 1}$, respectively). The action of the rule is described by the names of the nodes: As an example, the nodes identified by y and w, distinct in $d_L^{\sigma 1}$, are coalesced in $d_R^{\sigma 1}$. The node identifiers are of course arbitrary: They correspond to the actual elements of the set of nodes, and they are used just to characterize the span of functions.

Constraining the matches to be injective ensures that the production is not applied to a graph where nodes y and w are coalesced. Nevertheless, this turns out to be too restrictive, since a reduction step can be performed if name x coincides with either y or w. Hence, two additional productions are needed: They instantiate p_1^σ, coalescing the node x with either the node y or the node w. We leave these productions unnamed, since they play a minor rôle in the paper.

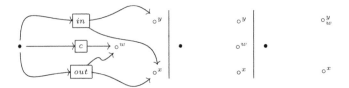

Fig. 12. The first production for $\mathcal{G}_\sigma$.

A similar situation occurs when the name y on the input operator, instead of the name w on the output operator, is bound: It suffices a production p_2^σ (together with two instances) mirroring p_1^σ. Most important, a production p_3^Σ is needed, where nodes y and w are already coalesced and the restriction operator is not required, as depicted in Figure 13. (Also an instance where the three names coincide, and the corresponding nodes are thus merged, has to be included.)

In fact, the presence of alternative productions is imposed by the possible constraints (as e.g. in the fusion of names) occurring in a reduction, and it is typical of other encodings of process calculi that we proposed [11, 13]. In general terms, a common constraint is that the match has to be injective on edges, since they represent different resources which have to be explicitly consumed: This condition is void in the encoding of solos, since for each production the two edges have different labels, but this is not always the case, as e.g. in the encoding for mobile ambients [13]. Concerning nodes, the reason for the match to be injective is linked with the politics which are involved with name replacing: Some of the possible substitutions among names are forbidden, usually by imposing structural constraints on the processes (as in the π-calculus, where the name occurring in the input has to be local [11]). Note also that the injectivity of the match is counterbalanced by the presence of rules with non-injective right-hand side, taking care of merging the names singled out by the match.

We now take the chance of summing up some advantages arising from the use of graphical encoding for nominal calculi. First of all, note that the reduction semantics of a calculus is usually given by a set of rules closed up-to structural congruence. Instead, no closure is required in our formalism: congruent processes are mapped to isomorphic graphs, and the graphical presentation accounts for a unique description of the normal form of a process, up-to α-conversion of bound names. Similarly, note that the reduction rules usually represent a schema, modulo the names occurring in the operators: The graphical representation is also basically up-to injective renaming of free names. It is also noteworthy that there is no need of explicit rules for closing with respect to the restriction and

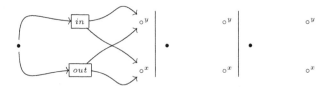

Fig. 13. The third production for $\mathcal{G}_\sigma$.

parallel operators: These structural rules are taken care of by the fact that graph morphisms allow for embedding the left-hand side of a production into a large graph, thus modeling the closure of reductions by context. Finally, we remark that, even if the search of a match can be considered as a global operation, rule application itself is a local operation, coalescing at most one node, and removing either two or three edges, so that also name fusion becomes local[2].

5.4 Causality via Fusion

Since the matching morphisms are required to be injective, we can apply the considerations on sequential independence provided in Section 2.

In fact, consider the process $(\nu w)(x(y) \mid \overline{x}w \mid w(z) \mid \overline{y}z)$, and its graphical depiction in Figure 11: A derivation is represented by the proof tree below.

$$\frac{y \neq w}{(\nu w)(x(y) \mid \overline{x}w \mid w(z) \mid \overline{y}z) \to (w(z) \mid \overline{y}z)\{^{y}/_{w}\} = y(z) \mid \overline{y}z \qquad y(z) \mid \overline{y}z \to 0}{(\nu w)(x(y) \mid \overline{x}w \mid w(z) \mid \overline{y}z) \to 0}$$

The reduction is obtained by first applying the rule removing the restriction operator for the name w occurring in the output operator $\overline{x}w$, coalescing it with the name y. Thanks to that fusion, the rule where no restriction operator occurs (since the names to be coalesced already coincide) can then be applied. Being the context rules immaterial, we end up by applying to the graph in Figure 11 first the rule p_1^σ, and then the rule p_3^σ. The derivation (namely, the derived graphs) is shown in Figure 14, and the associated colimit in Figure 15.

Now, the two steps are clearly *not* sequentially independent, since there is no suitable morphism leaving from $d_L^{\sigma3}$. This fact goes unnoticed in the graph process, since there is no dependence between the copies of the productions. Hence, the complete concurrency property does not hold, since only one of the two possible derivations obtained by linearizing the partial order actually exists.

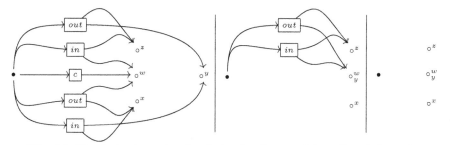

Fig. 14. The derived graphs of a derivation first applying p_1^σ and then p_3^σ.

---

[2] Even if it is not relevant here, we may add that dealing with reduction rules creating new names usually needs the presence of global checks on the whole process; in the DPO approach, this is easily dealt with by rules with non-surjective right-hand side.

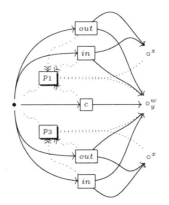

Fig. 15. The colimit of the derivation in Figure 14 (and of Figures 16 and 17 as well).

Let us now consider the process $(\nu w)(x(y) \mid \overline{x}w \mid w(z) \mid \overline{w}z)$, distinguished by the process above for the presence of an output operator on channel w. The same sequence of rule applications as for the derivation depicted in Figure 14 can be replicated, and the result is presented in Figure 16.

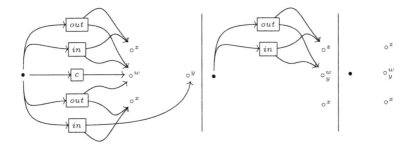

Fig. 16. The derived graphs of another derivation first applying p_1^σ and then p_3^σ.

However, now the components of the derivation are strongly sequentially independent, since the coalescing of nodes y and w has not enabled any derivation. The derivation obtained by applying the interchange operator is in Figure 17, while the graph underlying the graph process is the same presented in Figure 15.

6   Conclusions and Further Work

In our paper we proposed an extension of the classical process semantics for GTS's, in order to deal with injective derivations and productions with non-injective right-hand sides. We also proved that its correspondence with the concurrent semantics defined by shift equivalence still holds in this new setting.

Furthermore, we applied these results to the GTS obtained by the encoding of a simple nominal calculi. We argued about the relevance of similar encodings as

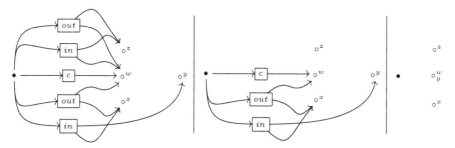

Fig. 17. The derivation obtained via interchange from the derivation in Figure 16.

an improvement with respect to the usual operational semantics for these calculi, given by a reduction system on terms up-to some structural congruence. In turn, the encoding motivated our research, since productions with non-injective right-hand side arise naturally for these graphical presentations of calculi.

Since shift-equivalent (injective) derivations are represented by the same graph process, our results provide a neat concurrent semantics for the reduction mechanism in nominal calculi. We leave as an open question the search of those conditions ensuring the classical *complete concurrency* property for the graph process semantics, as discussed in Section 5.4: Namely, the existence of a derivation for each possible total ordering, compatible with the derivation order, of the productions occurring in the graph process. A possible solution we foresee as viable is enriching the derivation order by recording the possible occurring of a fusion, as well as the nodes involved in the fusion itself.

References

1. P. Baldan, A. Corradini, H. Ehrig, M. Löwe, U. Montanari, and F. Rossi. Concurrent semantics of algebraic graph transformation. In H. Ehrig, H.-J. Kreowski, U. Montanari, and G. Rozenberg, editors, *Handbook of Graph Grammars and Computing by Graph Transformation*, volume 3, pages 107–187. World Scientific, 1999.
2. G. Berry and G. Boudol. The chemical abstract machine. *Theor. Comp. Sci.*, 96:217–248, 1992.
3. R. Bruni, F. Gadducci, and U. Montanari. Normal forms for algebras of connections. *Theor. Comp. Sci.*, 286:247–292, 2002.
4. L. Cardelli and A. Gordon. Mobile ambients. *Theor. Comp. Sci.*, 240:177–213, 2000.
5. A. Corradini and F. Gadducci. Rewriting on cyclic structures: Equivalence between the operational and the categorical description. *Informatique Théorique et Applications/Theoretical Informatics and Applications*, 33:467–493, 1999.
6. A. Corradini, U. Montanari, and F. Rossi. Graph processes. *Fundamenta Informaticae*, 26:241–265, 1996.
7. A. Corradini, U. Montanari, F. Rossi, H. Ehrig, R. Heckel, and M. Löwe. Algebraic approaches to graph transformation I: Basic concepts and double pushout approach. In G. Rozenberg, editor, *Handbook of Graph Grammars and Computing by Graph Transformation*, volume 1, pages 163–245. World Scientific, 1997.

8. V.-E. Căzănescu and Gh. Ştefănescu. A general result on abstract flowchart schemes with applications to the study of accessibility, reduction and minimization. *Theor. Comp. Sci.*, 99:1–63, 1992.

9. F. Drewes, A. Habel, and H.-J. Kreowski. Hyperedge replacement graph grammars. In G. Rozenberg, editor, *Handbook of Graph Grammars and Computing by Graph Transformation*, volume 1, pages 95–162. World Scientific, 1997.

10. Y. Fu. Variations on mobile processes. *Theor. Comp. Sci.*, 221:327–368, 1999.

11. F. Gadducci. Term graph rewriting and the π-calculus. In A. Ohori, editor, *Programming Languages and Semantics*, volume 2895 of *Lect. Notes in Comp. Sci.*, pages 37–54. Springer, 2003.

12. F. Gadducci, R. Heckel, and M. Llabrés. A bi-categorical axiomatisation of concurrent graph rewriting. In M. Hofmann, D. Pavlović, and G. Rosolini, editors, *Category Theory and Computer Science*, volume 29 of *Electr. Notes in Theor. Comp. Sci.* Elsevier Science, 1999.

13. F. Gadducci and U. Montanari. A concurrent graph semantics for mobile ambients. In S. Brookes and M. Mislove, editors, *Mathematical Foundations of Programming Semantics*, volume 45 of *Electr. Notes in Theor. Comp. Sci.* Elsevier Science, 2001.

14. P. Gardner and L. Wischik. Explicit fusion. In M. Nielsen and B. Rovan, editors, *Mathematical Foundations of Computer Science*, volume 1893 of *Lect. Notes in Comp. Sci.*, pages 373–382. Springer, 2000.

15. U. Golz and W. Reisig. The non-sequential behaviour of Petri nets. *Information and Control*, 57:125–147, 1983.

16. A. Habel, J. Müller, and D. Plump. Double-pushout graph transformation revisited. *Mathematical Structures in Computer Science*, 11:637–688, 2001.

17. C. Laneve, J. Parrow, and B. Victor. Solo diagrams. In N. Kobayashi and B. Pierce, editors, *Theoretical Aspects of Computer Science*, volume 2215 of *Lect. Notes in Comp. Sci.*, pages 127–144. Springer, 2001.

18. C. Laneve and B. Victor. Solos in concert. *Mathematical Structures in Computer Science*, 13:675–683, 2002.

19. R. Milner. *Communication and Concurrency*. Prentice Hall, 1989.

20. R. Milner, J. Parrow, and D. Walker. A calculus of mobile processes. Part I and II. *Information and Computation*, 100:1–77, 1992.

21. J. Parrow and B. Victor. The fusion calculus: Expressiveness and simmetry in mobile processes. In V. Pratt, editor, *Logic in Computer Science*, pages 176–185. IEEE Computer Society Press, 1998.

22. G. Plotkin. A structural approach to operational semantics. Technical Report DAIMI FN-19, Computer Science Department, Aarhus University, 1981.

Graph Transformation with Variables

Berthold Hoffmann

Technologiezentrum Informatik, Universität Bremen,
Bremen, Germany
hof@informatik.uni-bremen.de

Abstract. Variables make rule-based systems more abstract and expressive, as witnessed by term rewriting systems and two-level grammars. In this paper we show that variables can be used to define advanced ways of graph transformation as well. Taking the gluing approach to graph transformation [7, 3] as a basis, we consider extensions of rules with attribute variables, clone variables, and graph variables, respectively. In each case, the variables in a rule are instantiated in order to obtain a set of rule instances that in turn defines the transformation relation. By combining different kinds of variables, we define very expressive rules, and reduce them to plain rules by instantiation. Since gluing graph transformation has a well developed theory, this opens the door to lift results of that theory from instances to rules with variables.

1 Introduction

Rules are frequently used in computer science, for specifying the behavior of systems in an axiomatic way. Rules do often contain variables. *Term rewriting systems*, for instance, specify the evaluation of functions by rewrite rules such as

$$\mathsf{fib}(\mathsf{s}(\mathsf{s}(\mathsf{N}))) \to \mathsf{fib}(\mathsf{s}(\mathsf{N})) + \mathsf{fib}(\mathsf{N})$$

wherein the substitution of variables like N by terms yields ground rules like

$$\mathsf{fib}(\mathsf{s}(\mathsf{s}(\mathsf{s}(0)))) \to \mathsf{fib}(\mathsf{s}(\mathsf{s}(0))) + \mathsf{fib}(\mathsf{s}(0))$$

that define the term rewrite relation [19]. *Two-level grammars*, another example, derive languages of words by rules such as

$$\langle \mathsf{T_0\ expression} \rangle \ ::= \ \langle \mathsf{T_1\ to\ T_0\ operator} \rangle \ \langle \mathsf{T_1\ expression} \rangle$$

wherein variables like $\mathsf{T_0}$ and $\mathsf{T_1}$ are substituted by words of a context-free meta grammar in order to obtain context-free production rules like

$$\langle \mathsf{bool\ expression} \rangle \ ::= \ \langle \mathsf{int\ to\ bool\ operator} \rangle \ \langle \mathsf{int\ expression} \rangle$$

which in turn define the derivation relation of the grammar [2]. In both cases, rewriting is used twice: On the *meta level*, rule are instantiated by substituting variables, producing rule instances that generate the rewrite relation on the *object level*. Variables make rules more abstract and more expressive: term rewriting

H.-J. Kreowski et al. (Eds.): Formal Methods (Ehrig Festschrift), LNCS 3393, pp. 101–115, 2005.

systems define functions on infinite sets in a finite way, and two-level grammars derive recursively enumerable languages on the basis of context-free derivation.

This paper is about the use of variables in the area of graph transformation. Surprisingly, this concept has hardly been used in the major approaches of graph transformation that are documented in the handbook [27]. Early attempts by H. Göttler [11] and W. Hesse [17] to extend two-level word grammars to graphs were not successful. This work was inspired by three papers: D. Plump and A. Habel have devised variable hyperedges as placeholders for hypergraphs [25]; N. van Eetvelde and D. Janssens have introduced variable nodes as placeholders for graphs in [32]; and, D. Plump and S. Steinert have proposed variable labels as placeholders for attribute values [26]. These proposals follow the two-level model outlined above, but are based on different approaches to graph transformation. We catch on their ideas, but reformulate them so that they coherently use a single way of graph transformation on the object level. As a common basis, we choose the well-known gluing approach to graph transformation [7, 3] (also known as the algebraic, or double-pushout approach). In addition, we define clones, which are nodes that stand for sets of similar nodes within the same framework. The unified notions of variables allow to model several advanced concepts of graph transformation, such as the connection instructions of [10], and object set nodes and path expressions devised for programmed graph transformation [30]. The two-level model is modular so that different kinds of variables can also be combined easily. In this way, even the very advanced rules of [32] can be reduced to sets of simple gluing rules. Since variable instantiation is simply defined and easily understood, the resulting definitions are easily understood as well. And, since gluing transformation, the common basis of the instantiations, has a rich theory, results of this theory can help to prove properties of the rules as well.

The paper may raise a fundamental objection: *Need rules be that sophisticated?* Indeed, generative power is no issue, as gluing rules without variables already derive the recursively enumerated languages. However, in complex applications like software refactoring [22, 32], operations can be developed and verified more easily if they are expressed as a single rule – even a complex one – rather than as programs that control the application of simple rules in order to achieve the same effect.

The paper is structured as follows. We first recall graph transformation by gluing rules with relabeling in Section 2. This is the basis for discussing the use of variables in graph transformation in Section 3. Three kinds of variables are considered in Sections 4 to 6: attribute variables, clones, and graph variables. Section 7 discusses how instantiations can be combined, and how they may help to lift properties and results known from rule instances to rules. In Section 8 we conclude with pointers to related work and a discussion of further research.

2   Basic Graph Transformation

Rules in the gluing approach to graph transformation [7, 3] shall be used as rule instances on the object level. We have chosen this approach as it is not committed to a particular notion of graph (not even to graphs!), is widely used and has a rich theory. We confine the occurrences of rules to injective morphisms. It has been shown in [14] that this is no restriction; on the contrary, it permits a finer control of rule application. As in [15], we allow partially labeled graphs in rules so that nodes and edges may be relabeled in a transformation step.

Graphs. A *(partially labeled) graph* $G = \langle V_G, E_G, s_G, t_G, \ell_G \rangle$ over a set $\mathcal{C}$ of *labels* consists of disjoint finite sets V_G of *nodes* and E_G of *edges*, *source* and *target* functions $s_G, t_G \colon E_G \to V_G$ for edges, and a partial *labeling function* $\ell_G \colon V_G \cup E_G \to \mathcal{C}$ [1]. An edge $e \in E_G$ is called *incident* to its source and target $s_G(e)$ and $t_G(e)$ and makes these nodes *adjacent* with each other. G is called *totally labeled* if the function ℓ_G is total.

A *premorphism* $m \colon G \to H$ between two graphs G and H consists of two functions $m_V \colon V_G \to V_H$ and $m_E \colon E_G \to E_H$ that preserve sources and targets, i.e., $s_H \circ m_E = m_V \circ s_G$ and $t_H \circ m_E = m_V \circ t_G$. If m also preserves defined labels, i.e., if $\ell_H(m(n)) = \ell_G(n)$ for all $n \in \mathrm{Dom}(\ell_G)$, it is called a *morphism*. A morphism m is *injective (surjective)* if both m_V and m_E are injective (surjective, resp.), and it is an *inclusion* if $m(n) = n$ for all nodes and edges n in G. Two graphs G and H are *isomorphic*, written $G \cong H$, if there is an injective and surjective morphism $m \colon G \to H$ that preserves all labels, i.e., $\ell_H(m(n)) = \ell_G(n)$ for all $n \in V_G \cup E_G$.

Rules. A *rule* $t = (L \leftarrow I \to R)$ consists of two inclusions $I \to L$ and $I \to R$ between partially labeled graphs such that for all $n \in V_L \cup E_L$, $\ell_L(n) = \bot$ implies $n \in V_I \cup E_I$ and $\ell_R(n) = \bot$, and, vice versa, $\ell_R(n) = \bot$ implies $n \in V_I \cup E_I$ and $\ell_L(n) = \bot$.

Example 1 (Rule). Assuming that the set $\mathcal{C}$ of labels contains natural numbers, the rule in Fig. 1, taken from [26], relabels the target of an edge in a graph. In our examples, numbers attached to the nodes in the graphs of a rule define the morphisms between them.

Fig. 1. A rule.

Transformation. Let G and H be totally labeled graphs and $t = (L \leftarrow I \to R)$ a rule. We say that t *transforms G to H* and write $G \Rightarrow_t H$ if there exists a graph C with two natural pushouts

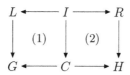

[1] The set $\mathrm{Dom}(f) = \{x \in A \mid f(a) \text{ is defined}\}$ denotes the *domain* of a partial mapping f; we write $f(x) = \bot$ if $f(x)$ is undefined.

so that the vertical morphisms are injective[2]. If $\mathcal{T}$ is a set of rules, we write $G \Rightarrow_{\mathcal{T}} H$ if $G \Rightarrow_t H$ for some rule $t \in \mathcal{T}$, and call $\Rightarrow_{\mathcal{T}} \subseteq \mathcal{G} \times \mathcal{G}$ the *transformation relation* induced by $\mathcal{T}$.

Rule Application. The above definition does not tell how a rule t is actually applied to a graph G in order to transform it into a graph H. For a totally labeled graph G and a rule $t = (L \leftarrow I \rightarrow R)$, an injective morphism $m\colon L \rightarrow G$ is called a *match* of t in G if it satisfies the following *dangling condition*: No node in $m(L \setminus I)$ is incident to an edge in $G \setminus m(L)$. Using this definition, a transformation $G \Rightarrow_t H$ is constructed as follows:

- Find a match $m\colon L \rightarrow G$ of t in G (if it exists).
- Remove all nodes and edges in $m(L \setminus I)$ from G, yielding the context graph C.
- Obtain H from the disjoint union of R and C by identifying the corresponding nodes and edges of $m(I)$ and R.

This construction defines H uniquely up to isomorphism.

3   A Framework for Graph Transformation with Variables

In the general setting for graph transformation with variables, a rule scheme will be instantiated to a set of rule instances that in turn defines the transformation relation. Below we outline general properties of rules, instantiation, and rule application that will be used in the following sections.

Rules. The set $\mathcal{C}$ of labels is extended by a set X of *variable names*. Graphs with labels from $\mathcal{C}$ and X are called *graph patterns* (or just *patterns*). In a pattern G, a variable name $x \in X$ may occur as a label, or be part of a label; it designates the label, or the so labeled node or edge as a placeholder. The *kernel* $\underline{G}$ of a pattern G is the graph obtained by removing all placeholders.

Then a *rule scheme* is a rule $t = (L \leftarrow I \rightarrow R)$ where L, I, and R are patterns.

Instantiation. A *substitution function* σ specifies how variable names occurring in a rule shall be substituted.

Instantiation of a rule scheme t according to some substitution σ defines a particular *rule instance* t^σ. Then $\mathcal{T}(t) = \{t^\sigma \mid \sigma \text{ is a substitution}\}$ defines the *set of rule instances* for t. $\mathcal{T}(t)$ is a set of rules without variables that defines a transformation relation as described in the previous section.

Rule Application. The application of a rule t cannot generate the set $\mathcal{T}(t)$ in order to apply one of the resulting rules because this set is infinite in general. Instead, rule application proceeds as follows.

Let G be a graph, and $t = (L \leftarrow I \rightarrow R)$ a rule scheme.

---

[2] A pushout is natural if it is a pullback as well. The construction of natural pushouts is described in [15].

1. Identify a *kernel match* $\underline{m}\colon \underline{L} \to G$ of the kernel $\underline{L}$ of L in G (if it exists).
2. Induce a substitution σ such that the kernel match $\underline{m}$ extends to a *full match* $m\colon L^\sigma \to G$ of t (if such a substitution exists).
3. Construct the instance R^σ and apply t^σ to construct the instance application $G \Rightarrow_{t^\sigma} H$.

Step 2 need not succeed in all cases; it may also be nondeterministic. We require that rule schemes used in the sequel satisfy the following two conditions:

– A rule scheme t is *left-linear* if every variable name occurs at most once in L.
– A rule scheme t is *closed* if every variable name occurring in R occurs in L as well.

These conditions make rule application easier. If a rule scheme is left-linear, the substitution σ can be induced (in step 2) by considering the unique occurrences of variable names in L one after the other. If a rule scheme is closed, the substitution σ induced in step 2 determines R^σ uniquely and completely.

4   Attribute Variables

In many applications of graph transformation, the nodes and edges of graphs have attributes like numbers or strings that shall be computed by functions during transformation. The attribute model of D. Plump and S. Steinert [26] represents attribute values as labels. The rule schemes on the meta-level are labeled with terms that specify how these values are computed. This is close to the models proposed in [28, 21].

Attributed Rules. *Attributed patterns* are graphs that are partially labeled with terms over a family $\mathcal{F} = (\mathcal{F}_n)_{n \geqslant 0}$ of graded *function symbols* that is disjoint to a set X of *variable names*. The set $\mathcal{T}(X)$ of *terms* is the least set satisfying (i) $x \in \mathcal{T}(X)$ for all variable names $x \in X$, (ii) $c \in \mathcal{T}(X)$ for all constant symbols $c \in \mathcal{F}_0$, and (iii) $f(t_1, \ldots, t_k) \in \mathcal{T}(X)$ for all function symbols $f \in \mathcal{F}_k$ and $k > 0$ terms $t_1, \ldots, t_k \in \mathcal{T}(X)$.

A rule scheme $t = (L \leftarrow I \to R)$ is an *attributed rule* if L, I, and R are attributed patterns. (Note that t is left-linear and closed, as every rule scheme.) We also require that t is *deterministic*, meaning that all terms used as labels in L are variables.

Instantiation. The meaning of function symbols $\mathcal{F}$ is given by an *algebra* A that consists of a *carrier set* A with elements $c_A \in A$ for all $c \in \mathcal{F}_0$, and functions $f_A\colon A^k \to A$ for all $f \in \mathcal{F}_k$ with $k > 0$.

A function $\alpha\colon X \to A$ is called an *assignment*. The extension $\hat{\alpha}\colon \mathcal{T}(X) \to A$ of α is defined by (i) $\hat{\alpha}(x) = \alpha(x)$ for all variable names $x \in X$, (ii) $\hat{\alpha}(c) = c_A$ for all constant symbols $c \in \mathcal{F}_0$, and (iii) $\hat{\alpha}(f(t_1, \ldots, t_k)) = f_A(\hat{\alpha}(t_1), \ldots, \hat{\alpha}(t_k))$ for all function symbols $f \in \mathcal{F}_k$ and all terms $t_1, \ldots, t_k \in \mathcal{T}(X)$.

For an attributed pattern G and an assignment $\alpha\colon X \to A$, its *instance* G^α is the partially labeled graph over A obtained by replacing the labeling function

Fig. 2. An attributed rule.

ℓ_G by $\hat{a} \circ \ell_G{}^3$. The instance of an attributed rule $t = (L \leftarrow I \rightarrow R)$ is the rule $t^\alpha = (L^\alpha \leftarrow I^\alpha \rightarrow R^\alpha)$ with partially labeled graphs over A.

Example 2 (Computing Path Weights). The attributed rule in Fig. 2 computes node attributes that represent the *weight* of paths in a graph. The assignment $\alpha = \{x \mapsto 3, y \mapsto 2, z \mapsto 7\}$ instantiates the attributed rule to the instance shown in Fig. 1 above (supposing that $+_A$ implements addition).

Rule Application. A transformation step $G \Rightarrow_t H$ via some attributed rule $t = (L \leftarrow I \rightarrow R)$ is constructed by finding a *pre-morphism* $m \colon L \rightarrow G$. Since t is left-linear and all defined labels in L are variables, mapping $\ell_L(n)$ onto $\ell_G(m(n))$ for every labeled node or edge n in L, uniquely defines a partial assignment $\alpha \colon X \rightarrow A$ that is defined for all $x \in X$ occurring in L. Then α determines the instantiation R^α completely since t is closed, and the transformation step $G \Rightarrow_t H$ is uniquely defined as well (up to isomorphism). The induction of assignments is thus deterministic and always successful.

Discussion. For simplicity, our definitions deal with untyped terms and algebra whereas the labels in [26] are many-sorted. In that paper, attribute values are also used to define *conditional rules* of the form $t = (L \leftarrow I \rightarrow R \text{ where } c)$ with a boolean expression c; instantiation yields an instance $t = (L^\alpha \leftarrow I^\alpha \rightarrow R^\alpha)$ if and only if $\hat{a}(c)$ evaluates to true. Then the induction of assignments is still deterministic, but partial, as some assignments α yield invalid rule instances t^α.

The algebra A can be implemented by some library in a programming language. It can also be defined by rewriting itself: Every confluent and terminating term rewriting system defines an algebra. The values of A may as well be graphs, and the operations can be defined by (confluent and terminating) graph transformations. This idea has been considered in [29] for the first time.

Attribution by instantiation is considerably simpler than other models, such as the one proposed in [16], where every graph is burdened by an infinite set of nodes that represents all values of A, and cluttered up with edges that point from nodes to their actual attribute values. It is also more general as edges may be labeled as well as nodes.

5   Clone Variables

Rules in programmed graph transformation [30] may contain "object set identifiers", which are nodes in a rule pattern that shall match the set of all nodes in a graph that are connected to the nodes of the rule in the same way. We call such nodes *clones*.

---

[3] The composition $\hat{a} \circ \ell_G$ is undefined at the nodes and wedges where ℓ_G is undefined.

Cloning Rules. Extend the label set $\mathcal{C}$ by the Cartesian product $\mathcal{C} \times X$, where X is a set of variable names disjoint to $\mathcal{C}$. A partially labeled graph G over $\mathcal{C} \cup (\mathcal{C} \times X)$ is a *clone pattern* if labels of the form (c, x) with $c \in \mathcal{C}$ and $x \in X$ are only used on nodes. A node v in a clone pattern G with $\ell_G(v) = (c, x)$ is called an *x-fold c-clone*, or just a *clone* if c and x do not matter. (Labels "$(\ ,x)$" indicate clones with undefined label.) All other nodes are called *constant*. The clone variable x in a label (c, x) stands for a number of c-nodes that are instantiated for v. Although this number is arbitrary for every variable name x, it restricts x-fold clones v and w to be instantiated by the same number of nodes.

A rule scheme $t = (L \leftarrow I \rightarrow R)$ consisting of clone patterns is a *cloning rule*. (It is left-linear and closed.)

Instantiation. Let $\mu \colon X \rightarrow \mathbb{N}$ be a *multiplicity function* which specifies how many instances of clones shall be inserted in a graph.

For a clone pattern G and a multiplicity function μ, the *instance* G^μ is defined as follows:

1. Replace every x-fold clone v by a set of $\mu(x)$ nodes, which are called the *instances* of v and are denoted by $v^{\mu(x)}$.
2. Replace every edge between a clone v and a constant node w by $\mu(x)$ edges between every instance in $v^{\mu(x)}$ and w, with the same label and direction.
3. Replace every edge between two x-fold clones v and w by $\mu(x)$ edges between corresponding instances in $v^{\mu(x)}$ and in $w^{\mu(x)}$, with the same label and direction. (See Fig. 3 on the left.)
4. Replace every edge between an x-fold clone v and a y-fold clone w (with different variables $x \neq y$) by $\mu(x) \times \mu(y)$ edges between every instance in $v^{\mu(x)}$ and every instance of $w^{\mu(y)}$, with the same label and direction. (See Fig. 3 on the right.)

The rule $t^\mu = (L^\mu \leftarrow I^\mu \rightarrow R^\mu)$ is the instance of a cloning rule for a multiplicity function μ.

Example 3 (A Pull-Down-Method Refactoring). The cloning rule in Fig. 4 has been adapted from [32]. It describes a transformation of graphs representing object-oriented programs where a method definition (denoted by the square δ-node) is pulled down from a class (represented by the constant C-node) to its subclasses (represented by the n-fold C-clone). For the transformation rule,

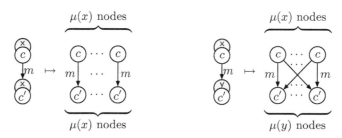

Fig. 3. Instantiation of edges between multiple nodes, case 3 (left) and case 4 (right).

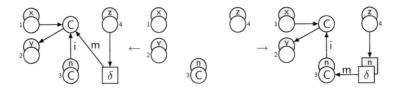

Fig. 4. Cloning rule for the pull-down-method refactoring.

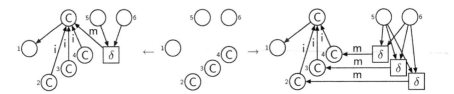

Fig. 5. Cloning rule instance for the pull-down-method refactoring.

the neighborhood of the class and the method definition have to be considered; this is done by the unlabeled clones with the clone variables x, y, and z.

Instantiating clone variables by $\mu = \{n \mapsto 3, x \mapsto 0, y \mapsto 1, z \mapsto 2\}$ yields the instance shown in Fig. 5. The i-edges are induced by case 2 of the instantiation, whereas m-edges are obtained according to case 3, and the unlabeled edges to the δ-clones are generated according to case 4.

Rule Application. The *kernel match* of a cloning rule $t = (L \leftarrow I \rightarrow R)$ is an injective morphism $\underline{m} \colon \underline{L} \rightarrow G$, where $\underline{L}$ is the kernel of L from which all clones and their incident edges have been removed. The kernel match uniquely determines the instances of the clones in L as follows: If $\{\bar{v}_1, \ldots, \bar{v}_k\}$ is the set of nodes adjacent with an x-fold c-clone $\bar{v}$ in L, search the adjacent nodes of their kernel matches $\underline{m}(\bar{v}_1), \ldots, \underline{m}(\bar{v}_k)$ for the set $\{v_1, \ldots, v_k\}$ of all c-nodes that are connected to $\underline{m}(\bar{v}_1), \ldots, \underline{m}(\bar{v}_k)$ in G in the same way as $\bar{v}$ is connected to $\bar{v}_1, \ldots, \bar{v}_k$ in L. This defines $\mu(x) = k$, $v^{\mu(x)} = \{v_1, \ldots, v_k\}$, and L^μ. The interface instance I^μ is included in L^μ, and the right hand side R can be instantiated by making $\mu(x)$ copies of all x-fold clones that are fresh in R, and connecting them accordingly. It is important that the clone variable x is defined by the left hand side of the rule in order to know how many copies shall be made for every fresh clone. (In Fig. 5, e.g., the clone variable n of the δ-clone on the right hand side is bound by the C-clone on the left hand side.)

The multiplicity function μ can be uniquely determined for every kernel match $\underline{m}$. However, the instance t^μ may not apply because it violates the dangling condition. For instance, the cloning rule in Fig. 4 cannot be applied to a graph where the match of the constant C-node has an incoming edge e that is not labeled with i because deletion of that node would leave e dangling.

Simulating Connecting Graph Transformation. Cloning rules can simulate connecting graph transformation in the sense of [22, 32]. The *connecting rules* in that paper take the form (L, R, in, out) where L is the graph to be matched (and deleted), R is the graph, a copy of which has to be replaced for the match of L, and the *connection instructions* $in, out \subseteq V_L \times V_R \times C_E \times C_E$ specify how the nodes incident to the match $m(L)$ are to be connected to the nodes in the copy of R:

- An instruction $(v, w, c, d) \in in$ says that every neighbor node v' pointing to v by a c-edge shall point to the node w in R by a d-edge.
- An instruction $(v, w, c, d) \in out$ says that if v points to some neighbor node v' by a c-edge, the node w in R shall point to the node v' by a d-edge.
- All other edges connecting nodes in L to neighbor nodes (with labels or directions not mentioned in connection instructions, that is) are deleted.

The cloning rule $t = (\tilde{L} \leftarrow \tilde{I} \rightarrow \tilde{R})$ simulating a connecting rule (L, R, in, out) is defined as follows:

- $\tilde{L}$ is obtained by extending L with a set of mutually distinct clones so that every node v in L is adjacent with two clones $x_{v,c,in}$ and $x_{v,c,out}$, for every edge label c in C_E.
- $\tilde{I}$ consists of all clones of $\tilde{L}$.
- $\tilde{R}$ is obtained by extending R with $\tilde{I}$, and by connecting the clones of $\tilde{I}$ to nodes w in R according to the connection instructions:
  - If $(v, w, c, d) \in in$, let the clone $x_{v,c,in}$ point to w by a d-edge.
  - If $(v, w, c, d) \in out$, let w point to the clone $x_{v,c,out}$ by a d-edge.

Admittedly, this simulation is rather clumsy because $2 \times |C|$ variables are needed for every node of a left hand side. Especially if C is infinite, we need a notation for clones that match "all other neighbor nodes" of some node v, i.e., all adjacent nodes that are connected by labels and directions not mentioned in the other clones at v.

6   Graph Variables

In [25], D. Plump and A. Habel have devised rules wherein variable hyperedges are placeholders for hypergraphs that are substituted by hyperedge replacement [5]. We "translate" hyperedge replacement to a simple way of node replacement in graphs.

Rules with Variable Nodes. The label set C is extended by a disjoint set X of *variable names*. Every variable name $x \in X$ comes with a *type* type$(x) \in C^* \times C^*$ that specifies the number and labels of its incident edges as follows. Let G be a graph over $C \cup X$. A node v in G is a *node variable* if $\ell_G(v) = x \in X$. A node variable v with $\ell_G(v) = x$ and type$(x) = (c_1 \cdots c_n, \bar{c}_1 \cdots \bar{c}_{\bar{n}})$ is *well-typed* if v is the target of n edges $e_1 \cdots e_n$ with $\ell_G(e_i) = c_i$ for $1 \leqslant i \leqslant n$, and the source of $\bar{n}$ edges $\bar{e}_1 \cdots \bar{e}_{\bar{n}}$ with $\ell_G(\bar{e}_j) = \bar{c}_j$ for $1 \leqslant j \leqslant \bar{n}$; v is *straight* if all sources of its ingoing edges and all targets of its outgoing edges are pairwise distinct, and finally, v is *apart* if there is no variable node among these adjacent nodes.

A graph G over $\mathcal{C} \cup X$ is a *graph pattern* if all edges have constant labels, and all variable nodes are well-typed, straight, and apart.

A rule scheme $t = (L \leftarrow I \rightarrow R)$ with graph patterns L, I, and R is a *rule with variable nodes* if I is constant. As usual, we also assume that t is left-linear and closed.

Instantiation. Variables in a graph pattern G are instantiated by replacing variable nodes by graphs.

With $\langle x^\star \rangle$ we denote the *star graph* of a variable name x, which is the graph with an x-labeled center node and incident edges according to type(x), plus unlabeled nodes at the other ends of these edges. The graph $\langle x^\circ \rangle$ is the discrete subgraph of $\langle x^\star \rangle$ that consists just of its border nodes.

A rule of the form $t = (\langle x^\star \rangle \leftarrow \langle x^\circ \rangle \rightarrow S)$ is called a *simple node replacement rule*[4]. (Fig. 6 shows such a rule.) A graph substitution γ maps variable names $x \in X$ onto node replacement rules $\gamma(x) = (\langle x^\star \rangle \leftarrow \langle x^\circ \rangle \rightarrow S)$.

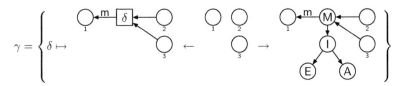

Fig. 6. A graph substitution.

The instantiation of a graph pattern G according to a graph substitution γ applies the simple node replacement rules $\gamma(\ell_G(v))$ to all variable nodes v in G in parallel.

In a graph pattern G, straightness guarantees that the simple node replacement rule $\gamma(x)$ applies to every x-node in G. Apartness ensures that the matches of two nodes replacement rules t and t' either overlap only in their border nodes, or they overlap completely, and $t = t'$. In the first case, the steps commute by the parallel independence results for gluing rules; in the second case, the definition of γ makes sure that the rules are equal. Thus simple node replacement via γ is strongly confluent for graph patterns, and the instance H^γ is unique up to isomorphism.

Example 4 (A Pull-Down-Method Refactoring). The substitution γ shown in Fig. 6 with Dom(γ) = $\{\delta\}$ specifies a simple node replacement rule for the variable name δ with type(δ) = $(\bot\bot, m)$ that replaces δ-stars by a method definition. In this case, the method body is a tree representing an if-statement **if** E **then** A with a condition E and a simple assignment A.

The substitution γ instantiates the rule with variable nodes in Fig. 5 to the rule shown in Fig. 7 below.

---

[4] This definition covers a restricted form of node replacement that corresponds to hyperedge replacement [5]. Full node replacement rules have connection instructions, and may be applied to variables that are not apart.

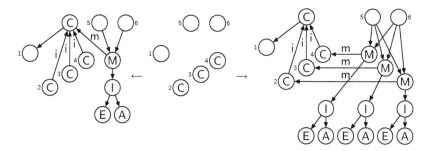

Fig. 7. Rule instance for the pull-down-method refactoring.

Rule Application. For constructing a transformation of a graph G via a rule t with variable nodes, we define the kernel of the left hand side pattern L in such a rule as the subgraph $\underline{L}$ where all variable nodes and their incident edges are removed, and determine a kernel match $\underline{m}: \underline{L} \to G$.

We then attempt to induce a substitution γ by starting, for every variable node v in L, at the matches of its adjacent nodes, say $v_1, \ldots, v_k$. Every candidate for $\gamma(\ell_L(v))$ has to be isomorphic to a subgraph $S \subseteq G$ that overlaps with $\underline{m}(\underline{L})$ and with the substitution candidates for other variables only in the nodes $\underline{m}(v_1), \ldots, \underline{m}(v_k)$.

One the one hand, there need not be such a graph, for instance if one v_i is not in the interface I, not adjacent with any other variable node in L, and if its kernel match is source or target of an edge e outside $\underline{m}(\underline{L})$ that is incident to the kernel match $\underline{m}(v_i)$ of a node $\bar{v}$ in $\underline{m}(\underline{L})$ other than $\underline{m}(v_1), \ldots \underline{m}(v_k)$. This edge cannot belong to S, it cannot belong to the substitution of another variable, and it may also not belong to the context of the transformation, since deletion of $\underline{m}(v_i)$ would leave e dangling. On the other hand, there may be several candidates for substituting a variable in general.

Instantiation of the right hand side R is unique once a graph substitution γ has been found. The transformation $\mathcal{G} \Rightarrow_{t^\gamma} H$ and the resulting graph H are then unique up to isomorphism.

Discussion. Variable nodes can also be substituted if they are not apart, and by applying full node replacement with connection instructions [10]. However, this requires some precaution as node replacement is not confluent in general.

Programmed graph transformation rules [30] feature *path expressions* by which the existence of paths can be specified. Path expressions can be considered as variable edges that may be substituted by chain-like graphs which are defined by edge replacement.

7   Combining Variables

It is fairly straight-forward to combine different kinds of variables in order to get even more powerful graph transformation rules. Simulation of the rules proposed in [32], for instance, requires at least two kinds of variables: The rules need clone variables as they use connection instructions, and variable nodes as they copy subgraphs of arbitrary size (like the method bodies). The rule in Fig. 4 contains clone variables and a variable node (δ) which are instantiated by making clones first, and instantiating the variable node δ afterwards. The combination requires some care in order to achieve the desired results: The clone variables like δ on the right hand side of Fig. 4 need to be cloned before the resulting set of variable nodes is substituted itself.

The rules in [32] might also take advantage of attribute variables if computations on primitive values shall be performed. In this case, attributes should be instantiated last.

Lifting Properties of Rule Instances. Given that instantiation is a simple concept, there is hope that some results for the underlying rule instances can be lifted to the level of the rule schemes. Confluence, for instance, is relevant for many applications. Fortunately, the theory of gluing graph transformation provides criteria for the parallel independence of transformations [7], and there is also a critical pair lemma [24]. In [13], parallel independence has been lifted to transformation with variable nodes. We think that confluence for the kind of rules used in [32] can be proved more easily in the two-level model than in the original definition.

8   Conclusions

In this paper we have studied how variables can be used in graph transformation. The general framework of two-level transformation – instantiation of rule schemes to rule instances which in turn define transformations – applies to several kinds of variables: attribute variables, clones, and graph variables. Since the model is modular, different kinds of variables can be combined. This yields very expressive rules that are still comprehensible, because variables are a familiar concept in specification and programming. Instantiation reduces rule schemes to standard gluing rules, and thereby allows to add important concepts of connecting and programmed graph transformation to the gluing approach – such as connection instructions and path expressions. Gluing graph transformation has a rich theory that is automatically available on the level of rule instances. Some of this theory can probably be lifted to the level of rule schemes as well.

Related Work. Several authors have studied advanced graph transformation rules. The early book [23] by M. Nagl defines a very general way of operational graph transformation. The structured rules of H.-J. Kreowski and G. Rozenberg [20] combine gluing rules with connection instructions. Both approaches do not consider graph variables.

Future Work. The work started in this paper can be continued in several directions.

The basic transformation model can be extended by types and shapes [18], by negative application conditions [12], and so on.

We would also like to loosen the standard conditions on rule schemes. For left-linearity, this is rather easy; but then, the induction of substitutions can no longer be done independently for every variable, since different occurrences of the same variable must have equal substitutions. Dropping the closedness condition is only feasible if transformation is lifted to graphs with variables. Then matching of graphs with variables against graphs has to be replaced by *unification* of two graphs with variables. However, it is unknown whether graph unification is decidable.

An important issue is to make the induction of substitutions for graph variables less nondeterministic, because variable nodes may have many different substitutions. The paper [6] gives rather strict conditions for the unique induction of substitutions. In [18], the nondeterminism of induction is reduced by requiring substitutions to be "shaped" graphs that are generated by a hyperedge replacement grammar. Then, substitution candidates can be found by parsing according to the grammar. In the rule in Fig. 5, for instance, the substitutions of δ could be restricted to the syntax trees of method definitions. A substitution for δ can be induced by parsing according to the syntactic rules.

Of course, we are also interested in evaluating how useful variables are for modeling realistic case studies. Refactoring seems to be an area where graph transformation can be applied successfully, and where advanced concepts are needed for the transformation rules. Such case studies could also show whether it is really possible to lift theorems from the underlying world of gluing rules to the high-level rules.

Acknowledgment

In 1978, Hartmut Ehrig urged I.-R. Schmiedecke and me to define extended affix-grammars (a variant of the two-level grammars mentioned in the introduction) so that he could understand them. This resulted not only in a graph grammar definition [9], but also introduced me to the field of graph transformation which has fascinated me ever since. Now the topics of this paper brought me back to two-level grammars where all this began. *Thank you, Hartmut!*

References

1. Volker Claus, Hartmut Ehrig, and Grzegorz Rozenberg, editors. *Proc. Graph Grammars and Their Application to Computer Science and Biology*, number 73 in Lecture Notes in Computer Science. Springer, 1979.
2. C.J. Cleaveland and R.C. Uzgalis. *Grammars for Programming Languages*. Elsevier, New York, 1977.
3. Andrea Corradini, Hartmut Ehrig, Reiko Heckel, Michael Löwe, Ugo Montanari, and Francesca Rossi. Algebraic approaches to graph transformation, part I: Basic concepts and double pushout approach. In Grzegorz Rozenberg, editor, *Handbook of Graph Grammars and Computing by Graph Transformation. Vol. I: Foundations*. World Scientific, 1997.

4. Andrea Corradini, Hartmut Ehrig, Hans-Jörg Kreowski, and Grzegorz Rozenberg, editors. *1st Int'l Conference on Graph Transformation (ICGT'02)*, number 2505 in Lecture Notes in Computer Science. Springer, 2002.
5. Frank Drewes, Annegret Habel, and Hans-Jörg Kreowski. Hyperedge replacement graph grammars. In Rozenberg [27], chapter 2, pages 95–162.
6. Frank Drewes, Berthold Hoffmann, and Mark Minas. Constructing shapely nested graph transformations. In Hans-Jörg Kreowski and Peter Knirsch, editors, *Proc. Int'l Workshop on Applied Graph Transformation (AGT'02)*, 2002. 107–118.
7. Hartmut Ehrig. Introduction to the algebraic theory of graph grammars. In Claus et al. [1], pages 1–69.
8. Hartmut Ehrig, Gregor Engels, Francesco Parisi-Presicce, and Grzegorz Rozenberg, editors. *2nd Int'l Conference on Graph Transformation (ICGT'04)*, number 3256 in Lecture Notes in Computer Science. Springer, 2004.
9. Hartmut Ehrig, Berthold Hoffmann, and Ilse-Renate Schmiedecke. A Graph-Theoretical Model for Multi-Pass Parsing. In J.R. Mühlbacher, editor, *Proc. 7th Conf. on Graph-Theoretical Concepts in Comp. Sci. (WG'81)*, pages 19–32, München-Wien, 1982. Hanser Verlag.
10. Joost Engelfriet and Grzegorz Rozenberg. Node replacement graph grammars. In Rozenberg [27], chapter 1, pages 1–94.
11. Herbert Göttler. *Zweistufige Graphmanipulationssysteme für die Semantik von Programmiersprachen*. Dissertation, Universität Erlangen-Nürnberg, 1977. [In German].
12. Annegret Habel, Reiko Heckel, and Gabriele Taentzer. Graph grammars with negative application conditions. *Fundamenta Informaticae*, 26:287–313, 1996.
13. Annegret Habel and Berthold Hoffmann. Parallel independence in hierarchical graph transformation. In Ehrig et al. [8], pages 178–193.
14. Annegret Habel, Jürgen Müller, and Detlef Plump. Double-pushout graph transformation revisited. *Mathematical Structures in Computer Science*, 11(5):637–688, 2001.
15. Annegret Habel and Detlef Plump. Relabelling in graph transformation. In Corradini et al. [4], pages 135–147.
16. Reiko Heckel, Jochen M. Küster, and Gabriele Taentzer. Confluence of typed attributed graph transformation systems. In Corradini et al. [4], pages 161–176.
17. Wolfgang Hesse. Two-level graph grammars. In Claus et al. [1], pages 255–269.
18. Berthold Hoffmann. Shapely hierarchical graph transformation. In *Proc. IEEE Symposia on Human-Centric Computing Languages and Environments*, pages 30–37. IEEE Computer Press, 2001.
19. Jan Willem Klop. Term rewriting systems. In S. Abramsky, Dov M. Gabbay, and T.S.E. Maibaum, editors, *Handbook of Logic in Computer Science*, volume 2, pages 1–116. Oxford University Press, 1992.
20. Hans-Jörg Kreowski and Grzegorz Rozenberg. On structured graph grammars, I and II. *Information Sciences*, 52:185–210 and 221–246, 1990.
21. Michael Löwe, Martin Korff, and Annika Wagner. An algebraic framework for the transformation of attributed graphs. In Sleep et al. [31], pages 185–199.
22. Tom Mens, Serge Demeyer, and Dirk Janssens. Formalising behaviour-preserving transformation. In Corradini et al. [4], pages 286–301.
23. M. Nagl. *Graph-Grammatiken: Theorie, Anwendungen, Implementierungen*. Vieweg-Verlag, Braunschweig, 1979. In German.
24. Detlef Plump. Hypergraph rewriting: Critical pairs and undecidability of confluence. In Sleep et al. [31], pages 201–213.

25. Detlef Plump and Annegret Habel. Graph unification and matching. In Janice E. Cuny, Hartmut Ehrig, Gregor Engels, and Grzegorz Rozenberg, editors, *Proc. Graph Grammars and Their Application to Computer Science*, number 1073 in Lecture Notes in Computer Science, pages 75–89. Springer, 1996.

26. Detlef Plump and Sandra Steinert. Towards graph programs for graph algorithms. In Ehrig et al. [8], pages 128–143.

27. Grzegorz Rozenberg, editor. *Handbook of Graph Grammars and Computing by Graph Transformation, Vol. I: Foundations*. World Scientific, Singapore, 1997.

28. Georg Schied. *Über Graphgrammatiken, eine Spezifikationsmethode für Programmiersprachen und verteilte Regelsysteme*. Dissertation, Universität Erlangen-Nürnberg, 1992. [In German].

29. Hans-Jürgen Schneider. On categorical graph grammars integrating structural transformations and operations on labels. *Theoretical Computer Science*, 109:257–274, 1993.

30. Andy Schürr. Programmed graph replacement systems. In Rozenberg [27], chapter 7, pages 479–546.

31. M. Ronan Sleep, Rinus Plasmeijer, and Marko van Eekelen, editors. *Term Graph Rewriting, Theory and Practice*. Wiley & Sons, Chichester, 1993.

32. Niels van Eetvelde and Dirk Janssens. Extending graph rewriting for refactoring. In Ehrig et al. [8], pages 399–415.

Graph Transformation in Molecular Biology[*]

Francesc Rosselló[1] and Gabriel Valiente[2]

[1] Department of Mathematics and Computer Science,
Research Institute of Health Science (IUNICS), University of the Balearic Islands,
Palma de Mallorca, Spain
[2] Department of Software, Technical University of Catalonia, Barcelona, Spain
cesc.rossello@uib.es, valiente@lsi.upc.es

Abstract. In the beginning, one of the main fields of application of graph transformation was biology, and more specifically morphology. Later, however, it was like if the biological applications had been left aside by the graph transformation community, just to be moved back into the mainstream these very last years with a new interest in molecular biology. In this paper, we review several fields of application of graph grammars in molecular biology, including: the modelling of higher-dimensional structures of biomolecules, the description of biochemical reactions, and the study of biochemical pathways.

1   Introduction

Once upon a time, biology was one of the main fields of application of graph transformation, as it is proved by the maiden name (back in 1978) "Workshop on Graph Grammars and Their Application to Computer Science and Biology" of the current "International Conference on Graph Transformation." Those early applications of graph rewriting in biology mostly belonged to the field of morphogenesis.

It is common knowledge that graphs describe structures in a simplified but explicit way. In such descriptions, nodes correspond to substructures and arcs represent relations among substructures. These arcs can be directed if the relation is so, labelled if one wants to record the kind of relation they stand for, and so on. On their turn, nodes may be labelled to make explicit what they symbolize, with labels that may be not only raw names, but also graphs themselves, or other higher-order objects that can be used to abstract the details of the substructure represented by the nodes in hierarchical structures. In any case, the actual meaning of the nodes and the arcs will depend on the actual application. Under this graphical representation of structures, the evolution of the latter can be described by graph rewriting mechanisms, where one or several subgraphs are replaced by other graphs in a way determined by evolution rules specified in a graph grammar.

It was soon noticed that the development states of an organism can be described as graphs in this way, with nodes representing for instance cells, body

---

[*] This work has been partially supported by the Spanish CICYT, project MAVERISH (TIC2001-2476-C03-01) and by the Spanish DGES and the EU program FEDER, project BFM2003-00771 ALBIOM.

H.-J. Kreowski et al. (Eds.): Formal Methods (Ehrig Festschrift), LNCS 3393, pp. 116–133, 2005.

segments, or tissues, and arcs representing spatial or biological relations among nodes. The nodes' labels may be used to denote their type and the arcs' labels the type of interaction they stand for. The rules governing some aspect of the development of such an organism can be described in this framework as graph rewriting rules and gathered in a graph grammar. In a given application, these rules can be fired simultaneously, in a synchronized way, or following some priority order. It was precisely the possibility of modelling the development of organisms where changes and segmentations take place simultaneously at different places that lead to the notion of parallel graph grammars, also called *graph L-systems*, as a generalization of string L-systems. They were introduced about thirty years ago by K. Culik and A. Lindenmayer [12], previously hinted by B. Mayoh [42], and they have been used since then in many applications of graph rewriting in morphogenesis.

This was the first kind of applications of graph rewriting in biology, and, as a matter of fact, the use of graph grammars as models of the development of organisms is still alive. For instance, Beck, Benkö et al [2] have proposed recently the use of graph transformation as an alternative to standard morphospace representations and geometric morphometrics in the field of theoretical morphology, while Tomita. Kurokawa and Murata [63] have introduced a new type of graph rewriting systems, *graph automata*, as an alternative to graph L-systems in the description of self-reproducing complex systems.

The success of graph grammars in the description of development pathways can be seen as a simple instance of their pattern handling power. According to D. Gernert [33], as soon as patterns are represented as graphs, graph grammars are a natural tool to describe the fundamental operations related to patterns: pattern generation, pattern transfer (the duplication of a certain subpattern and its insertion in a different location), pattern recognition, pattern interpretation (the influence of certain subpatterns on the behavior of whole system) and pattern application (the transmission of a certain pattern to another location). A type of graph grammars specifically tailored to handle patterns was proposed in [41].

Patterns that are conveniently modelled as graphs are found everywhere in biology, and not only in morphology. Molecular biology is no exception: the inner structure of chemical compounds [9], the tridimensional structure of nucleic acids and proteins [64], the chemical reactions [31], the biochemical networks [45], most formal components of molecular biology can be represented as graphs. This fact must be added to what is called in sociology of science "the phenomenon of the earlier tool" [33]: when some branch of mathematics reaches a high standard or it becomes fashionable, then it will be surely used in many other sciences[1].

---

[1] Historians of science put more emphasis on the converse phenomenon, when a problem in some science gives rise to new a branch of mathematics or gives new life to an already existing branch; for instance, the theory of Abstract Data Types gave a boost to universal algebra... and H. Ehrig [24] had his share of guilt! Graph grammars can also be seen as an example of this phenomenon, as they were born to solve the problem of specifying the transformation of non linear structures in software systems.

Therefore, it should not be a surprise that, with the recent thriving of computational molecular biology and computational systems biology, graph grammars have initiated what will probably become a second silver age of applications in biology.

The goal of this paper is to overview some applications of graph rewriting in molecular biology. In the next section we shall write about the modelling of tridimensional structures of nucleic acids and proteins. In Section 3 we will cover the modelling of chemical compounds and chemical reactions in artificial chemistries and then, in Section 4, the application of the latter in the analysis of biochemical pathways. With this short survey we want to celebrate H. Ehrig participation in the development of such a versatile specification and analysis tool.

2   Higher-Dimensional Structures of Biomolecules

A biomolecule can always be viewed as an oriented chain of monomers, which in turn can be mathematically described as a string over a suitable alphabet. This string is called the *primary structure* of the molecule. For instance, a DNA or an RNA molecule is a chain of nucleotides, each one of them characterized by the base attached to it: adenine, A, cytosine, C, guanine, G, or thymine, T, (in RNA, thymine is replaced by uracil, U). Thus, the primary structure of a DNA molecule is a string over $\{A, C, G, T\}$, while the primary structure of an RNA molecule is a string over $\{A, C, G, U\}$. In a similar way, proteins are chains of amino acids, and hence the primary structure of a protein is a string over a 20-letter alphabet, for instance

$$\{A, C, D, E, F, G, H, I, K, L, M, N, P, Q, R, S, T, V, W, Y\},$$

each letter representing an amino acid: A for Alanine, C for Cysteine, D for Aspartic acid, etcetera.

In the cell and *in vitro*, each RNA molecule and protein folds into a tridimensional structure, and this is structure what determines its biochemical function. The understanding of the folding process of these biomolecules and the prediction of their tridimensional structure from their primary structure are two of the main open questions in molecular biology.

As different levels of graining are suitable for different problems [51], we can sometimes forget about the detailed description of these tridimensional structures and consider only a simplified model of them, like for instance their contact structures. The *contact structure* [10] of a biomolecular tridimensional structure is the set of all pairs of monomers that are either consecutive in the chain or, in some specific sense, neighbors in the structure. Such a contact structure can be mathematically described as an undirected graph without multiple edges or self-loops, with sets of nodes representing the monomers numbered according to their position along the chain and with edges of two types: those that join pairs of consecutively numbered monomers, which are said to form the *backbone* of the contact structure, and the other ones, which are called *contacts*.

The secondary structures of RNA molecules form a special class of contact structures. In them, contacts represent the *hydrogen bonds* between pairs of non-consecutive bases[2] that hold together the tridimensional structure. A restriction, called the *unique bonds condition*, is added to the definition of RNA secondary structure [64]: a base can only pair with at most another base. It is usual to impose a final restriction on *RNA secondary structures*, by forbidding the existence of *knots*, i.e., of contacts that 'cross' each other. This restriction has its origin in the first dynamic programming methods to predict RNA secondary structures [64, 71], but real RNA structures can contain knots, which are moreover important structural elements of them. Contact structures with unique bonds and knots can also be used to represent the local basic building blocks of protein structures, like *α-helixes* or *β-sheets*, often called *protein secondary structures*.

Beyond secondary structures, the representation of the neighborhood in tri-dimensional structures of RNA molecules and proteins needs contact structures without unique bonds. The full contact structure of an RNA molecule may contain sets of contacts that violate the unique bonds condition, like base triplets and guanine platforms, and in the usually very complex contact structure of a protein, each amino acid uses to be a neighbor of several amino acids.

Although the theory of formal languages was born in the 1950s, and then almost simultaneously to modern molecular biology, (recall that F. Crick and J. Watson discovered DNA's double helix in 1953 and N. Chomsky published *Syntactic structures* in 1957), it was not until the 1980s that formal grammars methods started to be applied to biomolecular sequences [7]. A little later it was also noticed that string grammars could also be used to model and study not only the primary structure of biomolecules, but also certain aspects of their contact structures, as for instance secondary structures of RNA molecules [59, 58]. In these approaches, an RNA secondary structure is represented by a derivation tree of a certain context free grammar, while RNA contact structures with unique bonds and knots must be generated by new types of string grammars [53]. Many more works have focused on RNA secondary structures, or, more in general, contact structures with unique bonds, which can be easily described as strings over a complemented language. These include a few attempts to model and study simple aspects of the secondary structures of proteins using string grammar methods. Two typical examples are the SMART [57] and the TOPS [68] systems.

The goal of the representation of contact structures of biomolecules by means of grammars is to contribute to both main questions about contact structures of biomolecules mentioned above. From the theoretical point of view, one expects to deduce properties of their folding mechanisms from the performance of these grammars and the accuracy with which they generate real structures. In this way, these grammars would yield to a better understanding of the folding process of nucleic acids and proteins. From the practical point of view, stochastic versions of these grammars can be used to predict contact structures. Recall that a *stochastic grammar* specifies a probability for each production, and in

---

[2] Actually, a hydrogen bond can only form between bases that are at least four positions apart in the chain.

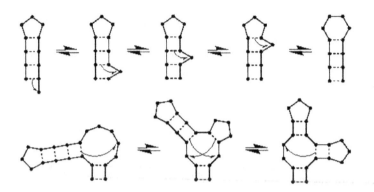

Fig. 1. Two RNA folding processes from [28].

this way it assigns a probability to every derivation. Once a grammar is *trained*, i.e., its probability parameters are tuned on a set of training examples, it can be used to predict the contact structure corresponding to a given primary structure as the most likely derivation of a structure with this primary structure. In the case of string regular grammars, this last step can also done using the very popular, and equivalent, formalism of Hidden Markov Models [21], while in the case of stochastic context-free grammars ad-hoc parsing methods are used [56]. Stochastic graph transformation was recently introduced [35].

Nevertheless, it is clear that it will be difficult to go beyond these results using string grammars in the study of protein structures, because of their high complexity [60]. As contact structures of proteins are graphs, the clear candidate to generate them are graph grammars.

There have been several important advances in the theoretical study of the protein folding problem using graph grammars in a hidden way. In these approaches, "rules take the form of local structure generators, from which structure evolves via iterative application of elementary steps" [48, p. 409]. Actually, the first rule-based approach to protein folding [52] dates back to 1977, and consists of several explicitly described composition rules for the formation, growth and coalescence of β-sheets that could perfectly be formalized as graph rewriting rules. Another description of the formation of protein domains[3] and their relative position as the result of the hierarchical application of explicit rules that are reminiscent of graph transformation rules is due to Lesk [40]. Rule-based descriptions of folding processes of RNA molecules have been proposed [38, 36, 28]; cf. Figure 1.

A paradigmatic, and very interesting, work in this line of research is Przytycka et al's rule-based description of a certain class of protein contact structures, the so-called *all-β proteins*, that admit a high variety of topologies and are difficult to predict from their primary structure. These researchers use a grammar consisting on four composition rules, or rather four families of composition rules, motivated by biophysical considerations that make them conjecture that their

---

[3] A *domain* of a protein is a piece that folds into a stable higher order contact structure.

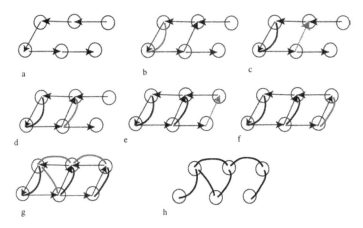

Fig. 2. A derivation of Przytycka et al's grammar.

rules have physical correlates in the actual mechanism of protein folding [48]. Contrary to all previous rule-based approaches to protein contact structures, their rules are explicitly presented as a graph grammar. For the purpose of this grammar, all-β proteins are represented as graphs with nodes corresponding to β-sheets and two types of edges: there are *domain edges*, that are generated by the application of the folding rules and combine the β-sheets to generate more complex folds, and *neighbor edges*, that represent the spatial juxtaposition of non-consecutive β-sheets after the application of a rule by means of a closure operation that can also be represented by a graph rewriting rule. The start graph has only neighbor edges between consecutive β-sheets and no domain edge, and successive applications of the rules group β-sheets by means of domain edges into more complex domains and connect β-sheets that are distant in the sequence but that become juxtaposed in space due to spatial restrictions. Figure 2 displays a derivation of this graph grammar, extracted from [48].

What can the graph grammar community bring to this line of research? To our opinion, it is the biologists' task to propose new rules of formation of contact structures of RNA molecules and proteins, but graph-grammarians could and should collaborate, among other tasks, in formalizing these rules and analyzing the redundancies that appear in the grammars; in determining the properties, for instance related to parallelism and concurrency, of the rewriting systems they define, which might lead to uncovering properties of the real folding process; and in developing general methods to characterize the sets of structures generated by any given set of biomolecules' folding rules, which might give new insights not only on folding processes but also on the evolution mechanisms.

As far as the prediction of protein contact structures goes, Abe and Mamitsuka [1] proposed in 1997 a stochastic tree grammar to predict β-sheets in a way similar to those developed for RNA secondary structures using string grammars that we recalled above. Stochastic versions of more general graph grammars will be necessary to predict contact structures of RNA molecules and proteins beyond secondary structures using this kind of methods. To do that, one should find a set of rules, perhaps in the spirit of those discussed in the previous paragraphs, that capture the formation of the different components of the contact structures of the target biomolecules as well as their relation; one should develop an efficient technique to estimate the grammar's parameters from a set of training graphs; and one should devise an efficient method to find the most probable structure of a protein given the grammar.

3 Artificial Chemistries

Roughly speaking, an *artificial chemistry* [20, 62] is a computational model of a chemical system. It consists of a set (a *soup*) of objects, called *molecules*, a set of *reaction rules* that produce new molecules from already existing molecules, and the definition of the *dynamics* of the system, that specifies the application conditions of the rules, the preference in their application, etc. Against other types of computational models, the goal of an artificial chemistry is to answer qualitative, rather than quantitative, questions: the existence of steady, or closed and self-maintaining, states, the size and diversity of the soup at some moment, etc.

The nature of the molecules, the reactions, and the dynamics of an artificial chemistry can be quite diverse. For instance, in one of the first artificial chemistries, Walter Fontana's *AlChemy* [29], objects were λ-terms, a reaction consisted of the application of the first λ-term to the second one, and the dynamics followed a combination of randomness (in the selection of the pair of molecules) and an explicit algorithm (to decide whether the reaction took place or not).

Now, although artificial chemistries can be, and have been, used to model many kinds of systems, their primary targets are 'real' chemistries, in which case molecules should be representations of chemical compounds, and reaction rules of chemical reactions. The chemical description of a 'real' molecule can be made at different levels of resolution:

- A *molecular descriptor* uniquely identifies a molecule in a biochemical database. For instance, beta-D-Glucose is entry number C00221 in the KEGG database [37].
- A *molecular formula* indicates the number of each type of atom in a molecule. For instance, beta-D-Glucose has the molecular formula $C_6H_{12}O_6$.
- A *constitutional formula* or *chemical graph* indicates which pairs of these atoms are bonded. For instance, beta-D-Glucose has the following chemical graph displayed in Figure 3 (left). Chemical graphs of molecules can be represented as strings using the SMILES language [65–67].
- A *structural formula* refines a chemical graph by indicating those stereochemical distinctions that are required to uniquely identify a molecule. For instance, Figure 3 (right) downloaded from the KEGG database, displays the structural formula of beta-D-Glucose; in it, plain lines depict bonds approximately in the plane of the drawing, bonds to atoms above the plane are shown with a bold wedge, and bonds to atoms below the plane are shown with short

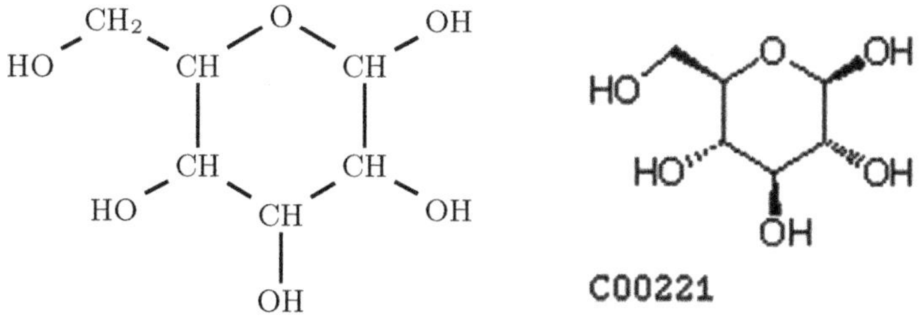

Fig. 3. Beta-D-Glucose's chemical graph (left) and structural formula (right).

parallel lines. This representation allows to distinguish beta-D-Glucose from other chemical compounds with the same chemical graph. D-glucose and L-glucose are mirror images and therefore they share the same chemical graph. Further, there are two possible orientations for the upper-right OH group, which is linked to the CH group number 7 in the ring structure: below the plane of the drawing (alpha-D-glucose) and above the plane of the drawing (beta-D-glucose).

A chemical description at the level of molecular descriptors or molecular formulas is useful for database retrieval purposes, and they can be used in an artificial chemistry when the knowledge of the structure of the chemical compounds is not necessary. In this case, molecular descriptors and formulas play the roles of simple labels, but then chemical reactions cannot be defined by means of local interactions of the atoms of the substrate's molecules.

Chemists have used chemical graphs to distinguish isomers since the second half of the nineteenth century. In first course Organic Chemistry classes, chemical reactions are explained in terms of constitutional formulas and a handful of reaction mechanisms, which corresponds to (chemical) graphs and rules to modify them by means of breaking, forming and changing the type of bonds. This leads in a natural way to artificial chemistries based on labelled graphs as molecules and graph transformation rules as reactions.

Several such artificial chemistries have been proposed so far. J. McCaskill and U. Niemann [44] proposed in 2000 an artificial chemistry for DNA and RNA processing based on graph transformation. In it, molecules are labelled graphs of a specific type, called *variable graphs*, that can represent nucleotides, nucleic acid single or double strands, or sets of all the latter. The reaction rules represent several types of chemical reactions: unimolecular (only one molecule is involved), bimolecular (two molecules react together) and enzymatic (a special type of unimolecular reaction that represents the attachment or the removal of an enzyme at a specific position of a molecule). These reaction rules are graph transformation rules that act following the usual matching-replacement-embedding schema, with some properties of the single-pushout approach [23]. All other reactions, including complex enzymatic reactions, can be decomposed into a series of applications of these reaction rules. The dynamics of the system simply consists of performing all possible reactions through a branching process to obtain all possible derivation paths. The final goal is to predict all libraries

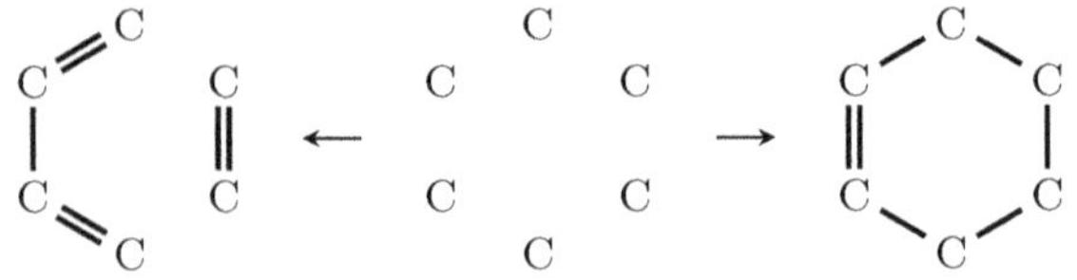

Fig. 4. The Toy model double-pushout rule for the Diels-Alder reaction.

of nucleic acids arising from a given set of strands by means of a given set of enzymatic reactions. The authors have implemented their artificial chemistry in a computer program called MOLGRAPH.

More recently, an artificial chemistry for organic chemistry called the *Toy Model* has been developed by G. Benkö, C. Flamm and P. Stadler [3, 5, 4]. In it, and following [46], molecules are *orbital graphs*: undirected graphs with nodes representing outer atom orbitals, labelled by the atomic element and the hybridization type of the orbital, and edges representing overlaps of adjacent orbitals. These orbital graphs represent sets of chemical compounds and they are uniquely determined by the chemical graphs of these chemical compounds, but they moreover incorporate chemically meaningful energy functions that allow the computation of reaction energies.

The reactions rules translate to this level of abstraction the basic organic reaction mechanisms as graph transformation rules that preserve the vertex labels and the total degrees of corresponding nodes, to capture the conservation of atoms and *valences* in organic reactions. These graph rewriting rules are actually double pushout production rules [11] over orbital graphs: the left-hand side, context, and right-hand side are orbital graphs with the same labelled nodes; the left-hand side graph represents the substrate, the right-hand side graph represents the product and the context graph has as edges those appearing in both the substrate and the product with the same type.

Consider[4], for example, the Diels-Alder reaction [19], one of the most important reactions in organic chemistry. The substrate of the reaction, 1,3-butadiene (C_4H_6) and ethylene (C_2H_4), is combined to form cyclohexene (C_6H_{10}), as described by the double-pushout transformation rule displayed in Figure 4.

A forward application of the previous double-pushout transformation rule to 1,3-butadiene (C_4H_6) and dihydro-2,5-furandione ($C_4H_4O_3$) to form 1,3-isobenzofurandione ($C_8H_8O_3$), corresponds to the double-pushout transformation in Figure 5.

In this artificial chemistry, the rules can be applied randomly or according to the reactivity index of the matching step, computed using suitable formulas from the energy functions, that can for instance be translated into a reaction rate. This graph rewriting system has been implemented in Maude as a client/server application. The final goal is again to compute extensively all possible results of any specific instance of this artificial chemistry, in this case under the form

---

[4] Usually, hydrogen atoms and the corresponding bonds are not represented explicitly in constitutional formulas.

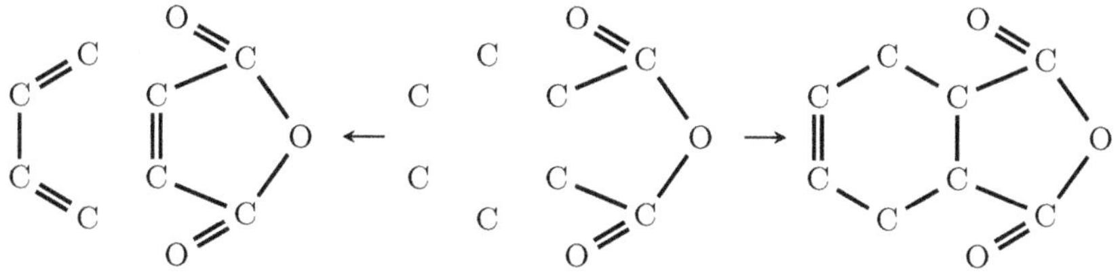

Fig. 5. A reaction in the Toy model.

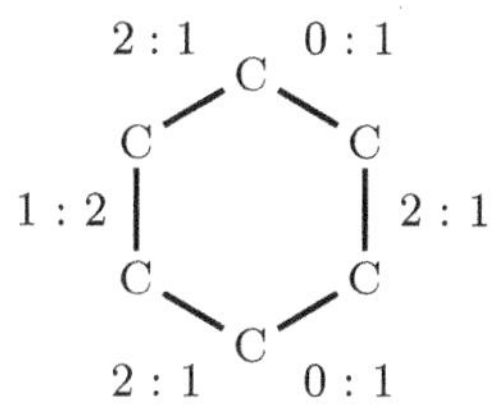

Fig. 6. The explicit chemical reaction for the Diels-Adler reaction.

of large chemical reaction networks defined by an initial set of molecules and the set of allowed reactions. We shall talk more on chemical networks, which are graphs themselves, in the next sections.

We have recently started to develop an artificial network based on graph relabeling grammars [55, 54], also with the final aim of studying biochemical networks. In our approach, and following Fujita's *imaginary transition structures* [30, 32, 31] to model chemical reactions, molecules are *generalized chemical graphs*, disjoint unions of chemical graphs with possibly some extra edges labelled 0. Reactions are then described as edge relabeling graph transformation rules. These reactions can be *explicit* and *implicit*.

An *explicit chemical reaction* is an undirected graph whose nodes are labeled by means of chemical elements and whose edges are labeled by the combination of two natural numbers: a *substrate weight* and a *product weight*. No edge can have both substrate and product weights equal to zero and, for all nodes, the total substrate weight (over all edges incident with the node) cannot be equal to zero and must coincide with the total product weight: it will be actually equal to the *valence* of the corresponding atom. An application of a generalized chemical reaction replaces the substrate weights by the product weights in the matching subgraph.

For instance, the aforementioned Diels-Alder reaction would be represented by the explicit chemical reaction shown in Figure 6, where a label of the form $x : y$ next to an edge means that the edge has substrate weight x and product weight y.

Then, the aforementioned reaction of 1,3-butadiene (C_4H_6) and dihydro-2,5-furandione ($C_4H_4O_3$) to form 1,3-isobenzofurandione ($C_8H_8O_3$) would correspond to the edge relabeling graph transformation described in Figure 7.

Explicit chemical reactions can be seen as edge relabeling versions of the Toy Model's double-pushout transformation rules. The novelty is that they admit implicit versions. An *implicit chemical reaction* is a compact representation of an explicit chemical reaction by means of a finite set of elementary edge relabeling operations that, when applied to a generalized chemical graph taking into account that the total degree of each node must remain constant and that no edge labeled 0 can still be labeled 0 after the application, determine uniquely the product chemical graph. Our conjecture is that, at least for reactions with molecules only involving hydrogen, oxygen, nitrogen and carbon, any one of the relabeling operations of the form 0 : 1 in an explicit chemical reaction forms an implicit chemical reaction for all its applications. For the moment we still have not found a counterexample (don't get fooled by Figure 7, which seems to be a counterexample: to simplify the representation, we omitted in it the weighted bonds between hydrogen atoms). We have implemented our generalized chemical graphs and reactions on top of `PerlMol`, and together with L. Félix, we are currently working on the dynamics and application of this artificial chemistry [27].

M. Yadav, B. Kelley and S. Silverman have recently written a paper [69] where they discuss several aspects of artificial chemistries based on graph transformation rules and designed to model 'real' chemistries. They argue that the reaction rules in any such artificial chemistry should carry associated relevant physical aspects (like temperatures, concentrations, nature and properties of the solvent, etc.), preconditions (like the necessary solvent or any other molecule that must be present or absent in order the reaction to take place), reactivity rates (it is already considered in the Toy Model), and inner mechanisms (similar to the implicit chemical reactions) of the chemical reactions. They also discuss several possible applications of these artificial chemistries: Computer Aided Organic Synthesis, the generation of exhaustive chemical compound libraries (from a given set of molecules through a permitted set of reaction rules and up to a fixed number of steps), and the modelling and analysis of chemical networks (the topic of the next section).

4   Symbolic Systems Biology

Cells are complex but organized chemical engines that carry out a large number of transformations, called *biochemical reactions*. The representation of a cellular process as an ordered set of biochemical reactions is usually called a *biochemical network* or a *biochemical pathway* [45]. There are many types of biochemical pathways, like *metabolic* (which use some molecules to produce others), *regulatory* (where proteins regulate the expression of genes that produce other

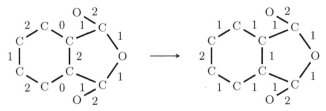

Fig. 7. An explicit chemical reaction in action.

proteins), or *signalling* (which transmit information from the environment to intracellular targets). Biochemical pathways can range in size from involving a few molecules to the complete network giving life to a complex organism, where billions of molecules interact.

The progress in molecular biology has made possible the detailed description of the main molecular components of living systems and their interactions. Nowadays' experimental molecular biology is producing vast amounts of quantitative data on biochemical networks that may support simulation-based research but also that need of theoretical models and the tools related to them, as well as their computer implementations, to interpret them and to make in a systematic way unambiguous descriptions and predictions that, by guiding new experiments, lead to a better understanding of the cellular processes [39].

The mathematical analysis of biochemical pathways has been approached through a large sort of techniques aiming at different goals. The traditional way of modelling them has been by means of quantitative methods based on differential equations. This is the formalism most widely used in science and engineering to model dynamical systems, and thus (recall the principle of the earlier tool mentioned in the introduction) also in molecular biology. In the simplest models of biochemical processes based on ordinary differential equations, the evolution of the concentrations of molecules is modeled by means of *rate equations* that express the rate of production of each molecule as a function of the concentrations of the other molecules, but other types of differential equations have also been used [6]. These equations can be solved when possible, or treated using numerical methods, and they are used to predict quantitative properties of the networks. They have shown to be fundamental in the understanding of cell processes, but they have also some drawbacks: it is difficult to obtain the input experimental data they need; the actual models are difficult to solve or simulate, as well as to understand intuitively; they assume the concentrations vary continuously, which at the cellular scale numbers of molecules is questionable; and they assume that these concentrations vary deterministically, which may be compromised by the stochastic behavior of the timing of cellular events, due again to the small numbers of molecules involved in the reactions [43]. Other methods include the use of discrete stochastic methods like stochastic master equations and bayesian networks.

The last years it has been observed that the biochemical reactions that take place in a cell behave like computation steps performed in a concurrent, nonlinear and asynchronous way [50]. This has motivated the introduction of advanced formal methods from computer science to model and study biochemical networks,from their automated reconstruction to the analysis of the networks obtained in this way. This has led to the new field of research of *symbolic systems biology* [39,47]. Some of the problems considered in this area can be attacked using graph transformation techniques

The most natural way to represent a biochemical network is, as the name 'network' hints, a directed graph with molecules as nodes and interactions as arcs. Many operations on these graphs can be carried out to analyze biochemical pathways: a search for a path between two genes in a regulatory pathway may reveal missing regulatory interactions; cycles in these networks mean feedback relations; clustering algorithms can be used to group together molecules with similar temporal reaction patterns.

Consider for instance metabolic pathways. In their graphical representation, nodes symbolize molecules and the arcs stand for applications of reaction rules in a suitable artificial chemistry [18]. If the molecules involved in a metabolic network are described as graphs and the reaction rules are graph rewriting rules, then graph transformation techniques can be used in a natural way.

One of the most important problems in metabolic pathway analysis is *pathway synthesis*: the construction of one or all pathways that can transform a given set of substrate molecules into a given product molecule. This problem is usually solved by means of retrosynthetic analysis using the reactions rules available. In order to synthesize meaningful metabolic pathways and to spot them in a graphical representation of the network, the so-called *combinatorially feasible reaction pathways* are defined by means of a set of axioms in [26, 61]. These axioms can be easily formalized in an artificial chemistry setting: for their translation in terms of generalized chemical graphs and explicit (or implicit) chemical reactions, see [55]. Then, the formal version of the pathway synthesis problem on a graphical representation of a metabolic network is to find one or all combinatorially feasible reaction pathways that, using as substrate as subset of the given set of substrates, produces the products.

In a graphical representation of a metabolic pathway with nodes symbolizing graphs and arcs symbolizing graph rewriting rules, the metabolic pathway synthesis problem can be solved using graph transformation by subjecting the target graph to a disconnection process by means of the reverses of the allowed graph rewriting rules and with application conditions that translate the satisfaction of the axioms of combinatorially feasible reaction pathways in this backwards construction. As a result, the target graph is transformed into a sequence of 'simpler' graphs in a stepwise manner. The aforementioned axioms guarantee that the repetition of this process eventually will result in a hierarchical synthesis tree rooted by the target graph, and then it only remains to check whether the set of graphs corresponding to the given substrate consists of leaves of this tree.

A graph transformation system for the analysis of metabolic pathways is being developed by G. Valiente at the UPC. The system is based on a database of explicit chemical reactions, a database of metabolic pathways, and a chemical graph transformation system. The efficient implementation of the latter system relies on the CANON method for labeling a molecular structure with canonical labels [65–67], in which a molecular structure is treated as a graph with nodes (atoms) and edges (bonds), and each atom is given a unique numerical label on the basis of the topology of the molecular structure.

The analysis of single metabolic, or, in general, biochemical networks must also be complemented with an investigation of the interactions between related networks. This could be done by composing them by means of rules abstracting these interactions. Biochemical networks being represented as graphs, their composition could be described as the application of graph rewriting rules, and hence graph transformation techniques could be used to analyze both the composition process and the resulting network. A particular example, the growth of metabolic networks obtained in a formal way from the repetition of a single biochemical reaction using graph transformation, is considered in [3]. There are already logic rewriting models that can be used to generate biochemical pathways, like the *Pathway Logic* [25], which has been designed to model and analyze signaling pathways. We mention specifically this formalism because we think that it can be translated into a graph transformation system.

Graph transformation can also be used in the formal modelling of cellular processes. Many formal approaches to develop formal computational models of the biochemical processes that take place in the cell have been based on process algebras: Petri nets [34, 70], CCS [14], π-calculus [49, 13], brane-calculi [8], κ-calculus [16]. All these approaches represent the molecules in the system by abstract symbols, not taking any care of their inner structure.

Several formal models of cellular biochemical processes specifically designed to model protein-protein interactions have been recently proposed. Pathway Logic has already been mentioned. Another very interesting one is the *graphical π-calculus* [15, 17]. In it, *proteins* can be seen as nodes with a set of *sites*, numbered hooks where edges can be attached to them. Then, *protein complexes* are seen as graphs obtained by connecting pairs of nodes by means of edges attached to sites; not all sites of proteins need to have edges attached, but even these orphan sites must be taken into account in the rules' application conditions. Sets of protein complexes are called *solutions*. These solutions evolve by means of reactions that reactions take place when a sub-solution, called a *reactant*, has a special shape: then, this reactant changes to a new reactant and binds to the rest of the solution in a way specified by the rewriting rules. Danos and Leneve carefully define the syntax and the rewriting rules for this system in a process-algebra style and, among other things, they give an implementation in π-calculus. The rewriting rules they propose can also be translated into graph rewriting rules. This surely would give a new semantic to this system and would bring new tools to study its properties. We believe that the graph transformation formalisms and techniques could be used in symbolic systems biology to model specific systems where the graphic nature of the molecules involved and the nature of the interaction between them ask for it.

5   Conclusion

Graph transformation was born more than 30 years ago with an eye on its applications, but since then it has become a very popular mathematical field of research in theoretical computer science. The obvious fact that graphs can be used to model many kind of structures and graph rewriting can be used to model the transformation of these structures, with some contribution of the phenomenon of the earlier tool, have made graph rewriting to find many applications, including in molecular biology.

We have overviewed several fields of application of graph rewriting techniques in molecular biology: the modelling of biomolecules's tridimensional structures, the definition of artificial chemistries and the analysis of biochemical networks. There are many other applications that have been omitted because of the lack of space. For instance, we are sure that everybody has in her or his mind the formal description of gene assembly in ciliates [22] as a graph rewriting system. We just wanted to call the attention of the graph transformation community, mostly oriented to software systems specification these days, and to invite them to catch a glimpse of a completely different world of possible applications.

References

1. Abe, N., Mamitsuka, H.: Predicting protein secondary structure using stochastic tree grammars. Machine learning **29** (1997) 275–301
2. Beck, M., Benkö, G., G. Eble, C.F., Müller, S., Stadler, P.: Graph grammars as models for the evolution of developmental pathways. In: The Logic of Artificial Life: Abstracting and Synthesizing the Principles of Living Systems (Proceedings of GWAL 2004), IOS Press (2004) 8–15
3. Benkö, G., Flamm, C., Stadler, P.F.: A graph-based toy model of chemistry. Journal of Chemical Information and Computer Sciences **43** (2003) 1085–1093
4. Benkö, G., Flamm, C., Stadler, P.F.: Multi-phase artificial chemistry. In: The Logic of Artificial Life: Abstracting and Synthesizing the Principles of Living Systems (Proceedings of GWAL 2004), IOS Press (2004) 16–22
5. Benkö, G., Flamm, C., Stadler, P.F.: Generic properties of chemical networks: Artificial chemistry based on graph rewriting. In: Proc. 7th European Conf. Advances in Artificial Life. Volume 2801 of Lecture Notes in Computer Science., Springer-Verlag (2003) 10–19
6. Bower, J.M., H. Bolouri, e.: Computational modeling of genetic and biochemical networks. MIT Press, Cambridge, MA (2001)
7. Brendel, V., Busse, H.G.: Genome structure described by formal languages. Nucleic Acid Research **12** (1984) 2561–2568
8. Cardelli, L.: Brane calculi. In: Proc. CMSB'03. Electronic Notes in Theoretical Computer Science, Elsevier (2003)
9. Cayley, A.: On the mathematical theory of isomers. Philosophical Magazine **47** (1874) 444–446
10. Chan, H.S., Dill, K.A.: Compact polymers. Macromolecules **22** (1989) 4559–4573
11. Corradini, A., Montanari, U., Rossi, F., Ehrig, H., Heckel, R., Löwe, M.: Algebraic approaches to graph transformation. Part I: Basic concepts and double pushout approach. In Rozenberg, G., ed.: Handbook of Graph Grammars and Computing by Graph Transformation, Volume 1: Foundations. World Scientific (1997) 163–246
12. Culik II, K., Lindenmayer, A.: Parallel rewriting on graphs and multidimensional development. Int. Journ. of General Systems **3** (1976) 53–66
13. Curti, M., Degano, P., Baldari, C.: Causal π-calculus for biochemical modelling. In: Proc. 1st Int. Workshop Computational Methods in Systems Biology. Volume 2602 of Lecture Notes in Computer Science., Springer-Verlag (2003) 21–33
14. Danos, V., Krivine, J.: Formal molecular biology done in CCS. In: Proc. CMSB'03. Electronic Notes in Theoretical Computer Science, Elsevier (2003)

15. Danos, V., Laneve, C.: Graphs for core molecular biology. In: Proc. 1st Int. Workshop Computational Methods in Systems Biology. Volume 2602 of Lecture Notes in Computer Science., Springer-Verlag (2003) 34–46

16. Danos, V., Laneve, C.: Core formal molecular biology. In: Proc. 12th European Symposium on Programming ESOP'03. Volume 2618 of Lecture Notes in Computer Science., Springer-Verlag (2003) 302–318

17. Danos, V., Laneve, C.: Formal molecular biology. Theoretical Computer Science (2004) in press

18. Deville, Y., Gilbert, D., van Helden, J., Wodak, S.J.: An overview of data models for the analysis of biochemical pathways. Briefings in Bioinformatics **4** (2003) 246–259

19. Fringuelli, F., Taticchi, A.: The Diels-Alder Reaction: Selected Practical Methods. John Wiley & Sons, Chichester, England (2002)

20. Dittrich, P., Ziegler, J., Banzhaff, W.: Artificial chemistries – a review. Artificial life **7** (2001) 225–275

21. Durbin, R., Krogh., A., Mitchison, G., Eddy, S.: Biological sequence analysis: Probabilistic models of proteins and nucleic acids. Cambridge Univ. Press, Cambridge (1998)

22. Ehrenfeucht, A., Harju, T., Petre, I., Prescott, D.M., Rosenberg, G.: Computation in Living Cells: Gene Assembly in Ciliates. Natural computing series. Springer-Verlag, Berlin (2004)

23. Ehrig, H., Heckel, R., Korff, M., Löwe, M., Ribeiro, L., Wagner, A., Corradini, A.: Algebraic approaches to graph transformation. part II: Single pushout approach and comparison with double pushout approach. In Rozenberg, G., ed.: Handbook of Graph Grammars and Computing by Graph Transformation. Volume 1: Foundations. World Scientific (1997) 247–312

24. Ehrig, H., Mahr, B.: Fundamentals of algebraic specification I: Equations and initial semantics. Springer Verlag (1985)

25. Eker, S., Knapp, M., Laderoute, K., Lincoln, P., Meseguer, J., Sonmez, K.: Pathway logic: symbolic analisys of biological signalling. In: Pacific symposium on Biocomputing 2001, World Scientific (2001) 400–412

26. Fan, L.T., Bertók, B., Friedler, F.: A graph-theoretic method to identify candidate mechanisms for deriving the rate law of a catalytic reaction. Computers & Chemistry **26** (2002) 265–292

27. Félix, L., Rosselló, F., Valiente, G.: Artificial chemistries and metabolic pathways. In Messeguer, X., Valiente, G., eds.: Proc. 5th Annual Spanish Bioinformatics Conference, Barcelona, Technical University of Catalonia (2004) 56–59

28. Flamm, C., Fontana, W., Hofacker, I., Schuster, P.: Kinetic folding of RNA at elementary step resolution. RNA **6** (2000) 325–338

29. Fontana, W.: Algorithmic chemistry. In: Artificial life II. Volume 47 of Santa Fe Institute Studies in the Sciences of Complexity., Addison-Wesley (1992) 159–210

30. Fujita, S.: Description of organic reactions based on imaginary transition structures. Part 1–5. Journal of Chemical Information and Computer Sciences **26** (1986) 205–242

31. Fujita, S.: Computer-Oriented Representation of Organic Reactions. Yoshioka Shoten, Kyoto (2001)

32. Fujita, S.: Description of organic reactions based on imaginary transition structures. Part 6–9. Journal of Chemical Information and Computer Sciences **27** (1987) 99–120

33. Gernert, D.: Graph grammars as an analytical tool in physics and biology. Biosystems **43** (1997) 179–187

34. Goss, P., Peccoud, J.: Quantitative modelling of stochastic systems in molecular biology using stochastic Petri nets. Proc. Nat. Acad. Sc. **95** (1998) 6750–6755
35. Heckel, R., Lajios, G., Menge, S.: Stochastic graph transformation systems. In: Proc. 2nd Int. Conf. Graph Transformation. Volume 3256 of Lecture Notes in Computer Science., Springer-Verlag (2004) 210–225
36. Hofacker, I., Fontana, W., Stadler, P., Bonhoeffer, L., Tacker, M., Schuster, P.: Fast folding and comparison of RNA secondary structures. Monatsh. Chem. **125** (1994) 167–188
37. Kanehisa, M., Goto, S.: KEGG: Kyoto encyclopedia of genes and genomes. Nucleic Acids Research **28** (2000) 27–30
38. Kister, A., Magarshak, Y., Malinsky, J.: The theoretical analysis of the process of RNA molecule self-assembly. BioSystems **30** (1993) 31–48
39. Kitano, H.: Computational systems biology. Nature **420** (2002) 206–210
40. Lesk, A.M.: Systematic representation of protein folding patterns. J. Mol. Graph. **13** (1995) 159–164
41. Mayoh, B.: On patterns and graphs. Preprint (1995)
42. Mayoh, B.: Multidimensional Lindenmayer organisms. In: L-systems. Volume 15 of Lecture Notes in Computer Science., Springer-Verlag (1974) 302–326
43. McAdams, H., Arkin, A.: It's a noisy business! Genetic regulation at the nanomolar scale. Trends in Genetics **15** (1999) 65–69
44. McCaskill, J., Niemann, U.: Graph replacement chemistry for DNA processing. In: DNA 2000. Volume 2054 of Lecture Notes in Computer Science., Springer-Verlag (2001) 103–116
45. Michal, G., ed.: Biological Pathways: An Atlas of Biochemistry and Molecular Biology. John Wiley & Sons, New York (1999)
46. Polanski, O.: Graphs in quantum chemistry. MATCH **1** (1975) 183–195
47. Priami, C., ed.: Computational methods in system biology. Proceedings first international workshop CMSM 2003. Volume 2602 of Lecture Notes in Computer Science., Springer-Verlag (2003)
48. Przytycka, T., Srinivasan, T., Rose, G.: Recursive domains in proteins. Protein Science **11** (2002) 409–417
49. Regev, A., Silverman, W., Shapiro, E.: Representation and simulation of biochemical processes using the π- calculus process algebra. In: Pacific symposium on Biocomputing 2001, World Scientific (2001) 459–470
50. Regev, A., Shapiro, E.: Cells as computation. Nature **419** (2002) 343
51. Reidys, C., Stadler, P.F.: Bio-molecular shapes and algebraic structures. Computers & Chemistry **20** (1996) 85–94
52. Richardson, J.: β-sheet topology and the relatedness of proteins. Nature **268** (1977) 495–500
53. Rivas, E., Eddy, S.R.: The language of RNA: a formal grammar that includes pseudoknots. Bioinformatics **16** (2000) 334–340
54. Rosselló, F., Valiente, G.: Chemical graphs, chemical reaction graphs, and chemical graph transformation. Electronic Notes in Theoretical Computer Science (2004) to appear
55. Rosselló, F., Valiente, G.: Analysis of metabolic pathways by graph transformation. In: Proc. 2nd Int. Conf. Graph Transformation. Volume 3256 of Lecture Notes in Computer Science., Springer-Verlag (2004) 70–82
56. Sakakibara, Y., Brown, M., Hughey, R., Mian, I., Sjolander, K., Underwood, R., Haussler, D.: Stochastic context-free grammars for tRNA modeling. Nucleic Acids Research **22** (1994) 5112–5128

57. Schultz, J., Milpetz, F., Bork, P., Ponting, C.: SMART, a simple molecular architecture research tool. PNAS **95** (1998) 5857–5864

58. Searls, D.: The computational linguistics of biological sequences. In: Artificial Intelligence and Molecular Biology, AAAI Press (1993) 47–120

59. Searls, D.: Formal language and biological macromolecules. In: Mathematical Support for Molecular Biology. Volume 47 of DIMACS Series in Discrete Mathematics and Theoretical Computer Science., AMS (1999) 128–141

60. Searls, D.: The language of genes. Nature **420** (2002) 211–217

61. Seo, H., Lee, D.Y., Park, S., Fan, L.T., Shafie, S., Bertók, B., Friedler, F.: Graph-theoretical identification of pathways for biochemical reactions. Biotechnology Letters **23** (2001) 1551–1557

62. Speroni, P.: Artificial chemistries. Bulletin EATCS **76** (2002) 128–141

63. Tomita, K., Kurokawa, H., Murata, S.: Graph automata: natural expression of self reproduction. Physica D **171** (2002) 197–210

64. Waterman, M.S., Smith, T.F.: RNA secondary structure: a complete mathematical analysis. Math. Biosci. **42** (1978) 257–266

65. Weininger, D.: SMILES, a chemical language and information system. 1. Introduction to methodology and encoding rules. Journal of Chemical Information and Computer Sciences **28** (1988) 31–36

66. Weininger, D., Weininger, A., Weininger, J.L.: SMILES. 2. Algorithm for generation of unique SMILES notation. Journal of Chemical Information and Computer Sciences **29** (1989) 97–101

67. Weininger, D.: SMILES. 3. DEPICT. Graphical depiction of chemical structures. Journal of Chemical Information and Computer Sciences **30** (1990) 237–243

68. Westhead, D., Slidel, T., Flores, T., Thornton, J.: Protein structural topology: automated analysis and diagrammatic representation. Protein Science **8** (1999) 897–904

69. Yadav, M.K., Kelley, B.P., Silverman, S.M.: The potential of a chemical graph transformation system. In: Proc. 2nd Int. Conf. Graph Transformation. Volume 3256 of Lecture Notes in Computer Science., Springer-Verlag (2004) 83–95

70. Zevedei-Oancea, I., Schuster, S.: Topological analysis of metabolic networks based on Petri net theory. In Silico Biology **3** (2003) 323–345

71. Zuker, M., Sankoff, D.: RNA secondary structures and their prediction. Bull. Math. Biol. **46** (1984) 591–621

Changing Labels in the Double-Pushout Approach Can Be Treated Categorically

Hans J. Schneider

Universität Erlangen-Nürnberg – Institut für Informatik (Lehrstuhl 2),
Erlangen, Germany
schneider@informatik.uni-erlangen.de

Abstract. In the double-pushout approach to graph transformations, most authors assume the left-hand side to be injective, since the noninjective case leads to ambiguous results. Taking into consideration productions that change labels, however, may add ambiguity even in the case of injective graph productions. A well-known solution to this problem is restricting the categorical treatment to the underlying graphs, whereas the labels on the derived graph are defined by other means. In this paper, we resume the detailed results on arbitrary left-hand sides that Ehrig and Kreowski have already given in 1976. We apply these results to the case of relabeling such that we can retain the elegant categorical constructions at the level of labeled graphs.

1 Introduction

Graph structures are ubiquitous in computer science as well as in many application areas. They are a very natural way to explain complex situations on an intuitive level, and they are used to define the syntax of structures according to given rules (graph grammars) as well as to describe dynamically deriving new situations from given ones (graph transformations). In 1973, we have introduced the double-pushout approach to formalize this process [9]. The approach allows replacing substructures in an intuitive way immediately instead of encoding the graphs into strings and then applying string transformations. Using concepts from category theory leads to elegant proofs, e.g., concerning local Church-Rosser property, parallelism, and concurrency, which can easily be applied to different categories [3]. Whereas the pushout construction is straightforward in the categories of interest, we need a pushout complement on the left-hand side of the double-pushout that may be ambiguous or not even exist.

The construction is well-understood in *Set* and *Graph*. We have already outlined in our first paper that the category *Lgraph* of labeled graphs with label preserving morphisms does not add any new difficulties to constructing pushouts or pushout complements. Many applications of interest, however, use morphisms that change not only the structure of the graph, but also change some labels. Examples can be found, e.g., in data bases [11, 16], compiler technique [15], term graph rewriting [12], asynchronous processes [17–19], and so on. To some extent, this is possible even in *Lgraph*: The node to be relabeled is not included in the

H.-J. Kreowski et al. (Eds.): Formal Methods (Ehrig Festschrift), LNCS 3393, pp. 134–149, 2005.

interface graph, and therefore, corresponding nodes on the left-hand and on the right-hand side can bear different labels. But, this node must have a fixed context given in the production, and it is not applicable in other contexts. In [9], we have overcome the limitations of label-preserving graph morphisms, by not labeling the interface graph and the corresponding nodes and edges in the context graph. The pushout construction is performed in $\mathcal{G}raph$, and a so-called "labeled gluing" ensures that the derivation includes only labeled graphs. *Rosen* has shown that this approach can be written in $\mathcal{L}graph$ using two different morphisms from the interface graph into the context graph [13]. With respect to term graph rewriting, *Habel* and *Plump* solve relabeling by considering partially labeled graphs [10]: If a node of the interface graph is not labeled, the corresponding nodes on the left-hand side and on the right-hand side may be labeled differently. Furthermore, their construction allows the left-hand side and the right-hand side to be partially labeled. In this case, the production represents a (possibly infinite) set of productions, which you get by labeling the unlabeled nodes with any element of a given set of symbols. Another approach that deserves special attention has been introduced by *Parisi-Presicce*, *Ehrig*, and *Montanari* [11]. With data base applications in mind, these authors define a structure on the labeling alphabet that allows the user to specify which changes are possible and which are not. Morphisms in the category $\mathcal{SL}graph$ are graph morphisms compatible with this structure. Both approaches define the labels on the derived graph by set-theoretic means.

In this paper, we show that the $\mathcal{SL}graph$-approach is able to model various applications that need relabeling or specifying sets of productions in a uniform way. Especially, it is possible to retain the elegant categorical constructions at the level of labeled graphs. Unfortunately, the structured alphabet adds a new difficulty in constructing the pushout complement: The solution may be unambiguous even in the case of injective graph morphisms. But this problem can be treated on a categorical level using the results by *Ehrig* and *Kreowski* concerning $\mathcal{E} - \mathcal{M}$-factorizable categories [6].

The rest of the paper is organized as follows. In the next section, we summarize the basic definitions of derivability in general and in the category of structurally labeled graphs. We recapitulate the decomposition theorem that allows us to characterize the pushout complements in the category of structurally labeled graphs (Section 3). Finally, Section 4 considers two special cases that are important in many applications, namely (infinite) sets of productions and productions changing some labels. The results, however, are formulated independent of these applications.

2   Background

In the double-pushout approach (see, e.g., [3]), the notion of derivability with respect to arbitrary categories is defined a symmetrical diagram:

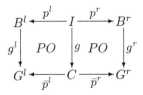

G^l and G^r are unambiguously defined if we know the *production* $p = (p^l, p^r)$ and the *embedding* g. Usually, however, the embedding g is not given, but the *handle* $g^l : B^l \to G^l$, and we have to look for a suitable context object C that allows an embedding $g : I \to C$ such that the given G^l is the pushout object of g and p^l. If we have found such an embedding, the right-hand part of the derivability diagram is unambiguously defined, and we can construct it in a straightforward manner. By this reason, the main problem in effectively constructing a derivation step is completing a pushout diagram backwards:

Definition 1 (Pushout complement). *If the diagram (a) can be completed such that diagram (b) is a pushout diagram:*

then, we call C together with morphisms g and $\bar{p}$ a pushout complement of $\bar{g} \cdot p$.[1]

Whereas the pushout is unambiguous up to isomorphism, the pushout complement need not exist in each case, and if it exists, it may be ambiguous. The different cases are well-understood in the categories *Set* and *Graph*. As usual, a *graph* is a quadruple $G = (E, V, s, t)$ with E and V being the sets of *edges* and *nodes*, respectively, and two set morphisms s and t assigning a *source node* and a *target node* to each edge. A *graph morphism* $f : G \to H$ is a pair $(f_E : E_G \to E_H, f_V :\to V_H)$ such that $f_V \cdot s_G = s_H \cdot f_E \wedge f_V \cdot t_G = t_H \cdot f_E$.

The category $\mathcal{L}graph$ of labeled graphs with label preserving morphisms does not add any new difficulties to constructing pushouts or pushout complements. In many areas of computer science as well as in application areas, however, it is necessary to change labels. We consider two examples illustrating different requirements.

Example 2. We consider a production removing common subexpressions from a term graph (Fig. 1). The nodes of a term graph are labeled with operator symbols, constants, and variables. In the productions, we use additional metavariables $(op, x, y, \ldots)$. Applying such a production, we have to replace these metavariables with operator symbols, and so on: The handle does not preserve the labels. In the figure, the nodes are represented by their labels. The different

---

[1] For reason of simplicity, we often call C the pushout complement without explicitly referring to the morphisms.

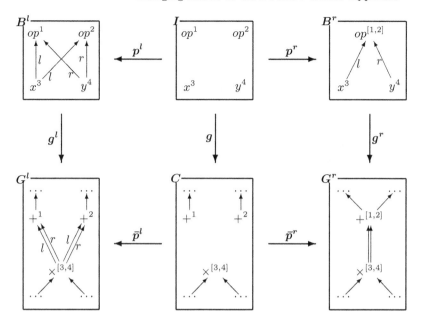

Fig. 1. Example of a derivation step removing common subexpressions.

nodes are distinguished by numbers that we write as exponents, and the mappings are given by these numbers, e.g., $p^r(3) = 3$. If the mapping is not injective, we use the notation $p^r(1) = p^r(2) = [1,2]$. Edge labels indicate the left-hand (l) and the right-hand (r) operand. In this example, we can omit the edge identifiers since they are unambiguously determined by the node mappings together with the labels. The figure shows only a part of the graph G^l explicitly, the rest is indicated by dots. We have chosen a noninjective g^l putting together the operand nodes 3 and 4, indicating that the left-hand operand and the right-hand operand of the operation are already detected to be identical. Of course, the production can also be applied to different operands; in this case, we have an injective g^l. Furthermore, we can apply it to a term graph in which several edges leave node 1 or node 2.

Example 3. We describe the problem of the dining philosophers in the following way: We have three places containing the thinking philosophers, the unused forks, and the eating philosophers, respectively. The transitions correspond to becoming hungry and satisfied. In the situation given in Fig. 2, philosophers p_2 and p_5 are eating using the forks f_2, f_3, f_5, f_1. Fork f_4 is not used. When p_2 is satisfied, he turns to thinking and puts back the forks f_2 and f_3. A derivation step modeling this transition is shown in Fig. 3. Please note that in the original net, the edge labels define the elements to be removed from a place or to be put onto it, whereas in the graph transformation approach, these changes are described by the transformation of the place labels. We introduce new edge labels such as ps (philosopher satisfied) or t (takes) to distinguish the edges. (We omit the identifying numbers to simplify the picture.)

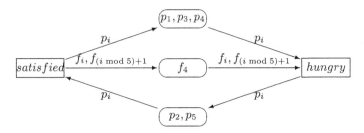

Fig. 2. Dining philosophers.

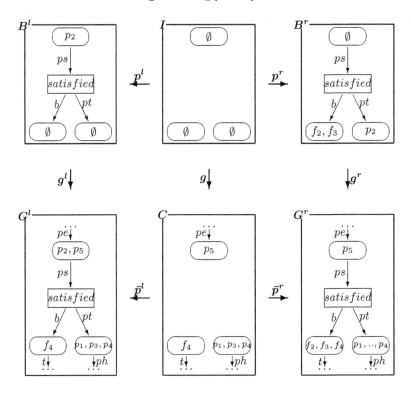

Fig. 3. Derivation step modeling the transition when p_2 is satisfied.

Although in these examples, we have different relabeling conditions, the approach introduced by *Parisi-Presicce, Ehrig,* and *Montanari* [11] is able to model both. With data base applications in mind, these authors define a structure on the labeling alphabet that allows the user to specify which changes are allowed and which are not:

Definition 4 (SLgraph). *Let L_E and L_V be two alphabets on which reflexive and transitive relations $\sqsubseteq_{L_E}$ and $\sqsubseteq_{L_V}$ are defined[2]. A structurally labeled graph*

---

[2] In the following, we almost always omit the indices, since the relations cannot be confused.

(SL-graph) is a labeled graph $G = (E, V, s, t, l_E, l_V)$ *with* $l_E : E \to L_E$ *and* $l_V : V \to L_V$. *An SL-graph morphism* $f : G \to H$ *is a graph morphism* $f = (f_E : E_G \to E_H, f_V : V_G \to V_H)$ *which additionally satisfies:*
$$(\forall v \in V_G)(l_{VG}(v) \sqsubseteq l_{VH}(f_V(v))) \text{ and } (\forall e \in E_G)(l_{EG}(e) \sqsubseteq l_{EH}(f_E(e))).$$

In Example 3, relation $\sqsubseteq_{L_V}$ is set inclusion and $\sqsubseteq_{L_E}$ is identity. But, the approach also allows to describe Example 2: We define that op can be replaced by a function symbol, whereas x and y can be replaced by constants, variables, or function symbols: $op \sqsubseteq +, -, \times, /$ and $x, y \sqsubseteq +, -, \times, /, v, c$, where v and c denote any variable or constant.

Constructing the pushout in $\mathcal{SL}graph$, some nodes (and edges) are put together. We have to label the resulting node with the least upper bound of the labels of the original nodes[3].

Lemma 5 ([11]). *If in the structured alphabet, the least upper bound exists, then $\mathcal{SL}graph$ has pushouts.*

Parisi-Presicce et al. restrict discussion to injective left-hand sides and additionally assume the right-hand side to be injective. Considering the underlying graphs, injective left-hand sides make the solution unambiguous if it exists. It is easy to see that structured labeling adds ambiguity to constructing pushout complements even in the case of an injective p^l, since different labels may lead to the same least upper bound. *Parisi-Presicce et al.*, however, make the definition unambiguous by an explicit condition written in a set-theoretic style. It requires existence of a minimal label that is chosen to be the solution.

In the next section, we resume this approach, but we do not make the pushout complement unambiguous by set-theoretic restrictions. Instead, we look for a categorical solution. In 1976, *Ehrig* and *Kreowski* have studied the construction of arbitrary pushout complements in a general setting based on $\mathcal{E}$–$\mathcal{M}$-factorizable categories [6]:

Definition 6 ($\mathcal{E} - \mathcal{M}$-factorizable category). *Given a category $\mathcal{K}$, let $\mathcal{E}$ be a class of epimorphisms that contains all the isomorphisms of $\mathcal{K}$ and is closed under composition, and let $\mathcal{M}$ be a class of monomorphisms that contains all the isomorphisms of $\mathcal{K}$ and is closed under composition. Then, $\mathcal{K}$ is called $\mathcal{E} - \mathcal{M}$-factorizable if and only if we can split each morphism $f \in \mathrm{Mor}_K$ unambiguously (up to ismorphism) into two morphisms such that*

$$(\exists e \in \mathcal{E})(\exists m \in \mathcal{M})(f = m \cdot e)$$

We call $\mathcal{E}$ ($\mathcal{M}$) *closed under construction of pushouts* if a given $\mathcal{E}$-morphism ($\mathcal{M}$-morphism) leads to an $\mathcal{E}$-morphism ($\mathcal{M}$-morphism) on the opposite side in constructing a pushout diagram.

Theorem 7 (Decomposition Theorem). *Let $\mathcal{K}$ be an $\mathcal{E}$–$\mathcal{M}$-factorizable category that has pushouts such that $\mathcal{E}$ and $\mathcal{M}$ are closed under construction of*

---

[3] *Parisi-Presicce et al.* use the inverse relation $\sqsupseteq$. Therefore, they need the greatest lower bound in constructing the pushout in $\mathcal{SL}graph$.

pushouts. Then, each pushout diagram $\bar{g} \cdot p = \bar{p} \cdot g$ *in* $\mathcal{K}$ *can be unambiguously split into four pushout diagrams such that each morphism with index e is in* $\mathcal{E}$ *and each morphism with index m is in* $\mathcal{M}$:

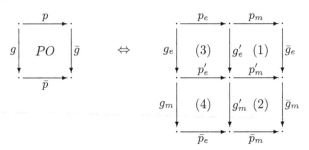

Obviously, the assumptions of the theorem are fulfilled with $\mathcal{E}$ being the set of all epimorphisms and $\mathcal{M}$ being the set of all coretractions. In the category of graphs, however, coretractions are too much a restriction. Therefore, we have formulated the theorem making explicit the assumptions the proof really needs.

As *Ehrig* and *Kreowski* have shown, this decomposition simplifies constructing pushout complements. The numbers in the figure indicate the order in which you can construct the subdiagrams. Subdiagrams (1), (2), and (4) can take advantage of the simpler concept of a coproduct complement:

Definition 8 (Coproduct complement). *The morphism* $\bar{f} : \bar{A} \to B$ *is called a coproduct complement of* $f : A \to B$ *if and only if* $A \xrightarrow{f} B \xleftarrow{\bar{f}} \bar{A}$ *is a coproduct.*

In *Set*, the situation is simple: If a morphism is injective, it has a unique coproduct complement, otherwise it has no coproduct complement at all.

Lemma 9 ([6,9]).

1. *If in Diagram (a), p is an arbitrary morphism and if* $\bar{g}_m$ *has a coproduct complement* $\bar{g}'_m$, *then the coproduct* (g_m, g'_m) *together with the morphism* $\bar{p}$ *factorizing* $\bar{g}'_m$ *and* $\bar{g}_m \cdot p$ *yields a pushout diagram* $\bar{g}_m \cdot p = \bar{p} \cdot g_m$.

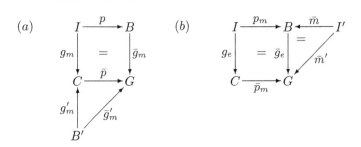

2. *If in Diagram (b),* $(p_m, \bar{m})$ *is a coproduct: then* $\bar{g}_e \cdot p_m = \bar{p}_m \cdot g_e$ *is a pushout if and only if* $(\bar{p}_m, \bar{m}')$ *is a coproduct.*

Subdiagrams (2) and (4) can be completed by the first part of this lemma. Restricting g_m and g'_m to injective morphisms makes the solution unambiguous. The second part of this lemma is not constructive, but in $\mathcal{S}et$, we get the solution by defining $g_e := (\bar{p}_m)^{-1} \cdot \bar{g}_e \cdot p_m$, where $(\bar{p}_m)^{-1}$ is the unique inverse of the restriction $\bar{p}_m : C \to \bar{p}_m[C]$.

Finally, subdiagram (3) can be completed in each case, since constructing the natural pushout complement is trivial:

Lemma 10 ([6]). *If in the following diagram*

$$
\begin{array}{ccc}
I & \xrightarrow{\ p\ } & B \\
{\scriptstyle \bar{g}\cdot p}\downarrow & & \downarrow{\scriptstyle \bar{g}} \\
G & \xrightarrow{\ \mathrm{id}\ } & G
\end{array}
$$

p and $\bar{g}$ are epimorphisms, then it is a pushout diagram.

In $\mathcal{S}et$, we have a fine situation: Proving existence is restricted to subdiagram (1), and ambiguity is restricted to subdiagram (3). All the morphisms indexed with m have a unique coproduct complement. If in addition, the identification condition is satisfied, $\bar{m}' := \bar{g}_e \cdot \bar{m}$ is injective, too, and has a coproduct complement. In this case, the solutions to subdiagrams (1), (2), and (4) can be constructed, unambiguously. Due to *Rosen*'s lemma [13, Lemma 4.1], we know all the solutions to subdiagram (3).

3   Ambiguous Pushout Complements in $\mathcal{SL}graph$

Discussing the situation in $\mathcal{S}et$, we have taken advantage of its special properties. Therefore, we could isolate the aspect to find all solutions in subdiagram (3) of the decomposition. If we, however, consider other categories, the situation may become more complicated. Fortunately, we can prove a lemma that is strongly related to *Rosen*'s lemma, but is not restricted to sets:

Lemma 11 (Ambiguous pushout complements). *Let $\bar{g} \cdot p = \bar{p} \cdot g$ be a pushout diagram.*

$$
\begin{array}{ccc}
I & \xrightarrow{\ p\ } & B \\
{\scriptstyle g}\downarrow\ \ \mathit{PO} & & \downarrow{\scriptstyle \bar{g}} \\
C & \xrightarrow{\ \bar{p}\ } & G
\end{array}
\qquad\qquad
\begin{array}{ccc}
I & \xrightarrow{\ p\ } & B \\
{\scriptstyle \alpha\cdot g}\downarrow\ \ \mathit{PO} & & \downarrow{\scriptstyle \bar{g}} \\
C' & \xrightarrow{\ \beta\ } & G
\end{array}
$$

If there is a factorization $\bar{p} = \beta \cdot \alpha$ with an epimorphic α and an arbitrary β, then the right-hand diagram $\bar{g} \cdot p = \beta \cdot (\alpha \cdot g)$ is a pushout diagram, too.

Proof. We consider the following diagram

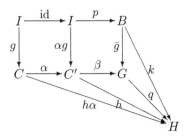

and assume $k \cdot p = h \cdot (\alpha \cdot g)$. We can rewrite this as $k \cdot (p \cdot \mathrm{id}) = (h \cdot \alpha) \cdot g$. Since the outer diagram is a pushout, we get a unique morphism $q : G \to H$ with $k = q \cdot \bar{g}$ and $h \cdot \alpha = q \cdot (\beta \cdot \alpha) = (q \cdot \beta) \cdot \alpha$ and with α assumed to be epimorphic: $h = q \cdot \beta$.   □

We call C together with morphisms g and $\bar{p}$ a *minimal pushout complement* if in every factorization $\bar{p} = \beta \cdot \alpha$ with an epimorphism α, this α is an isomorphism. Then, the lemma allows us to characterize ambiguous pushout complements: If we have a unique minimal solution C, we can find all the solutions by looking for factorizations of the morphism $\bar{p} : C \to G$. In general, however, we do not have a unique minimal pushout complement even in the category $\mathcal{S}et$.

Conversely, we call a pushout complement *maximal* if it is constructed according to Theorem 7 using the solution given in Lemma 10 as subdiagram (3). In this case, p'_e and therefore, $\bar{p}_e$ are isomorphisms, and $\bar{p}$ is in $\mathcal{M}$. *Rosen* has called this the *natural pushout complement*.

In the category $\mathcal{SL}graph$, we have ambiguous pushout complements even if we restrict discussion to injective graph morphisms. How to treat this situation in a categorical setting? *Parisi-Presicce et al.* have already observed that we can split an injective SL-graph morphism into a bijective SL-graph morphism followed by a label preserving graph morphism. We use this factorization to take advantage of the decomposition theorem.

Lemma 12 (Decomposition of $\mathcal{SL}graph$). *Given a structured alphabet, the category $\mathcal{SL}graph$ is $\mathcal{E} - \mathcal{M}$-factorizable with*

1. *$\mathcal{E}$ being the set of all epimorphisms of $\mathcal{SL}graph$ and*
2. *$\mathcal{M}$ being the set of all label preserving graph monomorphisms.*

Proof. Let $f : G \to H$ be a morphism in $\mathcal{SL}graph$. More precisely, we have $f_E : E_G \to E_H$, $f_V : V_G \to V_H$ and the labeling conditions

$$(\forall v \in V_G)(l_{VG}(v) \sqsubseteq l_{VH}(f_V(v)))$$
$$(\forall e \in E_G)(l_{EG}(e) \sqsubseteq l_{EH}(f_E(e)))$$

We can uniquely decompose the underlying graph morphism into an epimorphism and a monomorphism, and we label the intermediate graph with the labels of H: $G' := (f_E[E_G], f_V[V_G], s', t', l'_E, l'_V)$, where s', t', l'_E, and l'_V are the restrictions of s_H, t_H, l_{EH}, and l_{VH}, respectively. Now, we decompose f into $f = f_m \cdot f_e$:

$$f_e = (f_E : E_G \to f_E[E_G], f_V : V_G \to f_V[V_G])$$
$$f_m = (\mathrm{in}_{f_E[E_G]} : f_E[E_G] \to E_H, \mathrm{in}_{f_V[V_G]} : f_V[V_G] \to V_H)$$

where the morphism $\mathrm{in}_{f_E[E_G]}$ is the natural injection of $f_E[E_G]$ into E_H, etc. Trivially, f_e is an epimorphism satisfying the labeling conditions of an SL-graph morphism and f_m is a label preserving monomorphism.   □

This factorization satisfies the assumptions of Theorem 7. The proof is simple and left to the reader. (See, e.g., [20, Chapter 4].)

Lemma 13 (Decomposition of pushouts in SLgraph). *Decomposing the construction of a pushout complement in SLgraph as shown in Theorem 7 again restricts ambiguity to subdiagram (3).*

Proof. This assertion holds true for the underlying graphs (Lemma 9). Although coproduct complements in SLgraph are not unambiguous, we have no problems due to the special decomposition we have chosen[4]:

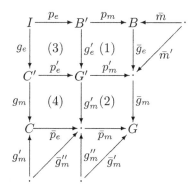

In subdiagrams (2) and (4), the construction of coproduct complements is applied to label preserving morphisms, and $\bar{p}_e$ is unambiguous because of the universal property of coproducts. In subdiagram (1), we construct the coproduct complement $\bar{m}$ of a label preserving morphism. But in the second step, we consider $\bar{m}' := \bar{g}_e \cdot \bar{m}$. Although $\bar{g}_e$ need not preserve labels, the composition does. Consider, e.g., a node v in B such that the label of $\bar{g}_e(v)$ is different from the label of v. Since the resulting diagram must be a pushout, this means that v is a gluing node and changing its label is caused by the label of the corresponding node in the graph G'. Therefore, v cannot be part of the coproduct complement, i.e., $\bar{m}'$ is label preserving, and there exists a unique p'_m.   □

---

[4] More precisely, we had to consider p_{Ee}, p_{Ve}, p_{Em}, p_{Vm}, etc., i.e., all formulaes hold for edges as well as for nodes. For reason of simplicity, we omit this distinction.

What about subdiagram (3)? Lemma 10 ensures existence of the maximal pushout complement. With respect to the underlying graphs, the well-known results presented in detail in [20, Chapter 4], can be used to find minimal solutions. We have to add the labeling:

Lemma 14. *In subdiagram* (3) *of the decomposition, the label of a node (or an edge) y must satisfy:*

(a) $l_I(v) \sqsubseteq l_{C'}(y) \sqsubseteq l_{G'}(g'_e(p_e(v)))$ for all v with $g_e(v) = y$,
(b) $l_{G'}(p'_e(y)) = \mathrm{lub}(\{l_{C'}(y') \mid p'_e(y') = p'_e(y)\} \cup \{l_{B'}(y'') \mid g'_e(y'') = p'_e(y)\})$.

The first inequality is a consequence of the commutativity of the diagram, the second follows from the pushout property. According to Lemma 10, the maximal solution is $l_{C'}(y) := l_{G'}(p'_e(y))$.

## 4	Results

In the case of an injective production, p_e and p'_e are bijective graph morphisms, and the underlying graph of C' is identical to that of G' and therefore, it is unambiguous if G' exists, i.e., if the identification condition and the dangling condition are satisfied [3]. Ambiguity can arise only from the labeling. But, condition (b) of Lemma 14 becomes simpler:

$$l_{G'}(p'_e(y)) = \mathrm{lub}(\{l_{C'}(y)\} \cup \{l_{B'}(y'') \mid g'_e(y'') = p'_e(y)\}).$$

Parisi-Presicce et al. have called the solutions $l_{C'}(y)$ to this equation g'_e-complements. In their definition of derivability, they add a condition on the labeling that is based on studying the g'_e-complements in detail. In our terminology, this additional property ensures existence of a unique minimal pushout complement which is used to complete the right-hand side of the derivation step.

We apply Lemma 14 to the examples of Section 2. We start with considering the simple case: Both the left-hand side of the production and the handle are injective, i.e., in the subdiagram of interest, the underlying graph morphisms p_e and g'_e as well as g_e and p'_e are bijections, especially, there is exactly one v with $y = g_e(v)$. The two conditions to label an element y become rather simple:

$$l_I(v) \sqsubseteq l_{C'}(y) \sqsubseteq l_{G'}(g'_e(p_e(v)))$$
$$l_{G'}(p'_e(y)) = \mathrm{lub}\{l_{C'}(y), l_{B'}(p_e(v))\}$$

This means that each solution of

$$l_I(v) \sqsubseteq x \sqsubseteq \mathrm{lub}\{x, l_{B'}(p_e(v))\} = l_{G'}(p'_e(y))$$

is a possible label of y. The following diagram makes clear what happens:

$$
\begin{array}{ccc}
l_I(v) & \xrightarrow{\ \sqsubseteq\ } & l_{B'}(p_e(v)) \\
\Big\downarrow{\scriptstyle\sqsubseteq} & & \Big\downarrow{\scriptstyle\sqsubseteq} \\
x & \xrightarrow[\ \sqsubseteq\ ]{} & l_{G'}(p'_e(y))
\end{array}
$$

Since the structured alphabet can be considered a category with $a \sqsubseteq b$ to be the unique morphism from a to b and with the least upper bound as the pushout construction, we have to find the pushout complements in the category of the structured alphabet. Since the monomorphic part of the decomposition is label preserving, we can summarize this case without restricting discussion to subdiagram (3):

Theorem 15 (Pushout complements of injective handles in $\mathcal{SL}$graph).
We assume a production with an injective left-hand side and an injective handle such that the pushout complement C in $\mathcal{G}$raph exists. Then, we can use any pushout complement of

$$
\begin{array}{ccc}
l_I(g^{-1}(y)) & \xrightarrow{\ \sqsubseteq\ } & l_{B^l}(p^l(g^{-1}(y))) \\
\sqsubseteq \downarrow & PO & \downarrow \sqsubseteq \\
l_C(y) & \xrightarrow{\ \sqsubseteq\ } & l_{G^l}(\bar{p}^l(y))
\end{array}
$$

to label a node (or an edge) y of C that has a pre-image in the interface graph, and $l_C(y) = l_{G^l}(\bar{p}^l(y))$ if it has not[5].

Whether or not a unique minimum exists depends on the structure of the alphabet. In Example 3, we have chosen set inclusion as $\sqsubseteq$. In this case, there is a unique minimal solution.

Corollary 16. *If the structured alphabet uses set inclusion as the $\sqsubseteq$-relation, the unique minimal solution to label a node or an edge that has a pre-image in the interface graph is*

$$
l_{G^l}(\bar{p}^l(y)) \setminus l_{B^l}(p^l(g^{-1}(y))).
$$

In Example 3, the solutions to label the input place must satisfy

$$
\{p_2, p_5\} = \text{lub}(l, \{p_2\}) \qquad \emptyset \sqsubseteq l
$$

The minimal solution is $l = \{p_5\}$ and the maximal is $l = \{p_2, p_5\}$. Restricting derivation steps to minimal pushout complements coincides with the usual definition of the token game. This example as well as some others (see, e.g., [17]) suggest that unique minimal pushout complements cannot only be used to characterize all solutions, but also are of special importance in many applications.

If we allow non-injective handles, the situation becomes more complicated. Let us consider a node (or an edge) y^C in C. Then, there may be several preimages y_i^I in the interface graph (Fig. 4). As we have shown in a more general setting [18], constructing the pushout in $\mathcal{SL}$graph in the case of noninjective morphisms leads to constructing general colimits in the category of the alphabet.

An interesting application using noninjective handles is term graph rewriting [12]. In this special case, the structure of the alphabet allows to characterize all the possible complements. We illustrate this structure in Fig. 5.

---

[5] Please note that this is the inverse of a theorem we have proved some years ago [18, Theorem 2.11].

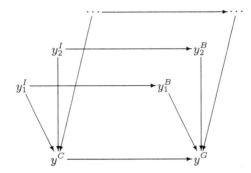

Fig. 4. The situation in the case of a non-injective handle.

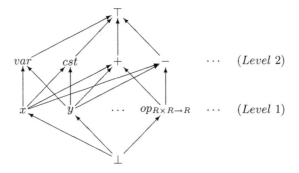

Fig. 5. Structure of the term graph alphabet.

In the graphs to be derived, we find only the labels given on the second level, where *var* and *cst* represent the variables and constants occurring in the application of interest. These two nodes are placeholders for a set of nodes with different identifiers. On the first level, we have metavariables that may occur in the productions. In order to simplify the figure, we have mentioned only one group of function symbols, namely the dyadic operators on the real numbers. We have arcs from the metalabel $op_{R \times R \to R}$ to all the operators of this type. The label $\perp$ is allowed only in the interface graph. Finally, the $\top$-symbol is necessary to ensure existence of the least upper bound. If it occurs in a derived graph, the production can be applied formally, but from the application point of view, the result is not valid.

Definition 17 (Term graph alphabet). *A term graph alphabet consists of the following:*
(1)   A set of (possibly typed) variables , constants, and operator symbols that are given by the application,
(2)   a set of (possibly typed) metalabels that may occur in the productions with $l \sqsubseteq v, c, op$ if and only if the type of the metalabel l and the type of the variable v, the constant c, or the operator symbol op agree,

(3)   *a special symbol ⊥ that may occur in the interface graph with ⊥ ⊑ l for
       each metalabel l,*
(4)   *a special symbol ⊤ with v, c, op ⊑ ⊤ for all variables v, constants c, and
       operator symbols op,*

If we do not have a typed alphabet, the second condition simply means that
the arity of the operator is the same as that of the metalabel.

Theorem 18 (Pushout complements in term graph rewriting). *In the
case of a term graph alphabet, the minimal solution always exists and is unique.
The minimal solution describes the usual interpretation of term graph rewriting.*

Proof. The structure of Fig. 5 allows us to find the possible labels of node y (or
edge y) in Fig. 4 by case discrimination. The labels on the y_i^B are either on level 1
or on level 2. Furthermore, the level-2 labels must be identical. (Otherwise, these
y_i^B can not be mapped onto the same y^G.) By Lemma 10, the maximal solution
is the label of y^G. *Case 1:* All the elements y_i^B bear metalabels. The label of y_i^I
can be either ⊥ or the label of the corresponding y_i^B. The label of y^C must be
the label of y^G. This solution is unique.

Case 2: Some y_i^B bear the label of y^G, and at least one of the y_i^I also takes
this label. y^C must be labeled with it, too.

Case 3: Some y_i^B bear the label of y^G, and all the y_i^I are labeled with ⊥. In
this case, the solution is ambiguous. ⊥ is the minimal label of y^C. The metalabel
between ⊥ and the label of y^G is the third solution.

Case 4: Some y_i^B bear the label of y^G, and some y_i^I are labeled with ⊥, others
with a metalabel. If all the metalabels are identical, this is the minimal label
of y^C. Otherwise, the label of y^G is the only solution.

Case 1 covers sets of productions. In Example 2, we get $\mathrm{lub}\{x, y\} \sqsubseteq l_C(v_{[3,4]})$
and $\times = \mathrm{lub}\{l_C(v_{[3,4]}), x, y\}$ resulting in $l_C(v_{[3,4]}) = \times$ as well as $op \sqsubseteq l_C(v_1)$
and $+ = \mathrm{lub}\{l_C(v_1), op\}$ resulting in $l_C(v_1) = +$.

Example 19. We consider the situation of Fig. 6. From Case 3, we get that the
question mark can be replaced by ⊥, fct_2, or f. Solution f would lead to ⊤
on the right-hand side that is not a legal label. The same situation results from
using fct_2 because this metalabel can be replaced only by a dyadic operation
symbol. Only the minimal label leads to a valid solution.

5   Conclusion

The categorical approach to graph transformations is very elegant and allows
to take advantage of rather general constructions. Relabeling, however, leads to
ambiguous results even in the case of injective graph morphisms. There are at
least two situations of practical importance that lead to relabeling: (1) The label
on the right-hand side of a production is different from the label on the corresponding node (edge) on the left-hand side. (2) The production describes a set of
productions using metalabels that must be replaced by real labels when applying
the production. Many authors have considered these applications restricting the

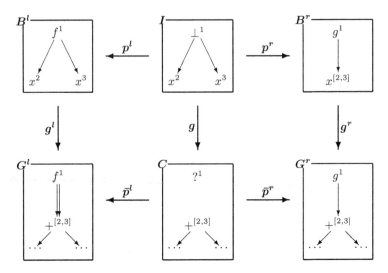

Fig. 6. Relabeling in the term graph example.

categorical treatment to the underlying graphs. Therefore, these approaches lead to different definitions of derivability in different applications, and novices entering the field of graph transformations feel annoyed by this diversity. Furthermore, the general categorical results, e.g., on parallel application of productions, cannot be applied to different applications immediately, but must be proved again and again.

Our approach defining derivability in the category $\mathcal{SL}graph$ allows to treat different applications in a uniform way and to avoid diversity. We have shown that the problem of ambiguity arising from changing labels can be controlled by a suitable $\mathcal{E} - \mathcal{M}$-decomposition and that we know all solutions if we know the minimal solutions. The applications under consideration suggest that the minimal solution is the solution of interest. What to do next? Now, we can reconsider the results on parallel independence, etc., which usually are proved using set-theoretic conditions (see, e.g., [3, Def. 3.4.1]). Applying such results to $\mathcal{SL}graph$ can be based on the purely categorical condition given in [18]. It seems that it is easy to modify that condition such that it uses a minimal context object C instead of a unique. This is left to the next paper. Then, we have a uniform description of parallel independence in term graph rewriting, Petri nets, data bases, and some other applications.

References

1. V. Claus/H. Ehrig/G. Rozenberg (Eds.): *Graph-Grammars and Their Application to Computer Science and Biology*, International Workshop, Bad Honnef, Germany, (Oct. 1978), Lect. Notes Comput. Science 73, Springer, Berlin, 1979
2. A. Corradini/H. Ehrig/H.-J. Kreowski/G. Rozenberg (Eds.): *Graph Transformation*, First International Conference, Barcelona, Spain, (Oct. 2002), Lect. Notes Comput. Science 2505, Springer, Berlin, 2002

3. A. Corradini/U. Montanari/F. Rossi/H. Ehrig/R. Heckel/M. Löwe: *Algrebraic approaches to graph transformation - Part I: Basic concepts and double pushout approach*, in [14], pp. 163-245

4. H. Ehrig: *Introduction to the algebraic theory of graph grammars (a survey)*, in: [1], pp. 1-69

5. H. Ehrig/G. Engels/H.J. Kreowski/G. Rozenberg (Eds.): *Handbook of Graph Grammars and Computing by Graph Transformation - Volume 2: Applications, Languages, and Tools*, World Scientific, Singapore, 1999

6. H. Ehrig/H.J. Kreowski: *Categorical approach to graphic systems and graph grammars*, Lect. Notes Economics Math. Systems 131 (1976), pp. 323-351

7. H. Ehrig/H.J. Kreowski/U. Montanari/G. Rozenberg (Eds.): *Handbook of Graph Grammars and Computing by Graph Transformation - Volume 3: Concurrency, Parallelism, and Distribution*, World Scientific, Singapore, 1999

8. H. Ehrig/M. Nagl/G. Rozenberg/A. Rosenfeld (Eds): *Graph-Grammars and Their Application to Computer Science*, 3rd International Workshop, Warrenton, Va., (Dec. 1986), Lect. Notes Comput. Science 291, Springer, Berlin, 1987

9. H. Ehrig/M. Pfender/H.J. Schneider: *Graph-grammars - An algebraic approach*, Proceed. Conf. Switching and Automata Theory 1973, Iowa, pp. 167-180

10. A. Habel/D. Plump: *Relabelling in graph transformation*, in [2], pp. 135-147

11. F. Parisi-Presicce/H. Ehrig/U. Montanari: *Graph rewriting with unification and composition*, in [8], pp. 496-514

12. D. Plump: *Term graph rewriting*, in: [5], pp. 3-61

13. B.K. Rosen: *Deriving graphs from graphs by applying a production*, Acta Informatica 4 (1975), pp. 337-357

14. G. Rozenberg (Ed.): *Handbook of Graph Grammars and Computing by Graph Transformation - Volume 1: Foundations*, World Scientific, Singapore, 1997

15. H.J. Schneider: *Syntax-directed description of incremental compilers*, Lecture Notes Computer Science 26, Springer, Berlin, 1975, pp. 192-201

16. H.J. Schneider: *Conceptual data base description using graph-grammars*, in: Graphen, Algorithmen, Datenstrukturen (Ed.: H. Noltemeier), Hanser, München, 1976, pp. 77-97

17. H.J. Schneider: *Describing distributed systems by categorical graph grammars*, Lecture Notes Computer Science 411, Springer, Berlin, 1990, pp. 121-135

18. H.J. Schneider: *On categorical graph grammars integrating structural transformations and operations on labels*, Theoretical Computer Science 109 (1993), pp. 257-274

19. H.J. Schneider: *Describing systems of processes by means of high-level replacement*, in [7], pp. 401-450

20. H.J. Schneider: *Graph Transformations – An Introduction to the Categorical Approach*, Preliminary version:
http://www2.informatik.uni-erlangen.de/~schneide/gtbook/index.xml
(Link checked on Nov. 2nd, 2004)

21. H.J. Schneider/H. Ehrig: *Grammars on partial graphs*, Acta Informatica 6, 2 (1976), pp. 297-316

Part II

Algebraic Specification

and

Logic

Modules, Brains and Schemas

Michael A. Arbib

Computer Science, Neuroscience and USC Brain Project
University of Southern California
Los Angeles, California 90089-2520, USA
`arbib@pollux.usc.edu`

Abstract. A short personal note briefly traces the author's interactions with Hartmut Ehrig. Where Ehrig has devoted much work to an algebraic theory of modules, the author has developed schema theory primarily as a tool for brain theory, but the author's version of schema theory has also been associated with algebraic theory and robotics. Topics presented in the present informal overview of schema theory include the role of schemas in bridging from action-oriented perception to knowledge, the notion of schema instances and their role in cooperative computation, learning in schemas, and ways of linking schemas to the study of the brain.

1 A Personal Introduction

My friendship with Hartmut Ehrig can be traced back to the volume "Universal Theory of Automata" which he co-authored with W. Kühnel, H.-J. Kreowski, and K.D. Kiermeier and published with Teubner, Stuttgart, in 1974. In this work, he applied category theory to provide a general framework for parts of automata theory. At around the same time, Ernie Manes – then my colleague at the University of Massachusetts at Amherst – had developed an alternative approach, first published in "Machines in a Category: An Expository Introduction" (in *SIAM Review*, 1974,16:163-192). In 1974, Ernie Manes arranged workshops in Amherst and San Francisco and we welcomed Hartmut to the United States, while in the following years he returned our hospitality in Germany. As the years went by, Hartmut's interests diverged from ours as he worked more and more on graph grammars and we continued in system theory, but as we moved on to apply our methods to the semantics of programming languages, we developed complementary algebraic approaches to the study of abstract data types.

Hartmut and I have only written one joint paper, "Linking schemas and module specifications for distributed systems", which Hartmut presented at the *2nd IEEE Workshop on Future Trends of Distributed Computing Systems*, in Cairo in 1990. It must be confessed that the paper was more a program for research than a presentation of results, and that research remains to be done. Since many readers of this volume will be well acquainted with the theory of module specifications (e.g., H. Ehrig and B. Mahr, 1990,*Fundamentals of Algebraic Specification 2: Module Specifications and Constraints*, volume 21 of *EATCS Monographs on Theoretical Computer Science*. Berlin: Springer Verlag), I present here

H.-J. Kreowski et al. (Eds.): Formal Methods (Ehrig Festschrift), LNCS 3393, pp. 153–166, 2005.

the *informal* background for schema theory in its motivation in the search for a high-level description of the ways in which brains function. Readers wishing a more formal treatment may find it in the paper "Port Automata and the Algebra of Concurrent Processes" written with Martha Steenstrup and Ernie Manes (*Journal of Computer and System Sciences*, 1983, 27:29–50) and its development with Damian Lyons of a schema-based framework for programming robots, "A Formal Model of Computation for Sensory-Based Robotics" (*IEEE Trans. on Robotics and Automation*, 1989, 5:280–293).

Finally, I want to acknowledge both Hartmut's continuing interest in brain theory, both as intellectual stimulation and for its potential yield of insights for new technology, and his support of my work in these areas. For example, his invitation to give a keynote address at *IDPT-2002: Integrated Design and Process Technology* in Pasadena in June 2002 was the stumulus for the perspective published as "Towards a neurally-inspired computer architecture" (*Natural computing*, 2003, 2:1–46).

I count myself fortunate to be among those many computer scientists who have benefited from both Hartmut's collaboration and his intellectual support, and I wish him many more years of intellectual excitement.

2   Basic Notions of Schema Theory

Many workers in cognitive science have little interest in brain or action, and much of their work focuses on linking Artificial Intelligence (AI) and cognition to symbol manipulation in general and to linguistics in particular. My own work, on the contrary, tries to see our cognitive abilities as rooted in our more basic capabilities to perceive and interact with the world. What, then, is this schema theory in which we are to give an account of the embodied mind, integrating an account of our mental representations with an account of the way in which we interact with the world?

There are many other approaches to schema theory, as outlined in the section "A Historical Sketch" below. My own version of *schema theory* [2,3] is an approach to cognitive neuroscience which explains behavior in terms of the concurrent interaction of many functional units called *schemas* (composable units of action, thought and perception). There are schemas for recognition of different objects, and the planning and control of different activities, and for more abstract operations besides. Schema theory now combines three distinct levels of theorizing:

1) **Basic Schema Theory:** Schema theory *simpliciter* provides a basic language which matches well with the "mental". It has its basic definition at a functional level which associates schemas with specific perceptual, motor, and cognitive abilities and other complex dispositions – and then stresses how our mental life results from the dynamic interaction, the competition and cooperation, of many schema instances. For example, one *perceptual* schema would let you recognize that a large structure is a house; in doing so,

it might provide strategies for locating the front door. The recognition of the door (activation of the perceptual schema for door) is not an abstract end in itself – it helps activate, and supply appropriate inputs to, *motor* schemas for approaching the door and for opening it (see Sec. 2.2 of [6], for examples). However, even at this functional level, a "computationally complete" explanation may involve schemas which are quite different from those that are suggested by introspection from conscious mental behavior.

2) **Neural Schema Theory:** Just as much human behavior can be explained by psychology without recourse to neurology, so can much successful schema theory proceed at a purely functional level. However, if we are to understand phenomena like episodic memory, emotion, consciousness, mental disorders, etc., it is clear that the details of schema function must make contact with data on brain localization and even neurochemistry. This motivates the "downward" extension of schema theory to form *neural schema theory*, in which we move from psychology and cognitive science as classically conceived (viewing the mind "from the outside") to cognitive neuroscience. The description of a schema can often be refined into a network of more detailed schemas. For a psychologist looking at overt behavior, the lowest level schemas employed may be relatively molar, themselves relatable to the subject's introspection or to the gross regional analysis of brain activity afforded by current human brain imaging techniques. For the neurophysiologist, further decomposition may be required until the schemas so defined are sufficiently fine-grain that their function may be played out across the detailed structures of specific neural networks of the brain. *Neural* schema theory provides a language for neuroscience appropriate to the analysis of data at the level of neuropsychology and human brain imaging, while at the same time showing that this "molar language" for neuroscience in no way precludes the relevance of finer-grain analysis in terms of neural circuitry and neurochemistry. Note that that our functional definition of a schema may change as we work out its implementation, revealing details that escaped our attention on superficial examination.

A schema in the basic sense of (1) is a functional notion (emphasizing the schema's causal role in some overall computation, never mind what implements it). It is only when we turn to neural schema theory that we seek to go the further step of studying the neural implementation of the schemas – thus linking schemas to the structural entities (brain regions or neural circuits, for example) which implement them. Interestingly, the language of schema theory is little used by neuroscientists. This is not because schema theory is irrelevant to neuroscience – but rather because few neuroscientists study large systems. Instead, they focus on one specific schema (for example, depth perception), the response of one neural circuit to specific patterns of stimulation, or fine details of neurochemistry and biophysics. I believe schema theory will become more widely accepted as more neuroscientists seek to link these details with larger cognitive systems, and relate them to the results of human brain imaging (which forces a more global view of interacting brain regions).

Neural schemas, then, are intermediate between behavior and neurons. How does a schema differ from an Edelman group [20], a Hebb assembly [25], or other notions that are similarly intermediate? They occupy the same "ecological niche" but my theory offers explicit analyses of perceptual-motor linkages and of the formation of assemblages/coordinated control programs that go beyond their theories. Below, I will strongly distinguish "schemas" from "modules" in the sense of Fodor's *Modularity of Mind* [22].

3) **Social Schema Theory:** In seeking to reconcile the "collective representations" of a community with the thought processes of individuals – creating an epistemology that integrates a sociology of knowledge with a psychology of knowledge – Arbib and Hesse [10] extended the basic schema theory "upward" to develop *social schema theory*. Their theory shows how "social schemas" constituted by collective patterns of behavior in a society may provide an external reality for a person's acquisition of schemas "in the head" in the sense of primary schema theory; conversely, it is the collective effect of behavior expressing schemas within the heads of many individuals that constitutes, and changes, this social reality. To understand the human individual we study the coherence and conflicts within a schema network that constitutes a personality, with all its contradictions. Social schema theory extends this to the holistic nets of social reality, of custom, language and religion. In this article, I concentrate on schema theory at the psychological and neural levels.

3   A Historical Sketch

Schema theory is designed to give an account of the embodied mind, an account which is to transcend mind-body dualism by integrating an account of our mental representations with an account of the way in which we interact with the world. To enrich the discussion of schemas, this section offers a brief historical review (see [8] for more). The history of schemas goes back to Immanuel Kant and beyond, but its links to neuroscience start with the work of the neurologists Head and Holmes [24] who discussed the notion of the *body schema* (see [35] for a current perspective). A person with damage to one parietal lobe of the brain may lose all sense of the opposite side of his body (see our earlier discussion of "neglect"), not only ignoring painful stimuli but neglecting to dress that half of the body; conversely, a person with an amputated limb but with the corresponding part of the brain intact may experience a wide range of sensation from the "phantom limb". Even at this most basic level of our personal reality – our knowledge of the structure of our own body – our brain is responsible for constructing that reality for us. Our growing scientific understanding of knowledge takes us far from what "common sense" will tell us is obvious. One of Head's students was Bartlett [13] noted that people's retelling of a story is based not on word-by-word recollection, but rather on remembering the story in terms of their own internal schemas, and then finding words in which to express this schema assemblage.

Such ideas prepare us for the work of Craik [18] who viewed the nature of the brain to be to "model" the world, so that when you recognize something, you "see" in it things that will guide your interaction with it. There is no claim of infallibility, no claim that the interactions will always proceed as expected. But the point is that you recognize things not as a linguistic animal, merely to name them, but as an embodied animal. I use the term "schema" for the building blocks of these models that guide our interactions with the world about us. To the extent that our expectations are false, our schemas can change, we learn. We then see many writers in the 1960s and 1970s [23,33,37] building upon this notion of an internal model of the world, at first in the cybernetic tradition, to develop the concept of representation so central to work in AI today.

One of the best-known users of the term "schema" is Piaget, whose *Biology and Knowledge* [42] gives an overview of his "genetic epistemology" which develops an embryological metaphor for the growth of a human's, and of human, knowledge. Piaget defines a schema as "the structure of interaction, the underlying form of a repeated activity pattern that can transcend the particular physical objects it acts on and become capable of generalization to other contexts". He traces the cognitive development of children, starting from basic schemas that guide their motoric interactions with the world, through stages of increasing abstraction that lead to language and logic, to abstract thought. We have already noted the importance of Piaget's concepts of *assimilation*, the ability to make sense of a situation in terms of the current stocks of schemas, and of *accommodation*, the way in which the stock of schemas may change over time as the expectations based on assimilation to current schemas are not met. These processes within the individual are reminiscent of the way in which a scientific community is guided by the *pragmatic criterion* of successful prediction and control [26]. We keep updating our scientific theories as we try to extend the range of phenomena they can help us understand. It is worth noting, however, that the increasing range of successful prediction may be accompanied by revolutions in ontology, in our understanding of what is real, as when we shift from the inherently deterministic reality of Newtonian mechanics to the inherently probabilistic reality of quantum mechanics.

Much work in brain theory and artificial intelligence contributes to schema theory, even though the scientists involved do not use this term. Schema theory provides a knowledge representation protocol which is part of the same theory-building enterprise as frames and scripts [38,45] but is distinguished in that, for example, schema theory has a grain size smaller than frames and scripts, but larger than neural models. Schema theory stresses the building up of new schemas; script theory stresses overarching organizational schemas for some family of behaviors. In its emphasis on the interaction of active computing agents (the schema instances), schema theory is related to studies in distributed artificial intelligence or multi-agent teams [50]. Since each schema combines knowledge with the processes for using it, schemas are more like actors than like frames or systems with unitary blackboards. Marvin Minsky espouses a *Society of Mind* analogy in which "members of society", the agents, are analogous to schemas

[39]. Rodney Brooks [17] controls robots with layers made up of asynchronous modules that can be considered as a version of schemas. This work shares with schema theory, with its mediation of action through a network of schemas, the point that no single, central, logical representation of the world need link perception and action; while sharing with Walter [51] and Braitenberg [15,16] the study of the "evolution" of simple "creatures" with increasingly sophisticated sensorimotor capacities – see [9] for further discussion). The term "schema theory", then, does not refer to one polished and widely accepted formalism. Aspects of schema theory have been within cognitive psychology [34,48] and motor control [47], for example.

For work within artificial intelligence, including work in machine vision and robotics, we ask how to define schemas as program units for cooperative computation that meet criteria for ease of implementation or for computational efficiency. For work within brain theory and cognitive psychology, schemas are designed to serve as units of complexity intermediate between behavior and neuron, and which help us "decompose" overall behavior in a fashion that gives us insight into the data of psychology and neuroscience. While my schema theory has been informed by that of Piaget – especially in its emphasis on the sensorimotor basis for mental development – and other work reviewed above, it is distinguished by:

(a) Its emphasis on the fact that our experience is usually mediated by an assemblage of schemas rather than a single schema. A situation is represented (consciously or unconsciously, repressed or not) by activation of a network of schemas that embody the significant aspects of a situation for the organism. Then, schemas determine a course of action by a process of analogy formation, planning, and schema interaction, in which formal deduction is not necessary implicated. Moreover, memory of a schema assemblage may be tuned to create a new schema.
(b) It relates perception to action in a unified representational framework (cf. [30]).
(c) A view of adaptation with links to Piaget's assimilation and accommodation, but which sees developmental stages as "emergent" rather than genetically prespecified.
(d) Related to (a), it introduces cooperative computation as a unifying style for cognitive science and neuroscience.

My approach is also distinguished by being structured in such a way that a schema may either be viewed purely as a functional unit in a network of interacting schemas (*basic* schema theory), or further analyzed in terms of its neural underpinnings (*neural* schema theory), and that individual schemas may be linked to social schemas (*social* schema theory [10]).

4   From Action-Oriented Perception to Knowledge

A schema is both a store of knowledge and the description of a process for applying that knowledge. As such, a schema may be instantiated to form multiple active copies called *schema instances*. For example, given a schema that represents generic knowledge about a chair, we may need several active instances of the chair schema, each suitably tuned, to subserve our perception of a scene containing several chairs. A schema is more like a molecule than an atom in that schema instances may well be linked to others to form *schema assemblages* which provide yet more comprehensive schemas.

Schema theory provides, inter alia, a language for the study of *action-oriented perception* [1,41] in which the organism's perception is in the service of current and intended action rather than (though not exclusive of) providing stimuli to which the organism provides unintended responses. According to schema theory, our minds comprise a richly interconnected network of schemas. Schema theory can also express models of language and other cognitive functions [11].

An assemblage of some instances of these schemas represents our current situation. A crucial notion is that of *dynamic planning*: the organism is continually making and remaking plans – in the form of schema assemblages called *coordinated control programs* which combine perceptual, motor, and coordinating schemas – but these are subject to constant updating as perception signals obstacles or novel opportunities. In particular, action-oriented perception involving passing parameters from perceptual to motor schemas: For example, perceiving a ball instructs the hand how to grasp it; perceiving obstacles adjusts one's navigation. However, schema assemblages and dynamic planning ensure that behavior seldom involves direct relationships of a behaviorist, stimulus-response simplicity; rather, context and plans help determine which perceptual clues will be sought and acted upon.

Schemas are modular entities whose instances can become *activated* in response to certain patterns of input from sensory stimuli or other schema instances that are already active. The *activity level* of an instance of a perceptual schema represents a "confidence level" that the object represented by the schema is indeed present; while that of a motor schema may signal its "degree of readiness" to control some course of action. The activity level of a schema may be but one of many parameters that characterize it. Thus a schema for "ball" might include parameters for its size, color, and velocity – in the sense of properties we might notice when we see a ball or play with it, i.e., with a level of detail appropriate to our skill and interest, rather than being highly precise measurements.

To make sense of any given situation we call upon hundreds of schemas in our current schema assemblage. Our lifetime of experience might be encoded in a personal "encyclopedia" of hundreds of thousands of schemas. As we act, we perceive; as we perceive, so we act. Perception is not passive, like a photograph. Rather it is active, as our current schemas determine what we take from the environment.

The essence of schema goes beyond the fact that we have concepts, for example, of a ball, because it makes explicit aspects of "concepts" that might be lost in other accounts. We first need to distinguish the "concepts" from the "schema". Are whales mammals? Now science says "Yes" – but "Do whales activate the mammal schema?" would be answered "No" for many individuals. Further, it integrates perceptual schemas (for example, how to recognize a ball) with motor schemas (such as what to do with a ball) through the parameter-passing mechanism, but also expresses likely and unlikely patterns of co-occurrence through the patterns of competition and cooperation that develop within the schema network. I discuss learning below – where perceptions lead (in an ongoing action-perception cycle) to actions with attendant expectations; failure of these expectations can lead to modifications of perceptual, motor and other schemas. Admittedly, this is an inadequate classification of schemas. Specific models introduce schemas whose role is to coordinate other schemas (for example, Hoff and Arbib [28] study the interaction of hand transport and preshape during visually guided reaching. Moreover, as assemblages or coordinated control programs are built up, they constitute compound schemas which are primarily neither perceptual nor motor.

One would like to have criteria (whether functional, neurological, phenomenological, conceptual, or behavioral) to individuate, or pick out, distinct schemas but none such exists at present. A schema analysis will often start with some overall function or phenomenon of interest and then refine the definition of the schema and its decomposition into other schemas in such a way as to match data on speed and error of behavior, or (if one studies schemas at the level of brain theory) the effects of lesions and other neural measurements and perturbations (see, for example, [6] or Chapter 3 of [12] for more details).

5   Schema Instances and Cooperative Computation

Schema theory sees behavior as based *not* on inferences from axioms nor on the operation of an inference engine on a passive store of knowledge. This moves us from the domain of serial computation to an understanding of how behavior results from *competition* and *cooperation* between schema instances (i.e., interactions which, respectively, decrease and increase the activity levels of these instances) which, due to the limitations of experience, cannot constitute a completely consistent axiom-based logical system.

Schema theory thus offers a new paradigm of computation, with "schemas" as the programs, and *cooperative computation* – a shorthand for "computation based on the competition and cooperation of concurrently active agents" – as their style of interaction. Cooperation yields a pattern of "strengthened alliances" between mutually consistent schema instances that allows them to achieve high activity levels to constitute the overall solution of a problem (as perceptual schemas become part of the current short-term model of the environment, or motor schemas contribute to the current course of action). It is as a result of competition that instances which do not meet the evolving (data-guided) consensus lose activity, and thus are not part of this solution (though their continuing subthreshold activity may well affect later behavior).

A schema network does not, in general, need a top-level executor since schema instances can combine their effects by distributed processes. This may lead to apparently emergent behavior, due to the absence of global control. For a very simple example: in my model of the frog, *Rana computatrix* (see, e.g., [5]), the decision on whether to feed or flee results from the interaction of schemas related to these two behaviors, not from explicit analysis of the relative merits of these two courses of action by higher level schemas. But the process does not stop there. A schema for hunger can shift the balance from "flee" to "feed" not by top-down control but by lowering the threshold for the "feed schema" to initiate behavior; schemas for recognition of obstacles can bias the chosen behavior to yield an appropriate detour, and this is expandable by learning.

To further see why a schema network may not need a top-level executor, think of schemas as linked in a network with two kinds of links: One kinds passes data, for example, the ball-schema might pass time-until-contact information to the catch-schema. The other passes activity levels so that, for example, perceptual schemas for two regions of an image may excite each other if the objects they represent are likely to occur in that spatial relationship; they might inhibit each other if such a juxtaposition is unlikely, as in seeing a snowball atop a fire. Since a surrealist painting *could* be seen to depict a snowball atop a fire, it is clear that these activity-links bias a dynamic process of interpretation rather than determining what can and cannot be seen. Similar considerations apply to other forms of integration of action, perception, and thought. Elsewhere ([6], Sec. 5.3), I provide a more fully developed example of cooperative computation in recognition in visual scene perception, which involves the continued interaction of bottom-up (more data-driven) and top-down (more hypothesis-driven) schemas.

6   Learning

Schema theory is a learning theory too. A schema provides us not only with abilities for recognition and guides to action, but also with expectations about what will happen. These may be wrong. We sometimes learn from our mistakes. Our schemas, and their connections within the schema network, change. In a general setting, there is no fixed repertoire of basic schemas. Rather, new schemas may be formed as assemblages of old schemas; but once formed a schema may be tuned by some adaptive mechanism. This tunability of schema-assemblages allows them to start as composite but emerge as primitive, much as a skill is honed into a unified whole from constituent pieces. My approach to schema theory thus adopts the idea of Jean Piaget (e.g., [42]), the Swiss developmental psychologist and genetic epistemologist, that the child has certain basic schemas and basic ways of *assimilating* knowledge to schemas, and that the child will find at times a discrepancy between what it experiences and what it needs or anticipates. On this basis, its schemas will change, *accommodation* will take place. It is an active research question as to what constitutes the initial stock of schemas. Much of Piaget's writing emphasizes the initial primacy of sensorimotor schemas, where other scientists study the interactions between mother and child to stress social and interpersonal schemas as part of the basic repertoire on which the child builds.

Another important concept in Piaget's work is that of *reflective abstraction*[1]. Piaget emphasizes that we do not respond to unanalyzed patterns of stimulation from the world. Rather, current stimuli are analyzed in terms of our current stock of schemas. It is the interaction between the stimulation – which provides variety and the unexpected – and the schemas already in place that provides patterns from which we can then begin to extract new operational relationships. These relationships can now be reflected into new schemas which form, as it were, a new plane of thought. And then – and this is the crucial point – since schemas form a network, these new operations not only abstract from what has gone before, but now provide an environment in which old schemas can become restructured. To the extent that we can form a general concept of an object, our earlier knowledge of a dog and a ball, and so on, become enriched.

7   Linking Schemas to the Brain

In brain theory, the analysis of schema instances is intermediate between the overall specification of some behavior and the neural networks that subserve it. A given schema, defined functionally, may be distributed across more than one brain region; conversely, a given brain region may be involved in many schemas. A top-down analysis may advance specific hypotheses about the localization of (sub)schemas in the brain, and these may be tested by lesion experiments, with possible modification of the model (for example, replacing one schema by several interacting schemas with different localizations) and further testing. Once a schema-theoretic model of some animal behavior has been refined to the point of hypotheses about the localization of schemas, we may then model a brain region by seeing if its known neural circuitry can indeed be shown to implement the posited schema. In some cases the model will involve properties of the circuitry that have not yet been tested, thus laying the ground for new experiments.

Schemas as "functional units" may be contrasted with the "structural units" of neuroanatomy and neurophysiology. The work of the nineteenth-century neurologists led us to think of the brain in terms of large interacting regions each with a more or less specified function, and this localization was reinforced by the work of the anatomists at the turn of the century who were able to subdivide the cerebral cortex on the basis of cell characteristics, cytoarchitectonics. It was at this same time that the discoveries of the neuroanatomist Ramón y Cajal [44] and the neurophysiologist Sherrington [49] helped establish the neuron doctrine, leading us to view the function of the brain in terms of the interaction of discrete units, the neurons. The issue for the brain theorist, then, is to map complex functions, behaviors, patterns of thought, upon the interactions of these rather large entities, anatomically defined brain regions, or these very small and numerous components, the neurons. This has led many neuroscientists to look for structures intermediate in size and complexity between brain regions

---

[1] The idea of reflective abstraction is developed [14]. I have argued [7] that Piaget pays insufficient attention to the role of social structures, including formal instruction, in the child's construction of logic and mathematics.

and neurons to provide stepping stones in an analysis of how neural structures subserve various functions. One early example was the Scheibels' [46] suggestion that the reticular formation could be approximated by a stack of "poker chips" each incorporating a large number of neurons receiving roughly the same input and providing roughly the same output to their environments. This modular decomposition provided the basis a model of the reticular formation [31].

In another direction, the theoretical ideas of Pitts and McCulloch [43] combined with the empirical observations of Lettvin and Maturana on the frog visual system to suggest that one might think of important portions of the brain in terms of interacting layers of neurons, with each layer being retinotopic in that the position of neurons in the layer was correlated with position on the retina, and thus in the visual field [32]. A neuron may participate in the implementation of multiple schemas. For example, in the toad brain we find that certain neurons in pretectum whose activity correlates with that of the perceptual schema for predators will also, via an inhibitory pathway to the tectum, contribute to the perceptual schema for prey (this is an explicit example of "cooperative computation"). A representation of some overtly defined concept or behavioral parameter will in general involve temporally coordinated activity of a multitude of neurons distributed over multiple brain regions. Moreover, each region will in general exhibit *coarse coding* of parameters: it is not the firing of a single cell that codes a value, but rather the averaged activity of a whole set of neurons that is crucial. In any case, the brain embodies many different schemas, some based on circuitry evolved for that purpose, others developed on the basis of experience with both social and nonsocial interactions.

Mountcastle and Powell [40] working in somatosensory cortex, followed by Hubel and Wiesel [29] working in visual cortex, established the notion of the column as a "vertical" aggregate of cells in cerebral cortex, again working on a common set of inputs to provide a well-defined set of outputs.

With this research, the notion of the brain as an interconnected set of "modules" – intermediate in complexity between neurons and brain regions – was well established within neuroscience[2], but here it may be useful to distinguish "neural modules" and "schemas" from "modules" in the sense of Fodor [22], a sense which has been excessively influential in recent cognitive science and related philosophizing. Rather than go into details, I simply list the key points from an earlier critique [4]: The fundamental point is that Fodor's modules – such as "language" or "vision" – are too large. It is clear from schema analyses of visual perception or motor control (e.g., in [6]) that a computational theory of cognition must use a far finer grain of analysis than that offered by Fodor. Fodor offers big modules (for example, one for all of language), argues vociferously that they are computationally autonomous, and despairs at the problem of explaining the central processes, since they are not informationally encapsulated. By contrast, my approach is to analyze the brain in terms of those smaller *functional* units called schemas, while stressing that each schema may involve

---

[2] For a recent overview, see Szentágothai's discussion of "modular architectonics of the brain" in Chapter 2 of [12].

the cooperative computation of many *structural* units ("modules" in the classical, medium-grain sense of neuroscience outlined above). Since the interactions between these schemas play a vital role in my models, the case for autonomy of large modules becomes less plausible. As a result, schema theory offers a continuity of theorizing between, say, vision and action and central processes, rather than recognizing the reality of the divide posited by Fodor.

The notion raised in the section on "Learning" that "schema assemblages may start as composite but emerge as primitive" [i.e., functionally grouped schemas can be described and/or activated by a single label] underlies the essential feature of hierarchical structuring. In fact, the issue of hierarchical structuring has been a central concern from, for example, the publication of Hebb's *The Organization of Behavior* to present day study of neural networks. The main ingredients are that patterns of neural activity become established ("attractors") as quasi-stable (i.e., until a certain amount of change of input activity), and then the formation of excitatory and inhibitory links which will encourage coactivation of several such patterns together (Hebb's "assemblies") or the activation of such patterns in some order (Hebb's "phase sequences"). However, many processes that we can now describe at the abstract schema level (as in the visual scene perception example mentioned above) still pose unanswered questions about whether and how they are realized in the brain's circuitry.

References

1. Arbib, M.A., 1972,The Metaphorical Brain: An Introduction to Cybernetics as Artificial Intelligence and Brain Theory, Wiley-Interscience: New York, p. 168.
2. Arbib, M.A., 1975, Artificial Intelligence and Brain Theory: Unities and Diversities, Ann. Biomed. Eng. 3:238–274.
3. Arbib, M.A., 1981, Perceptual Structures and Distributed Motor Control, in Handbook of Physiology, Section 2: The Nervous System, Vol. II, Motor Control, Part 1 (V.B. Brooks, Ed.), American Physiological Society , pp. 1449–1480.
4. Arbib, M.A., 1987a, Modularity and Interaction of Brain Regions Underlying Visuomotor Coordination, in Modularity in Knowledge Representation and Natural Language Understanding, (J.L. Garfield, Ed.), pp. 333–363.
5. Arbib, M. A. 1987b, Levels of Modelling of Visually Guided Behavior (with peer commentary and author's response), Behavioral and Brain Sciences, 10:407–465.
6. Arbib, M. A., 1989, The Metaphorical Brain 2: Neural Networks and Beyond, Wiley-Interscience
7. Arbib, M.A., 1990, A Piagetian Perspective on Mathematical Construction, Synthese, 84:43–58
8. Arbib, M.A., 1995, Schema Theory: From Kant to McCulloch and Beyond, in Brain Processes, Theories and Models . An International Conference in Honor of W.S. McCulloch 25 Years After His Death, (R. Moreno-Diaz and J. Mira-Mira, Eds.), Cambridge, MA: The MIT Press, pp.11–23.
9. Arbib, M.A., 2003, Rana computatrix to human language: towards a computational neuroethology of language evolution, Phil. Trans. R. Soc. Lond. A, 361: 2345–2379.
10. Arbib, M. A., and M. B. Hesse, 1986, The Construction of Reality, Cambridge University Press

11. Arbib, M.A., E.J. Conklin and J.C. Hill, 1987, From Schema Theory to Language, Oxford University Press.
12. Arbib, M. A., Érdi, P. and Szentágothai, J., 1998, Neural Organization: Structure, Function, and Dynamics, Cambridge, MA: The MIT Press.
13. Bartlett, F.C., 1932, Remembering, Cambridge University Press.
14. Beth, E.W., and Piaget, J., 1966, Mathematical Epistemology and Psychology, (Translated from the French by W. Mays), Reidel.
15. Braitenberg, V., 1965, Taxis, kinesis, decussation, Progress in Brain Research, 17:210–222.
16. Braitenberg, V., 1984, Vehicles: Experiments in Synthetic Psychology, Bradford Books/The MIT Press, Cambridge, MA.
17. Brooks, R.A., 1986, A robust layered control system for a mobile robot, IEEE Journal of Robotics and Automation, RA-2:14–23
18. Craik, K.J.W., 1943, The Nature of Explanation, Cambridge University Press.
19. Davis, R., and Smith, R.G., 1983, Negotiation as a metaphor for distributed problem solving, Artificial Intelligence, 20:63–109.
20. Edelman, G.M., 1987, Neural Darwinism: The Theory of Neuronal Group Selection, Basic Books.
21. Erman, L.D., Hayes-Roth, F.A., Lesser, V.R., and Reddy, D.R., 1980, The Hearsay-II Speech-Understanding System: Integrating Knowledge to Resolve Uncertainty, Computing Surveys, 12:213–253.
22. Fodor, J. , 1983, The Modularity of Mind, MIT Press/A Bradford Book.
23. Gregory, R.L., 1969, On How so Little Information Controls so much Behavior, in Towards a Theoretical Biology. 2, Sketches (C.H. Waddington, Ed.), Edinburgh University Press.
24. Head, H., and Holmes, G., 1911, Sensory Disturbances from Cerebral Lesions, Brain, 34:102–254.
25. Hebb, D.O., 1949, The Organization of Behavior, John Wiley & Sons.
26. Hesse, M.B., 1980, Revolutions and Reconstructions in the Philosophy of Science, Indiana University Press.
27. Hewitt, C.E., 1977, Viewing control structures as patterns of passing messages, Artificial Intelligence, 8:323–364.
28. Hoff, B., and Arbib, M.A., (1993) Simulation of Interaction of Hand Transport and Preshape During Visually Guided Reaching to Perturbed Targets, J .Motor Behav. 25: 175–192.
29. Hubel, D.H. and Wiesel, T.N., 1974, Sequence regularity and geometry of orientation columns in the monkey striate cortex. J. Comparative Neurology 158: 267–294.
30. Jeannerod, M., 1997, The Cognitive Neuroscience of Action, Oxford: Blackwell Publishers.
31. Kilmer, W.L., McCulloch, W.S., and Blum, J., 1969, A model of the vertebrate central command system, Int. J. Man-Machine Studies 1: 279–309.
32. Lettvin, J. Y., Maturana, H., McCulloch, W. S. and Pitts, W. H., 1959, What the frog's eye tells the frog's brain, Proc. IRE. 47: 1940–1951.
33. MacKay, D.M., 1966, Cerebral Organization and the Conscious Control of Action, in Brain and Conscious Experience (J.C. Eccles, Ed.), Springer-Verlag, pp.422–440.
34. Mandler, G. (1985): Cognitive Psychology: An Essay in Cognitive Science, Hillsdale, NJ: Lawrence Erlbaum Associates
35. Maravita, A., and Iriki, A., 2004, Tools for the body (schema), Trends in Cognitive Sciences, 8:79–86.

36. Minsky, M.L., 1975, A Framework for Representing Knowledge, In: The Psychology of Computer Vision, (P.H.Winston, Ed.), McGraw-Hill, pp.211–277.
37. Minsky, M.L., 1965, Matter, Mind and Models, In Information Processing 1965, Proceedings of IFIP Congress 65, Spartan Books, Vol.1, pp.45–59.
38. Minsky, M.L., 1975, A Framework for Representing Knowledge, In: The Psychology of Computer Vision, (P.H. Winston, Ed.), McGraw-Hill, pp.211–277.
39. Minsky, M.L., 1985, The Society of Mind, Simon and Schuster, New York, pp.244–250).
40. Mountcastle, V. B., and Powell, T.P.S., 1959, Neural mechanisms subserving cutaneous sensibility, with special reference to the role of afferent inhibition in sensory perception and discrimination, Bulletin of Johns Hopkins Hospital, 105:201–232.
41. Neisser, U., 1976, Cognition and Reality: Principles and Implications of Cognitive Psychology, W.H. Freeman.
42. Piaget, J., 1971, Biology and Knowledge, Edinburgh University Press.
43. Pitts, W.H., and McCulloch, W.S., 1947, How we know universals, the perception of auditory and visual forms. Bull. Math. Biophys., 9:127–147.
44. Ramón y Cajal, S., 1911, Histologie du systeme nerveux, Paris: A. Maloine, (English Translation by N. and L. Swanson, Oxford University Press, 1995;.
45. Schank, R., and Abelson, R., 1977, Scripts, Plans, Goals and Understanding: An Inquiry into Human Knowledge Structures, Erlbaum.
46. Scheibel, M.E. and Scheibel, A.B., 1958, Structural substrates for integrative patterns in the brain stem reticular core. In Reticular Formation of the Brain (H. H. Jasper et al., eds.), pp. 31–68, Little, Brown and Co.
47. Schmidt, R.A., 1976, The Schema as a Solution to Some Persistent Problems in Motor Learning Theory, in Motor Control: Issues and Trends (G.E. Stelmach, ed.), New York: Academic Press, pp.41–65.
48. Shallice, T. (1988): From Neuropsychology to Mental Structure, Cambridge: Cambridge University Press.
49. Sherrington, C.S., 1906, The integrative action of the nervous system, New Haven and London, Yale University Press.
50. Vidal, J.M., & Durfee, E.H., 2003, Multiagent systems, in The Handbook of Brain Theory and Neural Networks, (M.A. Arbib, Ed.), Second Edition, Cambridge, MA: A Bradford Book/The MIT Press, pp.707–711.
51. Walter, W.G., 1953, The Living Brain, Penguin Books, Harmondsworth.

From Conditional Specifications to Interaction Charts

A Journey from Formal to Visual Means to Model Behaviour

Egidio Astesiano and Gianna Reggio

DISI – Università di Genova, Italy
{reggio,astes}@disi.unige.it

Abstract. In this paper, addressing the classical problem of modelling the behaviour of a system, we present a paradigmatic journey from purely formal and textual techniques to derived visual notations, with a further attention first to code generation and finally to the incorporation into a standard notation such as the UML.

We show how starting from CASL positive conditional specifications with initial semantics of labelled transition systems, we can devise a new visual paradigm, the interaction charts, which are diagrams able to express both reactive and proactive/autonomous behaviour.

Then, we introduce the executable interaction charts, which are interaction charts with a special semantics, by which we try to ease the passage to code generation.

Finally, we present the interaction machines, which are essentially executable interaction charts in a notation that can be easily incorporated, as an extension, into the UML.

Keywords: design of visual notations, formal notations, behaviour modelling/specification, CASL, UML, interaction charts

1 Introduction

In a remarkable paper [10], celebrating and assessing a decade of TAPSOFT in 1995, Ehrig and Mahr, after admitting some disproportion between the original claims of formal methods and their real impact on software practices, were however insisting on the need of rooting engineering practices on "the contributions from theoretical and conceptual work". That call was taken up and expanded by the authors first in a talk at the last TAPSOFT (Lille,1997) [2, 3] and later on in some papers advocating the use of "well-founded methods" more than "formal methods" (see [5] for a general presentation). Well-founded methods are precisely rooted on theoretical and conceptual models, but presented in a way that is friendly for the user and more concerned with the practical engineering needs. In this paper, addressing the classical problem of modelling the behaviour of a system, we present in a sense a paradigmatic journey from purely formal and textual techniques to derived visual notations, with a further attention first to code generation and finally to the incorporation into a standard practical notation such as the UML [13]. A bit more precisely, we show how starting from

H.-J. Kreowski et al. (Eds.): Formal Methods (Ehrig Festschrift), LNCS 3393, pp. 167–189, 2005.

a formal specification technique, namely, positive conditional specifications with initial semantics of labelled transition systems, expressed using the CASL specification language [6, 12], we can devise a new visual paradigm, which can also be adopted for an extension of UML. The paradigm is centered on the interaction charts, which are diagrams able to express also a proactive/autonomous behaviour, in opposition to the only reactive behaviour of the state charts and of the UML state machines.

The first main new contribution of this paper is the introduction of the executable interaction charts, by which we try to tackle the problem of the treatment of the nondeterministic choice among various alternatives when moving from abstract formal notations to more practical notations that need a kind of operational/executable semantics in order to ease the passage to code. Whereas there are no needs to restrict the alternatives in the first case, namely, it is possible specify a system that may nondeterministically choose among a set of activities of any kind (internal, inputting, outputting, a mixture of inputting and outputting), in the latter either the sets of activities among which to choose are restricted (for example, only inputting and at most one internal, as in UML and Ada programming language) or some mechanism is introduced to be able to discover which alternatives are feasible in a certain situation (e.g., event queues/pools of UML). Executable interaction charts follow the second choice, by proposing the use of abstract buffers. The result is an abstract and executable visual notation to specify/model interactive behaviour, a kind of behaviour commonly found in "client" components, proactive agents and so on.

The second contribution of the paper is the introduction of the interaction machines, which are essentially executable interaction charts in a notation that can be easily incorporated, as an extension, in the UML notation.

We start in Sect. 2 by briefly summarizing the use of conditional specifications for modelling the behaviour of systems, then we introduce in Sect. 3 the interaction charts, showing how they have been derived from the corresponding conditional specifications. In Sect. 4 we present the executable interaction charts, and finally in Sect. 5 we show how they can be used to extend UML with a new kind of diagrams, the interaction machines.

2 Free Positive Conditional Specifications for Modelling Behaviour

Here, we use the word *system* to denote a dynamic entity of whatever kind, and so evolving along the time, without any assumption about other aspects; thus a system may be a communicating/nondeterministic/sequential/... process, a reactive/parallel/concurrent /distributed/... system, but also an object-oriented system (a community of interacting objects), and an agent or an agent system.

For modelling the behaviour of systems we adopt the well-known and accepted technique based on labelled transition systems (see [11, 16, 4]), which is today standard, widely used, and proven adequate in many cases, and there is a huge literature. For example, labelled transition systems are the basic formal models that we have used for giving the semantics to Ada [1] and to UML [18, 19].

A *labelled transition system* (shortly *lts*) is a triple $(STAT, LAB, \rightarrow)$, where $STAT$ and LAB are sets, the *states* and the *labels*, and $\rightarrow \subseteq STAT \times LAB \times STAT$ is the *transition relation*. A triple $(s, l, s') \in \rightarrow$ is said a *transition* and is usually denoted by $s \xrightarrow{l} s'$.

The behaviour of a system S may be represented by an lts $(STAT, LAB, \rightarrow)$ and an initial state $s_0 \in STAT$; then the states in $STAT$ reachable from s_0 represent the intermediate (interesting) situations of the life of S and the transition relation $\rightarrow$ the possibilities of S of passing from a situation to another one. It is important to note that here a transition $s \xrightarrow{l} s'$ has the following meaning: S in the state s has the *capability* of passing into the state s' by performing a transition whose interaction with the external (to S) world is represented by the label l. Thus the label l contains information on the conditions on the external world for the capability to become effective, and information on the transformation of such world induced by the execution of the action, i.e., it describes the interaction of S with the external world during such transition.

Labelled transition systems may be used also to model *structured systems* (i.e., systems built by putting together several subsystems, simple or in turn structured). The lts modelling a structured system is defined by composing the lts's describing its composing subsystems; the states of this lts are sets of states of the subsystems, and its transitions consist of the simultaneous execution of sets of transitions of the subsystems (at most one for each subsystem), see [4, 7].

Labelled transition systems may be specified by means of algebraic specifications having the form shown below. In this paper we present the algebraic specifications using the language CASL [6, 12]. CASL has been designed by CoFI[1], the international *Common Framework Initiative for algebraic specification and development*. It is based on a critical selection of features that have already been explored in various contexts, including subsorts, partial functions, first-order logic, and structured and architectural specifications.

A CASL specification may include the declarations of sorts, operations and predicates (together with their arity), and axioms that are first-order formulae with strong and existential equality and a *2-valued logics*. In CASL large and complex specifications are easily built out of simpler ones by means of (a small number of) specification building primitives, among them *union* (keyword '**and**') and *extension* can be used to structure specifications. Extensions, introduced by the keyword '**then**', may specify new symbols, possibly constrained by some axioms, or merely require further properties of old ones.

spec LTS = DATA$_1$ **and** ... **and** DATA$_r$ **then**
    sorts $State, Label, \ldots$
    ops ...
    preds $\_\_ \xrightarrow{\_\_} \_\_ : State \times Label \times State$
      ...
    axioms ...

---

[1] http://www.brics.dk/Projects/CoFI

where $\text{DATA}_1, \ldots, \text{DATA}_r$ are the names of the specifications of the basic data used to define the states and the labels.

Any algebra M that is a model of LTS defines a labelled transition system, precisely

$$(State^M, Label^M, \_\_ \xrightarrow{\_\_} \_\_^M).$$

By choosing appropriately the set of axioms of the above specification, it is possible to specify particular classes of lts, and thus particular classes of systems by characterizing their behaviour. However, first-order logic is not expressive enough to specify all relevant classes of lts, i.e., to express all relevant properties on them (see, e.g., [9, 4]); for example, using first-order logic it is not possible to require liveness conditions. A convenient solution is to extend the first-order logic with temporal combinators, as proposed by LTL (Labelled Transition Logic) presented in [9, 4], and its CASL version CASL-LTL [17].

If, instead, we want to specify a particular system with a given behaviour, that is a particular lts, we can use *positive conditional specifications* with free (initial) semantics. Such specifications in CASL have the form shown below. The CASL **free** construct defines free specifications, which are specifications having initial semantics. Such semantics avoids the need for explicit negation; indeed, in the models of free specifications, it is required that values of terms are distinct except when their equality follows from the specified axioms, and positive atoms built by predicates hold only when their truth follows from the specified axioms.

spec FCONDLTS = DATA_1 **and** ... **and** DATA_r **then**
free {   **sorts** $State, Label, \ldots$

  ops ...
  preds $\_\_ \xrightarrow{\_\_} \_\_ : State \times Label \times State$

  ...

  axioms *PosCond* **}** **end**

where $\text{DATA}_1, \ldots, \text{DATA}_r$ are the names of the free conditional specifications of the data used to define the states and the labels, and *PosCond* is a set of positive conditional formulae, which have the form $\bigwedge_{i=1,\ldots,n} \alpha_i \Rightarrow \beta$, where each α_i is a positive atom, i.e., either $pr(t_1, \ldots, t_m)$ or $t_1 = e = t_2$ (existential equation), and β is either $pr(t_1, \ldots, t_m)$ or $t_1 = t_2$ (strong equation).

The initial model I of FCONDLTS, unique up to isomorphism, defines the lts

$$(State^I, Label^I, \_\_ \xrightarrow{\_\_} \_\_^I).$$

Any element of $State^I$ and of $Label^I$ is the interpretation of a ground term, and we have that $I \models s \xrightarrow{l} s'$ iff $s \xrightarrow{l} s'$ follows from *PosCond*. Thus, any ground term *stat* of sort *State* represents a system, the one having as initial state the interpretation of *stat* in I.

We can specify algebraically also the structured systems, again by free conditional specifications, built by extending the union of the specifications of their subsystems. The transition predicates of the subsystems will appear in the premises of the axioms of these specifications, whereas the transition predicate of the structured system will appear in the consequences. For lack of room we do not further detail this topic, see, [4, 7].

3   Interaction Charts

In this section we introduce the *interaction charts* as the visual counterparts of the free conditional specifications of lts introduced in Sect. 2, and thus a visual notation to present lts, and so the behaviour of systems. A first version of interaction charts was presented in [20], recently refined in [7]; then a Java oriented version named *behaviour graph* was proposed as part of the notation JTN [8].

We restrict the considered class of free conditional specifications of lts's to be able to associate with them an interaction chart, by fixing the structure of the states and of the labels, and the form of the conditional axioms defining the transition predicate.

Here we consider two cases, which will result in two slightly different variants of interaction charts; which variant to use depends on the applications and on the specifier style.

Generator Variant, the states and the labels are defined by means of total generator operations.

Record Variant, the states have a record structure and the labels are defined by means of total generator operations.

3.1   Interaction Charts: Generator Variant

In this case we consider free conditional specifications of lts, written again in CASL, having the following form.

spec NAME = DATA$_1$ **and** ... **and** DATA$_r$ **then** **free {**
    sorts *State, Label*
    ops sg_1 : ... → *State* %% state generators
        ...
        sg_n : ... → *State*
        lg_1 : ... → *Label* %% label generators
        ...
        lg_m : ... → *Label*
    preds $\_\_ \xrightarrow{\ \_\_\ } \_\_$: *State* × *Label* × *State*
    axioms *GPosCond* **}** **end**

where NAME is an identifier, DATA$_1$, ..., DATA$_r$ are the names of the free conditional specifications (given elsewhere) of the datatypes used to define the states and labels, and each element of *GPosCond* has the form

$$(*)\ \ cond\ \Rightarrow\ sg(t_1,\ldots,t_k)\ \xrightarrow{\ lg(t''_1,\ldots,t''_p)\ }\ sg'(t'_1,\ldots,t'_h)$$

where sg and sg' are state generators, lg is a label generator, $t_1,\ldots,t_k$, $t''_1,\ldots,$ t''_p, t'_1, ..., t'_h are terms possibly with variables and *cond* is a conjunction of positive atoms, where $t_1,\ldots,t_k$, $t''_1,\ldots,$ t''_p, t'_1, ..., t'_h and their subterms may appear while the transition predicate $\_\_ \xrightarrow{\quad} \_\_$ cannot. Recall that the state and label generators are total operations[2].

---

[2] In CASL total operations are declared by ... : ... → ..., whereas the partial operation by ... : ... →? ....

Note that in the initial model of this specification the states/labels represented by different generators or by the same generator applied to different arguments are different.

The visual notation for presenting the above *system specification* is[3]

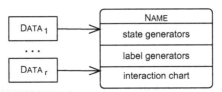

In the above picture, a label generator $lg : \ldots \rightarrow Label$ is written $lg(\ldots)$, and similarly a state generator $sg : \ldots \rightarrow State$ is written $sg(\ldots)$, since in both cases the result type may be omitted because it is implicit. The *interaction chart* is the visual presentation of the set *GPosCond* of the conditional axioms of the specification defining the transition predicate.

A conditional axiom having form (*) is visually represented as

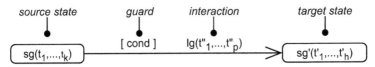

The visual presentations of all the axioms in *GPosCond* may then be put together building an oriented graph, as originally proposed in [20], by collecting together all rounded boxes related to states built by the same generator, and by writing only once repeated generator instantiations. The guards will be omitted when they are equivalent to true.

Example. We give, in Fig. 1, the specification of a simple process (component) operating a calculation over up to 100 negative integers and refusing any positive number. INTPLUS is the specification of integers extended with an operation *op*. To help understand the strong correspondence between the interaction chart and the conditional axioms of the corresponding specification, we report them below.

$run(100) \xrightarrow{null} stop$

$0 > N \Rightarrow run(CNT) \xrightarrow{receiveOk(N)} processing(N, CNT)$

$run(CNT) \xrightarrow{receiveKo(N)} ko$

$0 \leq N \Rightarrow run(CNT) \xrightarrow{receiveOk(N)} refusing(N, CNT)$

$refusing(N, CNT) \xrightarrow{refused(N)} run(CNT)$

$processing(N, CNT) \xrightarrow{result(op(N))} run(CNT + 1)$

Summarizing, an interaction chart is a labelled graph where

- nodes represent the relevant types/classes of situations in the life of the modelled system, during which some (usually implicit) invariant condition holds,

---

[3] Also the free conditional specifications of datatypes may be presented visually, see [20, 7].

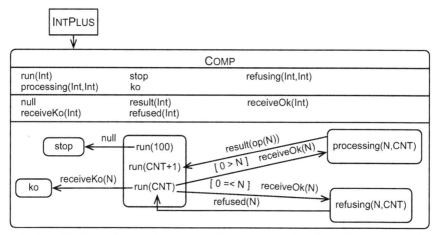

Fig. 1. Specification of a simple process with an interaction chart (generator variant)

- arcs represent the capabilities of the system of passing from a situation of one kind into another one of the same or of another kind and their labels describe the interaction of the system with the outside world during such move.

Thus, the interaction charts are a visual notation that follows the state-transition paradigm allowing to visually depict all the capabilities of interactions with the external environment of the modelled system, where transitions correspond to interaction capabilities, and interaction is intended as a description of the interchange between the modelled system and the external environment.

Notice that this is quite different from other visual notations, such as state-charts, where only the reactions to events coming from outside or from inside are visually depicted by the transitions. In some sense a statechart gives a picture of the reactive aspects of the behaviour of a system, whereas an interaction chart gives a picture of the interactive aspects of that behaviour.

3.2   Interaction Charts: Record Variant

In this case we consider free conditional specifications of lts, written again using CASL, having the following form.

spec NAME = DATA$_1$ **and** ... **and** DATA$_r$ **and** STRING **then**
free {   **sorts**   *State, Label*
         ops $< \_, \ldots, \_ >: s_1 \times \ldots \times s_n \times String \rightarrow State$ %% record generator
                  $lg_1 : \ldots \rightarrow Label$ %% label generators
                  ...
                  $lg_m : \ldots \rightarrow Label$
         preds $\_ \xrightarrow{\_\_} \_ : State \times Label \times State$
         vars $F_1 : s_1; \quad \ldots F_n : s_n;$
         axioms *RPosCond* **}** **end**

where NAME is an identifier, STRING is a specification of strings of characters, $DATA_1$, ..., $DATA_r$ are the names of the free conditional specifications (given elsewhere) of the datatypes used to define the states and the labels, and each element of $RPosCond$ has the form

$$(**) \quad cond \Rightarrow <F_1, \ldots, F_n, \text{``}ident_1\text{''}> \xrightarrow{lg(t_1, \ldots, t_m)} <t'_1, \ldots, t'_n, \text{``}ident_2\text{''}>$$

where $ident_1$ and $ident_2$ are two identifiers (and so "$ident_1$" and "$ident_2$" are ground terms of sort $String$), F_1, ..., F_n are variables of sorts s_1, ..., s_n respectively (always the same for all the axioms), t_1, ..., t_m, t'_1, ..., t'_n are terms possibly with variables and $cond$ is a conjunction of positive atoms, where $F_1, \ldots, F_n, t_1, \ldots, t_m, t'_1, \ldots, t'_n$ and their subterms may appear and the transition predicate cannot. Again, recall that the record and the label generators are total operations.

Note that in the initial model of this specification the states are records with n fields and the labels represented by different generators or by the same generator applied to different arguments are different.

The visual notation for presenting the above system specification is

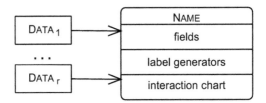

In the above picture, a label generator $lg : \ldots \to Label$ is written $lg(\ldots)$, as for the other variant. The fields are determined by the variables used to denote the state record components and are written $F_1 : s_1 \ldots F_n : s_n$.

The *interaction chart* is again a visual presentation of the set $RPosCond$ of the conditional axioms defining the transition predicate.

A conditional axiom having form (**) is visually represented as

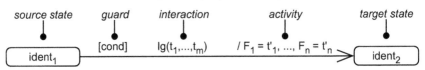

The visual presentations of all the conditional axioms may then be put together building an oriented graph by joining together all rounded boxes decorated by the same identifier.

The null field updates of the form $F = F$ will be omitted, as well as the guard when they are equivalent to true.

Example. We give, in Fig. 2, the specification of the same simple process used as example in Sect. 3.1. Again, to help understand the relationship of the interaction chart with the corresponding conditional axioms we report them below.

$$CNT = 100 \Rightarrow < CNT, N, \text{``}run\text{''} > \xrightarrow{null} < CNT, N, \text{``}stop\text{''} >$$

$$0 > X \Rightarrow < CNT, N, \text{``}run\text{''} > \xrightarrow{receiveOk(X)} < CNT, X, \text{``}processing\text{''} >$$

$$< CNT, N, \text{``run''} > \xrightarrow{receiveError(X)} < CNT, N, \text{``ko''} >$$
$$0 \leq X \Rightarrow < CNT, N, \text{``run''} > \xrightarrow{receiveOk(X)} < CNT, X, \text{``refusing''} >$$
$$< CNT, N, \text{``refusing''} > \xrightarrow{refused(N)} < CNT, N, \text{``run''} >$$
$$< CNT, N, \text{``processing''} > \xrightarrow{result(op(N))} < CNT + 1, N, \text{``run''} >$$

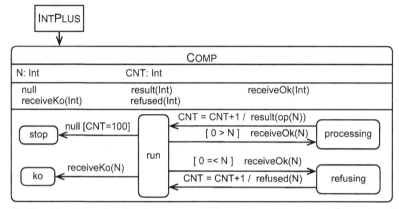

Fig. 2. Specification of a simple process with an interaction chart (record variant)

3.3 Specification of Structured Systems

We can visually present also the free conditional specifications of structured systems, by requiring they have the precise form described below.

- A transition of a structured system is made by the simultaneous execution of a group of transitions of its subsystems (obviously at most one for each subsystem), whose interactions form a set of *cooperations.*
- A cooperation is a set of complementary interactions, in the sense that the interactions being part of a cooperation can only be executed together; for example, sending and receiving a message along a channel, destroying a subsystem and being destroyed, sending a broadcast message and any number of reception of such message.
- There is a criterium for selecting which sets of cooperations will correspond to transitions; for example, interleaving (each transition corresponds to a unique cooperation), free parallel (each transition corresponds to a set of cooperations), and maximal parallelism (each transition corresponds to a maximal group of cooperations).

Here we do not have the room to present the details of the visual presentations of the cooperations and of the criteria; see [20, 7].

3.4 Interaction Charts: Additional Constructs

To effectively use the simple forms of interaction charts presented in Sect. 3.1 and 3.2, they have to be enriched with constructs allowing to easily present quite complex and large charts. Here we present some of them, those that we have found useful in the years; notice that some of them have been inspired by similar constructs of the UML state machines, whose introduction has been motivated by the needs of some of the proposers of the UML notation. Their semantics can be easily defined by transforming a chart using these features into a simpler one having the form defined in Sect. 3.1 or 3.2.

Syntactic Facilities. To help improve the layout of complex interaction charts.

- a state may be anonymous, i.e., the generator/the identifier is not written.
- a state may be depicted several times in a chart;
  - a generator chart may contain several rounded boxes decorated by patterns built by the same generator; such chart is equivalent to another one, where all those boxes are coalesced into a unique one including inside the decorations of all those boxes;
  - a record chart may contain several rounded boxes decorated by the same identifier; such chart is equivalent to another one, where all those boxes are coalesced into a unique one decorated by that identifier.
- a large system presentation may be split into several partial ones, where each one has the name compartment, and some of the other compartments (e.g., a representation containing the name, label and state generators, and another one containing the name and the interaction chart).

Initial/Final States. A node (at most one) of an interaction chart marked by ●——➤ is *initial* (only the states of the associated lts corresponding to that node may be used to determine the initial state of the specified system). The final states (any number), each one represented by ◉, explicitly show the end of the activity of the specified system; obviously no transition may leave a final state. A final state can be replaced by another one decorated by a zero-ary generator/identifier different from all those used in the chart.

For the record variant, the values of the fields in the initial state may be defined by decorating the arrow marking the initial state with $F_1 = t_1; \ldots F_n = t_n$.

Local Transitions (Null Interaction). A system may perform internal activity without any interaction with the external world, in this case we have transitions decorated by a null interaction. The null interaction is characterized by the fact that it takes part in a unique cooperation consisting just of itself. We assume that there is a unique predefined zero-ary label generator to represent a null interaction: null[4]. Moreover, null may be dropped from the transitions, to better depict the absence of interaction with the external environment.

Factorizing Transitions into Segments. It is useful to visually present a transition by joining many *transition segments* (that are not transitions) by a special symbol, the *junction*[5], visually presented by ●.

Technically, a *transition segment* is an arc either between two junctions or a junction and a state or a state and a junction annotated with a partial transition

---

[4] Similar to the τ label of Milner's CCS.
[5] We do not call a junction a pseudo-state as in UML [15], to stress that it is just a presentation mechanism without any special semantics in term of lts.

decoration (e.g., just a guard, an interaction, an activity, a guard and an interaction, ...). The meaning of junctions and segments is simply given by some
replacement rules: a junction may be eliminated by connecting any incoming arc
with any outgoing arc and annotating the resulting arc with the combination of
the two decorations (clearly, not all combination of segments are correct, e.g., a
guard cannot follow an activity).

The factorization of transitions improves the readability of the charts, by
splitting complex transitions into pieces, by avoiding to depict many times the
same part of decoration, and also by making more clear which are the differences
and the commonalities among some transitions.

For example the following fragment of interaction chart

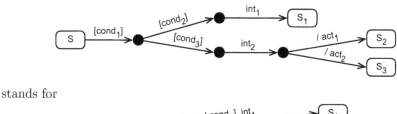

stands for

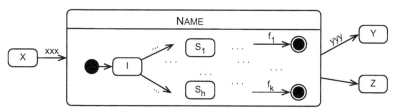

Composite (Sequential) States. A composite state is represented by a rounded
box with a compartment containing the name and another one containing an
interaction chart with a unique initial state and any number of final states. A
composite state may be the target or source state of a transition, and the source
of a unique special undecorated transition. Here we have a schematic generic
composite state.

A composite state can be replaced by

- dropping the state icon,
- making the initial state the target state of any incoming transition,
- adding to any internal state (neither initial nor final) any outgoing transition,
- replacing the final states by the state target of the undecorated outgoing
  transition.

The above schematic composite state stands for the following fragment of interaction chart.

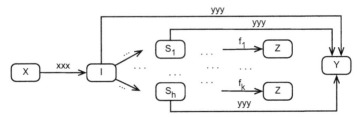

Because there is not a standard general well accepted way to define parallel composite states, and because the existing definitions are quite complicate always with subtle problematic points, we decided to avoid them in the interaction chart notation. Furthermore, this is not a big restriction; indeed, if we need to specify a system explicitly exhibiting a parallel behaviour it is always possible to see it as a structured system made by some subsystems cooperating among them in a parallel way.

Subcharts. Complex interaction charts may be modularly decomposed by defining and using subcharts. A *subchart* is an interaction chart with an initial state. Subcharts are declared in additional compartments of the system specifications that contain the name of the subchart (written in italic) and the interaction chart defining it. To include a subchart in the enclosing one it is sufficient to depict a rounded box with inside the name of the subchart, always written in italic. It stands for a composite state including the definition of that subchart.

Entry/Exit Actions and Internal Transition (Only for the Record Variant). Entry/exit actions and internal transition are associated to the states of an interaction chart. An *entry action* associated with a state is executed, as last thing, whenever a transition having that state as target is executed. An *exit action* associated with a state is executed, as first thing, whenever a transition having that state as source is executed. An *internal transition* presents an interaction capability that does not change the state.

The following picture shows a generic state, named Stat, with one entry action, one exit action and k internal transitions, plus a generic incoming and a generic outgoing transition.

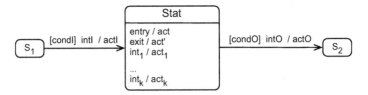

It stands for

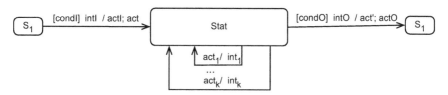

Example. We consider a variant of the simple process used as examples already in Sect. 3.1 and 3.2. In this case the counter is incremented also when a number is refused and the process may break down in any state, not only in the initial one. For simplicity, here we only give the new interaction chart using several of the additional constructs, among them initial, final and composite states, compound transitions and null interaction.

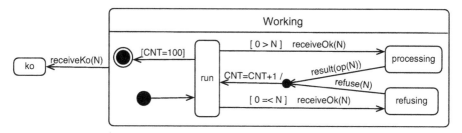

4   Executable Interaction Charts

The interaction charts presented in Sect. 3 specify the (simple) systems considered in isolation in an abstract formal way, by defining an lts determined by the corresponding free conditional specification. The concept of cooperation (finite sets of complementary interactions) together with a criteria to select among the possible groups of cooperations allow to specify the structured systems (i.e., systems built by several subsystems, simple or in turn structured) in an abstract formal way, again by determining a conditional specification defining an appropriate lts; however, for lack of room we cannot present the details and the corresponding visual notation here (see [20, 4, 7]).

Thus, the interaction charts are a rigorous/well-founded (i.e., based on a formal foundation) notation that is extremely flexible and powerful, since a very large class of systems may be specified/modelled using it, including almost any relevant case, as shown by the many applications made in the years, from Ada to OO systems. However, due to their extreme abstraction and generality interaction charts have less nice aspects, which may prevent their use in the current software development practice. Indeed,

- the lts modelling a simple system determined by an interaction chart may have infinite transitions leaving a state;
- the sets of transitions of the subsystems generating a transition of a structured system can be determined only by considering all the subsystems together and all their possible transitions.

As a consequence, it is very hard to develop software tools to support the use of interaction charts, such as a code generator. Thus, here we propose a less general version of interaction charts, which we call *executable*, characterized by the fact that they a have an executable (operational) semantics, similar to those of Petri nets and Harel's statecharts.

The executable interaction charts are based on the record variant (see Sect. 3.2) and must have an initial state. Syntactically only the form of the transitions is changed. A transition of an executable interaction chart has the form

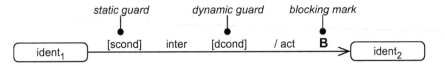

where scond and dcond are conjunctions of positive atoms, inter is a term built by a label generator, F_1, ..., F_n are the field names, act $= F_1 = t_1$; ...; $F_n = t_n$, *FreeVars*(scond) $\subseteq \{F_1, \ldots, F_n\}$, *FreeVars*(dcond) $\subseteq \{F_1, \ldots, F_n\} \cup$ *FreeVars*(inter), *FreeVars*(act) $\subseteq \{F_1, \ldots, F_n\} \cup$ *FreeVars*(inter). **B**, whenever present, denotes that the transition is blocking. As before, he null field updates of the form F = F will be omitted, as well as the guards when they are equivalent to true.

The operational semantics of an executable interaction chart is described below. The system goes on performing a basic-execution-step after another, where a *basic-execution-step* is defined in Fig. 3, where we write T.scond, T.inter, T.dcond, ... to denote the various parts of a transition T. Recall that at any time exactly one state is active (at the beginning the initial state is active).

(1) Let ET be the set of the transitions starting from the active state whose static
    guard holds;   if ET $= \emptyset$ then stop;
(2) let ETLIST be the list of the elements of ET in some order;
(3) if ETLIST is empty then go to (2);
    T = first(ETLIST); ETLIST = dropFirst(ETLIST);
(4) "attempt to execute T.inter";
    %%it can either fail or be successful returning a list of values VL instantiating the
    %%free variables of T.inter
    if it fails
    then
      if T is blocking then go to (4) else go to (3);
    else
      if T.dcond[VL/FreeVars(T.inter)] does not hold
      then
        go to (3)
      else
        execute T.act[VL/FreeVars(T.inter)];
        T.target becomes active and T.source, if different from T.target, becomes
        inactive; stop

Fig. 3. The basic-execution-step

In Fig. 3 we have a generic schema since step (4) "attempt to execute T.inter" must be defined case by case, i.e., the effect/meaning of performing an interaction must be defined, and it is not possible to simply say "there are other subsystems which have chosen to perform the transitions needed to build a cooperation". At this point we have two choices:

- to fix the interactions (e.g., reading and writing a buffer, sending and receiving messages along a channel, sending and receiving messages in a broadcasting way, ...), and thus the executable interaction charts are a unique notation;
- to propose a general schema for defining the meaning of executing a given set of interactions, and thus the executable interaction charts are a family of notations differing for the used interactions.

In this paper, we follow the second choice, and present it using, as example, the particular case of executable interaction charts, where subsystems communicate by sending and receiving asynchronous signals.

Syntactically the interactions are defined by a set of generators, as in Sect. 3. In this case we have two generators:

send(SysIdent,SignalName,ValueList) and rec(SysIdent,SignalName,ValueList).

For what concerns the semantics, we first define the cooperations among such interactions. In this case, send(si,n,vals) and rec(si',n',vals') form a cooperation whenever the arguments are identically, and these are all the possible cooperations.

Then, we introduce some *abstract buffers* that will be accessed by the subsystems by reading or writing information about their possibilities/willingness to perform some interactions. Thus, the attempt to perform an interaction will correspond to access one of these buffers, and depending on its content it can result in the buffer communicating either the failure or the success (together with the values needed to instantiate the free variables).

In this case, for each subsystem there is a buffer containing the set of signals received by it and not yet consumed. To attempt executing send(si,n,vals) consists in adding to the buffer of si the sent signal <n,vals>, and thus it will never fail; whereas to attempt executing rec(si,n,vars) by a subsystem with identity si consists in seeing whether its own buffer contains a signal having form <n,vals>, if the answer is positive the attempt is successful, the values vals are returned and <n,vals> is deleted from the buffer, otherwise the attempt fails.

Formally, the semantics of an executable interaction chart is given by transforming it into an equivalent normal one (presented in Sect. 3.2). Precisely, a specification of a structured system where the subsystems are modelled by executable interaction charts is transformed into an equivalent specification where the subsystems are modelled by normal interaction charts.

Let SP be a specification of a structured system whose n subsystems are specified respectively by $SP_1, \ldots, SP_n$. The specification SPeq equivalent to SP is defined in the following way. The specifications $SP_1, \ldots, SP_n$ using executable interaction charts are transformed in a standard way into specifications using normal interaction charts, say $SP'_1, \ldots, SP'_n$. The added buffers are defined by specifications of simple systems using normal interaction charts, say $B_1, \ldots,$ B_k. $SP'_1, \ldots, SP'_n$ and $B_1, \ldots, B_k$ are the specifications of the subsystems of SPeq, whereas all its non trivial cooperations are defined in a standard way, and have as participants one buffer and one original subsystem.

In the example, we show on a fragment of executable interaction chart using the rec/send interactions how to transform it into a normal one interacting with the buffers.

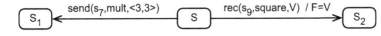

is transformed into

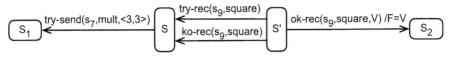

whereas the buffer for a subsystem identified by s_9 is modelled by the following interaction chart

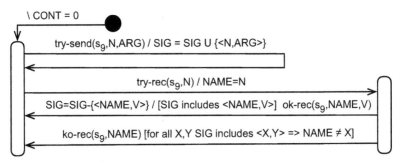

The cooperations between the subsystems and the buffers are just pairs of identical interactions (e.g., <try-inter(...),try-inter(...)>).

Notice that there is not a unique way to define the buffers realizing the cooperations, not even in this simple case of asynchronous signals exchange (in our example, we could have organized the buffer as a list instead of as a set).

We can summarize the tasks for defining a variant of executable interaction charts as follows:

(a) fix which are the interactions used by the variant, by giving their generators and defining the types of their arguments,
(b) fix which are the cooperations among them,
(c) define the buffers supporting the above cooperations. The possible interactions of these buffers are try-inter(...), ok-inter(...), ko-inter(...), where inter is one of the interaction generators defined at (a). Define also their behaviour by means of a normal interaction chart.

The point (3) of the definition of the basic execution step given in Fig. 3, concerning the choice of one among the various transitions, may be made less casual by offering the possibility to control it by decorating the transitions leaving a state with priorities, just integer numbers. Then it is sufficient to replace line (3) of the definition of the basic-execution-step in Fig. 3 by (3') let ETLIST be the list of the elements of ET ordered with respect to their priorities first those with the higher one (if several transitions have the same priority they are ordered in a casual way).

Then, a transition without priority stands for a transition with priority 0; and a transition decorated by else / act stands for a transition decorated by null / act i, where i is a number lower than the priorities of all the other transitions leaving the source state.

5   Extending UML with Interaction Machines

We propose to extend UML by adding a UML-like version of executable inter-action charts that we call *interaction machines*.

UML 2.0 (but versions 1.... are quite similar) offers three main ways to model the behaviour:

- (behaviour and protocol) state machines, or state charts or state diagrams, showing the reactive behaviour of objects,
- sequence/communication/interaction[6] overview diagrams showing sets of se-quences of events happening among a group of objects,
- and activity diagrams showing the control and dataflow aspect of the be-haviour.

Thus, there is no way to present the interactive aspects of an object in isolation; such aspects may be shown only by scenarios where its interactions are performed with a selected set of partners. As a consequence, to depict the behaviour in isolation of a proactive object, i.e., one which does not simply react to events, but instead mainly triggers events to which other objects will react, we can use only a state machine, which will be not very informing and readable, since it will have very few transitions with heavy decorations, mainly with huge activity part[7]. For these reasons, incorporating into UML the interaction charts may be seen as a real extension adding more notational power.

5.1   Interaction Machines

We define the interaction machines by changing as less as possible the definition of behaviour state machines[8] of UML 2.0 [15]. Here for lack of room we just show

---

[6] Note that in the UML world [15] the term interaction has a meaning different from the one used in this paper, a UML interaction is a set of sequences of event occur-rences among some objects.

[7] To overcome this problem UML 2.0 offers a very limited possibility to visually depict in a state machine some action either by enclosing it in a box, or, only for the send signal action, by a convex pentagon.

[8] Note, that it is possible to define also protocol interaction machines, since the dis-tinction between behaviour and protocol state machine is orthogonal with respect to depicting interactions or reactions.

how to define the basic form of the interaction machines (but there is no problem to incorporate the other more complex constructs of the state machines).

Recall that the abstract syntax of UML is given by means of an object-oriented description, a class diagrams, called *metamodel*, whose classes correspond to the abstract syntactic categories, presented inside [15]. At the meta-model level, the interaction machines may be added as a new subclass of the metaclass Behaviour, defined using the existing metaclasses whenever possible, see Fig. 4.

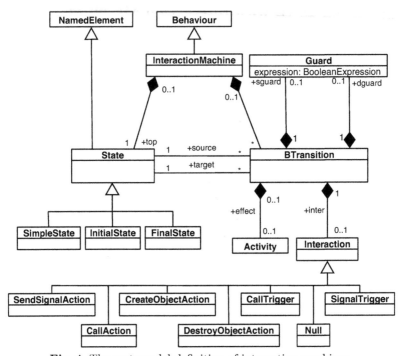

Fig. 4. The metamodel definition of interaction machines

Here we assume that the context of an interaction machines must be an active class.

Interaction machines are defined as the executable interaction charts, but the transition decorations are expressed using UML ingredients; precisely, the interactions are determined by the UML actions and events, and the guards and the activities are expressed by using the means offered by the UML. The generic form of the transitions of the interaction machines is

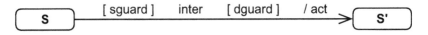

where

- sguard and dguard are boolean expressions (written using OCL[9]).
- inter is defined by the fragment of the UML metamodel in Fig. 4. It may be either a simple action requiring an interaction with some other object (here we only consider: operation call, signal sending, creation and destruction of objects), or an event that is the result of an interaction by some other object (call and signal trigger) or null that is no interaction with any other object, i.e., an activity purely internal to the context object.
- act, the *activity*, is an UML action.

In this case the "Constraints" defining the well-formed constructs are quite important and are as follows

- the evaluation of sguard and dguard cannot have any side effect, as already required for state machines [15];
- the evaluation of sguard, dguard and the execution of act must be possible without accessing anything outside the context object. For example, this means that, differently from UML state machines, a call of an operation of another object cannot appear in the activity part;
- the evaluation of any expression appearing in an interaction of the kind action must be possible without accessing anything outside the context object;
- only interactions of the form call and signal event may have formal parameters that may appear in dguard and act;
- all the attributes of the context active class are visible only inside the class itself and the interaction machine; thus no other object may access or modify them indirectly, except explicitly calling operations of the object (sending signals to it).

Notice that when we speak of objects, obviously we do not consider instances of UML datatypes.

The intuitive meaning of a transition of an interaction machine is that the object may perform (whenever possible) some interaction with some other object possibly followed by some *local* activity when some guard conditions are satisfied.

The precise meaning of an interaction machine is given by specializing the basic-execution-step of Fig. 3 to this particular case, that is essentially to define what means to attempt executing the particular interactions used here, as shown below.

CallAction: execute the call action; here we consider only asynchronous calls, thus it cannot fail;

CallTrigger: if there is matching call in the event pool, then take it instantiating the parameters, otherwise fail;

SendAction: execute the send action; sending signal is always asynchronous, thus it cannot fail;

SignalTrigger: if there is matching signal in the event pool, then take it instantiating the parameters, otherwise fail;

---

[9] The Object Constraint Language to specify constraints and other expressions appearing in UML models, see [14].

null: do nothing, clearly it cannot fail
CreateObjectAction: execute the create action; it cannot fail;
DestroyObjectAction: execute the destroy action, it cannot fail.

In this case, the abstract buffer supporting the cooperations are just the event pools associated with any UML object [15]. Notice, that the basic-execution-step in this case is a generalization of the run-to-completion-step used to describe the semantics of the state machines in [15], Sect. 15.3.12.

5.2   An Example: The Distributed Buffer Resetter

In this section we present a simple example of the use of the interaction machines to model a nonpurely-reactive active object, the *Distributed Buffer Resetter*. This example is quite paradigmatic of autonomous agents doing monitoring and maintenance over distributed systems or Internet. The resetter accesses some buffers one after another following some given ordering, and resets each buffer if its contents is "wrong". At each moment, the resetter can receive from some manager the list of the buffers to reset, or it can be stopped.

Using UML we model the buffer resetter using an active class with an associated interaction machine.

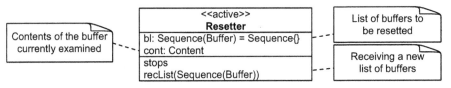

The class has two attributes cont and bl, and two operations stops and recList. The text enclosed by [comment symbol] is an UML comment.

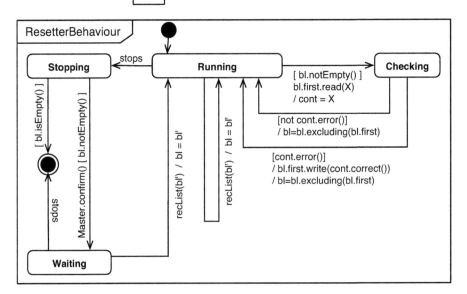

The resetter in the running state has three possible moves:

- when bl is not empty, it may access the first buffer of the list getting its content; if such content is an error (checked by the operation error) it resets such buffer by correcting its content;
- it may receive a new list of buffers to be reset by accepting a call of its operation recList;
- it may receive a request to stop (by a call of its operation stops); in such cases it passes in the state **stopping**. If bl is empty, it terminates by a transition into the final state, whose atomic interaction is the null one, otherwise it asks for a confirmation to its master by calling its operation confirm. Then, if it receives it as a new call of the stops operation, it terminates; if instead it receives a new list of buffers, it goes on to work again.

6   Conclusions

The main message of this paper is to witness the evolution from purely formal techniques to visual notations that are more friendly for the user, but still rooted in "theoretical and conceptual work", as advocated by Ehrig and Mahr [10] about ten years ago.

On the technical side, our journey was consisting of the following intermediate steps:

- a description of the behaviour of systems using CASL conditional specifications, with initial semantics, of labelled transition systems;
- their visual presentations by means of interaction charts;
- their specialization as executable interaction charts, i.e., interaction charts with a special operational semantics targeted at an easier passage to the code;
- finally a proposal of an extension of UML by introducing the interaction machines, which are essentially the UML version of the executable interaction charts.

The technical motivation of the introduction of the interaction machines in UML is their ability to represent both reactive and proactive behaviour of an object, where by proactive we mean autonomous behaviour of the kind required for example when modelling autonomous agents.

Note the difference with the approach taken in the UML [15], where interaction means a set of sequences of event occurrences among some objects. In other words in UML there is no provision to represent the interactions of an object in isolation.

The approach we have shown here allows to use visual notations where the formalities are completely hidden, though being amenable to a precise semantics; this is the essential meaning of the strategy that we call "well-founded methods" [5].

Note also that in the case that we have presented in this paper we have not given a semantics to an existing practical notation, but we have gone the opposite way: the interaction charts, with their ability to express autonomous behaviour, have been suggested by a purely formal specification technique (CASL specifications of labelled transition systems). This is one of the modalities appearing in what we have called "virtuous cycle" of the interaction between foundational and engineering work [5].

References

1. E. Astesiano, A. Giovini, F. Mazzanti, G. Reggio, and E. Zucca. The Ada Challenge for New Formal Semantic Techniques. In *Ada: Managing the Transition, Proc. of the Ada-Europe International Conference, Edimburgh, 1986*, pages 239–248. University Press, Cambridge, 1986.
2. E. Astesiano and G. Reggio. Formalism and Method. In M. Bidoit and M. Dauchet, editors, *Proc. TAPSOFT '97*, number 1214 in Lecture Notes in Computer Science, pages 93–114. Springer Verlag, Berlin, 1997.
3. E. Astesiano and G. Reggio. Formalism and Method. *T.C.S.*, 236(1,2):3–34, 2000.
4. E. Astesiano and G. Reggio. Labelled Transition Logic: An Outline. *Acta Informatica*, 37(11-12):831–879, 2001.
5. E. Astesiano, G. Reggio, and M. Cerioli. From Formal Techniques to Well-Founded Software Development Methods. In *Formal Methods at the Crossroads: From Panacea to Foundational Support, 10th Anniversary Colloquium of UNU/IIST the International Institute for Software Technology of The United Nations University, Lisbon, Portugal, March 18-20, 2002. Revised Papers.*, number 2757 in Lecture Notes in Computer Science, pages 132 – 150. Springer Verlag, Berlin, 2003.
6. M. Bidoit and P.D. Mosses. *CASL User Manual, Introduction to Using the Common Algebraic Specification Language*. Number 2900 in Lecture Notes in Computer Science. Springer-Verlag, 2004.
7. C. Choppy and G. Reggio. Towards a Formally Grounded Software Development Method. Technical Report DISI–TR–03–35, DISI, Università di Genova, Italy, 2003.
   Available at `ftp://ftp.disi.unige.it/person/ReggioG/ChoppyReggio03a.pdf`.
8. E. Coscia and G. Reggio. JTN: A Java-targeted Graphic Formal Notation for Reactive and Concurrent Systems. In Finance J.-P., editor, *Proc. FASE 99*, number 1577 in Lecture Notes in Computer Science. Springer Verlag, Berlin, 1999.
9. G. Costa and G. Reggio. Specification of Abstract Dynamic Data Types: A Temporal Logic Approach. *T.C.S.*, 173(2):513–554, 1997.
10. H. Ehrig and B. Mahr. A Decade of TAPSOFT: Aspects of Progress and Prospects in Theory and Practice of Software Development. In P.D. Mosses, M. Nielsen, and M.I. Schwartzbach, editors, *Proc. of TAPSOFT '95*, number 915 in Lecture Notes in Computer Science, pages 3–24. Springer Verlag, Berlin, 1995.
11. R. Milner. *A Calculus of Communicating Systems*. Number 92 in Lecture Notes in Computer Science. Springer Verlag, Berlin, 1980.
12. P.D. Mosses, editor. *CASL Reference Manual, The Complete Documentation of the Common Algebraic Specification Language*. Number 2960 in Lecture Notes in Computer Science. Springer-Verlag, 2004.
13. OMG. *UML Specification 1.3*, 2000.
    Available at `http://www.omg.org/docs/formal/00-03-01.pdf`.
14. OMG. *UML 2.0 OCL Specification*, 2003.
15. OMG. *UML 2.0 Superstructure Specification*, 2003.

16. G. Plotkin. An Operational Semantics for CSP. In D. Bjorner, editor, *Proc. IFIP TC 2-Working conference: Formal description of programming concepts*, pages 199–223. North-Holland, Amsterdam, 1983.

17. G. Reggio, E. Astesiano, and C. Choppy. CASL-LTL : A CASL Extension for Dynamic Reactive Systems Version 1.0– Summary. Technical Report DISI-TR-03-36, DISI – Università di Genova, Italy, 2003.
    Available at `ftp://ftp.disi.unige.it/person/ReggioG/ReggioEtAll03b.ps`.

18. G. Reggio, E. Astesiano, C. Choppy, and H. Hussmann. Analysing UML Active Classes and Associated State Machines – A Lightweight Formal Approach. In T. Maibaum, editor, *Proc. FASE 2000*, number 1783 in Lecture Notes in Computer Science. Springer Verlag, Berlin, 2000.

19. G. Reggio, M. Cerioli, and E. Astesiano. Towards a Rigorous Semantics of UML Supporting its Multiview Approach. In H. Hussmann, editor, *Proc. FASE 2001*, number 2029 in Lecture Notes in Computer Science. Springer Verlag, Berlin, 2001.

20. G. Reggio and M. Larosa. A Graphic Notation for Formal Specifications of Dynamic Systems. In J. Fitzgerald and C.B. Jones, editors, *Proc. FME 97 - Industrial Applications and Strengthened Foundations of Formal Methods*, number 1313 in Lecture Notes in Computer Science. Springer Verlag, Berlin, 1997.

Algebraic Properties of Interfaces

Michael Löwe, Harald König, and Christoph Schulz

Fachhochschule für die Wirtschaft (FHDW),
Hannover, Germany
michael.loewe@fhdw.de

Abstract. Interfaces became one of the most important features in modern software systems, especially in object-oriented systems. They provide for abstraction and detail hiding and, therefore, contribute to readable and reusable software construction. Formal specification methods have also addressed interfaces and provided formal semantics to them. These semantics are always based on some forgetful and/or restriction constructions. The main focus has been laid on integrating these constructions with free or behaviourable constructions in order to provide some combined semantics for specification modules.

In this paper, we investigate the algebraic properties of "forgetting" and "restricting". We define two different notions of model category for algebraic specification with explicit interface signatures, so-called interface specifications. The first one uses forgetful constructions only, the second one integrates the restriction to the part reachable by the interface.

We investigate closure properties of these categories w.r.t. subalgebras, products and directed limits. This analysis provides the first result, namely that interface specifications with Horn-axioms do not exceed the expressiveness of Horn-axiom specifications without interfaces. The inverse is more interesting. We show that each category specified by Horn-axioms can be specified by simple equations on possibly hidden operations. The constructive proof for this theorem leads to the next main result that every class specified by an interface specification with restriction semantics possesses an interface specification using forgetful semantics only.

1 Introduction and Preliminaries

Since the early days of algebraic specification research (see [4, 6]), it is well-known that some specification problems cannot be solved using a finite number of equations or Horn-axioms only ([2, 8, 9, 11]). There is no way to avoid some hidden operators for a finite solution. These results gave rise to a theory of implementation relations between abstract data types ([1]). The main mechanism here is a forgetful and/or restriction semantics in order to semantically model the realisation of target or export data types by some source or import data types. In these frameworks, the sorts and operations of the source types are considered hidden. In [3], specification and implementation methods are combined within one concept: algebraic module specifications. Semantics of these models consist of free constructions together with some forgetful and restriction steps. Algebraic

H.-J. Kreowski et al. (Eds.): Formal Methods (Ehrig Festschrift), LNCS 3393, pp. 190–203, 2005.

module specifications are able to provide semantics to access control concepts in object oriented languages as they are defined by "private"- or "protected"-keywords in Java ([5]) or UML ([10]) for example.

In this paper, we investigate forgetful and restriction semantics, as well. But we use a somewhat different approach. We are not interested in point-wise construction of export models from import models as it is done in [3]. We put the main emphasis on categories of algebras which are designed as model classes for algebraic specification with explicit export interfaces, so-called interface specifications. We show that, by the right choice of these model classes, data generating, forgetful, and restriction steps can be combined into a single free construction. Thus, the semantics of algebraic data types with hidden operations can be presented as a simple free construction, which possesses the well-known categorical compositional properties [7]. The paper is organised as follows.

The section "Interface Specifications" introduces syntax and semantics for algebraic specifications with explicit interfaces. We provide two kinds of semantics: forgetful semantics and restriction semantics. We show that forgetful model categories are not closed with respect to subalgebras. Therefore, they are no candidate for initial semantics. Restriction semantics is derived from forgetful semantics just by repairing this deficiency.

The section "Closure Properties" investigates algebraic closure properties of model classes with restriction semantics in the spirit of [12]. We address closure w.r.t. products, directed limits, and subalgebras using Horn-axioms as formulas[1]. The results of this section characterise restriction classes as classes which can be specified by Horn-axioms without hidden operators. They also guarantee the existence of initial semantics, which are explicitly constructed in section "Free Construction". Here the name restriction semantics is justified by showing that free constructions into restriction classes are composed of term-generating, equivalence factorisation, forgetful, and restriction steps.

Section "Interface Specifications with Equations" investigates the subset of interface specifications with restriction semantics where all axioms are equations. We show that for each Horn-specification without hidden operators there is one with hidden operators using equations only. The constructive proof provides another interesting corollary, namely that for each restriction class, there is an interface specification with forgetful semantics.

The "Summary and Outlook"-section provides an overview of all results contained in this paper and an outlook on parametric interface specifications as a future research topic.

We use the following notions and notations. A *signature* $SIG = (S, OP)$ consists of a set S of sorts and a family $OP = (OP_{w,s})_{w \in S^*, s \in S}$ of sets of operator symbols. With $op : s_1 s_2 \cdots s_n \to s$ we denote domain and codomain for operators $op \in OP_{s_1 s_2 \cdots s_n, s}$.

---

[1] We restrict ourselves to Horn-axioms, since they are the most expressive type of formulas that (1) are finite in all components and (2) provide an iterative access to the specified equivalence. Hence, they are the right choice on the design level of software systems, which is addressed by interface specifications, too.

$SIG = (S, OP)$ is a *subsignature* of $SIG' = (S', OP')$ (written $SIG \subseteq SIG'$), iff $S \subseteq S'$ and $OP_{w,s} \subseteq OP'_{w,s}$ for all $w \in S^*, s \in S$. The category of all algebras w.r.t. a signature SIG is denoted by $Alg(SIG)$. If $SIG \subseteq SIG'$ we obtain a functor $()_{SIG} : Alg(SIG') \to Alg(SIG)$, the so-called *forgetful functor*, where $((A)_{SIG})_s = A_s$ for each sort in SIG and $op^{(A)_{SIG}} = op^A$ for each operator in SIG. The application of this functor to an algebra A is denoted by $(A)_{SIG}$, the application to a homomorphism f by $(f)_{SIG}$. Whenever we mention categories of algebras, we assume them to be *full* and *abstract*, i.e. they contain all homomorphisms between objects, and with every object each isomorphic object is contained.

T_{SIG} means the *term algebra* w.r.t. a signature, $T_{SIG}(X)$ denotes the term algebra over some variable set $X = (X_s)_{s \in S}$, being a sort-indexed family of sets. Note, that the carrier sets of every SIG-algebra form a variable set.

A *Horn-axiom* $h = (X, P, e)$ consists of a finite variable set X (i.e. $\biguplus_{s \in S} X_s$ is finite), a finite set P of equations over X (premisses) and one equation e over X (conclusion). An *equation* (t_l, t_r) over X is a pair of terms $t_l, t_r \in (T_{SIG}(X))_s$ for some sort s. If P is empty, h reduces to an equation. Satisfaction is as usual. A *specification* $SPEC = (SIG, H)$ consists of a signature SIG and a set of Horn-axioms H. $Alg(SPEC)$ denotes the subcategory of those algebras in $Alg(SIG)$ that satisfy all axioms. Q_{SPEC} denotes (up to an isomorphism) the initial algebra in $Alg(SPEC)$.

A partially ordered set where every finite subset has an upper bound is called a *directed index set*. Let I be such a set. A *directed family of SIG-homomorphisms* on a family $(A_i)_{i \in I}$ of SIG-algebras is a familiy $\mathcal{H} = (h_{ij} : A_i \to A_j)_{i \leq j, i, j \in I}$ of homomorphisms with the following properties:

1. $h_{ii} = id_{A_i}$ for each $i \in I$
2. $h_{jk} \circ h_{ij} = h_{ik}$ for all $i, j, k \in I$ with $i \leq j \leq k$

We call $\vec{\mathcal{H}}$ with

$$(\vec{\mathcal{H}})_s = (\bigcup_{i \in I} (A_i)_s)/\equiv$$

the *directed limit*, where for any $x \in A_i, y \in A_j$: $x \equiv y$ iff there is $k \in I$ such that $h_{ik}(x) = h_{jk}(y)$. Moreover,

$$op^{\vec{\mathcal{H}}}([x_1], \ldots, [x_n]) = [op^{A_k}(h_{i_1 k}(x_1), \ldots, h_{i_n k}(x_n))] \tag{1}$$

for an upper bound k of $\{i_1, \ldots, i_n\}$ and $x_j \in A_{i_j}$.

2 Interface Specifications

Here we introduce the syntactical and semantical framework for specifications with auxiliary operators. Such operators do not occur in the description of the problem itself, but can provide additional structure that facilitates the formulation of the intended properties. Therefore we will distinguish between a public part and a *design* part which does not belong to the data type but is used internally. In the sequel we will refer to the public part as the *interface signature*.

Definition 1 (Interface specification). *An interface specification (a specification with auxiliary sorts and operators, resp.) $ISPEC = (SIG, DSPEC)$ consists of*

1. *a specification $DSPEC = (DSIG, H)$, where $DSIG = (DS, DOP)$ is the design signature and H is a set of Horn-axioms w.r.t. $DSIG$ and*
2. *a subsignature SIG of $DSIG$, called the interface[2].*

Interface specifications are written in the form $ISPEC = (S, OP, DS, DOP, H)$ with subset relations between the sort sets and operation sets or in one of the forms $ISPEC = (SIG, DSIG, H) = (SIG, DSPEC)$ dependent on what fits best our requirements. Moreover, in examples, we use the keywords *sorts*, *opns* for enumerating interface sorts and operator symbols, *Dsorts*, *Dopns* for declaring additional design sorts and operator symbols and *axioms* for Horn-axioms. If all these Horn-axioms reduce to equations, we write *eqns*.

The next definition introduces two alternatives for the category of all algebras that satisfy an interface specification. The question, what signature are these algebras of, can easily be answered: we want to describe algebras that are conform to the interface and not to the (hidden) design signature. Hence, we have

Definition 2 (Satisfaction). *A SIG-algebra A satisfies an interface specification $ISPEC = (SIG, DSPEC)$ if*

1. *there is $B \in Alg(DSPEC)$ such that $A \cong (B)_{SIG}$ (Forgetful Semantics).*
2. *there is $B \in Alg(DSPEC)$ such that $A \rightarrowtail (B)_{SIG}$ (Restriction Semantics).*

Here $C \cong D$ ($C \rightarrowtail D$) denotes the fact, that there is an isomorphism (a monomorphism) between two SIG-algebras C and D.

We use the abbreviations $Alg^F(ISPEC)$ or $Alg^R(ISPEC)$ for the category of all $ISPEC$-algebras with regard to forgetful semantics or restriction semantics, resp. Thus, there are two different possibilities for A to satisfy an interface specification. We deduce from the definition, that

$$Alg^F(ISPEC) \subseteq Alg^R(ISPEC). \qquad (2)$$

In general, we can replace $\subseteq$ by $\subset$ as is shown in the following examples.

Example 1. Consider the following interface specification:
$NEMPTY =$

$$\text{sorts} \quad S$$
$$\text{Dopns } c: \ \rightarrow S$$

Because the first alternative in Definition 2 requires an isomorphism, every $A \in Alg^F(ISPEC)$ contains at least one element in its carrier set. But the subalgebra $A_s = \emptyset$ is a SIG-algebra in $Alg^R(ISPEC)$. Hence, the subset relation is strict in (2) for $NEMPTY$ and we also deduce that, in general, $Alg^F(ISPEC)$ is not closed under subalgebras.

---

[2] The integration of the interface into the design signature could also be provided using general signature morphisms. In this contribution we restrict ourselves to inclusions but keep in mind that the following theory could also be presented in the general case.

Example 2. Consider the specification
$YOYO =$

$$
\begin{array}{lll}
\text{sorts} & S, T \\
\text{opns} & a, b : & \to S \\
& c, d : & \to T \\
\text{Dopns } f : & S \to T \\
& g : & T \to S \\
\text{eqns} & f(a) & = c \\
& f(b) & = c \\
& f(g(x)) = d
\end{array}
$$

Since f is constant and equal to c on the interface generated terms, any $DSPEC$-algebra needs a third element in the carrier set of S if $c \neq d$. So a minimal $B \in Alg^F(YOYO)$ is given by $B_S = \{a, b, z\}, B_T = \{c, d\}$. Indeed it is the forgetful image of the $DSPEC$-algebra C with $f^C = \{(a, c), (b, c), (z, d)\}$ and $g^C = \{(c, z), (d, z)\}$. But $T_{SIG}(\emptyset) \notin Alg^F(YOYO)$ because it lacks this third element.

Both cases underline that, in general, $Alg^F(ISPEC) \neq Alg^R(ISPEC)$ and that, using forgetful semantics, it might happen that hidden operations produce junk elements from the public point of view. Thus, forgetful semantics are not appropriate for initial semantics which is summarised by the following fact.

Proposition 1 (Main weakness of forgetful semantics). *In general, the inclusion functor* $I : Alg^F(ISPEC) \to Alg(SIG)$ *does not possess a left adjoint functor.*

Proof. We can easily deduce a counter example from Example 1. Assume that there is a free construction for $A \in Alg(SIG)$ with $A_S = \emptyset$. Then, the universal morphism $u = \emptyset$. Because $F(A)_S \neq \emptyset$, u cannot be an epimorphism, contradicting the universality of u. □

Restriction semantics does not possess these weaknesses as we will show in the next section. Moreover, we can point out another strength of these semantics: If $B \in Alg(DSPEC)$ with $A \rightarrowtail B$, we call B an *implementation* of A. The next proposition shows that the free construction $G : Alg(SIG) \to Alg(DSPEC)$ considered as left adjoint of the forgetful functor is consistent (in the sense of [11]) on $Alg^R(SIG, DSPEC)$.

Proposition 2. *Let* $A \in Alg^R(SIG, DSPEC)$ *and* $G : Alg(SIG) \to Alg(DSPEC)$ *be the free construction w.r.t. the forgetful functor* $Alg(DSPEC) \to Alg(SIG)$ *then* $G(A)$ *can be considered to be a* standard *implementation of* A, *i.e.* $A \rightarrowtail (G(A))_{SIG}$.

Proof. Let $B \in Alg(DSPEC)$, $i : A \to (B)_{SIG}$ be a monomorphism and let $u : A \to (G(A))_{SIG}$ denote the universal homomorphism. Hence, i extends to a unique homomorphism $i^* : G(A) \to B$ with $(i^*)_{SIG} \circ u = i$ on A. The injectivity of i forces u to be injective on A which yields the desired property. □

3   Closure Properties

The results of this section will provide sufficient conditions to classify the category $Alg^R(ISPEC)$. It turns out that hidden sorts and operations do not improve specification accuracy. A direct consequence of Definition 2 is

Proposition 3. *The category $Alg^R(ISPEC)$ is closed under subalgebras, i.e. if $A \in Alg^R(ISPEC)$ and $B \subseteq A$, then $B \in Alg^R(ISPEC)$.*   □

We also obtain

Proposition 4. *The category $Alg^R(ISPEC)$ is closed under products, i.e. if I is an index set, then*

$$A_i \in Alg^R(ISPEC) \text{ for all } i \in I \Rightarrow \prod_{i \in I} A_i \in Alg^R(ISPEC)$$

and

Proposition 5. *The category $Alg^R(ISPEC)$ is closed under directed limits, i.e. if I is a directed index set, $(A_i)_{i \in I}$ is a family of ISPEC-algebras, and $\mathcal{H} = (h_{ij} : A_i \to A_j)_{i \leq j \in I}$ is a directed family of ISPEC-homomorphisms, then $\vec{\mathcal{H}} \in Alg^R(ISPEC)$.*

For the proofs of Propositions 4 and 5 we need the following intermediate results.

Lemma 1. *Let $SIG \subseteq SIG'$ be two signatures.*

1. *If I is an index set and $\mathcal{A} = (A_i)_{i \in I}$ is a family of SIG'-algebras, then $(\prod \mathcal{A})_{SIG} = \prod_{i \in I}(A_i)_{SIG}$.*
2. *If I is an index set and $\mathcal{A} = (A_i)_{i \in I}, \mathcal{B} = (B_i)_{i \in I}$ are two families of SIG-algebras with $A_i \rightarrowtail B_i$ for all $i \in I$, then $\prod \mathcal{A} \rightarrowtail \prod \mathcal{B}$.*

Proof. 1 follows directly from the fact, that $()_{SIG}$ is a right adjoint functor and therefore preserves limits. To show 2, let $p_i : \prod \mathcal{A} \to A_i$ and $p'_i : \prod \mathcal{B} \to B_i$ be the projections and $m_i : A_i \to B_i$ monomorphisms for all $i \in I$. The product definition (for $\prod \mathcal{B}$) yields a unique homomorphism $h : \prod \mathcal{A} \to \prod \mathcal{B}$ such that

$$p'_i \circ h = m_i \circ p_i \text{ for all } i \in I. \tag{3}$$

We claim that h is a monomorphism. Indeed, let $f, g : X \to \prod \mathcal{A}$ be two arrows with $h \circ f = h \circ g$. Then $p'_i \circ h \circ f = p'_i \circ h \circ g$. By (3) we obtain

$$m_i \circ p_i \circ f = m_i \circ p_i \circ g \text{ for all } i \in I.$$

Because m_i is a monomorphism, this gives $p_i \circ f = p_i \circ g$ for all $i \in I$ and thus, by the universality of $\prod \mathcal{A}$: $f = g$.   □

Lemma 2. *Let $SIG \subseteq SIG'$ be two signatures.*

1. *If I is a directed index set, and $\mathcal{H} = (h_{ij} : A_i \to A_j)_{i \leq j \in I}$ is a directed family of of SIG'-homomorphisms, then $(\vec{\mathcal{H}})_{SIG} = \vec{\mathcal{H}_{SIG}}$ where $\mathcal{H}_{SIG} = ((h_{ij})_{SIG} : (A_i)_{SIG} \to (A_j)_{SIG})_{i \leq j \in I}$.*

2. *If I is a directed index set, and $\mathcal{H} = (h_{ij} : A_i \to A_j)_{i \leq j \in I}, \mathcal{K} = (k_{ij} : B_i \to B_j)_{i \leq j \in I}$ are two directed families of SIG-homomorphisms with $A_i \rightarrowtail B_i$ for all $i \in I$ via monomorphisms $m_i : A_i \to B_i$ and*

$$m_j \circ h_{ij} = k_{ij} \circ m_i \text{ for all } i \leq j \in I \tag{4}$$

then $\vec{\mathcal{H}} \rightarrowtail \vec{\mathcal{K}}$.

Proof. To derive 1 we observe that for each sort s in SIG and every homomorphism $((h_{ij})_{SIG})_s = (h_{ij})_s$. Or, equally, $\equiv' = \equiv$ on $(A_i)_s \times (A_j)_s$ where $\equiv$ is the SIG-congruence w.r.t. $((h_{ij})_{SIG})_{i \leq j \in I}$ and $\equiv'$ is the SIG'-congruence w.r.t. $(h_{ij})_{i \leq j \in I}$. Thus

$$((\vec{\mathcal{H}})_{SIG})_s = (\vec{\mathcal{H}})_s = (\bigcup_{i \in I}(A_i)_s)/_{\equiv'} = (\bigcup_{i \in I}((A_i)_{SIG})_s)/_{\equiv} = (\vec{\mathcal{H}_{SIG}})_s.$$

Using these arguments again, one easily derives $op^{(\vec{\mathcal{H}})_{SIG}} = op^{\vec{\mathcal{H}_{SIG}}}$ for all operator symbols op in SIG.

To show 2 we observe that the congruences $\equiv_h$ in $\vec{\mathcal{H}}$ and $\equiv_k$ in $\vec{\mathcal{K}}$ are coupled via the monomorphisms as follows: let $x \in A_i, y \in A_j$ and $x \equiv_h y$. Then, for some l: $m_l(h_{il}(x)) = m_l(h_{jl}(y))$. Using (4) and the fact that m_l is a monomorphism, we obtain for all $i, j \in I$:

$$x \equiv_h y \iff m_i(x) \equiv_k m_j(y) \text{ for all } x \in A_i, y \in A_j. \tag{5}$$

From (5) we easily deduce that the mapping $m : \vec{\mathcal{H}} \to \vec{\mathcal{K}}$ defined by

$$m([x]_{\equiv_h}) = [m_i(x)]_{\equiv_k}, \tag{6}$$

where $x \in A_i$, is well-defined and injective. We will show, that m is a homomorphism which will yield $\vec{\mathcal{H}} \rightarrowtail \vec{\mathcal{K}}$. Let $op : s_1 \cdots s_n \to s$ be an operator in SIG and $x_1 \in A_{i_1}, \ldots, x_n \in A_{i_n}$. If l is an upper bound of $i_1, \ldots, i_n$, then

$$op^{\vec{\mathcal{K}}}(m([x_1]_{\equiv_h}), \ldots, m([x_n]_{\equiv_h}))$$
$$\overset{(6)}{=} op^{\vec{\mathcal{K}}}([m_{i_1}(x_1)]_{\equiv_k}, \ldots, [m_{i_n}(x_n)]_{\equiv_k})$$
$$\overset{(1)}{=} [op^{B_l}(k_{i_1 l}(m_{i_1}(x_1)), \ldots, k_{i_n l}(m_{i_n}(x_n)))]_{\equiv_k}$$
$$\overset{(4)}{=} [op^{B_l}(m_l(h_{i_1 l}(x_1)), \ldots, m_l(h_{i_n l}(x_n)))]_{\equiv_k} \tag{7}$$

Since m_l is a homomorphism,

$$[op^{B_l}(m_l(h_{i_1 l}(x_1)), \ldots, m_l(h_{i_1 n}(x_n)))]_{\equiv_k}$$
$$= [m_l(op^{A_l}(h_{i_1 l}(x_1), \ldots, h_{i_n l}(x_n)))]_{\equiv_k}. \tag{8}$$

Moreover,

$$[m_l(op^{A_l}(h_{i_1 l}(x_1), \ldots, h_{i_n l}(x_n)))]_{\equiv_k}$$
$$\overset{(6)}{=} m([op^{A_l}(h_{i_1 l}(x_1), \ldots, h_{i_n l}(x_n))]_{\equiv_h})$$
$$\overset{(1)}{=} m(op^{\vec{\mathcal{H}}}([x_1]_{\equiv_h}, \ldots, [x_n]_{\equiv_h})). \tag{9}$$

Combining (7), (8), and (9) we obtain the desired result. $\qquad \square$

Proof (of Proposition 4). Let $(B_i)_{i \in I}$ be a family of $DSPEC$-algebras such that $A_i \rightarrowtail (B_i)_{SIG}$ for each $i \in I$. From Lemma 1, 2 we deduce that

$$\prod_{i \in I} A_i \rightarrowtail \prod_{i \in I} (B_i)_{SIG}. \tag{10}$$

By Lemma 1, 1 the right hand side of (10) equals $(\prod_{i \in I} B_i)_{SIG}$. Since implicational classes are closed under products (see [12]) we deduce that $\prod_{i \in I} B_i \in Alg(DSPEC)$ which gives $\prod_{i \in I} A_i \in Alg^R(ISPEC)$. $\qquad \square$

Proof (of Proposition 5). Let $G : Alg(SIG) \rightarrow Alg(DSPEC)$ be the free functor from Proposition 2. Thus, the family

$$u = (u_{A_i} : A_i \rightarrowtail (G(A_i))_{SIG})_{i \in I}$$

provides a natural transformation $u : Id_{Alg(SIG)} \Rightarrow (G)_{SIG}$ where the u_{A_i} are monomorphisms. Thus

$$(G(h_{ij}))_{SIG} \circ u_{A_i} = u_{A_j} \circ h_{ij}. \tag{11}$$

Moreover, the functor property of G shows that $(G(h_{ij}) : G(A_i) \rightarrow G(A_j))_{i \leq j \in I}$ is a directed familiy. Let $G(\vec{\mathcal{H}})$ be its directed limit. Then from Lemma 2, 2 together with (11) and Lemma 2, 1 we obtain

$$\vec{\mathcal{H}} \rightarrowtail G(\vec{\mathcal{H}})_{SIG} = (G(\vec{\mathcal{H}}))_{SIG}.$$

Since implicational classes are closed under directed limits (see [12]), we deduce that $G(\vec{\mathcal{H}}) \in Alg(DSPEC)$ and thus $\vec{\mathcal{H}} \in Alg^R(ISPEC)$. $\qquad \square$

It is shown in [12], Section 3.3, Theorem 25, that a necessary and sufficient condition for a class of algebras to possess a specification with Horn-axioms is that this class forms a quasi-variety, i.e. it is closed w.r.t. subalgebras (Proposition 3), products (Proposition 4), and directed limits (Proposition 5). Thus, the results of this chapter imply

Theorem 1 (Characterisation of restriction classes). *If C is a full and abstract subcategory of $Alg(SIG)$, then there is an interface specification $ISPEC$ for C, i.e. $C = Alg^R(ISPEC)$ if and only if C can be specified by Horn-axioms H, i.e. $C = Alg(SIG, H)$.* $\qquad \square$

4   Free Construction

Since $Alg^R(ISPEC)$ is closed w.r.t. subalgebras (Proposition 3) and products (Proposition 4), from [12], Section 3.3, Theorem 13 we obtain another important result.

Corollary 1 (Existence of free constructions). *Let $ISPEC$ be an interface specification then the inclusion functor $I : Alg^R(ISPEC) \rightarrow Alg(SIG)$ possesses a left adjoint.* $\qquad \square$

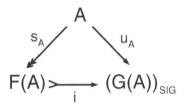

Fig. 1. Free construction w.r.t. inclusion.

In this chapter we will provide the explicit construction of this left adjoint. Let G be the standard implementation from Proposition 2. We will show that the free object for an algebra $A \in Alg(SIG)$ w.r.t. the inclusion $I : Alg^R(ISPEC) \to Alg(SIG)$ is the surjective part in the epi/mono-factorisation of the universal morphism $u_A : A \to I(G(A))$. The epi/mono-factorisation can be interpreted as a restriction step and, together with the well-known construction of G, completes the free construction into restriction classes as mentioned in chapter 1.

Theorem 2. *Let $ISPEC$ be an interface specification and $A \in Alg(SIG)$. If $u_A : A \to (G(A))_{SIG}$ is the universal morphism of the free construction $G : Alg(SIG) \to Alg(DSPEC)$ and $u_A = i \circ s_A$ is its epi/mono-factorisation (compare Figure 1), then $(F(A), s_A)$ is the free construction w.r.t. the inclusion functor $I : Alg^R(ISPEC) \to Alg(SIG)$.*

Proof. Obviously, $F(A) \in Alg^R(ISPEC)$ (because i is a monomorphism in Fig. 1). Let $B \in Alg^R(ISPEC)$ and $f : A \to B$ be a homomorphism. We have to show universality of the surjective homomorphism $s_A : A \to F(A)$. Since $B \in Alg^R(ISPEC)$ there is $C \in Alg(DSPEC)$ and a SIG-monomorphism $m : B \to (C)_{SIG}$. The universality of $u_A : A \to (G(A))_{SIG}$ provides a unique homomorphism $F^* : G(A) \to C$ such that

$$(F^*)_{SIG} \circ u_A = m \circ f \tag{12}$$

holds in $Alg(SIG)$. We define $f^* : F(A) \to B$ as

$$f^* = m^{-1} \circ (F^*)_{SIG} \circ i$$

and claim that this is a well-defined homomorphism and the unique extension of f. First of all we use the fact that for each $y \in F(A)$ there is $x \in A$ with $i(y) = u_A(x)$. Thus by (12)

$$(F^*)_{SIG}(i(y)) = (F^*)_{SIG}(u_A(x)) \in m(B),$$

which explains well-definedness. Moreover, using (12) a second time and the definition of f^*, we observe that

$$f^* \circ s_A = f \tag{13}$$

holds in A. Uniqueness of f^* among all arrows that fulfill (13) easily follows from the fact that s_A is an epimorphism. $\qquad\square$

Corollary 2 (Initial objects). *Let $ISPEC = (SIG, DSPEC)$ be an interface specification. Then the category $Alg^R(ISPEC)$ contains an initial object.*

Proof. Each object $F(T_{SIG})$ is initial in $Alg^R(ISPEC)$ because T_{SIG} is initial in $Alg(SIG)$ and free functors preserve initial objects. □

5   Interface Specifications with Equations

Theorem 1 shows that hidden operators do not add any specification power if Horn-axioms are used. We get a different situation if we restrict ourselves to equations, as the following example demonstrates.

Example 3. Consider the specification
$HIDINJ =$

$$
\begin{array}{lll}
\text{sorts} & S, T & \\
\text{opns} & c: & \to S \\
& f: & S \to T \\
\text{Dopns } g: & & T \to S \\
\text{eqns} & g(f(x)) = x &
\end{array}
$$

The hidden injectivity of f in every $DSPEC$-algebra is transported to every subalgebra and hence f must be injective in every $A \in Alg^R(HIDINJ)$. We consider the Algebra $I \in Alg^R(HIDINJ) : I_S = I_T = \{0,1\}, c^I = 0, f^I = id$. Constructing $A/\equiv$ where $\equiv$ is the equality on I_S and $I_T \times I_T$ on I_T, we obtain the surjective canonical homomorphism $[.] : A \to A/\equiv$. Thus, $A/\equiv$ is a homomorphism of A with non-injective f, since

$$
f_S^{A/\equiv}([0]_{\equiv_S}) = [0]_{\equiv_T} = [1]_{\equiv_T} = f_S^{A/\equiv}([1]_{\equiv_S}), \text{ but } [0]_{\equiv_S} \neq [1]_{\equiv_S}.
$$

Hence, $Alg^R(HIDINJ)$ is not closed under homomorphic images. □

Since all equational classes are closed w.r.t. homomorphic images (Birkhoff characterisation), Example 3 demonstrates that interface specifications with equations do not possess an equational specification in general. But the next theorem shows that hidden operators and equations are sufficient specification tools for any Horn-axiom specification.

The major step of the proof is the transformation of a Horn-axiom specification into an equational specification with hidden operators and was inspired by various examples of [11].

Theorem 3 (Characterisation of implicational classes). *If C is a full and abstract subcategory of $Alg(SIG)$, then there is an interface specification $ISPEC$ for C where all axioms are equations, i.e. $C = Alg^R(ISPEC)$ with $ISPEC = (SIG, DSIG, E)$, if and only if C can be specified by Horn-axioms H, i.e. $C = Alg(SIG, H)$.*

Proof. The first direction of the proof is shown easily, since Theorem 1 already stated that for each interface specification $ISPEC$, there exists a Horn-axiom specification H such that $Alg^R(ISPEC) = Alg(S, OP, H)$. This is obviously also true if the design axioms are equations only.

For the second direction, we observe that if (S, OP, H) is a specification with Horn-axioms, each axiom $h \in H$ has the form

$$h = (x_1 : s_{x_1}, \dots, x_n : s_{x_n}; t_1 = u_1, \dots, t_m = u_m \Rightarrow t_{m+1} = u_{m+1}) \qquad (14)$$

Assuming the sort of the equations $t_i = u_i$ is s_i for $i = 1, \dots, m+1$, we can define the following interface specification:
$ISPEC(H) =$

sorts	$(s)_{s \in S}$
opns	$(op)_{op \in OP}$
Dopns	$(h : s_{x_1} \cdots s_{x_n} s_1 s_1 \cdots s_m s_m s_{m+1} \to s_{m+1})_{h \in H}$

$$\text{eqns} \quad \left(\begin{array}{l} (e_{1,h}) \quad x_1 : s_{x_1}, \dots, x_n : s_{x_n}; \\ \qquad h(x_1, \dots, x_n, t_1, u_1, t_2, u_2, \dots, t_m, u_m, t_{m+1}) = u_{m+1} \\ (e_{2,h}) \quad x_1 : s_{x_1}, \dots, x_n : s_{x_n}, y_1 : s_1, \dots, y_m : s_m, z : s_{m+1}; \\ \qquad h(x_1, \dots, x_n, y_1, y_1, y_2, y_2, \dots, y_m, y_m, z) = z \end{array} \right)_{h \in H}$$

We are now going to prove $Alg^R(ISPEC(H)) = Alg(S, OP, H)$ by showing

$$Alg^R(ISPEC(H)) \subseteq Alg(S, OP, H) \qquad (15)$$

and

$$Alg(S, OP, H) \subseteq Alg^R(ISPEC(H)). \qquad (16)$$

To derive (15), we assume $A \in Alg^R(ISPEC(H))$. We want to show $A \models h$ for all $h \in H$. Let $ass : X \to A$ be an assignment for the set of variables $X = \{x_1, \dots, x_n\}$ of h in A, such that

$$\overline{ass}(t_i) = \overline{ass}(u_i) \qquad (17)$$

for all $i = 1, \dots, m$. Because $A \in Alg^R(ISPEC(H))$, there exists an algebra C for the design part of $ISPEC(H)$, such that there is a monomorphism $i : A \rightarrowtail (C)_{(S,OP)}$. Taking this monomorphism as a family of mappings, we obtain an assignment $ass_C = i \circ ass : X \to C$. For the extension $\overline{i \circ ass} : T_{SIG}(X) \to C$, we have

$$\overline{i \circ ass} = i \circ \overline{ass}. \qquad (18)$$

But C satisfies the equations, i.e.,

$$C \models (e_{1,h}) \qquad (19)$$

$$C \models (e_{2,h}) \qquad (20)$$

so we can conclude:

$$i \circ \overline{ass}(u_{m+1})$$
$$\overset{(18)}{=} \overline{ass}_C(u_{m+1})$$
$$\overset{(19)}{=} \overline{ass}_C\big(h(x_1,\ldots,x_n,t_1,u_1,\ldots,t_m,u_m,t_{m+1})\big)$$
$$= h^C\big(\overline{ass}_C(x_1),\ldots,\overline{ass}_C(x_n),\overline{ass}_C(t_1),\overline{ass}_C(u_1),\ldots,\overline{ass}_C(t_{m+1})\big)$$
$$\overset{(17)}{=} h^C\big(\overline{ass}_C(x_1),\ldots,\overline{ass}_C(x_n),\overline{ass}_C(t_1),\overline{ass}_C(t_1),\ldots,\overline{ass}_C(t_{m+1})\big)$$
$$= \overline{ass}_C\big(h(x_1,\ldots,x_n,t_1,t_1,\ldots,t_m,t_m,t_{m+1})\big)$$
$$\overset{(20)}{=} \overline{ass}_C(t_{m+1})$$
$$\overset{(18)}{=} i \circ \overline{ass}(t_{m+1})$$

Since i is a monomorphism, we obtain $\overline{ass}(u_{m+1}) = \overline{ass}(t_{m+1})$.

To show (16), we assume $A \in Alg(S,OP,H)$. Due to Definition 2, it is sufficient to construct an algebra $C \in Alg(DSIG,E)$, so that $(C)_{SIG} \supseteq A$. The construction is done as follows:

1. $C_s = A_s$ for all sorts $s \in S$
2. $op^C = op^A$ for all operators $op \in OP$
3. For each design operator $h \in DOP$, we define a corresponding operation $h^C : C_{s_{x_1}} \times \cdots \times C_{s_{x_n}} \times C_{s_1} \times C_{s_1} \cdots \times C_{s_m} \times C_{s_m} \times C_{s_{m+1}} \to C_{s_{m+1}}$ such that $h^C(c_1,\ldots,c_n,y_1,z_1,\ldots,y_m,z_m,x) =$

$$\begin{cases} x & \text{if } y_i = z_i \text{ for all } i = 1,\ldots,m \\ \overline{ass}(u_{m+1}) & \text{if } y_i = \overline{ass}(t_i) \text{ and } z_i = \overline{ass}(u_i) \text{ for all } i = 1,\ldots,m, \text{ and} \\ & x = \overline{ass}(t_{m+1}), \text{ where } ass: X_h \to C \text{ is defined as } ass(x_j) = \\ & c_j \text{ for all } j = 1,\ldots,n. \\ c \in C_{s_{m+1}} & \text{otherwise} \end{cases}$$

In the last of the three cases we could also have chosen x instead of an arbitrary $c \in C_{s_{m+1}}$, because $x \in C_{s_{m+1}}$ guarantees $C_{s_{m+1}} \neq \emptyset$.

The operations h^C are well-defined, even if both the first and second condition are met simultaneously. In such a case, we obtain for all $i = 1,\ldots,m$: $y_i = z_i$, $y_i = \overline{ass}(t_i)$, $z_i = \overline{ass}(u_i)$, and $\overline{ass}(t_{m+1}) = x$. From this, we can easily conclude $x = \overline{ass}(t_{m+1}) = \overline{ass}(u_{m+1})$ (since $A \models H$).

Now we have to show that C satisfies the equations of the design specification $ISPEC(H)$. It is obvious that there are two equations for each operation which are to be satisfied. But the operations have been defined deliberately so that these two equations are covered by the first two of the three cases in the definition of h^C. According to the definition of C, we deduce $A = (C)_{(S,OP)}$, completing the proof. $\qquad\square$

Corollary 3 (Forgetful and restriction semantics). *For each category Alg^R (ISPEC), there exists a design specification $ISPEC'$, such that $Alg^F(ISPEC') = Alg^R(ISPEC)$, i.e., each restriction class can be specified by forgetful semantics.*

Proof. Theorem 1 guarantees the existence of a specification with Horn-axioms (SIG, H) for $Alg^R(ISPEC)$, i.e., $Alg^R(ISPEC) = Alg(SIG, H)$. Due to the construction of the proof of Theorem 3, $Alg(SIG, H) = Alg^R(ISPEC(H))$. Additionally, for each $A \in Alg(SIG, H)$, there is a C satisfying the design equations of $ISPEC(H)$, such that $A = (C)_{SIG}$. From this, it follows directly that $Alg(SIG, H) = Alg^F(ISPEC(H))$. □

6   Summary and Outlook

In this paper, we introduced a notion of interface specification which integrates design specifications with explicit interfaces. The corresponding notion of satisfaction – also introduced here – provides a class of algebras for each interfaces specification. In this setting we can compare different specification mechanisms w.r.t. their expressive power. The results in this paper can be summarised as follows.

For each signature SIG, let $SIG(H)$ and $SIG(E)$ be the classes of SIG-algebras that can be specified using Horn-axioms or equations resp. without hidden sorts and operations, let $SIG^F(H)$ and $SIG^F(E)$ be the classes with an interface specification using forgetful semantics, and let $SIG^R(H)$ and $SIG^R(E)$ be the corresponding classes using restriction semantics. Then, by the results presented above, we obtain the following relations:

$$SIG(E) \subset SIG^R(E) = SIG(H) = SIG^R(H) \subset SIG^F(E) \overset{?}{\subseteq} SIG^F(H) \quad (21)$$

The only open question is $SIG^F(E) = SIG^F(H)$.

Most of the results we have achieved do not address finiteness of the specifications. For example, it is not guaranteed that the Horn-axiom specification that exists for a finite interface specification with equations is finite as well. Therefore, it is an interesting issue for future research to investigate the relationships of (21) for classes that can be specified by finite specifications[3] of the different types. One contribution to these questions can be found in the constructive proof of Theorem 3: If we have a finite Horn-axiom specification without hidden sorts and operations for a class C, then there is a finite interface specification for C using restriction semantics and equations only.

A second important result presented here that opens up a new research line is the existence of free constructions for interface specifications with restriction semantics. Theorem 2 demonstrates that this free construction integrates term generating, equivalence factorisation, and restriction steps into a single mechanism that is compositional in itself, since free constructions compose to free constructions.

On the basis of these constructions a module concept comparable to that presented in [3] can be built. A first step is to consider parametric interface specifications where the parameter is a subsignature of the interface. Initial semantics for this type of specifications is guaranteed by Theorem 2. A second step

---

[3] Finiteness means finite in all components.

is to allow interface specifications to formulate restrictions on the admissible parameter algebras. Again, we have free constructions. In a third step, a module concept that considers the implementation level explicitly can be obtained from this situation if we require specification morphisms from the parameter design to the target design, such that the resulting diagram of free functors between the related categories commutes. Now it should be possible to construct an explicit implementation for the target if the parameter comes equipped with an explicit implementation as well. What we get is a parametric specification with a parametric interface whose properties are worthwhile to be investigated by future research.

References

1. Hartmut Ehrig, Hans-Jörg Kreowski, Bernd Mahr, and Peter Padawitz. *Algebraic implementation of abstract data types.* Theoret. Comp. Sci. 20 (1982), 209-263.
2. Hartmut Ehrig and Bernd Mahr. *Fundamentals of Algebraic Specification 1: Equations and Initial Semantics*, volume 6 of *EATCS Monographs on Theoretical Computer Science.* Springer, 1985.
3. Hartmut Ehrig and Bernd Mahr. *Fundamentals of Algebraic Specification 2: Module Specifications and Constraints*, volume 21 of *EATCS Monographs on Theoretical Computer Science.* Springer, 1990.
4. Joseph A. Goguen, James W. Thatcher, and Eric G. Wagner. *An initial algebra approach to the specification, correctness and implementation of abstract data types.* IBM Research Report RC 6487, 1976. Also: Current Trends in Programming Methodology IV: Data Structuring, Prentice Hall (1978), 80-144.
5. James Gosling, Bill Joy, Guy Steele, and Gilad Bracha. *The $Java^{TM}$ Language Specification.* Second edition, Addison-Wesley, 2000.
6. John V. Guttag. *The specification and application to programming of abstract data types.* Ph.D. Thesis, University of Toronto, 1975.
7. Horst Herrlich and George E. Strecker. *Category Theory.* Second edition, Heldermann, 1979.
8. Hans-Jörg Kreowski. Internal communication. TU-Berlin, 1978.
9. Michael Löwe. *Algebra 2.* Technical Lecture Notes, TU-Berlin, 1991 (in german).
10. James Rumbaugh, Ivar Jacobson, and Grady Booch. *The Unified Modeling Language Reference Manual.* Addison-Wesley, 1999.
11. James W. Thatcher, Eric G. Wagner, J.B. Wright. *Data Type Specification. Parametrization and the Power of Specifiction Techniques.* ACM Transactions on Programming Languages and Systems, Vol. 4, No. 4, 1982, 711-732.
12. Wolfgang Wechler. *Universal Algebra for Computer Scientists.* Springer, 1992.

$\in_T$-Integration of Logics

Bernd Mahr and Sebastian Bab

Technische Universität Berlin,
Berlin, Germany
{mahr,bab}@cs.tu-berlin.de

Abstract. $\in_T$-logic was first designed by Werner Sträter as a first-order propositional logic with quantification, reference, and predicates for *true* and *false*. It is motivated by reconstruction of natural language semantics and allows, as a logic with self-reference and impredicativity, among others the treatment of the liar paradox despite the totality of its truth predicates. Its intensional models form a theory of propositions for which a correct and complete calculus is given.

$\in_T$-logic was picked up by Philip Zeitz to study the extension of abstract logics by the concepts of truth, reference and classical negation, thereby rebuilding the meta-level of judgements in a formal level of propositional logic. His parameterized $\in_T$-logic allows formulas from a parameter logic to become the constants in his $\in_T$-logic. Parameter-passing of logics with correct and complete calculus also admits, under certain conditions, the entailment of a calculus which is correct and complete for the extended logic.

Since in parameterized $\in_T$-logic Tarski Biconditionals not only apply for the truth of $\in_T$-logic sentences, but also for the meta-level truth of the parameter logic it is natural to view $\in_T$-logic as a theory of judgements whose propositions are expressed in the parameter logic.

We add a new interpretation to $\in_T$-logic as a theory of truth and judgements, and introduce $\in_T$-logic as a means for the integration of logics. Based on a particular choice of uniform view and treatment of logics we define $\in_T$-logics and $\in_T$-extensions as the foundation for $\in_T$-integration of logics and models.

Studies in $\in_T$-logic, which have started to deal with the difficulties of truth in natural language semantics, have evolved into a concept of logic integration where application oriented logics can be plugged in as parameters. This paper very much relies on the work of Philip Zeitz, but opens it for the new perspective of integration.

1 Introduction

There are various approaches to integration in the design and description of systems. Here integration of logics has the goal to define a logic formalism which admits compound propositions whose components come from different logics. For this task of integration $\in_T$-logic is used as a basis.

H.-J. Kreowski et al. (Eds.): Formal Methods (Ehrig Festschrift), LNCS 3393, pp. 204–219, 2005.
© Springer-Verlag Berlin Heidelberg 2005

1.1   Studies in $\in_T$-Logic

$\in_T$-logic, a theory of truth and propositions, was first defined by Werner Sträter in the context of reconstructing natural language semantics by means of self-referential structures (see [15, 14]). Based on propositional variables and constants formulas of $\in_T$-logic are built from classical propositional connectives together with quantification over propositional variables, propositional equality and predicates for truth and falsity. The resulting logic is intensional in that it has not just truth values but rather sets of propositions as its models. It has been proven free from antinomies despite its total truth-predicates and its ability to model self-referential sentences and impredicative quantification. This is shown in a model existence theorem. Also a correct and complete calculus of $\in_T$-logic is given by Sträter. The concept of $\in_T$-integration defined below is a reinterpretation and particular use of $\in_T$-logic.

Logic reconstruction of natural language semantics shows the need to simultaneously handle various kinds of logics as well as their extension by concepts of truth and reference. This need was taken as a motivation for a generalization of $\in_T$-logic, which admits propositional constants to be formulas from some other logic. The study of this generalization resulted in the definition of the so-called $\in_T$-extension by Philip Zeitz (in [26]). To deal with concrete extensions of arbitrarily given logics Zeitz studies different forms of abstract logics and introduces a particular form, in which the semantics is given by a system of sets, called basis. For extensions of such logics in abstract form Zeitz studies the existence of models and the conditions which allow to extend a correct and complete calculus of the parameter logic so that the resulting calculus is correct and complete for the extended logic. He also discusses basics of a model theory and the existence of intensional propositional interpretations.

$\in_T$-logic has proven to be a suitable concept for truth and reference. It avoids antinomies which necessarily appear with logics having total truth predicates and at the same time allow for representations of decidable relations and computable functions (Tarski, see [24]). In this respect $\in_T$-logic is not a rich language. The concept of $\in_T$-extension however circumvents limitations in expressive power by some kind of stratification in which language features that are rich enough for representations are being separated from their conceptual truth predicates.

1.2   Concepts of Integration

The task of specification has more and more become a task of dealing with the complexity of the system to be specified. Early approaches to cope with complex system structures were based on the concept of modularization, first studied in the context of programming (Parnas, see [18]) and later in the context of abstract data type specification (see Goguen, Tardo [12] and Ehrig, Mahr [4, 5]). Integration was here understood as composition of specifications based on some underlying logic. The need of a variety of logics for specification has motivated the study of abstract concepts of system description in appropriate formal or conceptual frameworks. Initially these frameworks were specification languages based on the categorical concept of institution (see for example Goguen, Burstall [10] and Ehrig, Mahr [5]). They have intensively been studied and have strongly influenced the design of later approaches in programming and specification. While specification languages over arbitrary institutions abstract away the underlying logic from specifications and allow thereby the use of different logics in system description it was still required that specifications in a given language are based on one and the same logic formalism.

Systematic description of systems and systems behaviour in a single formalism, however, has proven inadequate in many cases. In the context of application and technology development it was found that not only different tasks, but also different viewpoints and perspectives on the same object are required to deal with the many aspects involved.

Different tasks and different viewpoints however create the need for integration of different logic formalisms as well as frameworks of conformance and consistency. Computer systems have for a long time offered facilities for integration of programs written in different programming languages. While in the beginning these facilities were placed at the operating system level, using script languages for example, it was object-oriented approaches to system development, that first allowed integration at a higher language level. Middleware technologies like CORBA, J2EE, .NET or Sun.ONE are good examples. Integration is here achieved by service oriented architectures with interface conventions and definition languages.

An early example of specification under multiple views is the ANSA architectural model, a predecessor to the ISO reference model for open distributed processing (ODP, see [8] and Putman [19]). It distinguishes five viewpoints (enterprise, information, computation, engineering, technology), and, like RM-ODP, requires a system to be specified under each of these viewpoints.

In RM-ODP, integration of specification under the different viewpoints is not achieved by mere composition, but by the requirement of additional conformance specifications and activities of conformance assessment in which, among other properties, also consistency between specifications and compliance with standards is to be checked. Another example of integration of specifications under different viewpoints is found in the UML standard (see for example Born, Holz, Kath [1]) which provides several distinct diagrammatic specification formalisms to be applied in the specification of a single object.

Integration always requires concepts of abstraction and substitution. This is true for computer systems, programming languages, specification languages and logics in the same way. Abstraction in logic and other specification formalisms has indeed provided a good basis for the treatment of integration. In specification theory the study of integration beyond composition was first addressed in the merging of process and data type specifications (see Ehrig, Orejas [7] for a recent treatment of the matter) and was later also developed in the context of other formalisms.

General and more abstract approaches to integration have been studied in the framework of categories of institutions, dealing with mappings of logics and institutions (see Meseguer [16] and Goguen, Rosu [11]) and with so-called heterogeneous specifications and the moving between logical systems (see Tarlecki [22,23], Mossakowski [17] and Diaconescu [3]).

1.3   A Scenario for the Integration of Logics

The key feature of the parameterized $\in_T$-logic defined by Zeitz (see [26]) is found in the formal treatment of meta-level formulas which can be read as judgements about sentences from other logics. This feature of parameterized $\in_T$-logic is closely related to the integration of logics through the meta-level of their semantics. To elaborate this idea further we define the so-called $\in_T$-integration as an $\in_T$-logic which integrates families of logics through their meta-level.

Our approach to $\in_T$-integration is motivated by the following general *scenario of integration*:

> *Given a complex object A together with different views $V_1, \ldots, V_n$ on A, each view V_j is expressed in terms of some logic $\mathscr{L}_j$ and can be identified by a model M_j with respect to the logic $\mathscr{L}_j$. What we can say about these models M_j represents our knowledge about the object A. The goal of integration is now to create an integrated model in an appropriate integration logic, so that logical consequence and validity under the particular views are being preserved.*

In $\in_T$-integration, the integration logic, as being required in the scenario, will be an $\in_T$-logic whose parameters are the logics underlying the different views. And the integrated model in the scenario will in $\in_T$-integration be a model of propositions, which is constructed from the models representing the views. In this way $\in_T$-integration meets the integration scenario.

1.4   Goal of this Paper

The goal of this paper is to discuss the idea of reinterpreting and using $\in_T$-logic for the integration of logics. Many questions however will be left open and will be the subject of later studies. Nevertheless we can prove a number of results which show that $\in_T$-logic can well be used for logic integration and which serve as a starting point for further extensions. Among those may be the extension of $\in_T$-logic by modal operators to express propositions about time and possibility.

Even though integration of logics has never been a motivation for their study, it is becoming evident that $\in_T$-logics can well be used for a particular style of integration of languages and formalisms. The idea of integrating logics is not new. Extensive work of similar spirit is published by Gabbay in [9], where a fibred semantics methodology for combining logics and systems is studied. $\in_T$-integration of logics has yet to be analyzed in its relation to fibring logics as exposed in [9] and other approaches to logic integration.

With this paper we want to honor our colleague and friend Hartmut Ehrig by addressing questions in a field that has interested him for many years.

2   Uniform Views and Treatment of Logics

Before the concept of $\in_T$-integration can be defined, a suitable concept is needed which admits a uniform view and treatment of logics. In the study of logics a most natural way to express what is meant by a logic is to define systems $\mathscr{L}$ consisting of a set of formulas L and a proper consequence relation $\Vdash_{\mathscr{L}}$. We do not follow this tradition here, mainly for two reasons: first, because integration of logics is more convenient if logics are in abstract form, and second, because several logics like modal logic for example do not emphasise so strongly logical consequence, rather they emphasize validity in situations, which is closely related to the abstract form of a logic (see section 2.2).

Because of the abstract nature of this uniform view, it is conventional to state two axioms which have to be satisfied by every reasonable concept of logical consequence.

Definition 1 (Axioms of Logical Consequence). *Let $\mathscr{L}$ be a logic, L be its set of formulas and $\Vdash_{\mathscr{L}}$ its logical consequence relation. Then $\Vdash_{\mathscr{L}}$ is said to be* proper *if it satisfies the following two axioms: For all $\Phi, \Psi \subseteq L$ and all $\chi \in L$:*

1. *A1: If $\varphi \in \Phi$ then $\Phi \Vdash_{\mathscr{L}} \varphi$.*
2. *A2: If $\Phi \Vdash_{\mathscr{L}} \psi$ for all $\psi \in \Psi$ and $\Psi \Vdash_{\mathscr{L}} \chi$ then $\Phi \Vdash_{\mathscr{L}} \chi$.*

Here we present two other concepts for a uniform view and treatment of logics, logics in modeltheoretic and logics in abstract form. Both concepts have logical consequence as a derived concept and we will see that both give rise to a logical consequence relation which is proper.

2.1   Logics in Modeltheoretic and Abstract Form

The modeltheoretic form of a logic is defined as follows:

Definition 2 (Modeltheoretic Form, Logical Consequence, Validity). *A logic $\mathscr{L}$ is given in* modeltheoretic form *if $\mathscr{L} = (L, \mathbb{M}, \models_{\mathrm{M}})$ where L denotes a set of formulas, $\mathbb{M}$ denotes a class of models and $\models_{\mathrm{M}}$ denotes a validity relation saying that $\varphi \in L$ resp. $\Phi \subseteq L$ is* valid *in a model M if $M \models_{\mathrm{M}} \varphi$ resp. $M \models_{\mathrm{M}} \Phi$. For a logic in modeltheoretic form the following derived concepts are defined:*

1. *The* logical consequence relation *$\Vdash_{\mathrm{M}}$ is defined as: $\Phi \Vdash_{\mathrm{M}} \varphi$ iff for all $M \in \mathbb{M}$ it holds that $M \models_{\mathrm{M}} \Phi$ implies $M \models_{\mathrm{M}} \varphi$.*
2. *The* theory *of a model $M \in \mathbb{M}$ is defined as $Th_{\mathbb{M}}(M) := \{\varphi \in L \mid M \models_{\mathrm{M}} \varphi\}$.*
3. *The* theory *of a set of formulas Φ is defined as $Th_{\mathbb{M}}(\Phi) = \{\varphi \in L \mid \Phi \Vdash_{\mathrm{M}} \varphi\}$.*
4. *A set of formulas Φ is called* consistent *if there is an $M \in \mathbb{M}$ in which Φ is valid.*
5. *A set of formulas Φ is called* tautological *if Φ is valid in all $M \in \mathbb{M}$.*

The following are easy to prove facts:

Theorem 1 (Derived Concepts). *Given a logic $\mathscr{L} = (L, \mathbb{M}, \models_{\mathrm{M}})$ in modeltheoretic form. Then:*

1. *The logical consequence relation* $\Vdash_M$ *is proper.*
2. *For any* $M \in \mathbb{M}$ *and* $\Phi \subseteq L$ *the theories* $Th_M(M)$ *and* $Th_M(\Phi)$ *are closed under logical consequence.*

In the tradition of formal specification the concept of *institution* was defined (see Goguen, Burstall [10] and also Ehrig, Mahr [5]) to cover the logical core of specification languages. Institutions naturally extend the concept of a logic in modeltheoretic form by notions of signatures and morphisms in a categorical framework (see also Meseguer [16] for the concept of general logics). Our approach to integration is based on traditional logic notions rather than structural concepts like morphisms and universal properties of objects and therefore borrows its basic concepts from the Warschau School of logic (see Wójcicki [25] and Cleave [2]).

The abstract form of a logic is defined as follows:

Definition 3 (Abstract Form, Logical Consequence, Validity). *A logic* $\mathscr{L}$ *is given in* abstract form *if* $\mathscr{L} = (L, \mathscr{B})$ *where* L *denotes a set of formulas and* $\mathscr{B}$, *called the* basis *of* $\mathscr{L}$, *denotes a set of subsets of* L. *For a logic in abstract form the following derived concepts are defined:*

1. *The* logical consequence relation $\Vdash_{\mathscr{B}}$ *is defined as:* $\Phi \Vdash_{\mathscr{B}} \varphi$ *iff* $\Phi \subseteq B$ *implies* $\varphi \in B$ *for all* $B \in \mathscr{B}$.
2. *The* theory of a set of formulas Φ *is defined as* $Th_{\mathbb{B}}(\Phi) := \{ \varphi \mid \Phi \Vdash_{\mathbb{B}} \varphi \}$.
3. *A set of formulas* Φ *is called* consistent *if there is at least one* $B \in \mathscr{B}$ *with* $\Phi \subseteq B$.
4. *A set of formulas* Φ *is called* tautological *if* $\Phi \subseteq B$ *for all* $B \in \mathscr{B}$.

The following are easy to prove facts:

Theorem 2 (Derived Concepts). *Given a logic* $\mathscr{L} = (L, \mathscr{B})$ *in abstract form. Then:*

1. *The logical consequence relation* $\Vdash_{\mathscr{B}}$ *is proper.*
2. *For any* $\Phi \subseteq L$ *the theory* $Th_{\mathscr{B}}(\Phi)$ *is closed under logical consequence.*

Logics $\mathscr{L} = (L, \mathbb{M}, \models_M)$ in modeltheoretic form can easily be transformed into logics in abstract form by defining a basis to be the set of theories of models of $\mathbb{M}$. This transformation preserves logical consequence, tautologies and consistency which is stated in the following easy to prove theorem:

Theorem 3 (Conservative Transformation into Abstract Form). *Let* $\mathscr{L} = (L, \mathbb{M}, \models_M)$ *be a logic in modeltheoretic form and let* $\mathscr{B} := \{ Th_M(M) \mid M \in \mathbb{M} \}$. *Then* $\mathscr{L}' := (L, \mathscr{B})$ *is a logic in abstract form such that for all* $\varphi \in L$:

1. $\Vdash_M = \Vdash_{\mathscr{B}}$.
2. φ *is consistent in* $\mathscr{L}$ *iff* φ *is consistent in* $\mathscr{L}'$.
3. φ *is tautological in* $\mathscr{L}$ *iff* φ *is tautological in* $\mathscr{L}'$.

The concepts of logics in modeltheoretic and abstract form are in a way equivalent. While we have seen that logics in modeltheoretic form can be conservatively transformed into abstract form, it is also possible to define a modeltheoretic form from a logic in abstract form. Then the set $\mathscr{B}$ is taken as the class of models and the validity relation is just membership. But for the purposes of this paper there is no need to discuss the relationship between the two forms further.

2.2   Examples of Logics in Abstract Form

The following are three examples showing how well-known logics generate logics in abstract form.

Propositional Logic: Let V be a set of propositional variables and let $Form(V)$ be the set of propositional formulas over V defined as usual (see Ehrig, Mahr [6] for example): $V \subseteq Form(V)$ and if $a, b \in Form(V)$, then $\neg a \in Form(V)$ and $a \vee b \in Form(V)$. Let $L_1 := Form(V)$. A truth assignment is a function $B : V \to \{T, F\}$, and validity $B \models \varphi$ is defined as usual. The theory of an assignment is defined by $Th(B) := \{\varphi \in L_1 \mid B \models \varphi\}$.

Now let $\mathscr{B}_1 := \{Th(B) \mid B \text{ is a truth assignment}\}$. Then $\mathscr{L}_1 := (L_1, \mathscr{B}_1)$ denotes a logic in abstract form whose consequence relation and concepts of validity coincide to the corresponding concepts of propositional logic.

Propositional Modal Logic: The set of formulas of propositional modal logic $L_2 := Form(V)$ is defined over a set of propositional variables V and operators $\neg$, $\vee$ and $\square$ as follows: $V \subseteq Form(V)$ and if $a, b \in Form(V)$, then $\neg a \in Form(V)$, $a \vee b \in Form(V)$ and $\square a \in Form(V)$. Validation of formulas is defined on frames $\mathcal{F} = (W, R)$ where $W \neq \emptyset$ is a set of possible *worlds* and $R \subseteq W \times W$ is an *access relation* on W. An assignment to variables is defined as a function $\beta : V \to 2^W$, assigning to every variable the set of worlds in which this variable is considered a true proposition. An *interpretation* $Int = (\mathcal{F}, \beta, w)$ is then defined to consist of a frame $\mathcal{F} = (W, R)$, an assignment β and a world $w \in W$ denoting the perspective from which validation is checked. Validation is then inductively defined as follows:

- $Int \models v \ :\Leftrightarrow\ w \in \beta(v)\ (v \in V)$.
- $Int \models a \vee b \ :\Leftrightarrow\ Int \models a$ or $Int \models b$.
- $Int \models \neg a \ :\Leftrightarrow\ Int \not\models a$.
- $Int \models \square a \ :\Leftrightarrow\ (\mathcal{F}, \beta, w') \models a$ for all $w' \in W$ with $(w, w') \in R$.

Let $Th(Int) := \{a \in Form(V) \mid Int \models a\}$. Depending on the choice of the relation R in a frame $\mathcal{F}$ several variants of modal logic have been studied (see for example Hughes, Cresswell [13] and Rautenberg [20]).

Now let $\mathscr{B}_K := \{Th(Int) \mid Int \text{ is a modal logic interpretation}\}$. Then $\mathscr{L}_K := (L_2, \mathscr{B}_K)$ forms a logic in abstract form whose logical consequence and concepts of validity coincide with the propositional modal logic known as K in the literature. In a similar way we also get the well-known propositional modal logic S_4 in abstract form by $\mathscr{L}_{S_4} := (L_2, \mathscr{B}_{S_4})$ with $\mathscr{B}_{S_4} := \{Th(Int) \mid Int = (W, R, \beta, w)$ where R is reflexive and transitive$\}$.

First-Order Predicate Logic: The set of formulas of first-order predicate logic is defined with respect to a signature $\Sigma = (S, OP, R)$, providing names for sorts, functions and relations, and a family of variables $V = (V_s)_{s \in S}$. Formulas with respect to Σ and V are composed from operators $=$, $\neg$, $\vee$ and $\exists$ in the usual way. Traditionally first-oder predicate logic is given as a logic in modeltheoretic form $\mathscr{L} = (L_3, \mathbb{M}, \models_{\mathbb{M}})$ with L_3 being the set of formulas, $\mathbb{M}$ being the class of Σ-structures and $\models_{\mathbb{M}}$ being inductively defined (see Ehrig, Mahr [6] for example).

Then $\mathscr{L}_3 = (L_3, \mathscr{B}_3)$ with $\mathscr{B}_3 := \{ Th(A) \mid A$ is a Σ-structure$\}$ and $Th(A) := \{\varphi \mid A \models_M \varphi\}$ is a logic in abstract form whose concepts of logical consequence and validity coincide with the corresponding concepts of $\mathscr{L}$.

A further example of a logic in abstract form, however of trivial nature, is given by the logic of constants.

Logic of Constants: Let C be a nonempty set of constant symbols and $\mathscr{B} := 2^C$. Then $\mathscr{L}_C := (C, \mathscr{B})$ is a logic in abstract form. It can easily be shown that all constants $c \in C$ are consistent, while no constant $c \in C$ is tautological. The logical consequence relation $\Vdash_{\mathscr{L}_C}$ has the property that $\Phi \Vdash_{\mathscr{L}_C} \varphi$ iff $\varphi \in \Phi$. Thus all $B \in \mathscr{B}$ are closed under logical consequence.

For further elaboration and more examples of logics in abstract form see Zeitz [26].

2.3 Integration of Logics

Our intention is to study $\in_T$-*integration* as a particular way of integrating logics. The form of integration embodied by $\in_T$-integration however can be defined more abstractly. We therefore make no particular assumptions on the form in which logics are given:

Definition 4 (Integration of Logics). *Given two logics $\mathscr{L}_0$ and $\mathscr{L}_1$ with formulas L_0 and L_1, and logical consequence relations $\Vdash_{\mathscr{L}_0}$ and $\Vdash_{\mathscr{L}_1}$ respectively. Then:*

1. *$\mathscr{L}_1$ is said to integrate $\mathscr{L}_0$, written $\mathscr{L}_0 \leqslant \mathscr{L}_1$, if there exists a mapping $f : L_0 \to L_1$ such that for all $\Phi \subseteq L_0$ and all $\varphi \in L_0$ the following integration properties hold:*
   I1: $\Phi \Vdash_{\mathscr{L}_0} \varphi$ iff $f(\Phi) \Vdash_{\mathscr{L}_1} f(\varphi)$,
   I2: Φ is consistent in $\mathscr{L}_0$ iff $f(\Phi)$ is consistent in $\mathscr{L}_1$,
   I3: Φ is tautological in $\mathscr{L}_0$ iff $f(\Phi)$ is tautological in $\mathscr{L}_1$,
   where $f(\Phi)$ denotes the image of Φ under f, i.e. $f(\Phi) := \{f(\varphi) \mid \varphi \in \Phi\}$.
2. *If $L_0 \subseteq L_1$ and f is the inclusion, then $\mathscr{L}_1$ is called an extension of $\mathscr{L}_0$.*
3. *Given an I-indexed family $(\mathscr{L}_j)_{j \in I}$ of logics, then $\mathscr{L}$ is called an integration logic for $(\mathscr{L}_j)_{j \in I}$ if the logic $\mathscr{L}$ integrates $\mathscr{L}_j$ for all $j \in I$. We also say that $\mathscr{L}$ integrates the family $(\mathscr{L}_j)_{j \in I}$.*

Obviously integration can be iterated as stated in the following theorem.

Theorem 4 (Iterated Integration). *For any three logics $\mathscr{L}_0$, $\mathscr{L}_1$, $\mathscr{L}_2$ we have: if $\mathscr{L}_0 \leqslant \mathscr{L}_1$ and $\mathscr{L}_1 \leqslant \mathscr{L}_2$, then $\mathscr{L}_0 \leqslant \mathscr{L}_2$.*

For later use we state the easy to prove theorem saying that disjoint union logic is an integration logic for its components.

Definition 5 (Disjoint Union Logic). *Let $(\mathscr{L}_j)_{j \in I}$ be an I-indexed family of logics and let $(\mathscr{L}_j^*)_{j \in I}$ with $\mathscr{L}_j^* = (L_j^*, \mathscr{B}_j^*)$ be the corresponding family of logics in abstract form. Then the disjoint union logic $\mathscr{L}_\uplus = (L_\uplus, \mathscr{B}_\uplus)$ of $(\mathscr{L}_j)_{j \in I}$ is defined by:*

1. $L_{\uplus} := \biguplus_{j \in I} L_j^*$,

2. $\mathscr{B}_{\uplus} = \{ \biguplus_{j \in I} B_j \mid B_j \in \mathscr{B}_j^*, j \in I \}$,

where $\uplus$ denotes the disjoint union of sets.

Theorem 5 (Disjoint Union Integration). *Let $\mathscr{L}_{\uplus} = (L_{\uplus}, \mathscr{B}_{\uplus})$ be the disjoint union logic of an I-indexed family of logics $(\mathscr{L}_j)_{j \in I}$. Then for every $j \in I$ the logic $\mathscr{L}_{\uplus}$ integrates $\mathscr{L}_j$.*

3 $\in_T$-Logics and $\in_T$-Extension

Formulas of an $\in_T$-logic may be propositional variables, propositional parameters or compound expressions built up from the classical propositional connectives as well as quantification and assertions for truth, falsity and propositional equality. Accordingly, $\in_T$-logics are propositional logics. Their expressive power however is far beyond classical propositional logic since it allows quantification over propositional variables and propositional equality. Assertions of truth and falsity in $\in_T$-logics are explicit statements for truth predicates. This allows to deal with impredicativity, intensionality and truth (see Sträter [21] and Zeitz [26]). Our interest in $\in_T$-logics however is not in the study of these phenomena, but in the use of $\in_T$-logics as particular integration logics.

3.1   Syntax of $\in_T$-Logics

$\in_T$-formulas are inductively defined as follows:

Definition 6 ($\in_T$-Formulas). *Let $X := \{x_i \mid i \in \mathbb{N}\}$ be a well-ordered infinite set of variables and let P be an arbitrary set of parameters. Then the set $L := L(P)$ of $\in_T$-formulas with parameters P is the smallest set of expressions such that:*

$$
\begin{array}{lll}
(1) & x \in X & \Rightarrow x \in L \\
(2) & a \in P & \Rightarrow a \in L \\
(3) & \varphi \in L & \Rightarrow (\neg\varphi) \in L \\
 & & (\varphi{:}true) \in L \\
 & & (\varphi{:}false) \in L \\
(4) & \varphi, \psi \in L & \Rightarrow (\varphi \to \psi) \in L \\
 & & (\varphi \equiv \psi) \in L \\
(5) & \varphi \in L, \ x \in X & \Rightarrow (\forall x. \varphi) \in L
\end{array}
$$

Free and bound variables of $\in_T$-formulas as well as substitution of free variables are defined as usual. We write substitutions in the form of mappings $\sigma : X \to L$ and call σ *admissible* for a formula φ if for all free variables y in φ the following holds: if there is a free occurrence of y in the scope of a binding $\forall x.$ in the formula φ, then x is not a variable in $\sigma(y)$. Admissible substitutions do not cause unwanted binding of free variables.

Also renaming of bound variables is defined in the usual way. We call two formulas $\varphi, \psi \in L$ *α-congruent*, written $\varphi =_\alpha \psi$, if φ and ψ differ only in the choice of their bound variables. It holds that for every substitution σ and formula φ there is a formula ψ which is α-congruent to φ and σ is admissible for ψ (see for example Zeitz [26]). Hereby ψ can be obtained by renaming bound variables of φ.

On the basis of the propositional connectives $\neg$ and $\rightarrow$ other connectives like $\wedge$, $\vee$ and $\leftrightarrow$ can be defined in the usual way. Also existencial quantification $\exists x.\varphi$ can be defined by $\neg \forall x. \neg \varphi$. For economic reasons however we use the reduced set of connectives and quantifiers.

3.2 Semantics of $\in_T$-Logics

$\in_T$-logics are defined in modeltheoretic form. Since $\in_T$-logics are propositional logics their models are sets of propositions. In the semantics of $\in_T$-logics the nature and the structure of these propositions is not determined. Constructing new propositions from given propositions is expressed in the syntax of $\in_T$-logics and is semantically enforced by the properties of propositional interpretation of $\in_T$-formulas. In particular we define:

Definition 7 (Propositional Interpretation of $\in_T$-Formulas). *Let L be the set of $\in_T$-formulas with parameters P. A propositional interpretation of $\in_T$-formulas $\mathscr{I} = (M, T, \Gamma)$ is given by:*

1. *A set $M \neq \emptyset$ of propositions, called* propositional universe.
2. *A set $T \subsetneq M$ with $T \neq \emptyset$ of true propositions, called* truth domain.
3. *A mapping $\Gamma : L \times [X \rightarrow M] \rightarrow M$, called* propositional assignment, *such that the following truth and contextual properties hold:*
   (a) Truth Properties:
   * (1) $\Gamma((\varphi{:}true), \beta) \in T \;\Leftrightarrow \Gamma(\varphi, \beta) \in T$
   * (2) $\Gamma((\varphi{:}false), \beta) \in T \Leftrightarrow \Gamma(\varphi, \beta) \notin T$
   * (3) $\Gamma((\varphi \equiv \psi), \beta) \in T \;\Leftrightarrow \Gamma(\varphi, \beta) = \Gamma(\psi, \beta)$
   * (4) $\Gamma((\neg \varphi), \beta) \in T \qquad \Leftrightarrow \Gamma(\varphi, \beta) \notin T$
   * (5) $\Gamma((\varphi \rightarrow \psi), \beta) \in T \Leftrightarrow \Gamma(\varphi, \beta) \notin T \text{ or } \Gamma(\psi, \beta) \in T$
   * (6) $\Gamma((\forall x.\varphi), \beta) \in T \quad \Leftrightarrow \text{ for all } m \in M: \Gamma(\varphi, \beta[m/x]) \in T$
   (b) Contextual Properties:
   * (1) $\Gamma(x, \beta) \quad = \beta(x) \text{ for all } x \in X \text{ and all } \beta : X \rightarrow M$
   * (2) $\Gamma(\varphi, \beta_1) \quad = \Gamma(\varphi, \beta_2) \text{ if } \beta_1|_{Free(\varphi)} = \beta_2|_{Free(\varphi)}$
   * (3) $\Gamma(\varphi[\sigma], \beta) = \Gamma(\varphi, \langle \beta\sigma \rangle) \text{ for all substituions } \sigma$
   * (4) $\Gamma(\varphi, \beta) \quad = \Gamma(\psi, \beta) \text{ if } \varphi =_\alpha \psi$
   where $\beta[m/x]$ denotes the variant of β assigning m to x and being identical to β otherwise, $Free(\varphi)$ denotes the set of free variables in φ and $\langle \beta\sigma \rangle$ denotes the propositional assignment to variables, assigning to a variable x the proposition $\Gamma(\sigma(x), \beta)$.

We define $(\mathscr{I}, \beta) \models \varphi :\Leftrightarrow \Gamma(\varphi, \beta) \in T$ for validity and call $\mathscr{I}$ a model of a set of $\in_T$-formulas Φ if for all $\varphi \in \Phi$ and all $\beta : X \rightarrow M$ it holds that $(\mathscr{I}, \beta) \models \varphi$. We then write for short $\mathscr{I} \models \varphi$.

Definition 8 ($\in_T$-Logics). *Let $L(P)$ be the set of $\in_T$-formulas with parameter P, let $\mathbb{M}$ be a class of (not necessarily all) propositional interpretations of $\in_T$-formulas and let $\models$ be the validity relation for $\in_T$-formulas. Then $\in_T(P) := (L(P), \mathbb{M}, \models)$ is called an $\in_T$-logic.*

Note that an $\in_T$-logic is given in modeltheoretic form. Its logical consequence relation, denoted $\Vdash_\mathbb{M}$, is defined as in Definition 2. Its abstract form is defined as in Theorem 3.

Also note, that $x{:}true$ is a truth predicate in $\in_T(P)$. The propositional assignment of $(\varphi{:}true)$ falls into the truth domain of the propositional universe iff $\mathscr{I}$ is a model of φ. In other words the evaluation of the expression $(\varphi{:}true)$ coincides with the meta-level truth of the formula φ. This coincidence reflects the well-known Tarski Biconditionals (see Tarski [24]).

Note also that quantification in $\in_T$-logics is impredicative, since for example in a sentence $(\forall x.\varphi)$ the variable x ranges over all propositions, including the proposition expressed by $(\forall x.\varphi)$ itself. Sträter and Zeitz have shown that this has no effect on the existence of models (see [21, 26]). We will use their model construction later for the purpose of model integration.

Note further that $\in_T$-logic also allows to speak about the liar paradox saying that *"this sentence is false"*, where *this* refers to the sentence itself. Unlike other approaches (see Sträter [21] for a discussion), $\in_T$-logic has a total truth predicate. The liar paradox in $\in_T$-logic is phrased by $x \equiv (x{:}false)$ which is a formula that is false in all propositional interpretations for all values of x.

Finally note, that the semantics of $\in_T$-logic admits intensional models in the sense that the meaning of an $\in_T$-formula is not a truth value, but a proposition that may be true or false depending on it falling into the truth domain or not. Accordingly the negation of propositional equivalence $\neg(\varphi \equiv \psi)$ expresses that φ and ψ do not denote the same proposition, in contrast to $\neg(\varphi \leftrightarrow \psi)$ expressing that φ and ψ have different behaviours of truth. Sträter and Zeitz also show that intensional models for $\in_T$-logic can be constructed.

3.3 $\in_T$-Extension

In [26] Zeitz defines the concept of $\in_T$-extension as an $\in_T$-logic in the sense above which is an extension of some other logic. $\in_T$-extension is based on the observation that a parameter P in an $\in_T$-logic $\in_T(P)$, which is an arbitrary set, may well be the set of formulas of some other logic $\mathscr{P} = (P, \mathscr{B}_P)$. Zeitz gives conditions under which $\in_T(P)$ forms an extension of $\mathscr{P}$ (in the sense of Definition 4) and studies model- and prooftheoretic properties of such extensions. We adopt his notion, but rephrase it with only slight differences in their formal notation for our discussion of $\in_T$-integration below. The following theorem is easy to prove.

Theorem 6 ($\in_T$-Extension). *Let $\mathscr{P} = (P, \mathscr{B}_P)$ be a logic in abstract form, called parameter logic, and let $\in_T(P) = (L, \mathbb{M}, \models)$ be an $\in_T$-logic in modeltheoretic form such that for all propositional interpretations $\mathscr{I}$ the following extension property holds:*

$$E : \{a \in P \mid \mathscr{I} \models a\} \in \mathscr{B}_P.$$

Then $\in_T(P)$ is an extension of $\mathscr{P}$.

Note that in the semantics of $\in_T$-logics the assignment of propositions to parameters by the mapping Γ is not constrained in any way. $\in_T$-extensions however do constrain the propositional assignment to parameters as expressed in the extension property. We can, nevertheless, obtain every $\in_T$-logic as an $\in_T$-extension by choosing an appropriate logic of constants (see Section 2.2) as a parameter logic.

4 $\in_T$-Integration

The scenario of logic integration starts from different views on one and the same object and identifies a particular view with a model in some logic. Different views then correspond to different logics. The idea of integration was to provide an integration logic which allows statements which integrate other statements from different views. The concept of $\in_T$-logics, which was used by Zeitz to study $\in_T$-extensions of arbitrary logics in abstract form can now be used for the integration of logics. We model the scenario therefore by $\in_T$-integration of I-indexed families of logics which we assume without loss of generality to have mutually disjoint sets of formulas.

4.1 $\in_T$-Logic Integration

The integration of logics is generally defined in Section 2.3. $\in_T$-integration is seen as a particular form of integration where the expressiveness of $\in_T$-logic is added to the sum of logics to be integrated. This is achieved by defining the integration logic of a family of logics as an extension of the familiy's disjoint union. We state this in the following theorem.

Theorem 7 ($\in_T$-Integration Logic). *Given a family $(\mathscr{P}_j)_{j \in I}$ of logics and let $\mathscr{P}_{\uplus} = (P_{\uplus}, \mathscr{B}_{\uplus})$ be the disjoint union logic of $(\mathscr{P}_j)_{j \in I}$ as defined in Definition 5. Then $\in_T(P_{\uplus})$ integrates the family $(\mathscr{P}_j)_{j \in I}$ in the sense of Definition 4 if $\in_T(P_{\uplus})$ is an extension of $\mathscr{P}_{\uplus}$. We call $\in_T(P_{\uplus})$, also written as $\in_T((P_j)_{j \in I})$, the $\in_T$-integration logic of $(\mathscr{P}_j)_{j \in I}$.*

Proof. Let $\in_T(P_{\uplus})$ be an extension of $\mathscr{P}_{\uplus}$ which means that $\mathscr{P}_{\uplus} \leqslant \in_T(P_{\uplus})$. From Theorem 5 we know that $\mathscr{P}_{\uplus}$ integrates each of the component logics $\mathscr{P}_j$, in other words it holds $\mathscr{P}_j \leqslant \mathscr{P}_{\uplus}$ for all $j \in I$. From Theorem 4 it follows that $\in_T(P_{\uplus})$ integrates $\mathscr{P}_j$ for all $j \in I$. Thus from Definition 4 it follows that $\in_T(P_{\uplus})$ integrates the family $(\mathscr{P}_j)_{j \in I}$. ☐

4.2 $\in_T$-Model Integration

According to the above stated scenario of logic integration a view on some object corresponds with a model in some logic. Assertions which integrate statements from different views on some object will therefore need an interpretation in some model that integrates these views. The propositional meaning of $\in_T$-logics now allows to construct such an integrating model. One could think of different types of models for model integration, but we simply use the model construction given by Zeitz in [26].

Lemma 1 (Existence of $\in_T$-Interpretations). *Let $\mathscr{P} = (P, \mathscr{B}_P)$ be a logic in abstract form and let $\in_T(P) = (L, \mathbb{M}, \models)$ be the $\in_T$-extension of $\mathscr{P}$. Then for all $B \in \mathscr{B}_P$ there exists an interpretation $\mathscr{I}$ of $\in_T$-formulas such that $\{\varphi \in P \mid \mathscr{I} \models \varphi\} = B$.*

Proof. See Zeitz in [26], Theorem 3.17. The propositional interpretation $\mathscr{I} = (M, T, \Gamma)$ is defined by: $M = P \uplus \{0, 1\}$, the truth domain $T = B \uplus \{1\}$ and the propositional assignment $\Gamma : P \times [X \to M] \to M$ inductively defined as follows:

$$
\begin{array}{lll}
(1) & \Gamma(x, \beta) & := \beta(x) \text{ for all } x \in X \\
(2) & \Gamma(a, \beta) & := a \text{ for all } a \in P \\
(3) & \Gamma((\varphi{:}true), \beta) := 1 \text{ if } \Gamma(\varphi, \beta) \in T \\
(4) & \Gamma((\neg\varphi), \beta) & := 1 \text{ if } \Gamma(\varphi, \beta) \notin T \\
(5) & \Gamma((\varphi \equiv \psi), \beta) := 1 \text{ if } \Gamma(\varphi, \beta) = \Gamma(\psi, \beta) \\
(6) & \Gamma((\varphi \to \psi), \beta) := 1 \text{ if } \Gamma(\varphi, \beta) \notin T \text{ or } \Gamma(\psi, \beta) \in T \\
(7) & \Gamma((\forall x.\varphi), \beta) & := 1 \text{ if } \Gamma(\varphi, \beta[m/x]) \in T \text{ for all } m \in M \\
(8) & \Gamma(\varphi, \beta) & := 0 \text{ otherwise}
\end{array}
$$

In this interpretation all formulas of the parameter logic are considered to be propositions. In addition the propositions 0 and 1 are interpretations of $\in_T$-formulas, expressing truth and falsity. For this propositional interpretation truth and contextual properties as well as the extension property are shown in Zeitz [26]. □

Based on this model construction model integration can now be stated:

Theorem 8 ($\in_T$-Model Integration). *Given an I-indexed family $(\mathscr{P}_j)_{j \in I}$ of logics in modeltheoretic form with $\mathscr{P}_j = (P_j, \mathbb{M}_j, \models_j)$ and let $\in_T((P_j)_{j \in I})$ be the $\in_T$-integration logic of $(\mathscr{P}_j)_{j \in I}$. Then for all I-indexed families of models $(M_j)_{j \in I}$ with $M_j \in \mathbb{M}_j$ there is a model $\mathscr{I}$ of $\in_T((P_j)_{j \in I})$ such that:*

$$\text{For all } j \in I \text{ and all } \Phi_j \subseteq P_j : \mathscr{I} \models \Phi_j \text{ iff } M_j \models_j \Phi_j.$$

Proof. Given an I-indexed family $(\mathscr{P}_j)_{j \in I}$ of logics in modeltheoretic form with $\mathscr{P}_j = (P_j, \mathbb{M}_j, \models_j)$, let $\mathscr{P}_j^* = (P_j, \mathscr{B}_j)$ be the logic $\mathscr{P}_j$ after transformation into abstract form. Then for all $j \in I$, $M_j \in \mathbb{M}_j$ and $\Phi_j \subseteq P_j$

$$M_j \models_j \Phi_j \text{ iff } \Phi_j \subseteq Th(M_j) \in \mathscr{B}_j \tag{1}$$

This is obvious from the definition of the theory of a model and the transformation into abstract form (Theorem 3). Since the sets of formulas P_j for $j \in I$ are assumed to be pairwise disjoint we conclude directly from the definition of the disjoint union logic (see Definition 5) that for all $j \in I$

$$\Phi_j \subseteq Th(M_j) \in \mathscr{B}_j \text{ iff } \Phi_j \subseteq \biguplus_{j \in I} Th(M_j) \in \mathscr{B}_\uplus \tag{2}$$

Since $\in_T((P_j)_{j\in I})$ is an extension of $\mathscr{P}_{\uplus}$ we conclude from Lemma 1 the existence of an interpretation $\mathscr{I}$ of $\in_T$-formulas such that: $\biguplus_{j\in I} Th(M_j) = \{\varphi \in P_{\uplus} \mid \mathscr{I} \models \varphi\}$. We therefore have for $j \in I$

$$\Phi_j \subseteq \biguplus_{j\in I} Th(M_j) \in \mathscr{B}_{\uplus} \text{ iff } \mathscr{I} \models \Phi_j \tag{3}$$

Putting (1), (2) and (3) together we finally obtain for all $j \in I$: $M_j \models_j \Phi_j$ iff $\mathscr{I} \models \Phi_j$. Since this is true for all I-indexed families of models $(M_j)_{j\in I}$ and all families $(\Phi_j)_{j\in I}$ the theorem follows.   □

4.3   Discussion on $\in_T$-Calculus Integration

In [26] Zeitz proves (in Corollary 4.20) the following theorem.

Theorem 9 (Calculus Extension). *Let $\mathscr{P} = (P, \mathscr{B}_P)$ be a logic in abstract form having a Hilbert-type calculus $\mathscr{K}$ which is correct and complete for $\mathscr{P}$. Then $\in_T(\mathscr{K})$ is correct and complete for $\in_T(P)$ iff $\mathscr{B}_P$ is closed under intersection.*

Here a calculus of logic in abstract form is a pair consisting of a decidable set of axioms and a decidable set of rules. Provability is abstractly defined and correctness and completeness are defined in the usual way. The calculus $\in_T(\mathscr{K})$ is defined as an extension of the calculus $\mathscr{K}$ whose axioms contain besides all axioms of $\mathscr{K}$ and all rules of $\mathscr{K}$ being rewritten in axiomatic form the following list of $\in_T$-axioms

1. $\varphi \to (\psi \to \varphi)$
2. $(\varphi \to (\psi \to \chi)) \to ((\varphi \to \psi) \to (\varphi \to \chi))$
3. $(\neg\psi \to \neg\varphi) \to (\varphi \to \psi)$
4. $\varphi \to (\varphi{:}true)$
5. $(\varphi{:}true) \to \varphi$
6. $(\varphi \equiv \psi)$ (if $\varphi =_\alpha \psi$)
7. $(\varphi \equiv \psi) \to (\varphi \to \psi)$
8. $(\forall x.(\varphi \equiv \varphi')) \to ((\psi \equiv \chi) \to (\varphi[x := \psi] \equiv \varphi'[x := \chi]))$
9. $(\forall x.\varphi) \to \varphi[x := \psi]$
10. $(\forall x.(\varphi \to \psi)) \to (\varphi \to \forall x.\psi)$ (if $x \notin Free(\varphi)$)

as well as the two $\in_T$-rules

11. Modus Ponens: ψ can be derived from φ and $\varphi \to \psi$.
12. Generalization: $\forall x.\varphi$ can be derived from φ.

The assumption that $\mathscr{B}_P$ is closed under intersection is rather strong, since it implies that the set P of formulas is consistent, which is true for equational logic for example, but for no other logic with negation. Zeitz therefore studies extensions of the calculus $\in_T(\mathscr{K})$ and variants of conditions which admit correctness and completeness of the resulting calculi for $\in_T(P)$. Based on these results he can then show that numerous classical logics (among those the logics given in

Section 2.2) admit preservation of the correctness and completeness properties for their extended calculi (see Zeitz [26], Corollary 4.23 and 4.26, as well as Examples 4.24 and 4.27).

We claim here that these results can be inherited in the situation of $\in_T$-integration of logics. We expect no major difficulties to arise in the proofs. Because of a rather lengthy treatment in the transfer of these results, however, we do not check the details and leave this here as a claim.

5   Conclusion

Based on a minor generalization of $\in_T$-logic in the sense of Sträter and Zeitz we have shown that $\in_T$-logics can be used for integration of logics. This integration takes place at the meta-level of judgements and makes available the expressive power of elementary $\in_T$-logic for assertions about statements from parameter logics. It was our goal to explain how in a formal framework of abstract logics integration through $\in_T$-logic can work. We have so far not yet exploited all the potential of integration through $\in_T$-logic. Even though the disjoint union of parameter logics is not a limit construction in the category of logics in abstract form the use of limits and colimits in this category for combining logics might turn out to be much more appropriate to reflect the assumption in our integration scenario namely that parameter logics correspond to views on one and the same object.

Another example of fruitful study is the construction of models which reflect propositional equality between assertions from different parameter logics. Finally the extension of $\in_T$-logic by modal operators for truth and possibility seems to be a promising task. Along with such further studies the expressiveness of $\in_T$-integration has to be examined.

Acknowledgments

We would like to thank the anonymous referee for the positive suggestions and criticism.

References

1. Marc Born, Eckhardt Holz, and Olaf Kath. *Softwareentwicklung mit UML 2*. Addison-Wesley Verlag München, 2004.
2. J.P. Cleave. *A Study of Logics*. Clarendon Press, 1991.
3. R. Diaconescu. Grothendieck institutions, 2002.
4. Hartmut Ehrig and Bernd Mahr. *Fundamentals of Algebraic Specification 1*. Springer Verlag Berlin, 1985.
5. Hartmut Ehrig and Bernd Mahr. *Fundamentals of Algebraic Specification 2*. Springer Verlag Berlin, 1990.
6. Hartmut Ehrig, Bernd Mahr, Felix Cornelius, Martin Große-Rohde, and Philip Zeitz. *Mathematisch-strukturelle Grundlagen der Informatik, 2. Auflage*. Springer, 2001.

7. Hartmut Ehrig and Fernando Orejas. A generic component concept for integrated data type and process modeling techniques. Technical Report 2001/12, Technische Universität Berlin, 2001.

8. International Organization for Standardization. *Basic Reference Model of Open Distributed Processing*. ITU-T X.900 series and ISO/IEC 10746 series, 1995.

9. Dov M. Gabbay. *Fibring Logics*, volume 38 of *Oxford Logic Guides*. Oxford Science Publications, 1999.

10. J.A. Goguen and R.M. Burstall. Introducing institutions. In *Proceedings Logics of Programming Workshop, Carnegie-Mellon*, pages 221–256. LNCS 164, Springer, 1984.

11. J.A. Goguen and G. Rosu. Institution morphisms, 2001.

12. J.A. Goguen and J.J. Tardo. An introduction to OBJ: a language for writing and testing formal algebraic program specifications. In *Proceedings IEEE Conference on Specification for Reliable Software*, pages 170–189. IEEE Computer Society Press, 1979.

13. G. E. Hughes and M. J. Cresswell. *A New Introduction to Modal Logic*. Routledge, 1996.

14. Bernd Mahr. Applications of type theory. In *Proceedings of the International Joint Conference CAAP/FASE on Theory and Practice of Software Development*, pages 343–355. Springer-Verlag, 1993.

15. Bernd Mahr, Werner Sträter, and Carla Umbach. Fundamentals of a theory of types and declarations. Technical Report KIT-Report 82, Technische Universität Berlin, 1990.

16. José Meseguer. General logics. In H.-D. Ebbinghaus et al., editor, *Proceedings, Logic Colloquium, 1987*. North-Holland, 1989.

17. T. Mossakowski. Foundations of heterogeneous specification. In M. Wirsing, D. Pattinson, and R. Hennicker, editors, *Recent Trends in Algebraic Development Techniques, 16th International Workshop, WADT 2002*, pages 359–375. Springer London, 2003.

18. D.C. Parnas. A technique for software module specification with examples. In *CACM 15, 5*, pages 330–336, 1972.

19. Janis R. Putman. *Architecting with RM-ODP*. Prentice Hall PTR, 2000.

20. Wolfgang Rautenberg. *Klassische und nichtklassische Aussagenlogik*. Vieweg Verlag Braunschweig / Wiesbaden, 1979.

21. Werner Sträter. $\in_T$ *Eine Logik erster Stufe mit Selbstreferenz und totalem Wahrheitsprädikat*. Forschungsbericht, KIT-Report 98, 1992. Dissertation, Technische Universität Berlin.

22. A. Tarlecki. Moving between logical systems. In *COMPASS/ADT*, pages 478–502, 1995.

23. A. Tarlecki. Towards heterogeneous specifications. In D. Gabbay and M. van Rijke, editors, *Proceedings 2nd International Workshop on Frontiers of Combining Systems, FroCoS'98*. Kluwer, 1998.

24. Alfred Tarski. Der Wahrheitsbegriff in den formalisierten Sprachen. *Studia Philosophica 1*, pages 261–405, 1935.

25. R. Wójcicki. *Theory of Logical Calculi*. Kluwer, 1988.

26. Philip Zeitz. *Parametrisierte $\in_T$-Logik: Eine Theorie der Erweiterung abstrakter Logiken um die Konzepte Wahrheit, Referenz und klassische Negation*. Logos Verlag Berlin, 2000. Dissertation, Technische Universität Berlin, 1999.

Functorial Semantics of Rewrite Theories[*]

José Meseguer

University of Illinois at Urbana-Champaign, USA
meseguer@cs.uiuc.edu

Abstract. This paper develops a close analogy between Lawvere's functorial semantics of equational theories [21], and a similar 2-functorial semantics for rewrite theories, which specify concurrent systems and whose models are "true concurrency" models of such systems. This has the advantage of unifying within a single 2-functorial framework both models and rewrite theory morphisms. Such morphisms are used in Maude to "put rewrite theories together" in different constructions, including parameterized rewrite theory specifications.

1 Introduction

We owe to Lawvere [21], Bénabou [2], and to Eilenberg and Moore [14] the important insight that algebraic semantics can be developed not just on the category **Set** of sets and functions, but on general categories satisfying minimal requirements. This insight has been exploited in computer science since the 1970s, leading to many fruitful extensions of algebraic semantics beyond the traditional universal algebra of algebraic data types. Since Hartmut Ehrig has made fundamental contributions to both the original theory of algebraic data types and to extending algebraic semantics in new directions, including, for example, graph rewriting, and Petri-net based concurrent system specifications, it seems appropriate for this occasion to discuss the way in which the semantics of rewriting logic [25] is a form of universal algebra on the 2-category **Cat**, the same way that the semantics of traditional algebraic specifications is furnished by universal algebra on the category **Set**.

Of course, in traditional algebraic semantics we are not only interested in the categories of models (the algebras) but on "putting theories together" in a categorical way as originally proposed by Burstall and Goguen in [6] and further developed in the work of Hartmut Ehrig and other researchers (see his two-volume work with Mahr [12, 13] and references there). Theories are "put together" by means of categorical constructions in the category **EqtlTh** of equational theories and equational theory morphisms. Although the most common description of **EqtlTh** is in terms of theory *presentations*, the most elegant and presentation-independent way of describing **EqtlTh** is as the category of Lawvere-Bénabou theories [21, 2]. This has the conceptual advantage of unifying the semantics of the category of theories **EqtlTh** and that of the difference

---

[*] Research supported by ONR Grant N00014-02-1-0715 and NSF Grant CCR-0234524.

H.-J. Kreowski et al. (Eds.): Formal Methods (Ehrig Festschrift), LNCS 3393, pp. 220–235, 2005.

categories of algebras within a single *functorial semantics* framework, so that a theory morphism $H : T \longrightarrow T'$ becomes a product-preserving functor and can therefore be regarded as a T-algebra interpreted in the target Lawvere theory T'. In a completely similar way, rewrite theories can be viewed as Lawvere 2-theories and form a category **RWTh**, so that rewrite theory morphisms and models of rewrite theories (which intuitively correspond to "true concurrency" models of the concurrent system specified by the given rewrite theory) are again unified within a single 2-functorial semantics as 2-product preserving 2-functors. Furthermore, the semantics of a parameterized module as the left adjoint of the forgetful functor associated to the inclusion of the parameter theory into the body theory typical of algebraic specifications has an exact analogue for parameterized rewrite theories (supported for example by the Maude language [7, 11]) as a corresponding left adjoint of the forgetful functor in the 2-functorial semantics.

This paper is mostly based on two appendices of the SRI Technical report [24] now not easily available. The paper is organized as follows. The inference rules and model theory of (unconditional) rewriting logic is summarized in Section 2. The functorial semantics of rewrite theories is then presented in Section 3. Rewrite theory morphisms and parameterization are then discussed in Section 4. I then discuss related work and give some concluding remarks in Section 5.

2 Rewriting Logic and Its Models

2.1 Inference Rules and Their Meaning

A *signature* in rewriting logic is an equational theory[1] (Σ, E), where Σ is an equational signature and E is a set of Σ-equations. Rewriting will operate on equivalence classes of terms modulo E. In this way, we free rewriting from the syntactic constraints of a term representation and gain a much greater flexibility in deciding what counts as a *data structure*; for example, string rewriting is obtained by imposing an associativity axiom, and multiset rewriting by imposing associativity and commutativity. Of course, standard term rewriting is obtained as the particular case in which the set of equations E is empty. Techniques for rewriting modulo equations have been studied extensively [10] and can be used to implement rewriting modulo many equational theories of interest.

Given a signature (Σ, E), *sentences* of rewriting logic are sequents of the form

$$[t]_E \longrightarrow [t']_E,$$

where t and t' are Σ-terms possibly involving some variables X, and $[t]_E$, or $[t]$ for short, denotes the equivalence class of the term t modulo the equations E, that is, an element of the free (Σ, E)-algebra $T_{\Sigma,E}(X)$ generated by the variables X. A *rewrite theory* $\mathcal{R}$ is a 4-tuple $\mathcal{R} = (\Sigma, E, L, R)$ where Σ is a ranked alphabet

---

[1] Rewriting logic is parameterized by the choice of its underlying equational logic, that can be unsorted, many-sorted, order-sorted, membership equational logic, and so on. To ease the exposition I give an *unsorted* presentation.

of function symbols, E is a set of Σ-equations, L is a set of *labels*, and R is a set of pairs $R \subseteq L \times T_{\Sigma,E}(X)^2$ whose first component is a label and whose second component is a pair of E-equivalence classes of terms, with $X = \{x_1, \ldots, x_n, \ldots\}$ a countably infinite set of variables. Elements of R are called *rewrite rules*[2]. We understand a rule $(r, ([t], [t']))$ as a labeled sequent and use for it the notation $r : [t] \longrightarrow [t']$. To indicate that $\{x_1, \ldots, x_n\}$ is the set of variables occurring in either t or t', we write $r : [t(x_1, \ldots, x_n)] \longrightarrow [t'(x_1, \ldots, x_n)]$, or in abbreviated notation $r : [t(\overline{x})] \longrightarrow [t'(\overline{x})]$.

Given a rewrite theory $\mathcal{R}$, we say that $\mathcal{R}$ *entails* a sentence $[t] \longrightarrow [t']$, or that $[t] \longrightarrow [t']$ is a *(concurrent) $\mathcal{R}$-rewrite*, and write $\mathcal{R} \vdash [t] \longrightarrow [t']$ if and only if $[t] \longrightarrow [t']$ can be obtained by finite application of the following *rules of deduction* (where we assume that all the terms are well formed and $t(\overline{w}/\overline{x})$ denotes the simultaneous substitution of w_i for x_i in t):

1. **Reflexivity.** For each $[t] \in T_{\Sigma,E}(X)$, $\dfrac{}{[t] \longrightarrow [t]}$.

2. **Congruence.** For each $f \in \Sigma_n$, $n \in \mathbb{N}$,

$$\frac{[t_1] \longrightarrow [t_1'] \quad \cdots \quad [t_n] \longrightarrow [t_n']}{[f(t_1, \ldots, t_n)] \longrightarrow [f(t_1', \ldots, t_n')]}.$$

3. **Replacement.** For each rule $r : [t(x_1, \ldots, x_n)] \longrightarrow [t'(x_1, \ldots, x_n)]$ in R,

$$\frac{[w_1] \longrightarrow [w_1'] \quad \cdots \quad [w_n] \longrightarrow [w_n']}{[t(\overline{w}/\overline{x})] \longrightarrow [t'(\overline{w'}/\overline{x})]}.$$

4. **Transitivity.**

$$\frac{[t_1] \longrightarrow [t_2] \quad [t_2] \longrightarrow [t_3]}{[t_1] \longrightarrow [t_3]}.$$

Rewriting logic is a logic for reasoning correctly about *concurrent systems* having *states*, and evolving by means of *transitions*. The signature of a rewrite theory describes a particular structure for the states of a system – e.g., multiset, binary tree, etc. – so that its states can be distributed according to such a structure. The rewrite rules in the theory describe which *elementary local transitions* are possible in the distributed state by concurrent local transformations. The rules of rewriting logic allow us to reason correctly about which *general* concurrent transitions are possible in a system satisfying such a description. Thus, computationally, each rewriting step is a parallel local transition in a concurrent system. Alternatively, however, we can adopt a logical viewpoint instead, and regard the rules of rewriting logic as *metarules* for correct deduction in a *logical system*. Logically, each rewriting step is a logical *entailment* in a formal system.

---

[2] To simplify the exposition the rules of the logic are given for the case of *unconditional* rewrite rules. However, all the ideas presented here have been extended to conditional rules in [25] with very general rules of the form

$$r : [t] \longrightarrow [t'] \quad \textit{if} \quad [u_1] \longrightarrow [v_1] \wedge \ldots \wedge [u_k] \longrightarrow [v_k].$$

This increases considerably the expressive power of rewrite theories.

2.2   Models

I first sketch the construction of initial and free models for a rewrite theory $\mathcal{R} = (\Sigma, E, L, R)$. Such models capture the intuitive idea of a "concurrent system" in the sense that they describe systems whose states are E-equivalence classes of terms, and whose computations are concurrent rewritings using the rules in R. By adopting a logical instead of a computational perspective, we can alternatively view such models as "logical systems" in which formulas are validly rewritten to other formulas by concurrent rewritings which correspond to proofs for the logic in question. Such models have a natural *category* structure, with states (or formulas) as objects, computations (or proofs) as morphisms, and sequential composition as morphism composition, and in them dynamic behavior exactly corresponds to deduction.

Given a rewrite theory $\mathcal{R} = (\Sigma, E, L, R)$, for which we assume that different labels in L name different rules in R, the model that we are seeking is a category $\mathcal{T}_{\mathcal{R}}(X)$ whose objects are equivalence classes of terms $[t] \in T_{\Sigma,E}(X)$ and whose morphisms are equivalence classes of "proof terms" representing proofs in rewriting deduction, i.e., concurrent $\mathcal{R}$-rewrites. The rules for generating such proof terms, with the specification of their respective domains and codomains, are given below; they just "decorate" with proof terms the rules 1–4 of rewriting logic. Note that we always use "diagrammatic" notation for morphism composition, i.e., $\alpha; \beta$ always means the composition of α *followed by* β.

1. **Identities.** For each $[t] \in T_{\Sigma,E}(X)$, $\dfrac{}{[t] : [t] \longrightarrow [t]}$.

2. **Σ-structure.** For each $f \in \Sigma_n$, $n \in \mathbb{N}$,

$$\frac{\alpha_1 : [t_1] \longrightarrow [t_1'] \quad \cdots \quad \alpha_n : [t_n] \longrightarrow [t_n']}{f(\alpha_1, \ldots, \alpha_n) : [f(t_1, \ldots, t_n)] \longrightarrow [f(t_1', \ldots, t_n')]}.$$

3. **Replacement.** For each rewrite rule $r : [t(\overline{x}^n)] \longrightarrow [t'(\overline{x}^n)]$ in R,

$$\frac{\alpha_1 : [w_1] \longrightarrow [w_1'] \quad \cdots \quad \alpha_n : [w_n] \longrightarrow [w_n']}{r(\alpha_1, \ldots, \alpha_n) : [t(\overline{w}/\overline{x})] \longrightarrow [t'(\overline{w'}/\overline{x})]}.$$

4. **Composition.** $\dfrac{\alpha : [t_1] \longrightarrow [t_2] \quad \beta : [t_2] \longrightarrow [t_3]}{\alpha; \beta : [t_1] \longrightarrow [t_3]}.$

Each of the above rules of generation defines a different operation taking certain proof terms as arguments and returning a resulting proof term as its result. In other words, proof terms form an algebraic structure $\mathcal{P}_{\mathcal{R}}(X)$ consisting of a graph with nodes $T_{\Sigma,E}(X)$, with identity arrows, and with operations f (for each $f \in \Sigma$), r (for each rewrite rule), and $\_;\_$ (for composing arrows). Our desired model $\mathcal{T}_{\mathcal{R}}(X)$ is the quotient of $\mathcal{P}_{\mathcal{R}}(X)$ modulo the following equations[3]:

---

[3] In the expressions appearing in the equations, when compositions of morphisms are involved, we always implicitly assume that the corresponding domains and codomains match.

1. **Category**
   (a) *Associativity.* For all $\alpha, \beta, \gamma,$ $(\alpha; \beta); \gamma = \alpha; (\beta; \gamma)$.
   (b) *Identities.* For each $\alpha : [t] \longrightarrow [t'],$ $\alpha; [t'] = \alpha$ and $[t]; \alpha = \alpha$.
2. **Functoriality of the Σ-Algebraic Structure.** For each $f \in \Sigma_n$,
   (a) *Preservation of Composition.* For all $\alpha_1, \ldots, \alpha_n, \beta_1, \ldots, \beta_n$,

$$f(\alpha_1; \beta_1, \ldots, \alpha_n; \beta_n) = f(\alpha_1, \ldots, \alpha_n); f(\beta_1, \ldots, \beta_n).$$

   (b) *Preservation of Identities.* $f([t_1], \ldots, [t_n]) = [f(t_1, \ldots, t_n)]$.
3. **Axioms in E.** For $t(x_1, \ldots, x_n) = t'(x_1, \ldots, x_n)$ an axiom in E, for all $\alpha_1, \ldots, \alpha_n,$ $t(\alpha_1, \ldots, \alpha_n) = t'(\alpha_1, \ldots, \alpha_n)$.
4. **Exchange.** For each $r : [t(x_1, \ldots, x_n)] \longrightarrow [t'(x_1, \ldots, x_n)]$ in R,

$$\frac{\alpha_1 : [w_1] \longrightarrow [w_1'] \quad \ldots \quad \alpha_n : [w_n] \longrightarrow [w_n']}{r(\overline{\alpha}) = r(\overline{[w]}); t'(\overline{\alpha}) = t(\overline{\alpha}); r(\overline{[w']})}.$$

Note that the set X of variables is actually a parameter of these constructions, and we need not assume X to be fixed and countable. In particular, for $X = \varnothing$, we adopt the notation $\mathcal{T}_\mathcal{R}$. The equations in 1 make $\mathcal{T}_\mathcal{R}(X)$ a category, the equations in 2 make each $f \in \Sigma$ a functor, and 3 forces the axioms E. The exchange law states that any rewriting of the form $r(\overline{\alpha})$ – which represents the *simultaneous* rewriting of the term at the top using rule r *and* "below," i.e., in the subterms matched by the variables, using the rewrites $\overline{\alpha}$ – is equivalent to the sequential composition $r(\overline{[w]}); t'(\overline{\alpha})$, corresponding to first rewriting on top with r and then below on the subterms matched by the variables with $\overline{\alpha}$, and is also equivalent to the sequential composition $t(\overline{\alpha}); r(\overline{[w']})$ corresponding to first rewriting below with $\overline{\alpha}$ and then on top with r. Therefore, the exchange law states that rewriting at the top by means of rule r and rewriting "below" using $\overline{\alpha}$ are processes that are independent of each other and can be done either simultaneously or in any order.

Since each proof term is a description of a concurrent computation, what these equations provide is an equational theory of *true concurrency* allowing us to characterize when two such descriptions specify the same abstract computation.

Note that, since $[t(x_1, \ldots, x_n)]$ and $[t'(x_1, \ldots, x_n)]$ can both be regarded as functors $\mathcal{T}_\mathcal{R}(X)^n \longrightarrow \mathcal{T}_\mathcal{R}(X)$, from the mathematical point of view the exchange law just asserts that r is a *natural transformation*.

Lemma 1. *[25] For each rewrite rule $r : [t(x_1, \ldots, x_n)] \longrightarrow [t'(x_1, \ldots, x_n)]$ in R, the family of morphisms*

$$\{r(\overline{[w]}) : [t(\overline{w}/\overline{x})] \longrightarrow [t'(\overline{w}/\overline{x})] \mid \overline{[w]} \in T_{\Sigma,E}(X)^n\}$$

is a natural transformation $r : [t(x_1, \ldots, x_n)] \Rightarrow [t'(x_1, \ldots, x_n)]$ between the functors $[t(x_1, \ldots, x_n)], [t'(x_1, \ldots, x_n)] : \mathcal{T}_\mathcal{R}(X)^n \longrightarrow \mathcal{T}_\mathcal{R}(X)$.

The category $\mathcal{T}_\mathcal{R}(X)$ is just one among many *models* that can be assigned to the rewrite theory $\mathcal{R}$. The general notion of model, called an $\mathcal{R}$-*system*, is defined as follows:

Definition 1. *Given a rewrite theory* $\mathcal{R} = (\Sigma, E, L, R)$, *an* $\mathcal{R}$-*system* $\mathcal{S}$ *is a category* $\mathcal{S}$ *together with:*

- *a* (Σ, E)-*algebra structure given by a family of functors*

$$\{f_{\mathcal{S}} : \mathcal{S}^n \longrightarrow \mathcal{S} \mid f \in \Sigma_n, n \in \mathbb{N}\}$$

  satisfying the equations E, *i.e., for any* $t(x_1, \ldots, x_n) = t'(x_1, \ldots, x_n)$ *in* E *we have an identity of functors* $t_{\mathcal{S}} = t'_{\mathcal{S}}$, *where the functor* $t_{\mathcal{S}}$ *is defined inductively from the functors* $f_{\mathcal{S}}$ *in the obvious way.*
- *for each rewrite rule* $r : [t(\overline{x})] \longrightarrow [t'(\overline{x})]$ *in* R *a natural transformation* $r_{\mathcal{S}} : t_{\mathcal{S}} \Rightarrow t'_{\mathcal{S}}$.

An $\mathcal{R}$-*homomorphism* $F : \mathcal{S} \longrightarrow \mathcal{S}'$ *between two* $\mathcal{R}$-*systems is then a functor* $F : \mathcal{S} \longrightarrow \mathcal{S}'$ *such that it is a* Σ-*algebra homomorphism, i.e.,* $f_{\mathcal{S}} * F = F^n * f_{\mathcal{S}'}$, *for each* f *in* Σ_n, $n \in \mathbb{N}$, *and such that "F preserves R," i.e., for each rewrite rule* $r : [t(\overline{x})] \longrightarrow [t'(\overline{x})]$ *in* R *we have the identity of natural transformations*[4] $r_{\mathcal{S}} * F = F^n * r_{\mathcal{S}'}$, *where* n *is the number of variables appearing in the rule. This defines a category* $\mathcal{R}$-**Sys** *in the obvious way.*

This category has the additional property that the homsets $\mathcal{R}$-**Sys**$(\mathcal{S}, \mathcal{S}')$ are themselves categories with morphisms, called *modifications*, given by natural transformations $\delta : F \Longrightarrow G$ between $\mathcal{R}$-homomorphisms $F, G : \mathcal{S} \longrightarrow \mathcal{S}'$ satisfying the identities

$$\delta^n * f_{\mathcal{S}'} = f_{\mathcal{S}} * \delta$$

for each $f \in \Sigma_n$, $n \in \mathbb{N}$. This category structure actually makes $\mathcal{R}$-**Sys** into a 2-category [22, 20].

A detailed proof of the following theorem on the existence of initial and free $\mathcal{R}$-systems for the more general case of conditional rewrite theories is given in [25], where the soundness and completeness of rewriting logic for $\mathcal{R}$-system models is also proved.

Theorem 1. $\mathcal{T}_{\mathcal{R}}$ *is an initial object in the category* $\mathcal{R}$-**Sys**. *More generally,* $\mathcal{T}_{\mathcal{R}}(X)$ *has the following universal property: Given an* $\mathcal{R}$-*system* $\mathcal{S}$, *each function* $F : X \longrightarrow |\mathcal{S}|$ *extends uniquely to an* $\mathcal{R}$-*homomorphism* $F^{\natural} : \mathcal{T}_{\mathcal{R}}(X) \longrightarrow \mathcal{S}$.

3   Functorial Semantics of Rewrite Theories

If categories present difficulties of exposition for readers unfamiliar with the area, 2-categories are even more of a challenge. What follows is a quite informal exposition of ideas that I consider important for further work on rewriting logic's model theory. It explains in some detail the main intuitions about the basic concepts, and then builds up to the main goal of this section, namely obtaining a 2-categorical semantics for rewrite theories and a 2-functorial semantics for their models. To help the reader grasp the basic intuitions, I first discuss classical Lawvere theories, a fundamental concept that is then generalized to that of Lawvere 2-theories.

---

[4] Note that we use diagrammatic order for the *horizontal*, $\alpha * \beta$, and *vertical*, $\gamma; \delta$, composition of natural transformations [22].

3.1   Classical Lawvere Theories

Equational logic was the first instance of a categorical logic considered by Lawvere in his doctoral dissertation [21]. Lawvere restricted his analysis to classical set-theoretic models. Given an equational theory (Σ, E), he exhibited a category with finite products $\mathcal{L}_{\Sigma, E}$ such that Σ-algebras A that satisfy the equations E can be put in one-to-one correspondence with functors $\tilde{A} : \mathcal{L}_{\Sigma, E} \to \mathbf{Set}$ that strictly preserve products; i.e., chosen products in $\mathcal{L}_{\Sigma, E}$ are mapped to cartesian products in $\mathbf{Set}$.

The category $\mathcal{L}_{\Sigma, E}$ is easy to describe. Its objects are the natural numbers. A morphism $[t] : n \longrightarrow 1$ is the equivalence class modulo the equations E of a Σ-term t whose variables are among $x_1, \ldots, x_n$. A morphism $n \longrightarrow m$ is an m-tuple of morphisms $n \longrightarrow 1$. Morphism composition is term substitution. For example, $([x_7 * x_3], [x_4 + x_5]); [x_2 + x_1] = [(x_4 + x_5) + (x_7 * x_3)]$. It is then easy to see that the object n is the nth product of the object 1 with projections $[x_1], \ldots, [x_n]$; and, more generally, that the product of the objects n and m is $n + m$. The functor $\tilde{A}$ associated to the algebra A sends the morphism $[t] : n \longrightarrow 1$ to the derived operation $t_A : A^n \longrightarrow A$ associated to the term t. Under this correspondence between algebras and functors, an equation $t = t'$ is satisfied by a (Σ, E)-algebra A iff $\tilde{A}([t]) = \tilde{A}([t'])$.

Lawvere also showed that *any* small category with finite products and with objects the natural numbers such that n is the nth power of 1 is isomorphic to $\mathcal{L}_{\Sigma, E}$ for some equational theory (Σ, E). In fact, many equivalent *presentations* by operations and equations are possible for the same concept, e.g., groups, and what the Lawvere theory provides is a *presentation-independent* description of the concepts and, by taking product-preserving functors into $\mathbf{Set}$, also of the models.

The analogous case of many-sorted equational logic was studied by Bénabou in his thesis [2]. The category $\mathcal{L}_{\Sigma, E}$ is constructed as in the unsorted case, but now it has as its set of objects the free monoid S^* generated by the set S of sorts.

Note that, since $T_{\Sigma, E}(X)$ is the free (Σ, E)-algebra on X and n is the nth power of 1 in $\mathcal{L}_{\Sigma, E}$, we have the chain of bijections

$$\mathcal{L}_{\Sigma, E}(n, m) \cong \mathcal{L}_{\Sigma, E}(n, 1)^m = T_{\Sigma, E}(\{x_1, \ldots, x_n\})^m \cong$$
$$\cong \mathbf{Set}(\{x_1, \ldots, x_m\}, T_{\Sigma, E}(\{x_1, \ldots, x_n\})) \cong$$
$$\cong Alg_{\Sigma, E}(T_{\Sigma, E}(\{x_1, \ldots, x_m\}), T_{\Sigma, E}(\{x_1, \ldots, x_n\}))$$

and since composition of morphisms in $\mathcal{L}_{\Sigma, E}$ is given by term substitution, this bijection preserves compositions and is actually an isomorphism of categories between $\mathcal{L}_{\Sigma, E}^{op}$ and the full subcategory of $Alg_{\Sigma, E}$ whose objects are the algebras of the form $T_{\Sigma, E}(\{x_1, \ldots, x_n\})$ for $n \in \mathbb{N}$.

3.2   Enriched Categories and 2-Categories

Given a closed symmetric monoidal category $\mathcal{V}$ [22], a $\mathcal{V}$-*category* $\mathcal{A}$ is a class of *objects* together with an object $\mathcal{A}(A, B) \in \mathcal{V}$ for each pair of objects A, B, and morphisms in $\mathcal{V}$:

- $id_A : I \longrightarrow \mathcal{A}(A, A)$, called *identities*, for each object A, where I denotes the unit object for the tensor product $\otimes$ in $\mathcal{V}$,
- $\_ * \_ : \mathcal{A}(A, B) \otimes \mathcal{A}(B, C) \longrightarrow \mathcal{A}(A, C)$, called *compositions*, for each triple A, B, C of objects,

satisfying the usual category axioms for identities and for associativity of composition that are expressed in terms of commutative diagrams in $\mathcal{V}$ in the usual way (see [18] for a detailed and far-reaching exposition). The most basic example of a $\mathcal{V}$-enriched category is the category $\mathcal{V}$ itself, with hom objects the internal homs given by the closed structure.

2-Categories. The category **Cat** of small categories is cartesian closed. A *2-category* $\mathcal{A}$ is just a **Cat**-category. This can be equivalently described as an ordinary category $\mathcal{A}_0$, called the *underlying category* of $\mathcal{A}$, together with, for each pair of objects $A, B \in \mathcal{A}_0$, a small category $\mathcal{A}(A, B)$ whose set of objects is precisely the homset $\mathcal{A}_0(A, B)$; all this is defined in such a way that there exist composition and identity functors satisfying the usual axioms for categories. The morphisms in $\mathcal{A}(A, B)$ are called *2-cells*; given $f, g : A \longrightarrow B$ in $\mathcal{A}_0$ we use the notation

$$\phi : f \Longrightarrow g : A \longrightarrow B$$

to indicate a 2-cell from f to g. The example of 2-category *par excellence* is of course **Cat**, where the 2-cells are natural transformations. The graphical notation used for the calculus of natural transformations, with vertical and horizontal composition, can be used in any 2-category. *Vertical* composition just means composition *inside* the category $\mathcal{A}(A, B)$, and is denoted $\_; \_$, whereas *horizontal* composition is application of one of the composition functors

$$\_ * \_ : \mathcal{A}(A, B) \otimes \mathcal{A}(B, C) \longrightarrow \mathcal{A}(A, C)$$

and the so called "double law" is just the preservation of (vertical) composition due to the functoriality of $\_ * \_$.

Given 2-categories $\mathcal{A}$ and $\mathcal{B}$, a *2-functor*

$$F : \mathcal{A} \longrightarrow \mathcal{B}$$

is a mapping sending objects to objects, morphisms to morphisms and 2-cells to 2-cells in such a way that domains, codomains, identities and compositions are all preserved.

A *2-natural transformation*

$$\eta : F \Longrightarrow G : \mathcal{A} \longrightarrow \mathcal{B}$$

between 2-functors F and G is a morphism of 2-functors consisting of an ordinary natural transformation

$$\eta : F_0 \Longrightarrow G_0 : \mathcal{A}_0 \longrightarrow \mathcal{B}_0$$

of the underlying functors such that for any 2-cell $\alpha : f \Longrightarrow g : A \longrightarrow A'$ in $\mathcal{A}$ the identity

$$F(\alpha) * \eta(A') = \eta(A) * G(\alpha)$$

holds.

A *modification*

$$\rho : \eta \rightsquigarrow \nu : F \Longrightarrow G : \mathcal{A} \longrightarrow \mathcal{B}$$

between 2-natural transformations η and ν is a morphism of 2-natural transformations consisting of an assignment of a 2-cell

$$\rho(A) : \eta(A) \Longrightarrow \nu(A)$$

to each object A in $\mathcal{A}$, such that for each morphism $f : A \longrightarrow A'$ in $\mathcal{A}$ the identity

$$F(f) * \rho(A') = \rho(A) * G(f)$$

holds.

Limits and Colimits. Many concepts of ordinary category theory generalize to $\mathcal{V}$-categories; in fact, ordinary categories are $\mathcal{V}$-categories for the special case $\mathcal{V} = \mathbf{Set}$. It is for example possible to define $\mathcal{V}$-enriched notions of limit and colimit [18]. For the case of 2-categories, we shall call such limits and colimits 2-*limits* and 2-*colimits*. The point is that in an enriched context more subtle kinds of limits and colimits are possible.

For example, a 2-*product* of A and B in $\mathcal{A}$ is an object P such that there is a natural isomorphism

$$\mathcal{A}(X, P) \cong \mathcal{A}(X, A) \times \mathcal{A}(X, B)$$

where $\mathcal{A}(X, A) \times \mathcal{A}(X, B)$ denotes the product in $\mathbf{Cat}$ of the given categories. Similarly, an object F is a 2-*final* object in $\mathcal{A}$ if there is a natural isomorphism

$$\mathcal{A}(X, F) \cong 1$$

for 1 the category consisting of just one object and its identity morphism which is the final object in $\mathbf{Cat}$. We say that $\mathcal{A}$ has *finite 2-products* if it has binary 2-products and a 2-final object; if all finite 2-limits exist, we say that $\mathcal{A}$ has finite 2-*limits*. See [34] for a careful investigation of 2-limits. A 2-functor $F : \mathcal{A} \longrightarrow \mathcal{B}$ *preserves 2-products* if it preserves all existing 2-products and the 2-final object. Similarly, F is called *finitely (2-)continuous* if it preserves all the finite 2-limits that exist in $\mathcal{A}$.

3.3   Lawvere 2-Theories

A Lawvere 2-theory is a 2-category $\mathcal{L}$ with finite 2-products and having as objects the natural numbers[5] with n the nth 2-power of 1.

---

[5] The definition of a many-sorted Lawvere-Bénabou 2-theory is entirely analogous, and is left to the reader.

An important example is provided by the following Lawvere 2-theory $\mathcal{L}_\mathcal{R}$ associated to a rewrite theory $\mathcal{R}$. We define $\mathcal{L}_\mathcal{R}(n,1) = \mathcal{T}_\mathcal{R}(\{x_1, \ldots, x_n\})$ with horizontal composition of

$$\overline{\alpha}^n : \overline{[w]}^n \Longrightarrow \overline{[w']}^n : m \longrightarrow n$$

with

$$\beta : [t] \Longrightarrow [t'] : n \longrightarrow 1$$

given by

$$[t(\overline{\alpha}^n)]; \beta(\overline{w'}^n/\overline{x}^n) : [t(\overline{w}^n/\overline{x}^n)] \Longrightarrow [t'(\overline{w'}^n/\overline{x}^n)] : m \longrightarrow 1$$

where $\beta(\overline{w'}^n/\overline{x}^n)$ denotes the substitution of $\overline{w'}^n$ in β defined in Fact 3.12 of [25]. It follows easily from the equations defining $\mathcal{T}_\mathcal{R}(\{x_1, \ldots, x_n\})$ that this defines indeed a 2-category. It is the exact analogue of a classical Lawvere theory of the form $\mathcal{L}_{\Sigma,E}$ for (Σ, E) an equational theory, but now we have 2-cells corresponding to rewritings between $[t], [t'] \in T_{\Sigma,E}(\{x_1, \ldots, x_n\})$.

3.4 2-Functorial Semantics of Rewrite Theories

Given a Lawvere 2-theory $\mathcal{L}$, we define the 2-category $\mathbf{Mod}(\mathcal{L})$ of its models whose objects are finite product preserving 2-functors

$$M : \mathcal{L} \longrightarrow \mathbf{Cat},$$

whose morphisms are 2-natural transformations between such functors, and whose 2-cells are modifications. In a way entirely analogous to that of classical Lawvere theories we can now relate the model theory of unconditional rewrite theories developed in Section 2.2 with a (2-)functorial semantics as follows:

For $\mathcal{R}$ an unconditional rewrite theory, given an $\mathcal{R}$-system $\mathcal{S}$, the assignments:

- $n \mapsto \mathcal{S}^n$
- $[t(x_1, \ldots, x_n)] \mapsto t_\mathcal{S} : \mathcal{S}^n \longrightarrow \mathcal{S}$
- $\alpha : [t(x_1, \ldots, x_n)] \longrightarrow [t'(x_1, \ldots, x_n)] \mapsto \alpha_\mathcal{S} : t_\mathcal{S} \Longrightarrow t'_\mathcal{S}$

determine a 2-product preserving 2-functor

$$\tilde{\mathcal{S}} : \mathcal{L}_\mathcal{R} \longrightarrow \mathbf{Cat}$$

in such a way that $\mathcal{R}$-homomorphisms

$$F : \mathcal{S} \longrightarrow \mathcal{S}'$$

are in 1-1 correspondence with 2-natural transformations

$$\tilde{F} : \tilde{\mathcal{S}} \Longrightarrow \tilde{\mathcal{S}}',$$

and modifications

$$\rho : F \Longrightarrow G$$

are in 1-1 correspondence with modifications

$$\tilde{\rho} : \tilde{F} \rightsquigarrow \tilde{G}.$$

Requiring preservation of chosen 2-products "on the nose," this yields an iso-morphism of 2-categories

$$\mathcal{R}\text{-}\mathbf{Sys} \cong \mathbf{Mod}(\mathcal{L}_{\mathcal{R}}).$$

In a way analogous to classical Lawvere theories, we can discover $\mathcal{L}_{\mathcal{R}}^{op}$ "inside the belly" of $\mathcal{R}$-**Sys**. Indeed, using the freeness of $\mathcal{T}_{\mathcal{R}}(\{x_1, \ldots, x_n\})$ (where $\mathcal{T}_{\mathcal{R}}$ can be viewed as a 2-functor $\mathcal{T}_{\mathcal{R}} : \mathbf{Cat} \longrightarrow \mathcal{R}\text{-}\mathbf{Sys}$ 2-left adjoint to the obvious forgetful 2-functor, and $\{x_1, \ldots, x_n\}$ can be understood as a discrete category) and that n is the nth 2-power of 1 in $\mathcal{L}_{\mathcal{R}}$, we have the chain of isomorphisms of categories

$$\mathcal{L}_{\mathcal{R}}(n, m) \cong \mathcal{L}_{\mathcal{R}}(n, 1)^m = \mathcal{T}_{\mathcal{R}}(\{x_1, \ldots, x_n\})^m \cong$$
$$\cong \mathbf{Cat}(\{x_1, \ldots, x_m\}, \mathcal{T}_{\mathcal{R}}(\{x_1, \ldots, x_n\})) \cong$$
$$\cong \mathcal{R}\text{-}\mathbf{Sys}(\mathcal{T}_{\mathcal{R}}(\{x_1, \ldots, x_m\}), \mathcal{T}_{\mathcal{R}}(\{x_1, \ldots, x_m\}))$$

and since composition of morphisms and 2-cells in $\mathcal{L}_{\mathcal{R}}$ is given by substitution, this bijection preserves the 2-category structure and corresponds to an isomor-phism of 2-categories between $\mathcal{L}_{\mathcal{R}}^{op}$ and the full 2-subcategory of $\mathcal{R}$-**Sys** whose objects are the $\mathcal{R}$-systems of the form $\mathcal{T}_{\mathcal{R}}(\{x_1, \ldots, x_n\})$ for $n \in \mathbb{N}$.

More generally, we can define models of a Lawvere 2-theory $\mathcal{L}$ in a 2-category $\mathcal{A}$ with finite 2-products. Such models are defined as 2-product preserving 2-functors

$$M : \mathcal{L} \longrightarrow \mathcal{A}.$$

This, together with 2-natural transformations and modifications defines a 2-category $\mathbf{Mod}(\mathcal{L}, \mathcal{A})$. Similarly, given a rewrite theory $\mathcal{R}$, we can easily generalize the definition of the model 2-category $\mathcal{R}$-**Sys** by defining the models of $\mathcal{R}$ not on **Cat** but on any 2-category $\mathcal{A}$ with finite 2-products. In this way we obtain a 2-category that we denote $\mathcal{R}$-$\mathcal{A}$, so that the special case $\mathcal{R}$-**Sys** is simply $\mathcal{R}$-**Cat**. By requiring preservation of chosen 2-products "on the nose" we then obtain an isomorphism of 2-categories $\mathcal{R}$-$\mathcal{A} \cong \mathbf{Mod}(\mathcal{L}_{\mathcal{R}}, \mathcal{A})$.

4   Morphisms of Rewrite Theories and Parameterization

The 2-functorial semantics just developed allows us to investigate two notions of morphism between rewrite theories and also to give semantics to parameterized rewrite theories. The intuitions behind those two notions are as follows:

1. A *prescriptive* notion, in which a basic rewrite step in one theory is refined into a specified proof term, involving a possibly complex concurrent com-bination of rewritings. This generalizes the *implementation morphisms* that Ugo Montanari and I defined for Petri nets [27].

2. A notion of *implementable* morphism, where it is only required that the second theory should be capable of simulating any of the basic rewrites of the first, but no indication is given about how this should be done when several different ways are possible.

The precise definition of the prescriptive notion follows easily from our discussion of functorial semantics. To ease the exposition, this paper has presented the ideas in an unsorted context; however, everything has a straightforward generalization to the many-sorted case and I will assume from now on that we are dealing with many-sorted rewrite theories. Given two such rewrite theories, $\mathcal{R}$ and $\mathcal{R}'$ a *morphism*

$$H : \mathcal{R} \longrightarrow \mathcal{R}'$$

between them is an *interpretation* of $\mathcal{R}$ in the 2-category $\mathcal{L}'_{\mathcal{R}}$, i.e., an object H of the 2-category $\mathcal{R}\text{-}\mathcal{L}_{\mathcal{R}'}$. In addition – for considerations of syntactical ease – we impose on H the condition of mapping basic sorts of $\mathcal{R}$ to basic sorts of $\mathcal{R}'$. Therefore, what this interpretation H really means for $\mathcal{R} = (\Sigma, E, L, R)$ and $\mathcal{R}' = (\Sigma', E', L', R')$ is the specification of:

- An equational theory morphism for $H_0 : (\Sigma, E) \longrightarrow (\Sigma', E')$ mapping basic sorts to basic sorts, basic operations to (possibly derived) operations, and the axioms E to provable consequences of the axioms E' under the corresponding translation of terms, that is, we have $E' \vdash H(E)$.
- An assignment to each rewrite rule $r : [t(\overline{x}^n)] \longrightarrow [t'(\overline{x}^n)]$ in R (with $\overline{x}^n$ a sequence of many-sorted variables of sorts $\overline{s}^n$) of a 2-cell (that is, an equivalence class of proof terms)

$$H(r) : H_0([t]) \Longrightarrow H_0([t'])$$

Therefore, this notion corresponds to specifying a *simulation* of basic actions of the theory $\mathcal{R}$ as possibly complex actions of the theory $\mathcal{R}'$.

A nice consequence of the 2-functorial semantics of rewrite theories developed in Section 3.4 is that defining composition of morphisms is now trivial. Given morphisms

$$G : \mathcal{R} \longrightarrow \mathcal{R}' \qquad H : \mathcal{R}' \longrightarrow \mathcal{R}''$$

we can recall the isomorphisms

$$\mathcal{R}\text{-}\mathcal{L}_{\mathcal{R}'} \cong \mathbf{Mod}(\mathcal{L}_{\mathcal{R}}, \mathcal{L}'_{\mathcal{R}}) \qquad \mathcal{R}'\text{-}\mathcal{L}_{\mathcal{R}''} \cong \mathbf{Mod}(\mathcal{L}'_{\mathcal{R}}, \mathcal{L}''_{\mathcal{R}})$$

and define $G * H$ as the interpretation associated by the isomorphism

$$\mathcal{R}\text{-}\mathcal{L}_{\mathcal{R}''} \cong \mathbf{Mod}(\mathcal{L}_{\mathcal{R}}, \mathcal{L}''_{\mathcal{R}})$$

to the composition $\tilde{G} * \tilde{H}$ where $\tilde{G}$ and $\tilde{H}$ are the corresponding 2-product preserving 2-functors associated to G and H by the above isomorphisms. Therefore, rewrite theories and morphisms between them form a category that we denote

RWTh.

Another very important consequence of the functorial semantics is that we can give a model-theoretic semantics to parameterized rewrite theories. By a *parameterized rewrite theory* we mean a morphism

$$H : \mathcal{P} \longrightarrow \mathcal{R}.$$

In this terminology, we call $\mathcal{P}$ the *parameter* rewrite theory and $\mathcal{R}$ the *body*. Notice that composition along $\tilde{H}$ defines a 2-functor

$$H^{\flat} : \mathcal{R}\text{-}\mathbf{Sys} \longrightarrow \mathcal{P}\text{-}\mathbf{Sys}.$$

Thanks to general results about locally finitely presentable 2-categories (see [19], Prop. 9.13 and Section 10) $H^{\flat}$ has a left adjoint

$$H^{*} : \mathcal{P}\text{-}\mathbf{Sys} \longrightarrow \mathcal{R}\text{-}\mathbf{Sys}$$

which we adopt as our model-theoretic semantics for parameterized rewrite theories.

The definition of the category of *implementable rewrite theory morphisms* is as follows. A morphism of this kind between $\mathcal{R}$ and $\mathcal{R}'$ is of the form

$$H_0 : \mathcal{R} \longrightarrow \mathcal{R}'$$

with H_0 the underlying equational theory morphism (between the underlying equational theories (Σ, E) and (Σ', E')) of a morphism

$$H : \mathcal{R} \longrightarrow \mathcal{R}'.$$

Therefore, H_0 describes a *general way* in which a refinement of $\mathcal{R}$ by $\mathcal{R}'$ can *be implemented*, but we have forgotten the particular choice of implementation. Since there is a forgetful functor

$$\mathbf{RWTh} \longrightarrow \mathbf{EqtlTh}$$

sending $\mathcal{R} = (\Sigma, E, L, R)$ to the equational theory (Σ, E) and H to H_0, this defines a category of *implementable morphisms* which we can view as a quotient of $\mathbf{RWTh}$ under the equivalence relation on morphisms induced by such a forgetful functor.

5   Related Work and Concluding Remarks

For the case of a rewrite theory with $E = \varnothing$, a somewhat different construction of $\mathcal{L}_{\mathcal{R}}$ was given by D.E. Rydeheard and J.G. Stell in [32] who used 2-category presentations by generators and relations. With the aim of unifying the fix-point constructions of continuous functions and of recursive domain equations, A. Pitts [31] defined many-sorted Lawvere 2-theories with the additional structure of fixpoint operators and considered 2-product preserving 2-functors from such theories into 2-categories with 2-products and with fixpoint operators. In the same work, Pitts also introduced a logic of fixpoints whose fixpoint-free fragment essentially corresponds to unconditional rewriting logic.

In the classical case, in order to provide a functorial semantics in the more general setting of conditional theories and of "essentially algebraic" theories, Lawvere theories are generalized to categories with finite limits, presented by means of "sketches" in the sense of Ehresmann (see [1] for a good exposition) or by other means. The categories that are equivalent to categories of models for theories with finite limits were first characterized by Gabriel and Ulmer and are called *locally finitely presentable categories* [15]. The extension of the present work to *conditional* rewrite theories is sketched in [24] and has been further developed by Miyoshi in [30]. The key additional 2-limits needed beyond 2-products are *subequalizers* or, more generally, *inserters*, and models are then finitely continuous 2-functors from the corresponding 2-lim theories.

The point of view presented in this paper has been further generalized and exploited in Pisa to provide very useful connections with other concurrency theory models. Corradini, Gadducci and Montanari [9] provide a uniform construction for $\mathcal{L}_\mathcal{R}$ and for a *sesqui-category* model, similar to $\mathcal{L}_\mathcal{R}$ but satisfying fewer equations, that has been proposed by Stell [33]. They associate posets of partial orders of events to both models, and make the important observation that when a rewrite rule is not right linear – that is, when it has a repeated occurrence of a variable in its righthand side – then the poset associated to $\mathcal{L}_\mathcal{R}$ is not a prime algebraic domain, whereas the poset of the sesqui-category model is. In this way, the relationship between rewriting logic models and event structures is clarified. What happens is that, when rules are not right linear, $\mathcal{L}_\mathcal{R}$ is in a sense too abstract, because what is one event in one proof term may – because of repetition of variables – become several events in a proof term equivalent to it by the exchange axiom; in the sesqui-category model the exchange axiom does not hold, and therefore those computations are considered different.

Yet another important direction in which this work has been generalized is to provide models for *tile logic* [16, 17, 3], an algebraic model of synchronous concurrent systems. The key idea is to generalize Lawvere 2-theories to Lawvere double-theories, sometimes relaxing cartesian products to monoidal products as done in [28, 3, 5].

A theme common to the just-cited papers [28, 3, 5] is the use of membership equational logic [26] as a semantic framework to define categorical structures. This is the approach taken also in [4] to generalize the rewriting logic model theory of [25] to the case of rewrite theories such that: (1) their underlying equational theory is a theory in membership equational logic [26], and (2) some of the arguments in an operator f may be *frozen*, so that rewriting is forbidden in such arguments. A treatment of rewrite theories and rewrite theory morphisms along the same lines is currently under development in joint work with Roberto Bruni. Rewrite theory morphisms in this more general setting are important in several respects. First of all, they are used in the instantiation of parameterized modules in Maude [11, 8]. Furthermore, certain forms of rewrite theory morphism are very useful to describe *simulations* between different concurrent systems

[29, 23], which can then be used to define abstractions and implementation maps, and to preserve temporal logic properties which can be more easily verified on simpler and possibly finite state systems thanks to such simulations.

References

1. M. Barr and C. Wells. *Toposes, Triples and Theories*. Springer-Verlag, 1985.
2. J. Bénabou. Structures algébriques dans les catégories. *Cahiers de Topologie et Géometrie Différentielle*, 10:1–126, 1968.
3. R. Bruni. *Tile Logic for Synchronized Rewriting of Concurrent Systems*. PhD thesis, Dipartimento di Informatica, Università di Pisa, 1999. Technical Report TD-1/99. http://www.di.unipi.it/phd/tesi/tesi_1999/TD-1-99.ps.gz.
4. R. Bruni and J. Meseguer. Generalized rewrite theories. In J. Baeten, J. Lenstra, J. Parrow, and G. Woeginger, editors, *Proceedings of ICALP 2003, 30th International Colloquium on Automata, Languages and Programming*, volume 2719 of *Springer LNCS*, pages 252–266, 2003.
5. R. Bruni, J. Meseguer, and U. Montanari. Symmetric monoidal and cartesian double categories as a semantic framework for tile logic. *Mathematical Structures in Computer Science*, 12:53–90, 2002.
6. R. Burstall and J. Goguen. Putting theories together to make specifications. In R. Reddy, editor, *Proceedings, Fifth International Joint Conference on Artificial Intelligence*, pages 1045–1058. Department of Computer Science, Carnegie-Mellon University, 1977.
7. M. Clavel, F. Durán, S. Eker, P. Lincoln, N. Martí-Oliet, J. Meseguer, and J. Quesada. Maude: specification and programming in rewriting logic. *Theoretical Computer Science*, 285:187–243, 2002.
8. M. Clavel, F. Durán, S. Eker, P. Lincoln, N. Martí-Oliet, J. Meseguer, and C. Talcott. Maude 2.0 Manual. June 2003, http://maude.cs.uiuc.edu.
9. A. Corradini, F. Gadducci, and U. Montanari. Relating two categorical models of term rewriting. In J. Hsiang, editor, *Proc. Rewriting Techniques and Applications, Kaiserslautern*, pages 225–240, 1995.
10. N. Dershowitz and J.-P. Jouannaud. Rewrite systems. In J. van Leeuwen, editor, *Handbook of Theoretical Computer Science, Vol. B*, pages 243–320. North-Holland, 1990.
11. F. Durán and J. Meseguer. On parameterized theories and views in Full Maude 2.0. In K. Futatsugi, editor, *Proc. 3rd. Intl. Workshop on Rewriting Logic and its Applications*. ENTCS, Elsevier, 2000.
12. H. Ehrig and B. Mahr. *Fundamentals of Algebraic Specifications 1, Equations and Initial Semantics*, volume 6 of *EATCS Monographs on Theoretical Computer Science*. Springer-Verlag, 1985.
13. H. Ehrig and B. Mahr. *Fundamentals of Algebraic Specifications 2, Module Specifications and Constraints*, volume 21 of *EATCS Monographs on Theoretical Computer Science*. Springer-Verlag, 1990.
14. S. Eilenberg and J. C. Moore. Adjoint functors and triples. *Illinois J. Math.*, 9:381–398, 1965.
15. P. Gabriel and F. Ulmer. *Lokal präsentierbare Kategorien*. Springer Lecture Notes in Mathematics No. 221, 1971.
16. F. Gadducci. *On the Algebraic Approach to Concurrent Term Rewriting*. PhD thesis, Dipartimento di Informatica, Università di Pisa, Mar. 1996. Technical Report TD-2/96.

17. F. Gadducci and U. Montanari. The tile model. In G. Plotkin, C. Stirling, and M. Tofte, editors, *Proof, Language and Interaction: Essays in Honour of Robin Milner*. The MIT Press, 2000. http://www.di.unipi.it/~ugo/festschrift.ps.

18. G. Kelly. *Basic Concepts of Enriched Category Theory*. Cambridge Univ. Press, 1982.

19. G. Kelly. Structures defined by finite limits in the enriched context, I. *Cahiers de Topologie et Géometrie Différentielle*, 23:3–42, 1982.

20. G. Kelly and R. Street. Review of the elements of 2-categories. In G. Kelly, editor, *Category Seminar, Sydney 1972/73*, pages 75–103. Springer Lecture Notes in Mathematics No. 420, 1974.

21. F. W. Lawvere. Functorial semantics of algebraic theories. *Proceedings, National Academy of Sciences*, 50:869–873, 1963. Summary of Ph.D. Thesis, Columbia University.

22. S. MacLane. *Categories for the Working Mathematician*. Springer-Verlag, 1971.

23. N. Martí-Oliet, J. Meseguer, and M. Palomino. Theoroidal maps as algebraic simulations. To appear in Proc. WADT'04, Springer LNCS, 2004.

24. J. Meseguer. Rewriting as a unified model of concurrency. Technical Report SRI-CSL-90-02, SRI International, Computer Science Laboratory, February 1990. Revised June 1990.

25. J. Meseguer. Conditional rewriting logic as a unified model of concurrency. *Theoretical Computer Science*, 96(1):73–155, 1992.

26. J. Meseguer. Membership algebra as a logical framework for equational specification. In F. Parisi-Presicce, editor, *Proc. WADT'97*, pages 18–61. Springer LNCS 1376, 1998.

27. J. Meseguer and U. Montanari. Petri nets are monoids. *Information and Computation*, 88:105–155, 1990.

28. J. Meseguer and U. Montanari. Mapping tile logic into rewriting logic. In F. Parisi-Presicce, editor, *Recent Trends in Algebraic Development Techniques, 12th International Workshop, WADT'97, Tarquinia, Italy, June 3–7, 1997, Selected Papers*, volume 1376 of *Lecture Notes in Computer Science*, pages 62–91. Springer-Verlag, 1998.

29. J. Meseguer, M. Palomino, and N. Martí-Oliet. Equational abstractions. in Proc. CADE-19, Springer LNCS, Vol. 2741, 2–16, 2003.

30. H. Miyoshi. Modelling conditional rewriting logic in structured categories. In J. Meseguer, editor, *Proc. First Intl. Workshop on Rewriting Logic and its Applications*, volume 4 of *Electronic Notes in Theoretical Computer Science*. Elsevier, 1996.

31. A. Pitts. An elementary calculus of approximations. Unpublished manuscript, University of Sussex, December 1987.

32. D. Rydeheard and J. Stell. Foundations of equational deduction: A categorical treatment of equational proofs and unification algorithms. In *Proceedings of the Summer Conference on Category Theory and Computer Science, Edinburgh, Sept. 1987*, pages 114–139. Springer LNCS 283, 1987.

33. J. Stell. Modelling term rewriting systems by sesqui-categories. Technical Report TR94-02, Keele University, 1994. Also in shorter form in Proc. C.A.E.N., 1994, pp. 121–127.

34. R. Street. Limits indexed by category-valued 2-functors. *J. Pure Appl. Algebra*, 8:149–181, 1976.

Expander2

Towards a Workbench for Interactive Formal Reasoning

Peter Padawitz

University of Dortmund,
Dortmund, Germany
peter.padawitz@udo.edu

Abstract. Expander2 is a flexible multi-purpose workbench for interactive rewriting, verification, constraint solving, flow graph analysis and other procedures that build up proofs or computation sequences. Moreover, tailor-made interpreters display terms as two-dimensional structures ranging from trees and rooted graphs to a variety of pictorial representations that include tables, matrices, alignments, piles, partitions, fractals and turtle systems.

Proofs and computations performed with Expander2 follow the rules and the semantics of swinging types. Swinging types are based on many-sorted predicate logic and combine *visible* constructor-based types with *hidden* state-based types. The former come as *initial* term models, the latter as *final* models consisting of context interpretations. Relation symbols are interpreted as least or greatest solutions of their respective axioms. This paper presents an overview of Expander2 with particular emphasis on the system's prover capabilities.

1 Introduction

The following design goals distinguish Expander2 from many other proof editors or tools using formal methods:

- Expander2 provides several representations of formal expressions and allows the user to switch between linear, tree-like and pictorial ones when executing a proof or computation on formulas or terms.
- Proof and computation steps take place at three levels of interaction: the simplifier automates routine steps, axiom-triggered computations are performed by narrowing and rewriting, analytical rules like induction and coinduction are applied locally and stepwise.
- The underlying logic is general enough to cover a wide range of applications and to admit the easy integration of special structures or methods by adding or exchanging signatures, axioms, theorems or inference rules including built-in simplifications.
- Expander2 has an intelligent GUI that interprets user entries in dependence of the current values of global state variables. This frees the user from entering input that can be deduced from the context in which the system actually works.

H.-J. Kreowski et al. (Eds.): Formal Methods (Ehrig Festschrift), LNCS 3393, pp. 236–258, 2005.
© Springer-Verlag Berlin Heidelberg 2005

Proofs and computations performed with the system are correct with respect to the semantics of swinging types [16–19]. A swinging type is a functional-logic specification consisting of a many-sorted signature and a set of (generalized) Horn or co-Horn axioms (see section 3) that define relation symbols as least or greatest fixpoints and function symbols in accordance with the initial resp. final model induced by the specification.

Sortedness is only implicit because otherwise the proof and computation processes would become unnecessarily complicated. If used as a specification environment, the main purpose of Expander2 is *proof* editing and not *type* checking. Therefore, the syntax of signatures is kept as minimal as possible. The only explicit distinction between different types is the one between constants on the one hand and functions and relations on the other hand, expressed by the distinction between first-order variables (`fovars`) and higher-order variables (`hovars`). Proofs or computations that depend on a finer sort distinction can always be performed by introducing and using suitable *membership predicates*.

The prover features of Expander2 do not aim at the complete automation of proof processes. Instead, they support *natural* derivations, which humans can comprehend and thus control easily. Natural deduction avoids skolemization and other extensive normalizations that make formulas unreadable and thus inappropriate for interactive proving. For instance, the simplifier (see Section 5), which turns formulas into equivalent "simplified" ones, prefers implication to negation.

Of course, many conjectures can be proved both comprehensibly and efficiently without any human intervention into the proof process. Such proofs often follow particular schemas and thus may be candidates for derived inference rules. However, proofs of program correctness usually do not fall into this category, especially if induction or coinduction is involved and the original conjecture must be generalized in a particular way.

In fact, the simplifier of Expander2 performs certain normalizations. But they are in compliance with natural deduction and deviate from classical normalizations insofar as, for instance, implications and quantifiers are not eliminated by introducing negations and new signature symbols, respectively. On the contrary, the simplifier eliminates negation symbols by moving them to literal positions and then are removed completely by transforming negated (co)predicates into their complements. Axioms for relations and their complements can be constructed from each other: If P is a predicate specified by Horn axioms, then these axioms can be transformed systematically into co-Horn axioms for the copredicate *not_P*, and vice versa. This follows from the fact that relation symbols are interpreted by the least resp. greatest solutions of their axioms provided that these are negation-free and thus induce monotonic consequence operators [16–18].

Expander2 has been written in O'Haskell [12], an extension of Haskell [8] with object-oriented features for reactive programming and a typed interface to Tcl/Tk for developing GUIs. Besides providing a comfortable GUI the overall design goals of Expander2 were to integrate testing, proving and visualizing deductive methods, to admit several degrees of interaction and to keep the system open for extensions or adaptations of individual components to changing demands.

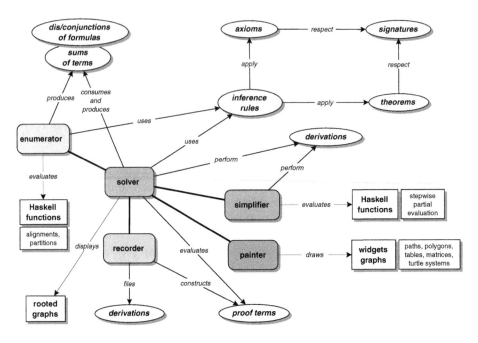

Fig. 1. Components of Expander2.

2   System Components

The main components of Expander2 are two copies of a **solver**, a **painter**, a **simplifier** an **enumerator** and a **recorder** that saves proofs and other computation sequences as well as executable proof terms. As Fig. 1 indicates, these components work together via several interfaces. For instance, the painter is used for drawing normal forms or solutions produced by the solver.

The **solver** is accessed via a window for entering, editing and displaying a list of trees (or graphs) that represents a disjunction or conjunction of logical formulas or a sum of algebraic terms (see Fig. 2). By moving the slider below the canvas of the solver window one selects the summand/factor to be shown on the canvas. If the *parse text* resp. *parse tree* button is pushed, the linear representation of a term or formula in the solver's text field is translated into an equivalent tree representation on the canvas and vice versa. Both representations are editable. As a linear representation is edited by selecting substrings, the tree representation is edited by selecting subtrees or nodes or redirecting edges.

The **painter** consists of several widget interpreters from which one is selected and applied to the current trees or parts of them. The resulting pictorial representations are displayed in a painter window. Pictures can be edited in the painter window and completed to *widget graphs*. Widgets are built up of path, polygon and *turtle action* constructors that admit the definition of a variety of pictorial representations ranging from tables and matrices via string alignments, piles and partitions to complex fractals generated by *turtle systems* [24]. The

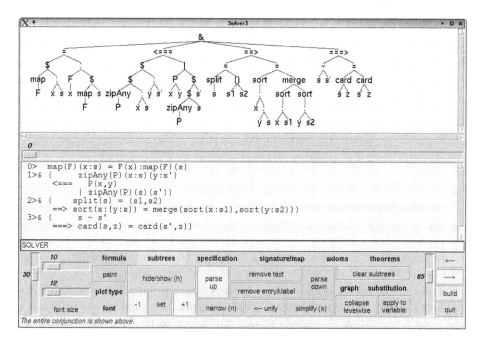

Fig. 2. The solver window.

latter define pictures in terms of sequence of basic actions that a turtle would perform when it draws the picture while moving over the canvas of a window. The turtle works recursively in two ways: it maintains a stack of positions and orientations where it may return to, and it may create trees whose pictorial representations are displayed at its current position.

The solver and its associated painter are fully synchronized: the selection of a tree in the solver window is automatically followed by a selection of the tree's pictorial representation in the painter window and vice versa. Hence rewriting, narrowing and simplification steps can be carried out from either window.

The **enumerator** provides algorithms that enumerate trees or graphs and pass their results both to the solver and the painter. Currently, two algorithms are available: a generator of all sequence alignments [5, 20] satisfying constraints that are partly given by axioms, and a generator of all nested partitions of a list with a given length and satisfying constraints given by particular predicates. The painter displays an alignment in the way DNA sequences are usually visualized. A nested partition is displayed as a rectangular dissection of a square where different levels are colored differently.

The user of Expander2 operates on specifications (consisting of signatures and axioms), theorems, substitutions, trees (representing algebraic terms, logical formulas or transition systems to be evaluated, solved, proved, or executed, respectively) via commands selected from the solver's menus (see Fig. 2). Sliders control the layout of a tree. With the slider in the middle of a solver window, one browses among several trees. All these actions yield input for the solver and may modify its **state variables**. Hence the solver can be regarded as a finite automaton whose actions are triggered not only by user input, but also by the actual system state. Here are the main state variables:

- The current **axioms** and **theorems** are applied to conjectures and build up the high- or medium-level steps of a computation or proof. Axioms and theorems are applied by narrowing or rewriting. A narrowing/rewriting step starts with unifying/matching a subtree (the redex) with/against an axiom. Narrowing applies (guarded) Horn or co-Horn clauses, rewriting applies only unconditional, but possibly guarded equations. The guard of an axiom is a subformula to be solved before the axiom is applied.

- The widget interpreter **pictEval** recognizes paintable terms or formulas and transforms them into their pictorial representations (see above).

- The current **proof** records the sequence of derivation steps performed since the last initialization of the list of current trees. Each element of the current proof consists of a description of a rule application, the resulting list of current trees and the resulting values of state variables.

- The current **proof term** represents the current proof as an executable expression for the purpose of later proof checking. It is built up automatically when a derivation is carried out and can be saved to a user-defined file. A saved proof term is loaded by writing its name into the entry field and pushing *check proof term from file*. This action overwrites the current proof term. The proof represented by the loaded proof is carried out stepwise (and thus checked) on the displayed tree by pushing only the ---> button. Each click triggers a proof step.

- The current **signature** consists of symbols denoting *basic specifications* consisting of signatures and axioms, *predicates* interpreted as the least solutions of their (Horn) axioms, *copredicates* interpreted as the greatest solutions of their (co-Horn) axioms, *constructors* and *cofunctions* for building up data, *defined functions* specified by (Horn) axioms or implemented as Haskell functions called by the simplifier, first-order variables that may be instantiated by terms or formulas, and higher-order variables that may be instantiated by functions or relations. Most built-in signature symbols have the syntax and semantics as the synonymous Haskell functions (see [21]).

- The current **substitution** maps the variables of its domain to terms over the current signature. It is generated, modified and applied by particular commands.

- **treeMode** indicates whether the list **trees** of current trees (or other rooted graphs) is a singleton or represents a disjunction or conjunction of formulas or a sum (= disjoint union) of terms. *True*, *False* and *()* are the respective zero elements. The slider between the canvas and the text field of a solver window allows one to browse among the current trees and to select the one to be displayed on the canvas.

- The list **treePoss** consists of the positions of selected subtrees of the actually displayed tree. Subtrees are selected (and moved) by pushing the left mouse button while placing the cursor over their roots.

- **varCounter** maps a variable x to the maximal index i such that x_i occurs in the current proof. *varCounter* is updated when new variables are needed.
- Further state variables may occur as function symbols in terms or formulas. Their values are stored by rewriting steps, retrieved and modified by simplification steps and represented pictorially in a painter window by applying suitable widget interpreters. The purpose of these state variables is to hide complex function parameters from the screen whose current values are needed by the simplifier for evaluating built-in iterative Haskell functions [21].

Expander2 allows the user to control proofs and computations at three levels of interaction. At the high level, analytic and synthetic inference rules and other syntactic transformations are applied individually and locally to selected subtrees. The rules cover single axiom applications, substitution or unification steps, Noetherian, Hoare, subgoal or fixpoint induction and coinduction. Derivations are correct if, in the case of trees representing terms, their sum is equivalent to the sum of their successors or, in the case of trees representing formulas, their disjunction/conjunction is implied by the disjunction/conjunction of their successors. The underlying models are determined by built-in data types and the least/greatest interpretation of Horn/co-Horn axioms. Incorrect deduction steps are recognized and cause a warning. All proper tree transformations are recorded, be they correct proofs or other transformations.

At the medium level, rewriting and narrowing realize the iterated and exhaustive application of all axioms for the defined functions, predicates and copredicates of the current signature. Rewriting terminates with **normal forms**, i.e. terms consisting of constructors and variables. Terminating narrowing sequences end up with the formula `True`, `False` or *solved formulas* that represent solutions of the initial formula (see section 3). Since the axioms are functional-logic programs in abstract logical syntax, rewriting and narrowing agree with program execution. Hence the medium level allows one to test such programs, while the inference rules of the high level provide a "tool box" for program verification. In the case of finite data sets, rewriting and narrowing is often sufficient even for program verification. Besides classical relations or deterministic functions, non-deterministic functions (e.g. state transition systems) and "distributed" transition systems like *Maude programs* [10] or algebraic nets [26] may also be axiomatized and verified by Expander2. The latter are executed by applying associative-commutative rewriting or narrowing on *bag terms*, i.e. multisets of terms (see section 3).

At the low level, built-in Haskell functions simplify or (partially) evaluate terms and formulas and thereby hide most routine steps of proofs and computations. The functions comprise arithmetic, list, bag and set operations, term equivalence and inequivalence and logical simplifications that turn formulas into **nested Gentzen clauses** (see section 5). Evaluating a function f at the medium level means narrowing upon the axioms for f, Evaluating f at the low level means running a built-in Haskell implementation of f. This allows one to test and debug algorithms and visualize their results. For instance, translators between different representations of Boolean functions were integrated into Expander2 in this way. In addition, an execution of an iterative algorithm can be split into its loop traversals such that intermediate results become visible. Currently, the computation steps of Gaussian equation solving, automata minimization, OBDD optimization, LR parsing, data flow analysis and global model checking can be carried out and displayed.

Section 3 presents the syntax of the axioms and theorems that can be handled by Expander2 and describes how they are applied to terms or formulas and how the applications build up proofs. Section 4 shows how axiom applications are combined to narrowing or rewriting steps. Section 5 goes into the logical details of the simplifier and lists the simplification rules for formulas. Section 6 provides induction, coinduction and other rules that Expander2 offers at the high level of interaction. The correctness of the rules presented in Sections 4, 5 and 6 follows almost immediately from corresponding soundness results given in [15–18]. The concluding section 7 focuses on future work.

3   Axioms, Theorems, and Derivations

Axioms and theorems to be applied in derivations are **Horn clauses** ((1)-(7)) or **co-Horn clauses** ((8)-(13)):

$$
\begin{array}{llll}
(1) & \{guard \Rightarrow\} \ \underline{(f(t) = u} & \{\Longleftarrow prem\}) \\
(2) & \{guard \Rightarrow\} \ \underline{(t_1 \ ^\wedge \ldots \ ^\wedge t_k \ ^\wedge!^\wedge t_{k+1} \ ^\wedge \ldots \ ^\wedge t_n = u} & \{\Longleftarrow prem\}) \\
(3) & \{guard \Rightarrow\} \ \underline{(p(t)} & \{\Longleftarrow prem\}) \\
(4) & \underline{t = u} & \{\Longleftarrow prem\} \\
(5) & \underline{q(t)} & \{\Longleftarrow prem\} \\
(6) & \underline{at_1 \wedge \ldots \wedge at_n} & \{\Longleftarrow prem\} \\
(7) & \underline{at_1 \vee \ldots \vee at_n} & \{\Longleftarrow prem\} \\
(8) & \{guard \Rightarrow\} \ \underline{(f(t) = u} & \Longrightarrow conc) \\
(9) & \{guard \Rightarrow\} \ \underline{(q(t)} & \Longrightarrow conc) \\
(10) & \underline{t = u} & \Longrightarrow conc \\
(11) & \underline{p(t)} & \Longrightarrow conc \\
(12) & \underline{at_1 \wedge \ldots \wedge at_n} & \Longrightarrow conc \\
(13) & \underline{at_1 \vee \ldots \vee at_n} & \Longrightarrow conc \\
\end{array}
$$

Curly brackets enclose optional parts. f, p and q denote a defined function, a predicate and a copredicate, respectively, of the current signature. In the case of a higher-order symbol f, p or q, (t) may denote a "curried" tuple $(t_1) \ldots (t_n)$. Usually, $at_1, \ldots, at_n$ are atoms, but may also be more complex formulas (see section 6).

The underlined terms or atoms are called **anchors**. Each application of a clause to a **redex**, i.e. a subterm or subformula of the current tree, starts with the search for a most general unifier of the redex and the anchor of the clause. If the unification is successful and the unifier satisfies the guard, then the redex is replaced by the **reduct**, i.e. the instance of *prem*, u or *conc*, respectively, by the unifier. Moreover, the reduct is augmented with equations that represent the restriction of the unifier to the redex variables (see section 4). If the current trees are terms, then the reducts must be terms and thus only premise-free, but possibly guarded clauses of the form (1) or (2) can be applied.

A guarded clause is applied only if the instance of the guard by the unifier is solvable. The derived (most general) solution extends the unifier. Guarded axioms are needed for efficiently evaluating ground, i.e. variable-free, formulas. Axioms or theorems used as lemmas in proofs, however, should be unguarded. Otherwise the search for a solution of the guard may block the proof process.

Axioms of form (2) are called **AC equations** because they take into account that $^\wedge$ is an associative-commutative function (see section 2). (2) can be applied to a bag term $u_1 \, ^\wedge \ldots \, ^\wedge u_m$ if a list $L_1 = [u_{i_1}, \ldots, u_{i_n}]$ of elements of $L = [u_1, \ldots, u_m]$ unifies with $[t_1, \ldots, t_n]$ and if the unifier satisfies the guard. At first, a substitution f that unifies $L' = [t_1, \ldots, t_k]$ with members of L is looked for. Then f must be extendable to a substitution g that satisfies the guard and unifies $[t_{k+1}, \ldots, t_m]$ with a permutation of the list $L_2 = [v_1, \ldots, v_{n-k}]$ that consists of all elements of L that were not unified with elements of L'. The search is performed by traversing the permutations of L_2 in reverse lexicographic order. If a suitable permutation has been found, the elements of L_1 are replaced by the instance of t by g, while the remaining elements of L are replaced with their instances by g. At most 720 permutations of L_2 are checked. If this case is reached without achieving a unifier, then the application of (2) consists of replacing L_2 by the permutation achieved at last. Further permutations may then be tried by re-applying the AC equation.

For applying a clause of type (6), (7), (12) or (13), n subformulas $at'_1, \ldots, at'_n$ must be selected in the displayed tree such that for all $1 \leq i \leq n$, at'_i is unifiable with at_i. at'_i is replaced by the corresponding instance of *prem* resp. *conc*. The resulting reducts are combined conjunctively in the case of a Horn clause and disjunctively in the case of a co-Horn clause (see section 6).

Axioms represent functional-logic programs and thus are of the form (1), (2), (3) or (9). Axioms determine the least/greatest fixpoint model of a specification (see section 1). Theorems are supposed to be valid in this model. Narrowing and rewriting consist of automatic axiom applications (see section 4). Applications of individual axioms are restricted to the high level of interaction (see section 6).

Example 1. An Expander2 specification of finite lists with a defined function *flatten* for flattening lists of lists and a predicate *part* for generating list partitions reads as follows:

```
specs:     NAT
defuncts:  flatten
preds:     part
fovars:    x y s s' p

axioms:    part([x],[[x]])                                &
           (part(x:y:s,[x]:p) <=== part(y:s,p))           &
           (part(x:y:s,(x:s'):p) <=== part(y:s,s':p))     &
           flatten[] = []                                 &
           flatten(s:p) = s++flatten(p)
```

Example 2. An Expander2 specification of streams (infinite lists) with defined functions *head*, *tail* and *eq* and, given a Boolean function *f*, for a predicate *exists(f)* and a copredicate *fair(f)*, which check whether *f* holds for one element resp. infinitely many elements of a stream reads as follows:

```
constructs:  blink
defuncts:    head tail eq
preds:       exists
copreds:     fair
fovars:      x y s
hovars:      f

axioms:      head(x:s) = x                                            &
             tail(x:s) = s                                            &
             head(blink) = 0                                          &
             tail(blink) = 1:blink                                    &
             eq(x)(x) = true                                          &
             (x =/= y ==> eq(x)(y) = false)                           &
             (f(head(s)) = true  ==> exists(f)(s))                    &
             (f(head(s)) = false
                ==> (exists(f)(s) <=== exists(f)(tail(s))))           &
             (fair(f)(s) ===> exists(f)(s) & fair(f)(tail(s)))
```

A **derivation** with Expander2 is a sequence of successive values of the state variable *trees* (see Section 2). It is stored in the state variables *proof* and *proof term*. All three variables are initialized when the contents of the text field is parsed and the resulting tree *t* is displayed on the canvas. Then the state variable *trees* is set to the singleton $[t]$.

A derivation is **correct** if the derived disjunction/conjunction (resp. sum) of the current trees implies (resp. is a possible result of) the original one. The underlying semantics is described in section 1. Built-in symbols are interpreted by the simplifier. Expander2 checks the correctness of each derivation step and delivers a warning if the step may be incorrect.

A correct derivation that ends up with the formula *True* or *False* is a proof resp. refutation of the original formula φ. Further possible results are **solved formulas**, which are conjunctions of existentially quantified equations or universally quantified inequations that represent a substitution of the free variables of φ by normal forms (see section 2). The substitution is a solution of φ if the derivation of the solved formula is correct.

The correctness of a derivation step depends on the **polarity** of the redex with respect to its position within the current trees. The polarity is *positive* if the number of preceding negation symbols or premise positions is even. Otherwise it is *negative*. A rule is **analytical** or **expanding** if the reduct implies the redex. Here the redex must have positive polarity if the derivation step shall be correct. A rule is **synthetical** or **contracting** if the redex implies the reduct. Here the redex must have negative polarity if the derivation step shall be correct. Expander2 checks these applicability conditions automatically. Of course, *both* analytical and synthetical rules transform a redex into an *equivalent* formula and thus may be applied regardless of the polarity.

4   Narrowing and Rewriting

The narrowing procedure of Expander2 applies axioms and simplification rules repeatedly from top to bottom and from left to right, first to the currently displayed tree and then to other current trees.

Usually, the narrowing procedure applies all applicable axioms for the anchor of a redex simultaneously. Hence narrowing steps within a proof provide case distinctions. The axioms for a relation or defined function may be **eager** or **lazy**. Eager axioms are applied first. Lazy axioms are applied only if, within a complete top-down traversal of the current trees, no eager axiom is applicable. AC equations are typical candidates for lazy axioms (see Section 2). The repeated application of an AC equation to the same AC term may prevent other axioms from being applied to subredices. Hence other axioms are preferred to AC equations.

Applying all applicable (Horn) axioms for a predicate or defined function simultaneously results in the replacement of the redex by the *disjunction* of their *premises* together with equations representing the computed unifiers (see Section 3). Applying all applicable (co-Horn) axioms for a copredicate simultaneously results in the replacement of the redex by the *conjunction* of their *conclusions*. The narrowing rules read as follows:

- **Narrowing upon a Predicate p**

$$\frac{p(t)}{\bigvee_{i=1}^{k} \exists Z_i : (\varphi_i \sigma_i \wedge \boldsymbol{x} = \boldsymbol{x} \sigma_i)} \; \updownarrow$$

  where $\gamma_1 \Rightarrow (p(u_1) \Longleftarrow \varphi_1), \dots, \gamma_n \Rightarrow (p(u_n) \Longleftarrow \varphi_n)$ are the (Horn) axioms for p,

  $(*)$ $\boldsymbol{x}$ is a list of the variables of t,
  for all $1 \le i \le k$, $t\sigma_i = u_i\sigma_i$, $\gamma_i\sigma_i \vdash True$[1] and $Z_i = var(u_i, \varphi_i)$,
  for all $k < i \le n$, t is not unifiable with u_i.

- **Narrowing upon a Copredicate p**

$$\frac{p(t)}{\bigwedge_{i=1}^{k} \forall Z_i : (\boldsymbol{x} = \boldsymbol{x} \sigma_i \Rightarrow \varphi_i \sigma_i)} \; \updownarrow$$

  where $\gamma_1 \Rightarrow (p(u_1) \Longrightarrow \varphi_1), \dots, \gamma_n \Rightarrow (p(u_n) \Longrightarrow \varphi_n)$ are the (co-Horn) axioms for p and $(*)$ holds true.

- **Narrowing upon a Defined Function f**

$$\frac{\varphi(f(t))}{\begin{array}{c} \bigvee_{i=1}^{k} \exists Z_i : (\varphi(v_i\sigma_i) \wedge \varphi_i\sigma_i \wedge \boldsymbol{x} = \boldsymbol{x}\sigma_i) \; \vee \\ \bigvee_{i=k+1}^{l} (\varphi(f(t\sigma_i)) \wedge \boldsymbol{x} = \boldsymbol{x}\sigma_i) \end{array}} \; \updownarrow$$

  where $\gamma_1 \Rightarrow (f(u_1) = v_1) \Longleftarrow \varphi_1, \dots, \gamma_n \Rightarrow (f(u_n) = v_n) \Longleftarrow \varphi_n$ are the (Horn) axioms for f,

---

[1] Hence σ_i solves the guard γ_i. Expander2 tries to solve γ_i by applying at most 500 narrowing steps.

(∗) x is a list of the variables of t,
for all $1 \leq i \leq k$, $t\sigma_i = u_i\sigma_i$, $\gamma_i\sigma_i \vdash True$ and $Z_i = var(u_i, \varphi_i)$,
for all $k < i \leq l$, σ_i is a partial unifier of t and u_i,
for all $l < i \leq n$, t is not partially unifiable with u_i.

$u_1, \ldots, u_n$ may be tuples of terms. In the case of narrowing upon a defined function, the unification of t with u_i may fail because at some position, the root symbols of t and u_i are different and one of them is a defined function f. Since the unification may succeed later, when subsequent narrowing steps have replaced f by a constructor or a variable, we save the already obtained *partial* unifier σ_i and construct a reduct that consists of the σ_i-instance of the redex and equations that represent σ_i. This version of the narrowing rule has been derived from the **needed narrowing** strategy [1, 15]. If the underlying specification is *functional*, the strategy of applying these narrowing rules iteratively from top to bottom to a formula φ leads to a set S of solutions of φ such that each solution of φ is an instance of some $s \in S$ [16, 17]. Hence, in the context of this strategy, the narrowing rules are not only expansions (or analytical rules), but even equivalence transformations. This fact is indicated by the symbol ↕ attached to the rules.

Non-narrowable logical atoms $p(t)$ with normal form t are simplified to by *False* if p is a predicate and by *True* if p is a copredicate. This complies with the semantics of p as the least/greatest solution of the axioms for p. Non-narrowable terms $f(t)$ with normal form t are simplified to the undefinedness constant (). Of course, this transformation may lead to undesired results if some function or relation occurring in a (simplified) conjecture has not been specified completely!

If the current trees are terms, only rewriting steps can be applied. Rewriting is the special case of narrowing upon defined functions where the unifiers σ_i do not instantiate redex variables:

- **Rewriting upon a Defined Function f**

$$\frac{u(f(t))}{u(v_1\sigma_1)<+>\ldots<+>u(v_k\sigma_k)}$$

where $\gamma_1 \Rightarrow f(u_1) = v_1, \ldots, \gamma_1 \Rightarrow f(u_n) = v_n$ are the axioms for f and

(∗)   for all $1 \leq i \leq k$, $t = u_i\sigma_i$ and $\gamma_i\sigma_i \vdash True$,
for all $k < i \leq n$, t does not match u_i.

If a rewriting step delivers a proper sum of terms, the applied axioms specify a non-deterministic function: each $v_i\sigma_i$ is a possible value. Non-rewritable terms $f(t)$ with normal form t are simplified to the undefinedness constant (), which is neutral with respect to the sum operator <+>.

5   Simplification

Narrowing removes predicates, copredicates and defined functions from the current trees. The simplifier does the same with logical operators, constructors and symbols of the built-in signature. Simplifications realize the highest degree of automation and the lowest level of interaction (see section 2). The reducts of rewriting or narrowing steps are simplified automatically.

The simplifier turns formulas into minimal *nested Gentzen clauses* of the form[2]

$$\forall \boldsymbol{x} \ (\exists \boldsymbol{y} \ (t_1 \wedge \ldots \wedge t_m) \Rightarrow \forall \boldsymbol{z} \ (u_1 \vee \ldots \vee u_n)).$$

"Nested" means that the clause is derivable by the following grammar:

$$
\begin{aligned}
S &\longrightarrow A \mid B & A &\longrightarrow A_1 \mid C \\
B &\longrightarrow B_1 \mid C & A_1 &\longrightarrow \exists \boldsymbol{x} \ (B \wedge \ldots \wedge B) \\
B_1 &\longrightarrow \forall \boldsymbol{x} \ (A \vee \ldots \vee A) & C &\longrightarrow \forall \boldsymbol{x} \ (A_1 \Rightarrow B_1) \mid atom
\end{aligned}
$$

The evaluation rules used by the simplifier are equivalence transformations. Besides the partial evaluation of built-in predicates and functions, the following rules are applied:

- *Elimination of Zero Elements*

$$\frac{t_1 <+> \ldots <+> () <+> \ldots <+> t_n}{t_1 <+> \ldots <+> t_n}$$

$$\frac{\varphi_1 \wedge \ldots \wedge True \wedge \ldots \wedge \varphi_n}{\varphi_1 \wedge \ldots \wedge \varphi_n} \qquad \frac{\varphi_1 \vee \ldots \vee False \vee \ldots \vee \varphi_n}{\varphi_1 \vee \ldots \vee \varphi_n}$$

- *Disjunctive Normal Form*

$$\frac{f(\ldots, t_1 <+> \ldots <+> t_n, \ldots)}{f(\ldots, t_1, \ldots) <+> \ldots <+> f(\ldots, t_n, \ldots)}$$

$$\frac{p(\ldots, t_1 <+> \ldots <+> t_n, \ldots)}{p(\ldots, t_1, \ldots) \vee \ldots \vee p(\ldots, t_n, \ldots)}$$

$$\frac{\varphi \wedge \forall \boldsymbol{x}(\psi_1 \vee \ldots \vee \psi_n)}{\forall \boldsymbol{x}((\varphi \wedge \psi_1) \vee \ldots \vee (\varphi \wedge \psi_n))} \quad \text{if no } x \in \boldsymbol{x} \text{ occurs freely } \varphi$$

- *Equation Splitting*

$$\frac{c(t_1, \ldots, t_n) = c(u_1, \ldots, u_n)}{t_1 = u_1 \wedge \ldots \wedge t_n = u_n} \qquad \frac{c(t_1, \ldots, t_n) = d(u_1, \ldots, u_n)}{False}$$

$$\frac{c(t_1, \ldots, t_n) \neq c(u_1, \ldots, u_n)}{t_1 \neq u_1 \vee \ldots \vee t_n \neq u_n} \qquad \frac{c(t_1, \ldots, t_n) \neq d(u_1, \ldots, u_n)}{True}$$

- *Quantifier Distribution*

$$\frac{\forall \boldsymbol{x}(\varphi_1 \wedge \ldots \wedge \varphi_n)}{\forall \boldsymbol{x}\varphi_1 \wedge \ldots \wedge \forall \boldsymbol{x}\varphi_n} \qquad \frac{\exists \boldsymbol{x}(\varphi_1 \vee \ldots \vee \varphi_n)}{\exists \boldsymbol{x}\varphi_1 \vee \ldots \vee \exists \boldsymbol{x}\varphi_n} \qquad \frac{\exists \boldsymbol{x}(\varphi \Rightarrow \psi)}{\forall \boldsymbol{x}\varphi \Rightarrow \exists \boldsymbol{x}\psi}$$

$$\frac{\exists \boldsymbol{x}(\varphi_1 \wedge \ldots \wedge \varphi_n)}{\exists \boldsymbol{x}_1\varphi_1 \wedge \ldots \wedge \exists \boldsymbol{x}_n\varphi_n} \qquad \frac{\forall \boldsymbol{x}(\varphi_1 \vee \ldots \vee \varphi_n)}{\forall \boldsymbol{x}_1\varphi_1 \vee \ldots \vee \forall \boldsymbol{x}_n\varphi_n}$$

if for all $1 \leq i \leq n$, no variable of $\boldsymbol{x}_i$ occurs freely in φ_j, $1 \leq j \leq n$, $j \neq i$.

---

[2] The binding-priority ordering of logical operators is given by $\{\neg, \forall, \exists\} > \wedge > \vee > \Rightarrow$.

- *Removal of Quantifiers.* Unused **bounded variables** are removed. **Successive quantifiers** are merged.
- *Removal of Negation.* Negation symbols are moved to literal positions where they are replaced by complement predicates: $\neg P(t)$ is reduced to $not\_P(t)$, $\neg not\_P(t)$ is reduced to $P(t)$.
  Co-Horn/Horn axioms for not_P can be generated automatically from Horn/Co-Horn axioms for P.
- *Flattening of Conjunctions and Disjunctions*

$$\frac{\varphi_1 \wedge \ldots \wedge (\psi_1 \wedge \ldots \wedge \psi_k) \wedge \ldots \wedge \varphi_n}{\varphi_1 \wedge \ldots \wedge \psi_1 \wedge \ldots \wedge \psi_k \wedge \ldots \wedge \varphi_n} \qquad \frac{\varphi_1 \vee \ldots \vee (\psi_1 \vee \ldots \vee \psi_k) \vee \ldots \vee \varphi_n}{\varphi_1 \vee \ldots \vee \psi_1 \vee \ldots \vee \psi_k \vee \ldots \vee \varphi_n}$$

$$\frac{\varphi_1 \wedge \ldots \wedge \mathit{False} \wedge \ldots \wedge \varphi_n}{\mathit{False}} \qquad \frac{\varphi_1 \vee \ldots \vee \mathit{True} \vee \ldots \vee \varphi_n}{\mathit{True}}$$

- *Subsumption I.* The set of factors of a conjunction or summands of a disjunction is reduced to its minimal elements with respect to the subsumption relation (see below).
- *Removal of Equations*

$$\frac{\exists \boldsymbol{x}(x = t \wedge \varphi)}{\exists \boldsymbol{x}\varphi[t/x]} \qquad \frac{\forall \boldsymbol{x}(x \neq t \vee \varphi)}{\forall \boldsymbol{x}\varphi[t/x]} \quad \text{if } x \in \boldsymbol{x} \setminus var(t)$$

$$\frac{\forall \boldsymbol{x}(x = t \wedge \varphi \Rightarrow \psi)}{\forall \boldsymbol{x}(\varphi \Rightarrow \psi)[t/x]} \quad \text{if } x \in \boldsymbol{x} \setminus var(t)$$

- *Creation of Narrowing Redices*

$$\frac{x = t \wedge \varphi}{x = t \wedge \varphi[t/x]} \qquad \frac{x \neq t \vee \varphi}{x \neq t \vee \varphi[t/x]}$$

if x is a free variable of φ that does not occur in t and t consists of constructors and codefined functions

- *Modus Ponens*

$$\frac{\varphi \wedge (\psi \Rightarrow \theta)}{\varphi \wedge \theta} \quad \text{if } \varphi \text{ subsumes } \psi$$

- *Subsumption II.* An implication $\varphi \Rightarrow \psi$ is reduced to *True* if φ **subsumes** ψ. Subsumption is the least binary relation on terms and formulas that satisfies the following implications:

$$\exists\, 1 \leq i \leq n : \varphi \text{ subsumes } \psi_i \qquad \Longrightarrow \varphi \text{ subsumes } \psi_1 \vee \ldots \vee \psi_n$$
$$\forall\, 1 \leq i \leq n : \varphi \text{ subsumes } \psi_i \qquad \Longrightarrow \varphi \text{ subsumes } \psi_1 \wedge \ldots \wedge \psi_n$$
$$\exists\, \boldsymbol{t} : \varphi(\boldsymbol{t}) = \psi(\boldsymbol{y}) \text{ and } \exists\, \boldsymbol{u} : \psi(\boldsymbol{u}) = \varphi(\boldsymbol{x}) \Longrightarrow \exists \boldsymbol{x}\varphi(\boldsymbol{x}) \text{ subsumes } \exists \boldsymbol{y}\psi(\boldsymbol{y})$$
$$\exists\, \boldsymbol{u} : \psi(\boldsymbol{u}) = \varphi \qquad \Longrightarrow \varphi \text{ subsumes } \exists \boldsymbol{y}\psi(\boldsymbol{y})$$
$$\forall\, 1 \leq i \leq n : \varphi_i \text{ subsumes } \psi \qquad \Longrightarrow \varphi_1 \vee \ldots \vee \varphi_n \text{ subsumes } \psi$$
$$\exists\, 1 \leq i \leq n : \varphi_i \text{ subsumes } \psi \qquad \Longrightarrow \varphi_1 \wedge \ldots \wedge \varphi_n \text{ subsumes } \psi$$
$$\exists\, \boldsymbol{t} : \varphi(\boldsymbol{t}) = \psi(\boldsymbol{y}) \text{ and } \exists\, \boldsymbol{u} : \psi(\boldsymbol{u}) = \varphi(\boldsymbol{x}) \Longrightarrow \forall \boldsymbol{x}\varphi(\boldsymbol{x}) \text{ subsumes } \forall \boldsymbol{y}\psi(\boldsymbol{y})$$
$$\exists\, \boldsymbol{t} : \varphi(\boldsymbol{t}) = \psi \qquad \Longrightarrow \forall \boldsymbol{x}\varphi(\boldsymbol{x}) \text{ subsumes } \psi$$
$$\varphi = \psi \text{ modulo reorderings of bounded variables, factors or summands}$$
$$\Longrightarrow \varphi \text{ subsumes } \psi$$

- *Implication Splitting*

$$\frac{\forall\boldsymbol{x}(\varphi_1 \vee \ldots \vee \varphi_n \Rightarrow \psi)}{\forall\boldsymbol{x}(\varphi_1 \Rightarrow \psi) \wedge \ldots \wedge \forall\boldsymbol{x}(\varphi_n \Rightarrow \psi)} \qquad \frac{\forall\boldsymbol{x}(\varphi \Rightarrow \psi_1 \wedge \ldots \wedge \psi_n)}{\forall\boldsymbol{x}(\varphi \Rightarrow \psi_1) \wedge \ldots \wedge \forall\boldsymbol{x}(\varphi \Rightarrow \psi_n)}$$

- *Universal Quantification of Implications*

$$\frac{\exists\boldsymbol{x}\varphi \Rightarrow \psi}{\forall\boldsymbol{x}(\varphi \Rightarrow \psi)} \qquad \frac{\psi \Rightarrow \forall\boldsymbol{x}\varphi}{\forall\boldsymbol{x}(\psi \Rightarrow \varphi)}$$

if no variable of $\boldsymbol{x}$ occurs freely in ψ.

- *Uncurrying*

$$\frac{\varphi \Rightarrow (\theta \Rightarrow \psi_1) \vee \psi_2}{\varphi \wedge \theta \Rightarrow \psi_1 \vee \psi_2}$$

Besides being an essential part of proof processes, simplification in Expander2 may be used for testing algorithms, especially iterative ones, which change values of state variables during loop traversals [21]. Several such algorithms have been integrated into the simplifier by translating a loop traversal into a simplification step. Consequently, intermediate results can be visualized in a painter window (see Section 2). The respective state variables are initialized as a side-effect of applying particular axioms that rewrite constants.

Similarly to narrowing and rewriting, the simplifier pursues a top-down strategy that ensures termination and the eventual application of all applicable rules. This is necessary because it usually works in the background. For instance, narrowing reducts are simplified automatically before they are submitted to further narrowing steps.

The notion of simplification differs from prover to prover. For instance, Isabelle [13] subsumes rewriting upon equational axioms under simplification.

6   Rules at the High Level of Interaction

Narrowing steps and simplifications are both analytical and synthetical and thus turn formulas into semantically *equivalent* ones. Instances of the rules that are accessible via the solver's subtrees menu (see Fig. 2), however, may be strictly analytical or strictly synthetical. Hence they can be applied only individually and only to subtrees with positive resp. negative polarity (see Section 3). We describe the main rules in terms of the actions to be taken by the user in order to apply them.

- **Instantiation.** Select an existentially/universally quantified variable x. If the scope of x has positive/negative polarity, then all occurrences of x in the scope are replaced by the term in the solver's entry field. Alternatively, the replacing term t may be taken from the dispalyed tree and moved to a position of x in the scope. Again, all occurrences of x in the scope are replaced by t.

- **Generalization.** Select a subformula φ and enter a formula ψ into the solver's entry field. If φ has positive/negative polarity, then φ is combined conjunctively/disjunctively with ψ.

- **Unification.** Select two factors of a conjunction $\varphi = \exists \boldsymbol{x}(\varphi_1 \wedge \ldots \wedge \varphi_n)$ or two summands of a disjunction $\psi = \forall \boldsymbol{x}(\varphi_1 \vee \ldots \vee \varphi_n)$. If they are unifiable and the unifier instantiates only variables of $\boldsymbol{x}$, then one of them is removed and the unifier is applied to the remaining conjunction/disjunction. The transformation is correct if φ/ψ has positive/negative polarity.
- **Copy.** Select a subtree φ. A copy of φ is added to the children of the subtree's parent node. The transformation is correct if the parent node holds a conjunction or disjunction symbol.
- **Removal.** Select subtrees $\phi_1, \ldots, \phi_n$. $\phi_1, \ldots, \phi_n$ are removed from the displayed tree. The transformation is correct if $\phi_1, \ldots, \phi_n$ are summands/factors of the same disjunction/conjunction with positive/negative polarity.
- **Reversal.** The list of selected subtrees is reversed. The transformation is correct if all subtrees are arguments of the same occurrence of a *permutative* operator. Currently, the permutative operators are:

$$\&, |, =, = / =, \sim, \sim\sim, \sim/\sim, +, *, \ \hat{}\ , \{\}.$$

- **Congruence.** Select (1) an atom tRt' with positive polarity such that $R \in \{=, \sim\}$ or (2) an atom tRt' with negative polarity such that $R \in \{\neq, \not\sim\}$ or (3) an atom tRt' with positive polarity such that

$$R \in Trans =_{def} \{<, \leq, >, \geq, =, \sim, \sim\sim\}$$

or (4) $n - 1$ factors

$$t_1 R t_2, \ t_2 R t_3, \ \ldots, \ t_{n-1} R t_n$$

of a conjunction with negative polarity such that $R \in Trans$. The selected atoms are composed resp. decomposed in accordance with the assumption that R is compatible with function symbols (cases 1 and 2) or transitive (cases 3 and 4).
- **Constrained Narrowing.** Select subtrees $\phi_1, \ldots, \phi_n$ and write axioms into the text field or a signature symbol f into the solver's entry field. Then narrowing/rewriting steps upon the axioms in the text field or the axioms for f, respectively, are applied to $\phi_1, \ldots, \phi_n$.
- **Axiom/Theorem Application.** Select subtrees $\phi_1, \ldots, \phi_n$ and write the number of an axiom or theorem into the solver's entry field. The selected axiom or theorem is applied from left to right or from right to left to $\phi_1, \ldots, \phi_n$. If the applied clause is of the form (1), (2), (4), (8) or (10), then left/right refers to the respective side of the leading equation, otherwise to the respective side of implication that is given by the clause (see Section 3). The transformation is correct if the conclusion/premise is applied to a subformula with positive/negative polarity. A clause of type (6), (7), (12) or (13) is applied to atoms $at'_1, \ldots, at'_n$ each of which is part of a conjunction or disjunction: Let V be the set of variables of *prem* resp. *conc* that do not occur in $at_1, \ldots, at_n$.

$$\text{application of (6)} \quad \frac{\varphi_1(at'_1) \wedge \ldots \wedge \varphi_n(at'_n)}{(\bigwedge_{i=1}^{n} \varphi_i(\exists V(prem\sigma \wedge \bigwedge_{x \in dom(\sigma)} x \equiv x\sigma)))} \ \Uparrow$$

where for all $1 \leq i \leq n$, $at'_i \sigma = at_i \sigma$ and φ_i does not contain existential quantifiers or negation or implication symbols.

$$\text{application of (7)} \quad \frac{\varphi_1(at'_1) \vee \ldots \vee \varphi_n(at'_n)}{(\bigwedge_{i=1}^{n} \varphi_i(\exists \, V(prem\sigma \wedge \bigwedge_{x \in dom(\sigma)} x \equiv x\sigma)))} \Uparrow$$

where for all $1 \leq i \leq n$, $at'_i \sigma = at_i \sigma$ and φ_i does not contain universal quantifiers or negation or implication symbols.

$$\text{application of (12)} \quad \frac{\varphi_1(at'_1) \wedge \ldots \wedge \varphi_n(at'_n)}{(\bigvee_{i=1}^{n} \varphi_i(\forall \, V(\bigwedge_{x \in dom(\sigma)} x \equiv x\sigma \Rightarrow conc\sigma)))} \Downarrow$$

where for all $1 \leq i \leq n$, $at'_i \sigma = at_i \sigma$ and φ_i does not contain existential quantifiers or negation or implication symbols.

$$\text{application of (13)} \quad \frac{\varphi_1(at'_1) \vee \ldots \vee \varphi_n(at'_n)}{(\bigvee_{i=1}^{n} \varphi_i(\forall \, V(\bigwedge_{x \in dom(\sigma)} x \equiv x\sigma \Rightarrow conc\sigma)))} \Downarrow$$

where for all $1 \leq i \leq n$, $at'_i \sigma = at_i \sigma$ and φ_i does not contain universal quantifiers or negation or implication symbols.

- **Noetherian Induction.** Select a list of free or universal induction variables $x_1, \ldots, x_n$ in the displayed tree. If $\varphi = (prem \Rightarrow conc)$, then the *induction hypotheses*

$$conc' \Longleftarrow (x_1, \ldots, x_n) \gg (x'_1, \ldots, x'_n) \wedge prem'$$
$$prem' \Longrightarrow ((x_1, \ldots, x_n) \gg (x'_1, \ldots, x'_n) \Rightarrow conc')$$

are added to the current theorems. If φ is not an implication, then

$$conc' \Longleftarrow (x_1, \ldots, x_n) \gg (x'_1, \ldots, x'_n)$$

is added. Primed formulas are obtained from unprimed ones by priming the occurrences of $x_1, \ldots, x_n$. $\gg$ denotes the induction ordering. Each left-to-right application of an added theorem corresponds to an induction step and introduces an occurrence of $\gg$. After axioms for $\gg$ have been added to the current axioms, narrowing steps upon $\gg$ should remove the occurrences of $\gg$ because the transformation is correct only if φ can be derived to *True* [14, 15].

- **Factor Shift.** Select an implication $\varphi = (prem_1 \wedge \ldots \wedge prem_n \Rightarrow conc)$ and premise indices $i_1, \ldots, i_k$. φ is turned into the equivalent implication

$$prem_{j_1} \wedge \ldots \wedge prem_{j_r} \quad \Rightarrow \quad (prem_{i_1} \wedge \ldots \wedge prem_{i_k} \Rightarrow conc')$$

where $j_1, \ldots, j_r = \{1, \ldots, n\} \setminus i_1, \ldots, i_k$. This transformation may be necessary for submitting φ to a proof by fixpoint induction.

- **Summand Shift.** Select an implication $\varphi = (prem \Rightarrow conc_1 \vee \ldots \vee conc_n)$ and conclusion indices $i_1, \ldots, i_k$. φ is turned into the equivalent implication

$$prem \wedge \neg conc_{i_1} \wedge \ldots \wedge \neg conc_{i_k} \quad \Rightarrow \quad conc_{j_1} \vee \ldots \vee conc_{j_r}$$

where $j_1, \ldots, j_r = \{1, \ldots, n\} \setminus i_1, \ldots, i_k$. This transformation may be necessary for submitting φ to a proof by coinduction.

The following rules are correct if the selected subformulas have positive polarity.

- **Coinduction on a Copredicate p.** Select subformulas

$$\{prem_1 \Rightarrow\} \, p(t_1)$$
$$\wedge \ldots \qquad\qquad (A)$$
$$\wedge \{prem_k \Rightarrow\} \, p(t_k)$$

such that p does not depend on any predicate or function occurring in $prem_i$. (A) is turned into

$$p(\boldsymbol{x}) \quad\Longleftarrow\quad \{prem_1 \wedge\} \, \boldsymbol{x} = t_1$$
$$\wedge \ldots \qquad\qquad (A')$$
$$\wedge \{prem_k \wedge\} \, \boldsymbol{x} = t_k$$

where $\boldsymbol{x}$ is a list of variables. Moreover, a new predicate p' is added to the current signature and

$$p'(\boldsymbol{x}) \quad\Longleftarrow\quad \{prem_1 \wedge\} \, \boldsymbol{x} = t_1$$
$$\wedge \ldots \qquad\qquad (*)$$
$$\wedge \{prem_k \wedge\} \, \boldsymbol{x} = t_k$$

becomes the axiom for p'. All occurrences of p in the axioms for p are replaced by p'. Then (*) is applied to all occurrences of p' in the transformed axioms for p. The conjunction of the clauses resulting from these applications replaces the original conjecture (A).

- **Fixpoint Induction on a Predicate p.** Select subformulas

$$p(t_1) \Rightarrow conc_1$$
$$\wedge \ldots \qquad\qquad (B)$$
$$\wedge \, p(t_k) \Rightarrow conc_k$$

such that p does not depend on any predicate or function occurring in $conc_i$. (B) is turned into

$$p(\boldsymbol{x}) \quad\Longrightarrow\quad (\boldsymbol{x} = t_1 \Rightarrow conc_1)$$
$$\wedge \ldots \qquad\qquad (B')$$
$$\wedge \, (\boldsymbol{x} = t_k \Rightarrow conc_k)$$

where $\boldsymbol{x}$ is a list of variables. Moreover, a new predicate p' is added to the current signature and

$$p'(\boldsymbol{x}) \quad\Longrightarrow\quad (\boldsymbol{x} = t_1 \Rightarrow conc_1)$$
$$\wedge \ldots \qquad\qquad (*)$$
$$\wedge \, (\boldsymbol{x} = t_k \Rightarrow conc_k)$$

becomes the axiom for p'. All occurrences of p in the axioms for p are replaced by p'. Then (*) is applied to all occurrences of p' in the transformed axioms for p. The conjunction of the clauses resulting from these applications replaces the original conjecture (B).

- **Fixpoint Induction on a Function f.** Select subformulas

$$f(t_1) = u_1 \;\Rightarrow\; conc_1$$
$$\wedge \ldots \tag{C}$$
$$\wedge\; f(t_k) = u_k \;\Rightarrow\; conc_k$$

or

$$f(t_1) = u_1 \;\{\wedge\; conc_1\}$$
$$\wedge \ldots \tag{D}$$
$$\wedge\; f(t_k) = u_k \;\{\wedge\; conc_k\}$$

such that f does not depend on any predicate or function occurring in u_i or $conc_i$. (C) is turned into

$$f(\boldsymbol{x}) = z \quad\Longrightarrow\quad (\boldsymbol{x} = t_1 \wedge z = u_1 \Rightarrow conc_1)$$
$$\wedge \ldots \tag{C'}$$
$$\wedge\; (\boldsymbol{x} = t_k \wedge z = u_k \Rightarrow conc_k),$$

(D) is turned into

$$f(\boldsymbol{x}) = z \quad\Longrightarrow\quad (\boldsymbol{x} = t_1 \Rightarrow z = u_1\{\wedge\; conc_1\})$$
$$\wedge \ldots \tag{D'}$$
$$\wedge\; (\boldsymbol{x} = t_k \Rightarrow z = u_k\{\wedge\; conc_k\})$$

where $\boldsymbol{x}$ is a list of variables and z is a variable. Moreover, a new predicate f' is added to the current signature and

$$f'(\boldsymbol{x}, z) \quad\Longrightarrow\quad ((\boldsymbol{x} = t_1 \wedge z = t_1) \Rightarrow conc_1)$$
$$\wedge \ldots \tag{*}$$
$$\wedge\; ((\boldsymbol{x} = t_k \wedge z = t_k) \Rightarrow conc_k)$$

resp.

$$f'(\boldsymbol{x}, z) \quad\Longrightarrow\quad (\boldsymbol{x} = t_1 \Rightarrow (z = t_1\{\wedge\; conc_1\}))$$
$$\wedge \ldots \tag{*}$$
$$\wedge\; (\boldsymbol{x} = t_k \Rightarrow (z = t_k\{\wedge\; conc_k\}))$$

becomes the axiom for f'. All occurrences of f in the flattened axioms for f are replaced by f'. Replacing f actually means replacing equations $f(t) = u$ by logical atoms $f'(t, u)$. Then (*) is applied to all occurrences of f' in the transformed axioms for f. The conjunction of the clauses resulting from these applications replaces the original conjecture (C)/(D).

- **Hoare Induction.** Select a subformula of the form (C) or (D) such that $k = 1$ and f has a single axiom of the form $f(\boldsymbol{x}) = loop(\boldsymbol{v})$. (C)/(D) is turned into (C')/(D') and then transformed into the following conjectures, which characterize INV as a Hoare invariant:

$$INV(\boldsymbol{x}, \boldsymbol{v}) \tag{INV 1}$$
$$loop(\boldsymbol{y}) = z \;\wedge\; INV(\boldsymbol{x}, \boldsymbol{y}) \;\Rightarrow\; conc_1 \tag{INV 2}$$

- **Subgoal Induction.** Same as Hoare induction except that the following conjectures are created, which characterize INV as a subgoal invariant:

$$INV(\boldsymbol{v}, z) \;\Rightarrow\; conc_1 \tag{INV 1}$$
$$loop(\boldsymbol{y}) = z \Rightarrow INV(\boldsymbol{y}, z) \tag{INV 2}$$

Example 1 (continued). A proof by fixpoint induction as Expander2 records it is presented. The conjecture says that *part* is correct insofar as it only computes partitions of the given list. All, Any, & and | denote $\forall$, $\exists$, $\wedge$ and $\vee$, respectively.

```
part(s,p) ==> s = flatten(p)
```

Applying fixpoint induction w.r.t.

```
      part([x0],[[x0]])
   & (part(x1:(y0:s3),[x1]:p0) <=== part(y0:s3,p0))
   & (part(x2:(y1:s4),(x2:s'0):p1) <=== part(y1:s4,s'0:p1))
```

to the preceding tree leads to the formula

```
All x0 x1 y0 s3 p0 x2 y1 s4 s'0 p1:
(  [x0] = flatten[[x0]]
   & (x1:(y0:s3) = flatten([x1]:p0) <=== y0:s3 = flatten(p0))
   & (x2:(y1:s4) = flatten((x2:s'0):p1) <=== y1:s4 = flatten(s'0:p1)))
```

Simplifying (6 steps) the preceding tree leads to the factor

```
All x0:([x0] = flatten[[x0]])
```

Narrowing the preceding factor leads to the factor

```
[] = flatten[]
```

Narrowing the preceding factor leads to new ones.
The current factor is given by

```
All x1 y0 s3 p0:
(y0:s3 = flatten(p0) ==> x1:(y0:s3) = flatten([x1]:p0))
```

Applying the axiom

```
   flatten(s:p) = s++flatten(p)
```

at position [1,1] of the preceding factor leads to the factor

```
All x1 y0 s3 p0:(y0:s3 = flatten(p0) ==> x1:(y0:s3) = [x1]++flatten(p0))
```

Simplifying (12 steps) the preceding factor leads to a new formula.
The current formula is given by

```
All x2 y1 s4 s'0 p1:
(y1:s4 = flatten(s'0:p1) ==> x2:(y1:s4) = flatten((x2:s'0):p1))
```

Applying the axiom

```
   flatten(s:p) = s++flatten(p)
```

at positions [0,1,1],[0,0,1] of the preceding tree leads to the formula

```
All x2 y1 s4 s'0 p1:
(y1:s4 = s'0++flatten(p1) ==> x2:(y1:s4) = x2:s'0++flatten(p1))
```

Simplifying (11 steps) the entire formula leads to the formula
True

A proof by Noetherian induction (see above) of the same conjecture is less straightforward and more than twice as long as the above proof by fixpoint induction (see [21], *Examples*, PARTPROOF2).

Example 2 (continued). A proof by coinduction as Expander2 records it is presented. The conjecture says that *blink* is a fair stream insofar as it contains infinitely many zeros. At first, the conjecture must be generalized. *blink* and *1:blink* have to be fair streams.

```
fair(eq(0))(blink) & fair(eq(0))(1:blink)
```

Adding the other factors leads to

```
fair(eq(0))(blink) & fair(eq(0))(1:blink)
```

Applying coinduction w.r.t.

```
    fair(f0)(s0)  ===>  exists(f0)(s0) & fair(f0)(tail(s0))
```

to the preceding tree leads to the formula

```
All f0 s0:
(         f0 = eq(0) & s0 = blink
    |     f0 = eq(0) & s0 = 1:blink
  ===>    exists(f0)(s0)
       & (    f0 = eq(0) & tail(s0) = blink
          |   f0 = eq(0) & tail(s0) = 1:blink))
```

Simplifying (49 steps) the preceding tree leads to the summand

```
  exists(eq(0))(blink) & tail(blink) = blink
& exists(eq(0))(1:blink) & tail(1:blink) = blink
```

Adding the other summands leads to

```
    exists(eq(0))(blink) & tail(blink) = blink
  & exists(eq(0))(1:blink) & tail(1:blink) = blink
|   exists(eq(0))(blink) & tail(blink) = blink
  & exists(eq(0))(1:blink) & tail(1:blink) = 1:blink
|   exists(eq(0))(blink) & tail(blink) = 1:blink
  & exists(eq(0))(1:blink) & tail(1:blink) = blink
|   exists(eq(0))(blink) & tail(blink) = 1:blink
  & exists(eq(0))(1:blink) & tail(1:blink) = 1:blink
```

```
Narrowing (9 steps) the preceding tree leads to the formula
True
```

7   Conclusion

We have given an overview of Expander2 with special focus on the system's prover capabilities. Other features, such as the generation, editing and combination of pictorial term representations or the use of state variables by the simplifier are described in detail in [21]. Future work on Expander2 and on the underlying Swinging Types approach will concentrate on the following:

➤ Representation of *coalgebraic* data types in terms of coinductively defined functions and of corresponding subtypes defined in terms of co-Horn clauses for membership predicates or *coequalities*. First steps towards this extension can be found in [19]. Coalgebraic specifications are also dealt with in, e.g., [6, 23, 9, 11]. O'Haskell records [12] may be suitable for embedding standard coalgebraic data types into the simplifier.

➤ Compilers that translate functional or relational programs written in, e.g., Haskell, Maude [10], Prolog or Curry [7] into simplification rules. This might involve the combination of particular programming language constructs and their semantics with the pure algebraic-logic semantics of Expander2 specifications. Related work has been done by combining the algebraic specification language CASL [3] with Haskell [25].

➤ A compiler of UML class diagrams and OCL constraints into Expander2 specifications has been developed in a students' project. This yields a basis for proving invariants, reachabilities and other safety or liveness properties of object-oriented specifications within Expander2.

➤ Commands for the automatic generation of particular axioms, theorems or simplification rules. Such commands are already available for specifying complement predicates, deriving "generic" lemmas from the least/greatest fixpoint semantics of relations and for turning co-Horn axioms into equivalent Horn axioms (see [21], *Axioms menu*).

➤ Simplification rules that cooperate with other theorem provers [2, 22, 27–29] or constraint solvers [4] via tailor-made interfaces.

➤ Narrowing and fixpoint (co)induction complement each other with respect to the direction axioms are combined with conjectures: In the first case, axioms are applied to conjectures, and the proof proceeds by transforming the modified conjectures. In the second case, conjectures are applied to axioms and the proof proceeds by transforming the modified axioms. Moreover, narrowing on a predicate p is, at first, a computation rule, i.e. a rule for evaluating p, while fixpoint induction on p is a proof rule, i.e. a rule for proving something about p. Strinkingly, the situation turns upside down for copredicates: narrowing on a copredicate q is rather a proof rule, whereas coinduction on q is used as a computation rule [18]. This observation makes it worthwhile to look for a uniform proof/computation strategy that uses fixpoint (co)induction already at the medium level of interaction.

➤ The range of applications of Expander2 will be investigated and extended by further case studies. Most specifications designed and proofs and computations performed with the system up to now are listed and classified in the *Examples* section of the manual [21]. So far, the above-mentioned students' project for translating UML/OCL specifications into Expander2 has led to the most extensive examples.

References

1. S. Antoy, R. Echahed, M. Hanus, *A Needed Narrowing Strategy*, Journal of the ACM 47 (2000) 776-822
2. *Automated Reasoning Systems*, http://www-formal.stanford.edu/clt/ARS/systems.html
3. M. Bidoit, P.D. Mosses, *CASL User Manual*, Springer LNCS 2900 (2004)
4. Th. Frühwirth, S. Abdennadher, *Essentials of Constraint Programming*, Springer 2003
5. R. Giegerich, *A Systematic Approach to Dynamic Programming in Bioinformatics. Parts 1 and 2: Sequence Comparison and RNA Folding*, Report 99-05, Technical Department, University of Bielefeld 1999
6. J. Goguen, G. Malcolm, *A Hidden Agenda*, Theoretical Computer Science 245 (2000) 55-101
7. M. Hanus, ed., *Curry: A Truly Integrated Functional Logic Language*, http://www.informatik.uni-kiel.de/~curry
8. *Haskell: A Purely Functional Language*, http://haskell.org
9. B. Jacobs, J. Rutten, *A Tutorial on (Co)Algebras and (Co)Induction*, EATCS Bulletin 62 (1997) 222-259
10. *The Maude System*, http://maude.cs.uiuc.edu
11. Till Mossakowski, Horst Reichel, Markus Roggenbach, Lutz Schröder, *Algebraic-coalgebraic specification in CoCASL*, Proc. WADT 2002, Springer LNCS 2755 (2003) 376-392
12. J. Nordlander, ed., *The O'Haskell homepage*, http://www.cs.chalmers.se/~nordland/ohaskell
13. T. Nipkow, L.C.Paulson, M. Wenzel, *Isabelle/HOL*, Springer LNCS 2283 (2002)
14. P. Padawitz, *Deduction and Declarative Programming*, Cambridge University Press 1992
15. P. Padawitz, *Inductive Theorem Proving for Design Specifications*, J. Symbolic Computation 21 (1996) 41-99
16. P. Padawitz, *Proof in Flat Specifications*, in E. Astesiano, H.-J. Kreowski, B. Krieg-Brückner, eds., *Algebraic Foundations of Systems Specification*, IFIP State-of-the-Art Report, Springer 1999
17. P. Padawitz, *Swinging Types = Functions + Relations + Transition Systems*, Theoretical Computer Science 243 (2000) 93-165
18. P. Padawitz, *Structured Swinging Types*, http://ls5-www.cs.uni-dortmund.de/~peter/SST.ps.gz
19. P. Padawitz, *Dialgebraic Swinging Types*, http://ls5-www.cs.uni-dortmund.de/~peter/Dialg.ps.gz
20. P. Padawitz, *Swinging Types At Work*, http://ls5-www.cs.uni-dortmund.de/~peter/BehExa.ps.gz
21. P. Padawitz, *Expander2: A Formal Methods Presenter and Animator*, http://ls5-www.cs.uni-dortmund.de/~peter/Expander2/Expander2.html
22. *The QPQ Database of Deductive Software Components*, http://www.qpq.org
23. H. Reichel, *An Approach to Object Semantics based on Terminal Coalgebras*, Math. Structures in Comp. Sci. 5 (1995) 129-152
24. G. Rozenberg, A. Salomaa, eds., *Handbook of Formal Languages, Vol. 3: Beyond Words*, Springer 1997
25. L. Schröder, T. Mossakowski, *Monad-Independent Dynamic Logic in HasCASL*, Proc. WADT 2002, Springer LNCS 2755 (2003) 425-441

26. M.-O. Stehr, J. Meseguer, P.C. Ölveczky, *Rewriting Logic as a Unifying Frame-work for Petri Nets*, in: H. Ehrig et al., eds., Unifying Petri Nets, Springer LNCS 2128 (2001)

27. G. Sutcliffe, *Problem Library for Automated Theorem Proving*,
    http://www.cs.miami.edu/~tptp

28. F. Wiedijk, ed., *The Digital Math Database*,
    http://www.cs.kun.nl/~freek/digimath

29. *The Yahoda Verification Tools Database*, http://anna.fi.muni.cz/yahoda

Relationships Between Equational and Inductive Data Types

Eric G. Wagner

Wagner Mathematics, 1058 Old Albany Post Road,
Garrison, NY 10524, USA
wagner@highlands.com

Abstract. This paper explores the relationship between equational algebraic specifications (using initial algebra semantics) and specifications based on simple inductive types (least fixed points of equations using just products and coproducts, e.g. $N \cong 1 + N$). The main result is a proof that computable data type (one in which the corresponding algebra is computable in the sense of Mal'cev) can be specified inductively. This extends an earlier result of Bergstra and Tucker showing that any computable data type can be specified equationally.

1 Introduction

For the purposes of this paper, an equational data type is one specified as the initial Σ-algebra for a given signature, Σ and set of equations E over Σ; and a simple inductive data type is one defined as the least fixpoint solution of equations using just products and coproducts, e.g., $N \cong 1 + N$.

The early papers on equational data type specifications [5–8], and inductive specifications [12], were written in the seventies. (See [11] for other earlier references relevant to inductive types). The two approaches seem rather different:

1. Early papers on inductive data types used $\rightarrow$ in addition to $+$ and $\times$ and thus introduced higher level types not found in the equational approach.
2. In the equational approach it is easy to specify types such as finite sets of natural numbers, but this can not be done directly in the inductive approach, that is, finite sets do not appear as a fixpoint of equations using just $+$ and $\times$.
3. Papers on inductive types frequently also consider co-inductive types: greatest fixpoints of equations in $+$ and $\times$ (and $\rightarrow$). This also leads to data types, such as possibly-infinite strings and possibly-infinite trees; I do not know of any equational specifications for such types.
4. Equational specifications are not necessarily implementable as is shown by the existence of unsolvable word problems; but simple inductive data types are always implementable [14].

We could continue listing differences, but the real point of this paper is to argue that, for all "practical purposes", the equational and inductive approaches are equally powerful. Of course this is only true when we make certain "practical" restrictions:

H.-J. Kreowski et al. (Eds.): Formal Methods (Ehrig Festschrift), LNCS 3393, pp. 259–274, 2005.
© Springer-Verlag Berlin Heidelberg 2005

1. We restrict ourselves to computable data types.
2. We restrict ourselves to simple inductive types (no $\rightarrow$, no greatest fixpoints).

The first restriction is justified on the grounds that, after all, we do want to use the specified data types in real programs which means they must be implementable, i.e., computable. This does not justify the second restriction, but without the second restriction we are clearly faced with insurmountable differences.

Section 2 of the paper reviews the relevant concepts from the theories of equational specifications and recursive functions, and the work on computable data types (or algebras) from [13] and [1]. Section 3 explores simple inductive specifications and ways to enrich them with additional operations. In particular, it is shown how we can generalize primitive recursion and the While-do operator to inductive types. Section 4 applies the results of Section 3 to traditional recursion theory; showing that both primitive recursion and minimalization can be expressed using While-do. The main result is given in Section 5, where we show that computable data types can be specified both equationally and inductively following a particular specification strategy.

2   Preliminaries

2.1   Signatures and Algebras

The following is meant only to clarify our notation, we assume the reader is already familiar with the concepts.

Definition 1. *A* signature *consists of the following data*

> *S, a finite set, the* set of sorts.
> *F, a finite set, the* set of operators.
> $\mathrm{dom} : F \rightarrow S^*$
> $\mathrm{cod} : F \rightarrow S$

> *We write* $\Sigma = \lceil S, F, \mathrm{dom}, \mathrm{cod} \rceil$ *to indicate that* Σ *is a signature with the given data.*

> *Given a signature* $\Sigma = \lceil S, F, \mathrm{dom}, \mathrm{cod} \rceil$ *we define a* Σ-algebra, *A, as consisting of*

1. *For each* $s \in S$ *a set* $A(s)$, *called the* carrier of sort *s.*
2. *For each* $\sigma \in F$ *a function* $A(\sigma) : A(\mathrm{dom}(\sigma)) \rightarrow \mathrm{cod}(\sigma)$ *where, for any* $w \in S^*$, *if* $w = s_1 \cdots s_n$, *then* $A(w) = A(s_1) \times \cdots \times A(s_n)$.

We will sometimes denote a Σ-algebra *as* $A = \lceil \langle A(s) \mid s \in S \rangle, \langle A(\sigma) \mid \sigma \in F \rangle \rceil$.

If A and B are Σ-algebras *then a* Σ-homomorphism, $h : A \rightarrow B$, *consists of an S-indexed family of mappings,* $h = \langle h(s) : A(s) \rightarrow B(s) \mid s \in S \rangle$, *such that for any* $\sigma \in F$,

$$B(\sigma) \circ h(\mathrm{dom}(\sigma)) = h(\mathrm{cod}(\sigma)) \circ A(\sigma),$$

where, for any $w \in S^*$, *if* $w = s_1 \cdots s_n$ *then* $h(w) = h(s_1) \times \cdots \times h(s_n)$.

Definition 2. *Let $\Sigma = \lceil S, F, \mathrm{dom}, \mathrm{cod} \rceil$ be a signature, then for each $s \in S$ we define the set of Σ-terms as follows:*

1. *If $\sigma \in F$, $\mathrm{dom}(\sigma) = \epsilon$ (the empty string), and $\mathrm{cod}(\sigma) = s$ then σ is a Σ-term of sort s.*
2. *If $\sigma \in F$, $\mathrm{dom}(\sigma) = w = s_1 \cdots s_n \in S^*$, $\mathrm{cod}(\sigma) = s$, and, for $i = 1, \ldots, n$, t_i is a Σ-term of sort s_i, then $\sigma(t_1, \ldots, t_n)$ is a Σ-term of sort s.*

We write $T(\Sigma)_s$ for the set of all Σ-terms of sort s. We write $T(\Sigma)$ for the S-indexed set $T(\Sigma) = \langle T(\Sigma)_s \mid s \in S \rangle$.

Given a signature, Σ, we also write $T(\Sigma)$ to denote the Σ-term algebra, where $T(\Sigma)(s) = T(\Sigma)_s$ and where for any $\sigma \in F$ and $\langle t_1, \ldots, t_n \rangle \in T(\Sigma)(\mathrm{dom}(\sigma))$,

$$T(\Sigma)(\sigma)(t_1, \ldots, t_n) = \sigma(t_1, \ldots, t_n).$$

Recall that $T(\Sigma)$ is "the" initial Σ-algebra, that is, for any Σ-algebra, A, there is exactly one Σ-homomorphism $h_A : T(\Sigma) \to A$.

2.2 Recursive Functions and Recursive Sets

We follow the definitions given in [4] for the definitions of recursive functions on the natural numbers, ω, and for recursive sets of natural numbers.

Definition 3. *Given functions $\psi : \omega^n \to \omega$ and $\theta : \omega^{1+1+n} \to \omega$, the operation of primitive recursion yields the unique function $\phi : \omega^{1+n} \to \omega$ such that, for $y \in \omega$ and $\overline{x} \in \omega^n$,*

$$\phi(0, \ \overline{x}) = \psi(\overline{x})$$
$$\phi(y + 1, \ \overline{x}) = \theta(y, \ \phi(y, \overline{x}), \ \overline{x}).$$

A function is primitive recursive *if it can be obtained by a finite number of applications of the operations of composition[1] and primitive recursion starting from the constant 0-function, the successor function, and the projection functions.*

Given a total function $\theta : \omega^{n+1} \to \omega$ the operation of minimalization *yields the unique function $\phi : \omega^n \to \omega$, whose value for given $\overline{x}$ is the least value of y, if such exists, for which $\theta(y, \overline{x}) = 0$, and is undefined if no such y exists. The function θ is said to* Min-regular *if ϕ is total.*

A class of functions is recursive *if it can be obtained by a finite number of applications of composition, primitive recursion, and minimalization of Min-regular functions.*

We say that a set is (primitive) recursive if its characteristic function is. Thus, a subset $P \subseteq \omega \times \omega$ is (primitive) recursive iff the function

$$C_P(x, y) = \begin{cases} 0 & \langle x, y \rangle \in P \\ 1 & \langle x, y \rangle \notin P \end{cases}$$

is (primitive) recursive.

---

[1] In this definition, as in [4], composition may have more than two arguments: given a function $g : \omega^n \to \omega$ and n functions $f_i : \omega^p \to \omega$, $i = 1, \ldots, n$, this composition operation gives us the function $g(f_1, \ldots, f_n) : \omega^p \to \omega$. However, in most of the paper we will use binary composition.

The above definition of a recursive function is convenient for theoretical purposes but is not necessarily convenient for writing specifications of functions even when we restrict ourselves to functions on the natural numbers. For example, the specification of the equality predicate on natural numbers given in [4] using primitive recursive and composition requires a sequence of five functions to be defined, but we can give a succinct definition using double recursion

$$eq(0,\ 0) = 0$$
$$eq(succ(n),\ 0) = 1$$
$$eq(0,\ succ(p)) = 1$$
$$eq(succ(n),\ succ(p)) = eq(n,\ p)$$

The original treatments of recursion allowed such double recursions among many others; indeed a set of equations was defined as recursive if it could always be applied and evaluated "recursively", see [10]. That is, the emphasis is on how the equations are applied rather than on specific forms of equations schemata. It is straightforward to show that even with double recursion we can define functions which are not primitive recursive even though they are recursive.

2.3   Gödelizations of Σ-Terms

In order to treat Σ-algebras in the context of traditional recursive function theory it is necessary to be able to represent Σ-terms by natural numbers. What we want is to do this in a way in which the translation from term to number, and its inverse from number to term, are, in some sense, computable, and in which the set of numbers representing terms is a recursive set. Such a representation is called a Gödelization. Rather than spell out the desired properties of Gödelizations in detail, we shall always use the following specific Gödelization, GN, in this paper.

Definition 4. *For each $i \in \omega$ let $Pr(i)$ denote the ith prime (take $Pr(0) = 0$). Given a signature $\Sigma = \lceil S, F, \mathrm{dom}, \mathrm{cod} \rceil$, let $\langle \sigma_1, \ldots, \sigma_n \rangle$ be an enumeration of the elements of F (without repetitions). Then for each $t \in T(\Sigma)$ let its Gödelization, $GN(t)$, be given as follows:*

1. *If $t = \sigma_i$ where $\mathrm{dom}(\sigma_i) = \epsilon$ (the empty string), then $GN(t) = 2^{Pr(i)}$.*
2. *If $t = \sigma_j(t_1, \ldots, t_p)$ then $GN(t) = 2^{Pr(j)} \cdot 3^{GN(t_1)} \cdots Pr(p+1)^{GN(t_p)}$.*

Definition 5. *A function $f : \bigcup \langle T(\Sigma)_s \mid s \in S \rangle \to \omega$ is called GN-computable if there is a recursive function $g : \omega \to \omega$ such that $g \bullet GN = f$.*

A set $W \subseteq \bigcup \langle T(\Sigma)_s \mid s \in S \rangle$ is said to be GN-computable if the set $GN(W)$ $=_{def} \{GN(w) \mid w \in W\}$ is recursive.

Example 1. For each $s \in S$, the set $T(\Sigma)_s$ is GN-computable. The set $\bigcup \langle T(\Sigma)_s \mid s \in S \rangle$ is GN-computable.

2.4   Computable Algebraic Specifications

The following definition comes from [13] via [2] and has been rephrased to reflect our needs and notation:

Definition 6. *Let* $\Sigma = \lceil S, F, \mathrm{dom}, \mathrm{cod} \rceil$ *be a signature and let* A *be a* Σ-algebra. *We say that* A *is* effectively presented *if for each* $s \in S$ *there is a recursive set* $\Omega(s)$ *and a surjective function* $\alpha_s : \Omega(s) \to A_s$, *and, for each* $\sigma \in F$, *there exists a recursive function* $\Omega(\sigma) : \Omega(\mathrm{dom}(\sigma)) \to \Omega(\mathrm{cod}(\sigma))$ *such that,*

$$A(\sigma) \circ \alpha^{\mathrm{dom}(\sigma)} = \alpha_{\mathrm{cod}(\sigma)} \circ \Omega(\sigma)$$

where, for $w = s_1 \cdots s_n \in S^*$, $\alpha^w = \alpha_{s_1} \times \cdots \times \alpha_{s_n}$.

We say that a Σ-algebra, A, *effectively presented as above, is* computable *if for each* $s \in S$ *the relation* $\equiv_{\alpha,s}$ *on* $\Omega(s)$, *such that* $n \equiv_{\alpha,s} p$ *iff* $\alpha_s(n) = \alpha_s(p)$ *for all* $n, p \in \Omega(s)$, *is recursive.*

Note that $\Omega = \lceil \langle \Omega_s \mid s \in S \rangle, \langle \Omega(\sigma) \mid \sigma \in F \rangle \rceil$ is a Σ-algebra. Furthermore, α is a surjective homomorphism from Ω to A.

Theorem 1. *Every computable* Σ-algebra, A, *finitely generated by* $a_1, \ldots, a_n$ *has a specification consisting of a finite set,* E, *of equations over some signature* $\Sigma' = \lceil S, F \cup F', \mathrm{dom}', \mathrm{cod}' \rceil$ *where, for each* $\sigma \in F$, $\mathrm{dom}'(\sigma) = \mathrm{dom}(\sigma)$, $\mathrm{cod}'(\sigma) = \mathrm{cod}(\sigma)$, *and the* Σ-reduct of $T(\Sigma', E)$ *is isomorphic to* A.

Proof. See Theorem 3.1. in [2].

3   Inductive Specifications Using Only + and ×

In contrast to the approach taken in [12], where $+$, $\times$, $\to$ are used to construct inductive types, we will only consider simple inductive types constructed using $+$ and $\times$. Our main effort in this section will be directed to exploring the extension of inductive types to include operations other than those corresponding directly to injections and projections. Such extensions can be thought of as embedding the inductive types in some form of programming language [9, 3, 17, 16].

The restriction to $+$ and $\times$ has the additional advantage that it simplifies the mathematics since everything (that we will do in this paper) can be carried out within the category, **Set**, of sets and total functions rather than in a more general CPO. It is worthwhile, however, to point out that **Set**

1. Has terminal objects (which we denote by **1**), **1** is the empty product of sets.
2. Is distributive, so, in particular, for any sets, A, B and C, the function

$$\Delta = [1_C \times i_A, \ 1_C \times i_B] : (C \times A) + (C \times B) \cong C \times (A + B)$$

is an isomorphism, where i_A and i_B are the evident coproduct injections, and 1_C is the identity on C.

3.1   Carrier Specifications

Notation: Given a set S and strings $u_1, \ldots, u_n \in S^*$ we write $(u_1) \cdots (u_n)$ to denote the string-of-strings whose i^{th} element is u_i. Note, $()$ denotes the empty string.

Definition 7. *A* carrier specification: *A carrier specification consists of*

1. K, *a finite set (of* class names*)*
2. $\iota : K \to (K^*)^*$, *assigning each class name a string-of-strings of class names.*

Given a set, K, of class names we can interpret a string-of-strings, $\bar{u} = (u_1)\cdots(u_n)$, as a sum of products, or, more precisely, as a polynomial in the algebra with operations $+$ and $\times$ generated by the set K. Such a polynomial then defines (up-to-isomorphism) a function $|\mathbf{Set}|^K \to |\mathbf{Set}|$ with $+$ interpreted as coproduct, and with $\times$ interpreted as product. The function $\iota : K \to (K^*)^*$ then defines an endofunctor $\bar{\iota} : \mathbf{Set}^K \to \mathbf{Set}^K$. Such a functor has a least fixed point (with respect to inclusion and up-to-isomorphism). We say then that the semantics of the carrier specification, $\lceil K, \iota \rceil$, is the algebra consisting of the sets making up this least fixpoint together with the associated coproduct injections and product projections. (Note, we have not formally presented a signature for this algebra, but see below.)

Example 2. : Let $K = \{NAT, STACK\ BOOL\}$ and let $\iota : K \to (K^*)^*$ be such that $\iota(NAT) = ()(NAT)$, $\iota(STACK) = ()(NAT \cdot STACK)$ and $\iota(BOOL) = ()()$. The corresponding set of least points consists of sets, $\underline{NAT}$, $\underline{STACK}$ and $\underline{BOOL}$ such that

$$\underline{NAT} \cong 1 + \underline{NAT}$$
$$\underline{STACK} \cong 1 + (\underline{NAT} \times \underline{STACK})$$
$$\underline{BOOL} \cong 1 + 1$$

That is, $\underline{NAT}$ is isomorphic to the set of natural numbers, $\underline{STACK}$ is isomorphic to the set of stacks of natural numbers (isomorphic to the set of strings over the natural numbers), and $\underline{BOOL}$ is "the" two-element set.

If we now look at the injections and projections we see that they all correspond to natural operations on the corresponding sets. E.g., the injection $1 \to \underline{NAT}$ picks out 0, while the other injection is naturally interpreted as the successor function. For $\underline{BOOL}$ it is natural to name the injections *true* and *false*. Going further, it is suggestive to name the remaining injections and projections as in the following diagram:

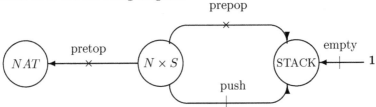

in which we have used special arrows to denote injections and projections: $\longmapsto$ for injections, and $\longrightarrow\!\!\times\!\!\longrightarrow$ for projections. It is tempting to replace the labels *prepop* and *pretop* by *pop* and *top* respectively; but the domains are wrong.

However, it is reasonable to say that the coproduct injections and product projections "always correspond to the basic underlying operations of the corresponding data type".

3.2   Operation Specifications

Only a very limited class algebras can be defined in the above manner. Two particular limitations are:

1. The only operations are the injections and projections. Thus while the above definition of NAT gives us zero and successor, it does not give addition and multiplication to say nothing of the other operations readily definable in equational specifications.
2. We don't have any mechanism for defining congruences on the inductive types.

In this section we address the first problem.

Defining Functions Using Source, or Target, Tupling

Definition 8. *Given sets $A_0, A_1, \ldots, A_n$ and a function $f_i : A_i \to A_0$ for each $i = 1, \ldots, n$, then the* source tupling of the f_i, *denoted $[f_1, \ldots, f_n]$, is the function*

$$[f_1, \ldots, f_n] : A_1 + \cdots + A_n \to A_0$$

where, for all $j = 1, \ldots, n$, if $i_j : A_j \to A_1 + \cdots + A_n$ is the evident coproduct injection, then $[f_1, \ldots, f_n] \circ i_j = f_j$.

Example 3. Two simple examples of a function defined by means of source tupling are the definition of the identity and predecessor functions on the natural numbers: $1_{NAT} = [zero, \ succ]$, and $pred = [zero, 1_{NAT}]$. Two other examples would be to define the stack operations as $pop = [empty, \ prepop]$ and $top = [zero, \ pretop]$. Note that we could, equally well, define $top = [s^{17}(zero), pretop]$, the choice of the first value is no more restricted here than in the equational case.

Definition 9. *Given sets $A_0, A_1, \ldots, A_n$ and a function $f_i : A_0 \to A_i$ for each $i = 1, \ldots, n$, then the* target tupling of the f_i, *denoted $\langle f_1, \ldots, f_n \rangle$, is the function*

$$\langle f_1, \ldots, f_n \rangle : A_0 \to A_1 \times \cdots \times A_n$$

where, for all $j = 1, \ldots, n$, if $p_j : A_1 \times \cdots \times A_n \to A_j$ is the evident product projection, then $p_j \circ \langle f_1, \ldots, f_n \rangle = f_j$.

Target tupling is important for several reasons; a major one being that it allows us to define composition as a binary operation. See the proof of Proposition 3 for another use of target-tupling.

Generalizing Primitive Recursion

Definition 10. *Let $\lceil K, \iota \rceil$ be a carrier specifications with corresponding carrier algebra A. Then given*

1. $k \in K$ with $\iota(k) = (u_1)(u_2) \cdots (u_n)$
2. *a function ρ_i, for each $i = 1, \ldots, n$ such that, if u_i contains exactly $p_i \geq 0$ occurrences of k, then they are enumerated by $\rho_i : \{1, \ldots, p_i\} \to \{1, \ldots, |u_i|\}$, where $|u_i|$ is the length of u_i. That is, $u_{i, \rho_i(j)} = k$ and is the jth occurrence of k in u_i,*
3. $k' \in K$

4. $v \in K^*$

5. *for each* $i = 1, \ldots, n$, *a function* $g_i : A(v) \times A(u_i) \times A(k')^{p_i} \to A(k')$

we say that the operation of primitive recursion associates with the above data a new function $h : A(v) \times A(k) \to k'$ *such that for each* $j = 1, \ldots, n$, *if* $\bar{a} \in A(v)$ *and* $\bar{b} \in A(u_j)$, *then*

$$h(\bar{a}, i_{u_j}(\bar{b})) = g_j(\bar{a}, \bar{b}, \ h(\bar{a}, \bar{b}_{\rho(1)}), \ldots, h(\bar{a}, \bar{b}_{\rho(p_j)})),$$

where $i_{u_j} : A(u_j) \to A(k)$ *is the indicated coproduct injection.*

Definition 11. *Let* $\lceil K, \iota \rceil$ *be a carrier specification with corresponding carrier algebra* A. *We say that a function* $f : A(v) \times A(k) \to A(k')$ *is primitive recursive with respect to* $\lceil K, \iota \rceil$ *if there exists a finite sequence of functions* $g_1, \ldots, g_p$ *such that each function* g_i *is either*

1. *An injection or projection function from* A, *or*
2. *A source-tupling or target tupling of functions earlier in the sequence, or*
3. *A binary composite of functions of preceding it in the sequence (i.e., we compose functions* $f : A \to B$ *and* $g : C \to D$ *to get* $f \bullet g : A \to D$ *only when* $B = C$, *rather than using the multi-argument version of composition given in the earlier footnote),or*
4. *A result of applying the operation of primitive recursion to functions preceding it in the sequence.*

Proposition 1. *For the carrier specification*

$$nat = \lceil \{NAT\}, \iota(NAT) = ()(NAT) \rceil$$

of the natural numbers, primitive recursion with respect to NAT *is equivalent to that for primitive recursion as in Definition 3.*

Example 4. Consider the presentation $\lceil \{T\}, \iota(T) = ()(T \cdot T) \rceil$ for binary trees with the projections and injections named as in the diagram:

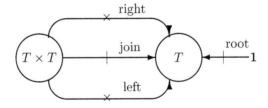

The primitive recursive function for the depth of such a tree is

$$depth(root) = 0$$
$$depth(join(\langle T_1, T_2 \rangle)) = max(depth(T_1), \ depth(T_2))$$

Generalized While-Do

Definition 12. *Let* $\lceil K, \iota \rceil$ *be a carrier specifications with corresponding carrier algebra A. Then given*

1. *Strings* $t, u, v, w \in K^*$ *such that* $A(u) = A(v) + A(w)$,
2. *A function* $f : A(t) \to A(u)$ *(the* initialization *function).*
3. *a function* $g : A(u) \to A(u)$ *(the* iterated *function)*

The operation of while-do *associates with the above data a partial function* $W(g) :$ $A(u) {\dashrightarrow} A(u)$ *that is a fix point for the diagram*

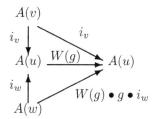

where i_v *and* i_w *are the indicated coproduct injections. Equationally:* $W(g) = [i_v, \ W(g) \bullet g \bullet i_w]$.

The operation of initialized while-do *associates with the above data the function* $W(f, g) =_{def} W(g) \bullet f$. *We say that* : $W(f, g)$ *is* WD-regular *if* $W(f, g)$ *is total.*

Our claim is that this captures what we would intuitively expect as the meaning of "While in $A(w)$ do g starting with initialization f". More precisely:

Proposition 2. *Given* $A(u) = A(v) + A(w)$ *and a function* $g : A(u) \to A(u)$ *then there is a least partial function[2]* $h : A(u) {\dashrightarrow} A(u)$ *which is a fixpoint for the equation* $h = [i_v, \ h \bullet g \bullet i_w]$. *In particular, if* $\bar{a} \in A(u)$ *and* $g^n(\bar{a}) \in A(v)$ *then* $h(\bar{a}) = g^n(\bar{a})$, *and, if no such n exists for* $\bar{a}$ *then* $h(\bar{a})$ *is undefined.*

Proof. See [15], where the proof is given in a slightly more general setting.

In the review of recursion theory, given in section 2, we employed the operation of minimalization. The minimalization operation does not carry over to arbitrary inductive data types defined using $+$ and $\times$ since such algebras lack the "natural" total ordering enjoyed by NAT. Of course we can (and, in some of our proofs, we will) employ minimalization after first introducing a Gödelization to provide the necessary ordering; but this does not seem to me to be the road to intuitively clear specifications.

4   Recursive Functions on NAT via While-Do

The following results show that we can replace primitive recursion, and minimalization of min-regular functions, with while-do (starting from a small, finite

---

[2] Given partial functions $f, g : A {\dashrightarrow} B$ we say f *is less-than* g if for every $a \in A$, if $f(a)$ is defined the $g(a) = f(a)$.

set of given primitive recursive functions). Of course this does not say anything about either the convenience or efficiency of relying on while-do. But it does show that we can implement recursive functions in a reasonably conventional programming language without complex recursions.

Proposition 3. *If $\phi : \omega^n \to \omega$ is defined by minimalization from the Min-regular function $\theta : \omega^{1+n} \to \omega$, then it is definable in the inductive specification $\lceil \{NAT\}, \iota(NAT) = ()(NAT) \rceil$, as $\phi = \pi_{(1)}^{1+n+1} \bullet W(f,g)$, from the WD-regular iterative while-do function, $W(f,g)$ given by the data*

- $A(t) = \omega^n$, $A(u) = \omega^{1+n+1}$, $A(v) = \omega^{1+n} \times \mathbf{1}$, $A(w) = \omega^{1+n} \times \omega$, with coproduct injections as shown in the diagram below.
- With initializing function $f = \langle zero, 1_{\omega^n}, \theta \bullet \langle zero, 1_{\omega^n} \rangle \rangle$
- With iterated function $g = \langle succ \bullet \pi_{(1)}^{1+n+1}, \pi_{(2)}^{1+n+1}, \theta \bullet \langle succ \bullet \pi_{(1)}^{1+n+1}, \pi_{(2)}^{1+n+1} \rangle \rangle$

giving the fixpoint diagram

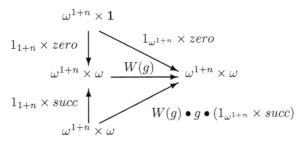

Proposition 4. *If $\phi : \omega^{1+n} \to \omega$ is defined by primitive recursion from the functions $\psi : \omega^n \to \omega$ and $\theta : \omega^{1+1+n} \to \omega$, then it is definable in the inductive specification $\lceil \{NAT\}, \iota(NAT) = ()(NAT) \rceil$, as $\phi = \pi_{(4)}^{1+n+1+1} \bullet W(f,g)$, from the WD-regular iterative while-do function, $W(f,g)$ given by the data*

- $A(t) = \omega^{1+n}$, $A(u) = \omega^{1+n+1+1} \times \omega$, $A(v) = \omega^{1+n+1+1} \times \mathbf{1}$, $A(w) = \omega^{1+n+1+1} \times \omega$, with coproduct injections $(1_{\omega^{1+n+1+1}} \times zero) : A(v) \to A(u)$ and $(1_{\omega^{1+n+1+1}} \times succ) : A(w) \to A(u)$.
- With initializing function

$$f = \langle \pi_{(1)}^{1+n}, \pi_{(2)}^{1+n}, zero, \psi \bullet \langle zero, \pi_{(2)}^{1+n} \rangle, eq \bullet \langle \pi_1^{1+n}, zero \rangle \rangle$$

- With iterated function

$$g = \langle \pi_{(1)}, \pi_{(2)}, succ \bullet \pi_{(3)}, \theta \bullet \langle \pi_{(3)}, \pi_{(4)}, \pi_{(2)} \rangle, eq \bullet \langle succ \bullet \pi_{(3)}, \pi_{(1)} \rangle \rangle$$

where, for $i = 1, \ldots, 5$, $\pi_{(i)}$ is an abbreviation for the projection $\pi_{(i)}^{1+n+1+1}$.

Corollary 1. *All the usual recursive functions on ω can be defined as inductive functions on NAT.*

5   Congruences and Computable Data Types

Let $\Sigma = \lceil S, F, \text{dom}, \text{cod} \rceil$ be a signature. Nothing we have proposed so far allows us to go from an inductively defined Σ-algebra, R, to a quotient algebra, $A = (R/\equiv_R)$, where $\equiv_R$ is a congruence on R. One approach to this problem is to require the congruence, $\equiv_R$ to by inductively definable, that is, that its characteristic function, $\mathcal{C}$, be inductively definable in R. Note that since $\equiv_R$ is an S-indexed set, $\equiv_R = \langle \equiv_{R,s} \mid s \in S \rangle$, what we want is an S-indexed set of functions $\mathcal{C} = \langle \mathcal{C}_s : T(\Sigma)_s \to NAT \rangle$ (or, equivalently, $\mathcal{C} = \langle \mathcal{C}_s : T(\Sigma)_s \to BOOL \rangle$). This is sufficient in that, for each $s \in S$, $\mathcal{C}_s$ tells us precisely when two elements of $R(s)$ represent the same element of A. See Example 6 below, for a familiar example, namely that of fractions as representatives of rational numbers.

Definition 13. *Let R be an inductively defined algebra in the sense of being an extension of an algebra given by a carrier presentation $\lceil K, \iota \rceil$ to include additional defined inductively operations, and let $\equiv_R$ be a congruence on R, then we say $(R/\equiv_R)$ is inductively defined providing that $\equiv_R$ can be inductively defined (using "hidden" sorts and operators).*

The question then is, when can this be done? The answer is that it can be done providing that the algebra A is computable (see Definition 6). The following theorem provides the key. Note that it is in a more general context than the above discussion in that it starts from a Σ-algebra A and provides an inductively defined algebra $R(\Gamma)$ for each choice, Γ, of constructors for A. The theorem does not show that R is inductively defined, that is shown by the propositions following the theorem.

Theorem 2. *Let $\Sigma = \lceil S, F, \text{dom}, \text{cod} \rceil$ be a signature and let A be a computable Σ-algebra (data type with signature Σ) with an effective presentation $\lceil \Omega, \alpha \rceil$. Let $\Gamma = \lceil S, F_\Gamma, \text{dom}|_{F_\Gamma}, \text{cod}|_{F_\Gamma} \rceil \subseteq \Sigma$ consisting of constructors, that is, of operators, σ, whose corresponding operations, $A(\sigma)$, generate (the carriers of) A. Let $T(\Gamma)$ denote the initial Γ-term algebra. Then*

1. *Let $\gamma : T(\Sigma) \to \Omega$ be the unique homomorphism given by the initiality of $T(\Sigma)$. Then γ is GN-computable.*
2. *There exists a computable Σ-algebra $R(\Gamma)$ such that*
   (a) $R(\Gamma)(s) = T(\Gamma)(s)$ for each $s \in S$.
   (b) For each $\sigma \in F_\Gamma$, and $\langle t_1, \ldots, t_n \rangle \in T(\Gamma)(\text{dom}(\sigma))$, we have

$$R(\sigma)(t_1, \ldots, t_n) = \sigma(t_1, \ldots, t_n),$$

3. *Let $\equiv_R$ be the congruence on $R(\Gamma)$ such that $(R(\Gamma)/\equiv_R) \cong A$. Then $\equiv_R$ is GN-computable.*

The proof is given in the Appendix.

Proposition 5. *Given a signature $\Sigma = \lceil S, F, \text{dom}, \text{cod} \rceil$ then there exists an inductive presentation $\lceil \{T\} \cup S, \iota \rceil$, where $T \notin S$, defining an inductive algebra, A, of signature $\lceil \{T\} \cup S, S \cup F, \cdots \rceil$ with*

carriers:
$$A(s) = T(\Sigma)_s$$
$$A(T) = \coprod \langle T(\Sigma)_s \mid s \in S \rangle$$
operations:
$$A(\sigma) : A(\text{dom}(\sigma)) \to A(\text{cod}(\sigma)) \ \textit{where} \ A(\sigma)(t_1, \ldots, t_n) = \sigma(t_1, \ldots, t_n)$$
$$A(s) : T(\Sigma)_s \to \coprod \langle T(\Sigma)_s \mid s \in S \rangle = \bigcup \langle T(\Sigma)_s \mid s \in S \rangle, \ \textit{the inclusion}$$
function

Corollary 2. *The carrier algebra for $R(\Gamma)$ is inductively definable by applying the above proposition to Γ.*

Proposition 6. *Extending the above presentation to include ;the carrier specification $\lceil \{NAT\}, \iota(NAT) = ()(NAT) \rceil$, we can inductively define the function $GN : \bigcup \langle T(\Sigma)_s \mid s \in S \rangle \to NAT$ and, for each $s \in S$, the restriction $GN|_{T(\Sigma)_s} : T(\Sigma)_s \to NAT$.*

Proposition 7. *Every GN-recursive function and set (see Definition 5) is inductively definable.*

Theorem 3. *If A is computable then $R(\Gamma)$ and $\equiv_R$ are inductively definable and so $A = (R(\Gamma)/\equiv_R)$ is inductively specifiable. Furthermore, by Theorem 1 (originally from [1]), A is equationally specifiable.*

Example 5. In STACKS-OF-NAT the above approach works very neatly, There we could take

$$\Sigma = \lceil \{NAT, STACK\}, \{zero, succ, plus, times, empty, push, pop, top\}, \ldots \rceil$$

and $\Gamma = \lceil \{NAT, STACK\}, \{zero, succ, empty, push\}, \ldots \rceil$ (we leave dom and cod to the reader). We immediately get $T(\Gamma)(NAT) \cong \omega \cong A(NAT)$ and $T(\Gamma)(STACK) \cong \omega^* \cong A(STACK)$. Then, assuming $A(pop)(A(empty)) = A(empty)$, and $A(top)(A(empty)) = A(zero))$ it is easy to write equations for the remaining operators. The desired congruences are both trivial, i.e., the diagonal (equality).

Example 6. A more interesting example is provided by the positive rational numbers. Let NAT, POS and RAT have signatures:

$$\Sigma_{NAT} = \lceil \{NAT\}, \{zero, \ succ_N, +_N, \times_N\}, \ \cdots \rceil$$
$$\Gamma_{NAT} = \lceil \{NAT\}, \{zero, \ succ_N\}, \cdots \rceil$$
$$\Sigma_{POS} = \lceil \{POS\}, \{one, \ succ_P, +_P, \times_P\}, \ \cdots \rceil$$
$$\Gamma_{POS} = \lceil \{POS\}, \{one, \ succ_P\}, \cdots \rceil$$
$$\Sigma_{RAT} = \lceil \{NAT, \ POS, \ RAT\}, F_{NAT} \cup F_{POS} \cup \{make, +_R, \times_R\}, \ \cdots \rceil$$
$$\Gamma_{RAT} = \lceil \{NAT, \ POS, \ RAT\}, \{zero, one, succ_N, succ_P, make\}, \cdots \rceil.$$

where $\text{dom}(make) = NAT \cdot POS$, $\text{cod}(make) = RAT$.

There is a "coercion"

$$\mathcal{C} : T(\Gamma_{POS}) \rightarrow T(\Gamma_{NAT})$$
$$\mathcal{C}(one) = succ_N(zero)$$
$$\mathcal{C}(succ_P(n)) = succ_N(\mathcal{C}(n))$$

which we could either add to the signature or regard as a "hidden operator". Then the equations defining the operations are:

$$+_N(n,\ zero) = n \qquad\qquad +_P((n,\ one) = succ_P(n)$$
$$+_N(n,\ succ_N(p)) = succ_N(+_N(n,p)) \qquad +_P(n,\ succ_P(p)) = succ_P(+_P(n,p))$$

$$\times_N(n,\ zero) = zero \qquad\qquad \times_P(n,\ one) = n$$
$$\times_N(n,\ succ_N(p)) = +_N(\times_N(n,p),\ n) \quad \times_P(n,\ succ_P(p)) = +_P(\times_P(n,p),\ n)$$

$$+_R(make(n,p),\ make(q,r))$$
$$= make(+_N(\times_N(n,\mathcal{C}(r)),\ \times_N(\mathcal{C}(p),q)),\ \times_P(p,r))$$

$$\times_R(make(n,p),\ make(q,r)) = make(\times_N(n,q),\ \times_P(p,r))$$

With the equations given so far we get an algebra of positive fractions. To get the positive rationals we need to introduce a congruence on $T(\Gamma_{RAT})$. In the equational specification framework it suffices to add one more axiom, namely

$$make(\times_N(n,\mathcal{C}(q)),\ \times_N(p,\mathcal{C}(q)) = make(n,\ p),$$

where n is of sort NAT and p and q are of sort POS. But it is equally easy to show that the desired congruence is recursive. Let eq_N be the equality on NAT (which we defined earlier), then the characteristic function, eq_S, of the desired congruence is given by the equation

$$eq_S(make(n,p),\ make(q,r)) = eq_N(\times_N(n,\mathcal{C}(r)),\ \times_N(\mathcal{C}(p),q));$$

the familiar definition for the equality of rational numbers written as fractions.

References

1. J. A. Bergstra and J. V. Tucker. Algebraic specifications of computable and semi-computable data structures. *TCS*, 50:137–181, 1987.
2. J. A. Bergstra and J. V. Tucker. *Algebraic Specifications of Computable and Semicomputable data structures*. Technical Report 115, Mathematical Centre, Department of Computer Science, Amsterdam, 1979.
3. J. R. B. Cockett, H. G.Chen, and L. R. Smith. *Preliminary User Manual for CHARITY*. Technical Report CS-89-82, University of Tennessee, 1989.
4. Martin Davis. *Computability and Unsolvability*. McGraw-Hill, New York, 1958.
5. J. A. Goguen, J. W. Thatcher, and E. G. Wagner. An initial algebra approach to the specification, correctness, and implementation of abstract data types. In R. T. Yeh, editor, *Current Trends in Programming Methodology, IV, Data Structuring*, pages 80–149, Prentice-Hall, 1978.

6. J. A. Goguen, J. W. Thatcher, E. G. Wagner, and J. B. Wright. Abstract data types as initial algebras and the correctness of data representations. In *Proc. Conference on Computer Graphics, Pattern Recognition and Data Structures*, 1975.
7. J. V. Guttag. Abstract data types and the development of data structures. *Communications of the ACM*, 20(6):396–404, 1977.
8. K. V. Guttag and J. J. Horning. The algebraic specification of data types. *Acta Mathematica*, 10:27–52, 1978.
9. T. Hagino. *A categorical programming language*. PhD thesis, University of Edinburgh, 1988.
10. S. C. Kleene. *Introduction to Metamathematics*. D. Van Nostrand, New York, 1952.
11. D. J. Lehmann and M. B. Smyth. Algebraic specification of data types: a synthetic approach. *Mathematical Systems Theory*, 14:97–139, 1981.
12. D. J. Lehmann and M. B. Smyth. Data types. In *18th Annual Symposium on Foundations of Computer Science, Providence, RI, IEEE, New York, NY*, pages 7–12, IEEE, New York, NY, 1977.
13. A. I. Mal'cev. Constructive algebra, i. *Russian Mathematical Surveys*, 16:77–129, 1961.
14. Eric G. Wagner. All recursive types defined using products and sums can be implemented using pointers. In *Proceedings of Conference on Algebraic Logic and Universal Algebra in Computer Science, Jun1-June4 1988, Iowa State University, Ames, Iowa*, pages 111–132, Springer – Verlag LNCS 425, 1990.
15. Eric G. Wagner. Categorical semantics, or extending data types to include memory. In H.-J. Kreowski, editor, *Recent Trends in Data Type Specification: 3rd Workshop on Theory and Applications of Abstract Data Types Selected Papers*, pages 1–21, Informatik-Fachberichte 116, Springer-Verlag, 1984.
16. Eric G. Wagner. Generic types in a language for data directed design. In *Recent Trends in Data Type Specification: Proceedings of the 7th Workshop on Specification of Abstract Data Types*, pages 341–361, LNCS 534, Springer Verlag, 1990.
17. R. F. C. Walters. *An imperative language based on distributive categories*. Research Report 89-26, Department of Pure Mathematics, The University of Sydney, December 1989.

A   The Proof of Theorem 2

Proof. Let $\lceil \Omega,\ \alpha \rceil$ be an effective presentation for A.

1) Let $\gamma : T(\Sigma) \to \Omega$ be the unique homomorphism given by the initiality of $T(\Sigma)$. To see that γ is GN-computable let $g_{\gamma,s} : \omega \to \omega$, for each $s \in S$, such that

$$
g_{\gamma,s}(n) = \begin{cases} 0 & \text{if } n \notin GN(T(\Sigma)_s) \\ \Omega(\sigma) & \text{if } n = GN(\sigma) \text{ where } \mathrm{dom}(\sigma) = \epsilon \\ & \text{and } \mathrm{cod}(\sigma) = s \\ \Omega(\sigma)(\gamma(t_1), \ldots, \gamma(t_n)) & \text{if } n = GN(\sigma(t_1, \ldots, t_n)) \text{ with } \mathrm{cod}(\sigma) = s. \end{cases}
$$

That $g_{\gamma,s}$ is recursive follows from the recursiveness of $GN(T(\Sigma)_s)$ and $\Omega(\sigma)$. But then we see, immediately, that $\gamma_s = g_{\gamma,s} \circ GN$ and so γ_s is GN-computable by Definition 5.

2) Let $R(\Gamma)$ be the Σ-algebra such that

1. For each $s \in S$, $R(\Gamma)_s = T(\Gamma)_s$,
2. For each $\sigma \in F_\Gamma$, and $\langle t_1, \ldots, t_n \rangle \in T(\Gamma)(\mathrm{dom}(\sigma))$,

$$R(\Gamma)(\sigma)(t_1, \ldots, t_n) = \sigma(t_1, \ldots, t_n) \in T(\Gamma)_{\mathrm{cod}(\sigma)}.$$

3. For each $\sigma \in (F_\Sigma \backslash F_\Gamma)$, and $\langle t_1, \ldots, t_n \rangle \in T(\Gamma)(\mathrm{dom}(\sigma))$,

$$R(\Gamma)(\sigma)(t_1, \ldots, t_n) =$$
$$GN^{-1}(\text{the least } n \in GN(T(\Gamma)_s) \text{ such that}$$
$$g_{\gamma,s}(n) = \Omega(\sigma)(\gamma(t_1), \ldots, \gamma(t_n)).$$

To show that $R(\Gamma)$ is computable we must give an appropriate effective presentation, $\lceil \Delta, \delta \rceil$. To this end, let Δ be the Σ-algebra where

1. For each $s \in S$, $\Delta(s) = GN(T(\Gamma)_s)$
2. For each $\sigma \in F_\Gamma$, and $\langle p_1, \ldots p_n \rangle \in \Delta(\mathrm{dom}(\sigma))$,

$$\Delta(\sigma)(p_1, \ldots, p_n) = 2^{Pr(i)} \cdot 3^{p_1} \cdots Pr(n+1)^{p_n}$$

where $\sigma = \sigma_i$ in the enumeration $\sigma_1, \ldots, \sigma_{|F|}$ of F_Σ in Definition 4.
3. For each $\sigma \in F_\Sigma \backslash F_\Gamma$ and $\langle p_1, \ldots, p_n \rangle \in \Delta(\mathrm{dom}(\sigma))$

$$\Delta(\sigma)(p_1, \ldots, p_n) =$$
$$\text{the least } n \in GN(T(\Gamma)_s) \text{ such that } g_{\gamma,s}(n) = \Omega(\sigma)(g_\gamma(p_1), \ldots, g_\gamma(p_n)).$$

Now let $\delta = \langle \delta_s \mid s \in S \rangle$, where, for each $s \in S$, $\delta_s = (GN|_{T(\Gamma)_s})^{-1}$.

What we need to show, for each $\sigma \in F_\Sigma$, is that $R(\Gamma)(\sigma) \bullet \delta^{\mathrm{dom}(\sigma)} = \delta_{\mathrm{cod}(\sigma)} \bullet \Delta(\sigma)$ and that the congruence $\equiv_{\delta_s}$ on $\Delta(s)$ is recursive for each $s \in S$.

Let $\sigma \in F_\Sigma$ and let $\langle p_1, \ldots, p_n \rangle \in \Delta(\mathrm{dom}(\sigma))$.

If $\sigma \in F_\Gamma$ then

$$
\begin{aligned}
&(R(\Gamma) \bullet \delta^{\mathrm{dom}(\sigma)})(p_1, \ldots, p_n) \\
&= \sigma(GN^{-1}(p_1), \ldots, GN^{-1}(p_n)) \\
&\text{which, if } \sigma = \sigma_i \\
&= GN^{-1}(2^{Pr(i)} \cdot 3^{p_1}, \ldots, Pr(n+1)^{p_n}) \\
&= GN^{-1}(\Delta(\sigma)(p_1, \ldots, p_n)) \\
&= \delta_{\mathrm{cod}(\sigma)}(\Delta(\sigma)(p_1, \ldots, p_n)) \\
&= (\delta_{\mathrm{cod}(\sigma)} \bullet \Delta(\sigma))(p_1, \ldots, p_n)
\end{aligned}
$$

as desired,.

While, if $\sigma \in F_\Sigma \backslash F_\Gamma$ then

$$
\begin{aligned}
&(R(\Gamma) \bullet \delta^{\mathrm{dom}(\sigma)})(p_1, \ldots, p_n) \\
&= R(\Gamma)(\sigma)(GN^{-1}(p_1), \ldots, GN^{-1}(p_n)) \\
&= GN^{-1}(\text{the least } n \in GN(T(\Gamma)_s) \text{ such that} \\
&\qquad g_{\gamma,s}(n) = \Omega(\sigma)(\gamma(GN^{-1}(p_1)), \ldots, \gamma(GN^{-1}(p_n))) \\
&= GN^{-1}(\text{the least } n \in GN(T(\Gamma)_s) \text{ such that} \\
&\qquad g_{\gamma,s}(n) = \Omega(\sigma)(g_\gamma(p_1), \ldots, g_\gamma(p_n)) \\
&= (\delta)_{\mathrm{cod}(\sigma)} \bullet \Delta(\sigma))(p_1, \ldots, p_n)
\end{aligned}
$$

as desired.

That $\equiv_\delta$ is recursive follows immediately from δ_s being bijective for each $s \in S$. Thus $R(\Gamma)$ is a computable algebra with effective presentation $\lceil \Delta, \delta \rceil$.

3) Let $\equiv_R$ be the congruence on $R(\Gamma)$ such that $(R(\Gamma)/ \equiv_R) \cong A$. We want to show that $\equiv_R$ is GN-computable. By definition, if $t, t' \in R(\Gamma)_s =_{def} T(\Gamma)_s$, then $t \equiv_R t'$ if, and only if, $A(t) = A(t')$. Now let $\beta : T(\Sigma) \to A$ and $\rho : T(\Sigma) \to R(\Gamma)$ be the unique homomorphisms given by the initiality of $T(\Sigma)$. By the construction of $R(\Gamma)$ it follows that $\rho|_{T(\Gamma)}$ is the identity and thus that $\kappa : R(\Gamma) \to A$ such that $\kappa_s(t) = A(t)$ is a homomorphism. But then, for any $s \in S$ and $t, t' \in T(\Gamma)_s$

$$t \equiv_R t'$$
$$\Leftrightarrow A(t) = A(t')$$
$$\Leftrightarrow \beta(t) = \beta(t')$$
$$\Leftrightarrow \alpha \bullet \gamma(t) = \alpha \bullet \gamma(t')$$
$$\Leftrightarrow \alpha(g_{\gamma,s}(GN(t))) = \alpha(g_{\gamma,s}(GN(t')))$$
$$\Leftrightarrow g_{\gamma,s}(GN(t)) \equiv_\alpha g_{\gamma,s}(GN(t'))$$

But, by Definition 6, the congruence $\equiv_\alpha$ is recursive, and, by an earlier part of this proof, the function $g_{\gamma,s}$ is recursive, and so $\equiv_R$ is GN-computable.

Cofree Coalgebras for Signature Morphisms*

Uwe Wolter

Department of Informatics, University of Bergen,
Bergen, Norway
wolter@ii.uib.no

Abstract. The paper investigates the construction of cofree coalgebras for 'unsorted signature morphisms'. Thanks to the perfect categorical duality between the traditional concept of equations and the concept of coequations developed in [14] we can fully take profit of the methodological power of Category Theory [2] and follow a clean three step strategy: Firstly, we analyse the traditional BIRKHOFF construction of free algebras and reformulate it in a systematic categorical way. Then, by dualizing the BIRKHOFF construction, we obtain, in a second step, corresponding results for cofree coalgebras. And, thirdly, we will interpret the new "abstract" categorical results in terms of more familiar concept. The analysis of a sample cofree construction will provide, finally, some suggestions concerning the potential rôle of cofree coalgebras in System Specifications.

1 Introduction

It is an old observation that unsorted signatures used in Universal Algebra and Algebraic Specifications can be coded by functors $\mathcal{F} : \mathbf{Set} \to \mathbf{Set}$. $\mathcal{F}$-algebras are given in this setting by a carrier A and a map $\alpha : \mathcal{F}(A) \to A$. A corresponding unsorted signature morphism is modeled by a natural transformation $\tau : \mathcal{F} \Rightarrow \mathcal{G}$ and gives rise to a forgetful functor $U_\tau : \mathbf{Alg}(\mathcal{G}) \to \mathbf{Alg}(\mathcal{F})$, where $\mathbf{Alg}(\mathcal{F})$ denotes the category of all $\mathcal{F}$-algebras and all homomorphisms between them.

Free algebras w.r.t. U_τ and the corresponding free functor $T_\tau : \mathbf{Alg}(\mathcal{F}) \to \mathbf{Alg}(\mathcal{G})$, i.e., the functor left-adjoint to U_τ, play an important rôle in Algebraic Specifications, especially for parametrization, structuring, and modularization [2–4, 8, 9].

On the other hand, it has become evident, in the last few years, that coalgebras, the categorical dual of algebras, provide a unifying framework for formal specification of dynamical and behavioural aspects of systems [6, 7, 10, 15]. $\mathcal{F}$-coalgebras are given by a carrier A and a (reversed) map $\alpha : A \to \mathcal{F}(A)$. An unsorted signature morphism $\tau : \mathcal{F} \Rightarrow \mathcal{G}$ gives rise to a functor $U_\tau^c : \mathbf{Alg}^c(\mathcal{F}) \to \mathbf{Alg}^c(\mathcal{G})$ between the corresponding categories of coalgebras. Those "co-forgetful" functors have been, e.g., successfully used in establishing a hierarchy of probabilistic system types [1]. In contrast the corresponding cofree functors have not been investigated and applied up to now, as far as we know.

---

* Research partially supported by the Norwegian NFR project MoSIS/IKT.

H.-J. Kreowski et al. (Eds.): Formal Methods (Ehrig Festschrift), LNCS 3393, pp. 275–290, 2005.

But, taking into account the importance of free algebras in Algebraic Specifications, we should be curious about the rôle of cofree coalgebras in structuring
and modularizing (dynamic) System Specifications.

In the paper we take a first step and investigate the construction of cofree
coalgebras for unsorted signature morphisms. Due to the perfect categorical duality between the traditional concept of equations and the concept of coequations
developed in [14] we can fully take profit of the methodological power of Category
Theory [2] and follow a clean three step strategy: The first, most demanding,
step will be to analyse the traditional BIRKHOFF construction of free algebras
and to develop a systematic categorical description of this construction. In a second, easy step we will dualize the categorical construction in a quite formal way.
And, in a third step, we will try to interpret the dual construtions and results in
terms of known concepts, i.e., in terms of concepts we are familiar with because
of our today education.

The analysis of a sample (co)free construction will, hopefully, give some hints
for a future, more comprehensive investigation of the potential rôle of cofree coalgebras in System Specification. We will close the paper with a short discussion
of possible generalizations and extensions of the results presented here.

2   Sets, Algebras, Termalgebras, and Equations

We summarize the necessary concepts, constructions, and results from [14].

A *(generalized) subset* $(S, i)^1$ of a set A is a set S together with a mono
(injective map) $i : S \rightarrow A$. We write $(S_1, i_1) \subseteq_A (S_2, i_2)$ (or simply $(S_1, i_1) \subseteq
(S_2, i_2)$), if there is a map $m : S_1 \rightarrow S_2$ such that $i_1 = i_2 \circ m$, and we write
$(S_1, i_1) \cong_A (S_2, i_2)$ in case $(S_1, i_1) \subseteq_A (S_2, i_2)$ and $(S_2, i_2) \subseteq_A (S_1, i_1)$.Note,
that S_1 and S_2 are isomorphic if $(S_1, i_1) \cong_A (S_2, i_2)$.

The category **Set** has all limits thus we can construct for any family $\mathcal{S} =
((S_j, i_j) \mid j \in J)$ of subsets of A the multiple pullback $(\bigcap \mathcal{S} \xrightarrow{m_j} S_j \mid j \in J)$.
The m_j are mono since multiple pullbacks preserve mono's in any category.
In such a way, we obtain a new subset $(\bigcap \mathcal{S}, i_\cap)$ of A with $i_\cap = i_j \circ m_j$, i.e.,
$(\bigcap \mathcal{S}, i_\cap) \subseteq_A (S_j, i_j)$, for all $j \in J$, called the *intersection* of $\mathcal{S}$.

Using intersection we can define for any map $f : A \rightarrow B$ the *image of A under
f* : We build the intersection of all subsets (S, i) of B with $f = i \circ l$ for some
map $l : A \rightarrow S$ and obtain a subset $(f(A), m_f)$ of B and a map $e_f : A \rightarrow f(A)$
with $f = m_f \circ e_f$. In **Set** holds the *axiom of choice*: Every epi $e : A \rightarrow B$ in **Set**
is a *split epi*, i.e., there exists at least one $r : B \rightarrow A$ such that $e \circ r = id_B$. This
axiom ensures that the image construction provides an epi-mono factorization:

Proposition 1. *Any map* $f : A \rightarrow B$ *in* **Set** *can be factorized as* $f = m_f \circ e_f$
with $m_f : f(A) \rightarrow B$ *a mono and* $e_f : A \rightarrow f(A)$ *a (split) epi.*

A *relation* between two sets A and B is a subset (R, i_R) of the product $A \times B$,
where we take here for $A \times B$ the cartesian product $\{(a, b) \mid a \in A, b \in B\}$.

---

[1] Note, that the position of S indicates that S is the domain of i.

Kernels play an important technical role in Universal Algebra. Categorically, the *kernel* $ker(f) = \{(a, a') \in A \times A \mid f(a) = f(a')\}$ of a map $f : A \to B$ is the equalizer of the parallel pair $(f \circ \pi_1, f \circ \pi_2)$ of maps

$$ker(f) \xrightarrow{\ i_f\ } A \times A \underset{\pi_2}{\overset{\pi_1}{\rightrightarrows}} A \xrightarrow{\ f\ } B$$

$(ker(f), i_f)$ is a relation in A since equalizer are always mono. Limit reasoning shows that intersection of kernels is related to tupling of maps as expressed in

Proposition 2. *For each family* $(f_j : A \to B_j \mid j \in J)$ *of maps in* **Set** *we have*
$$(ker(\langle f_j \rangle_{j \in J}), i_{\langle f_j \rangle}) \cong_{A \times A} (\textstyle\bigcap_{j \in J} ker(f_j), i_\cap).$$

Moreover, kernels are compatible with composition:

Proposition 3. *For any map* $f : A \to B$ *and any mono* $m : B \to C$ *in* **Set** *we have* $(ker(f), i_f) \subseteq_{A \times A} (ker(g \circ f), i_{g \circ f}).$

Split epi's are regular epi's in any category thus the axiom of choice also provides

Proposition 4. *Each (split) epi* $e : A \to B$ *in* **Set** *is the coequalizer of the pair* $\pi_1 \circ i_e, \pi_2 \circ i_e : ker(e) \to A$ *of maps.*

Given a functor $\mathcal{F} : \textbf{Set} \to \textbf{Set}$ an *$\mathcal{F}$-algebra* (α, A) consists of a set A, called the *carrier*, and a map $\alpha : \mathcal{F}(A) \to A$, called the *(algebraic) structure map*[2]. An *$\mathcal{F}$-homomorphism* $f : (\alpha, A) \to (\beta, B)$ between $\mathcal{F}$-algebras is a map $f : A \to B$ such that $\beta \circ \mathcal{F}(f) = f \circ \alpha$.

$$
\begin{array}{ccc}
A & \xrightarrow{\ f\ } & B \\
\alpha \uparrow & & \uparrow \beta \\
\mathcal{F}(A) & \xrightarrow{\mathcal{F}(f)} & \mathcal{F}(B)
\end{array}
$$

An *$\mathcal{F}$-subalgebra* of an $\mathcal{F}$-algebra (α, A) is a subset (S, i) of A with an $\mathcal{F}$-algebraic structure (σ, S) such that the map $i : S \to A$ defines an $\mathcal{F}$-homomorphism $i : (\sigma, S) \to (\alpha, A)$.

By $\textbf{Alg}(\mathcal{F})$ we denote the category of all $\mathcal{F}$-algebras and all homomorphisms between them. Obviously, the assignments $(\alpha, A) \mapsto A$ and $(f : (\alpha, A) \to (\beta, B)) \mapsto (f : A \to B)$ extend to a functor $U_{\mathcal{F}} : \textbf{Alg}(\mathcal{F}) \to \textbf{Set}$. Straightforward categorical reasoning shows, that only the fact that the functor $\mathcal{F}$ appears in the domain of the algebraic structure maps ensures that $U_{\mathcal{F}}$ creates limits. This means, $\textbf{Alg}(\mathcal{F})$ has all limits and both the carriers of limit $\mathcal{F}$-algebras and the mediating $\mathcal{F}$-homomorphisms are obtained by the corresponding limit constructions on carriers and on maps in **Set** [10, 14]. The product of two $\mathcal{F}$-algebras (α, A) and (β, B), for instance, is given by $(\alpha \otimes \beta, A \times B)$ with $\alpha \otimes \beta := (\alpha \times \beta) \circ \langle \mathcal{F}(\pi_1), \mathcal{F}(\pi_2) \rangle$, thus the projections π_1 and π_2 become indeed $\mathcal{F}$-homomorphisms:

---

[2] Note, that the position of A indicates now that A is the codomain of α.

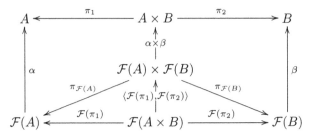

And, since kernels are equalizer, we obtain for any $\mathcal{F}$-homomorphism $f : (\alpha,A) \to (\beta,B)$ an $\mathcal{F}$-subalgebra $(\alpha_f, ker(f))$ of $(\alpha \otimes \alpha, A \times A)$.

Example 1. We consider the identity functor $\mathcal{I} := Id_{\mathbf{Set}} : \mathbf{Set} \to \mathbf{Set}$ and a functor $\mathcal{D} : \mathbf{Set} \to \mathbf{Set}$ with $\mathcal{D}(A) := A \times A$ for all sets A and $\mathcal{D}(f) := f \times f : A \times A \to B \times B$ for all maps $f : A \to B$. An $\mathcal{I}$-algebra is then given by a set A and a unary operation $\alpha : A \to A$, and $\mathcal{G}$-algebras are just a set A together with a binary operation $\alpha : A \times A \to A$.

There are three crucial technical results in Algebraic Specifications. Firstly, the image factorization in **Set** can be transferred to algebras:

Proposition 5. *Any $\mathcal{F}$-homomorphism $f : (\alpha, A) \to (\beta, B)$ can be factorized as $f = m_f \circ e_f$ with $m_f : (f(A), \hat{\beta}) \to (\beta, B)$ a mono in $\mathbf{Alg}(\mathcal{F})$ (and $m_f : f(A) \to B$ a mono in **Set**) and with $e_f : (\alpha, A) \to (f(A), \hat{\beta})$ an epi in $\mathbf{Alg}(\mathcal{F})$ (and $e_f : A \to f(A)$ a (split) epi in **Set**).*

Secondly we have different versions of *homomorphism theorems* as, for instance:

Theorem 1. *For any $\mathcal{F}$-homomorphisms $e : (\alpha, A) \to (\beta, B)$, $g : (\alpha, A) \to (\gamma, C)$ with $e : A \to B$ (split) epi in **Set** there exists a unique $\mathcal{F}$-homomorphism $h : (\beta, B) \to (\gamma, C)$ with $g = h \circ e$ iff $(ker(e), i_e) \subseteq_{A \times A} (ker(g), i_g)$.*

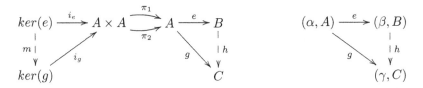

Another useful and more categorical formulation of this statement is given by a generalization of Proposition 4

Proposition 6. *Any $\mathcal{F}$-homomorphism $e : (\alpha, A) \to (\beta, B)$ with $e : A \to B$ (split) epi in **Set** is the coequalizer of the pair $\pi_1 \circ i_e, \pi_2 \circ i_e : (\alpha_e, ker(e)) \to (\alpha, A)$ of $\mathcal{F}$-homomorphisms.*

Thirdly, we can construct terms for all ω-continous functors $\mathcal{F} : \mathbf{Set} \to \mathbf{Set}$ [11], i.e., especially for all *polynomial functors*, i.e., functors that can we build from the constant functors $A : \mathbf{Set} \to \mathbf{Set}$, the identical functor $I : \mathbf{Set} \to \mathbf{Set}$, the product functor $\times : \mathbf{Set} \times \mathbf{Set} \to \mathbf{Set}$, and the coproduct functor

$+ : \mathbf{Set} \times \mathbf{Set} \to \mathbf{Set}$ [6, 10]. Note, that functors $\mathcal{F} : \mathbf{Set} \to \mathbf{Set}$ related to "algebraic signatures" are special polynomial functors assigning to a carrier A a corresponding coproduct of products of A:

Let X be a set (of *variables*). An $\mathcal{F}$-algebra $(\iota_{\mathcal{F},X}, T_{\mathcal{F}}(X))$ together with a (variable) *assignment* $\eta_{\mathcal{F},X} : X \to T_{\mathcal{F}}(X)$ is *free over* X w.r.t. $U_{\mathcal{F}}$ if for every $\mathcal{F}$-algebra (α, A) and for every assignment $a : X \to A$ there exists a unique $\mathcal{F}$-homomorphism $a^{\mathcal{F}} : (\iota_{\mathcal{F},X}, T_{\mathcal{F}}(X)) \to (\alpha, A)$ such that $a^{\mathcal{F}} \circ \eta_{\mathcal{F},X} = a$.

$$
\begin{array}{ccc}
X \xrightarrow{\eta_{\mathcal{F},X}} T_{\mathcal{F}}(X) & \qquad & (\iota_{\mathcal{F},X}, T_{\mathcal{F}}(X)) \\
\searrow \quad \big\downarrow a^{\mathcal{F}} & & \big\downarrow a^{\mathcal{F}} \\
a \qquad A & & (\alpha, A)
\end{array}
$$

The 'elements' of $T_{\mathcal{F}}(X)$ are usually called $\mathcal{F}$-terms and $(\iota_{\mathcal{F},X}, T_{\mathcal{F}}(X))$ is called the $\mathcal{F}$-termalgebra over X. Moreover, we will call the universal assignment $\eta_{\mathcal{F},X} : X \to T_{\mathcal{F}}(X)$ the $\mathcal{F}$-*unit for* X and $a^{\mathcal{F}}$ the *(unique)* $\mathcal{F}$-*extension of* a.

Standard categorical arguments show that the assignments $X \mapsto (\iota_{\mathcal{F},X}, T_{\mathcal{F}}(X))$ and $(s : X \to Y) \mapsto (\eta_{\mathcal{F},Y} \circ s)^{\mathcal{F}} : (\iota_{\mathcal{F},X}, T_{\mathcal{F}}(X)) \longrightarrow (\iota_{\mathcal{F},Y}, T_{\mathcal{F}}(Y))$ define a functor $T_{\mathcal{F}} : \mathbf{Set} \to \mathbf{Alg}(\mathcal{F})$ left-adjoint to $U_{\mathcal{F}}$ and called the 'free functor for $\mathcal{F}$'.

Example 2. For $X = \{0, 1\}$ we can identify $T_{\mathcal{I}}(X)$ with the set $\mathbb{N} \times X$ thus we have $\iota_{\mathcal{I},X}(n, b) = (n + 1, b)$. And $T_{\mathcal{D}}(X)$ consists of all binary trees with no symbols at the branching nodes but with 0 or 1 at the leafs, where $\iota_{\mathcal{D},X}$ just makes a new binary tree out of two given one.

An *equational* $\mathcal{F}$-*specification over* X is a relation $(E, spec)$ in $T_{\mathcal{F}}(X)$, where the 'elements' $\mathbf{1} \xrightarrow{eq} T_{\mathcal{F}}(X) \times T_{\mathcal{F}}(X)$ are called *equations*. An assignment $a : X \to A$ is a *solution* of $(E, spec)$ in a $\mathcal{F}$-algebra (α, A), if $(E, spec) \subseteq (ker(a^{\mathcal{F}}), i_{a^{\mathcal{F}}})$. Since $(ker(a^{\mathcal{F}}), i_{a^{\mathcal{F}}})$ is an equalizer this condition is equivalent to the condition: $a^{\mathcal{F}} \circ \pi_1 \circ spec = a^{\mathcal{F}} \circ \pi_2 \circ spec$

$$
\begin{array}{ccccc}
ker(a^{\mathcal{F}}) & \xrightarrow{i_{a^{\mathcal{F}}}} & T_{\mathcal{F}}(X) \times T_{\mathcal{F}}(X) & \overset{\pi_1}{\underset{\pi_2}{\rightrightarrows}} T_{\mathcal{F}}(X) \xrightarrow{a^{\mathcal{F}}} A \\
\big\uparrow{\scriptstyle i} & \nearrow{\scriptstyle spec} & & \\
E & & &
\end{array}
$$

3   Free Algebras

In our categorical setting an "unsorted signature morphism" is represented by a natural transformation and provides, as usual, a *forgetful* functor

Definition 1. *Any natural transformation* $\tau : \mathcal{F} \Rightarrow \mathcal{G} : \mathbf{Set} \to \mathbf{Set}$ *gives rise to a functor* $U_{\tau} : \mathbf{Alg}(\mathcal{G}) \to \mathbf{Alg}(\mathcal{F})$ *defined for any* $\mathcal{G}$-*algebra* (α, A) *and any* $\mathcal{G}$-*homomorphism* $f : (\alpha, A) \to (\beta, B)$ *as* $U_{\tau}(\alpha, A) := (\alpha \circ \tau_A, A)$ *and* $U_{\tau}(f) := f$.

$$\begin{array}{ccccc}
\mathcal{F}(A) & \xrightarrow{\tau_A} & \mathcal{G}(A) & \xrightarrow{\alpha} & A \\
\downarrow{\scriptstyle\mathcal{F}(f)} & & \downarrow{\scriptstyle\mathcal{G}(f)} & & \downarrow{\scriptstyle f} \\
\mathcal{F}(B) & \xrightarrow{\tau_B} & \mathcal{G}(B) & \xrightarrow{\beta} & B
\end{array}$$

Remark 1. To model, e.g., morphisms between many-sorted signatures, we would need a more general construction: Let be given functors $\mathcal{F} : \mathbf{C} \to \mathbf{C}$, $\mathcal{G} : \mathbf{D} \to \mathbf{D}$, a functor $\mathcal{V} : \mathbf{D} \to \mathbf{C}$, and a natural transformation $\tau : \mathcal{F} \circ \mathcal{V} \Rightarrow \mathcal{V} \circ \mathcal{G} : \mathbf{C} \to \mathbf{C}$. Then we obtain a functor $U_{\mathcal{V},\tau} : \mathbf{Alg}(\mathcal{G}) \to \mathbf{Alg}(\mathcal{F})$ defined for any $\mathcal{G}$-algebra (α, A) and any $\mathcal{G}$-homomorphism $f : (\alpha, A) \to (\beta, B)$ as $U_{\mathcal{V},\tau}(\alpha, A) := (\mathcal{V}(\alpha) \circ \tau_A, \mathcal{V}(A))$ and $U_{\mathcal{V},\tau}(f) := \mathcal{V}(f)$.

Example 3. Definition 1 covers the traditional unsorted signature morphisms, but is slightly more general: Obviously, the assignments $A \mapsto \Delta_A : A \to A \times A$ with $\Delta_A(a) := (a, a)$ for all $a \in A$ define a natural transformation $\Delta : \mathcal{I} \Rightarrow \mathcal{D}$. Given a $\mathcal{D}$-algebra $(\alpha : A \times A \to A, A)$ the $\mathcal{I}$-algebra $U_\Delta(\alpha, A) = (\alpha \circ \Delta_A, A)$ will forget all the applications of α to non-identical input pairs.

The objective of the paper is to analyse the construction of free algebras:

Definition 2. *Let be given an $\mathcal{F}$-algebra (α, A). A $\mathcal{G}$-algebra $(\iota_{\tau,\alpha}, T_\tau(A))$ together with an $\mathcal{F}$-homomorphism $\eta_\alpha : (\alpha, A) \to U_\tau(\iota_{\tau,\alpha}, T_\tau(A))$ is* free over *(α, A) w.r.t. U_τ if for any $\mathcal{G}$-algebra (β, B) and for any $\mathcal{F}$-homomorphism $h : (\alpha, A) \to U_\tau(\beta, B)$ there exists a unique $\mathcal{G}$-homomorphism $h^\tau : (\iota_{\tau,\alpha}, T_\tau(A)) \to (\beta, B)$ such that $U_\tau(h^\tau) \circ \eta_\alpha = h$.*

$$\begin{array}{ccc}
(\alpha, A) & \xrightarrow{\eta_\alpha} & U_\tau(\iota_{\tau,\alpha}, T_\tau(A)) \\
 & \searrow{\scriptstyle h} & \big\downarrow{\scriptstyle U_\tau(h^\tau) = h^\tau} \\
 & & U_\tau(\beta, B)
\end{array}
\qquad
\begin{array}{c}
(\iota_{\tau,\alpha}, T_\tau(A)) \\
\big\downarrow{\scriptstyle h^\tau} \\
(\beta, B)
\end{array}$$

Usually, the free algebra $(\iota_{\tau,\alpha}, T_\tau(A))$ is obtained by constructing an appropriate quotient of the $\mathcal{G}$-termalgebra over the carrier of (α, A). To do this, we have to "syntactify" the $\mathcal{F}$-algebra (α, A). Traditionally, this is done either by "signature extensions" [3] or by "generators" [9, 13]. Categorically, these approaches are reflected as follows: We consider the $\mathcal{F}$-termalgebra over the carrier of (α, A). Then, the $\mathcal{F}$-extension of the trivial assignment $id_A : A \to A$ gives us an *evaluation* $ev_\alpha := id_A^{\mathcal{F}} : (\iota_{\mathcal{F},A}, T_{\mathcal{F}}(A)) \to (\alpha, A)$ of the "arithmetical expressions over (α, A)" thus we have

$$ev_\alpha \circ \eta_{\mathcal{F},A} = id_A \text{ in } \mathbf{Set}. \tag{1}$$

Note, that, in such a way, ev_α becomes split epi and $\eta_{\mathcal{F},A}$ split mono in **Set**.

The *(internal) $\mathcal{F}$-theory* of (α, A) is given now by the kernel $(ker(ev_\alpha), i_{ev_\alpha})$ (see the diagram in Proposition 7). To translate this $\mathcal{F}$-theory into a $\mathcal{G}$-theory we need a translation of $\mathcal{F}$-terms over A into $\mathcal{G}$-terms over A: We consider the $\mathcal{G}$-unit for A $\eta_{\mathcal{G},A} : A \to T_{\mathcal{G}}(A)$ and take the corresponding $\mathcal{F}$-extension $\eta_{\mathcal{G},A}^{\mathcal{F}} : (\iota_{\mathcal{F},A}, T_{\mathcal{F}}(A)) \to U_\tau(\iota_{\mathcal{G},A}, T_{\mathcal{G}}(A))$ with

$$\eta_{\mathcal{G},A}^{\mathcal{F}} \circ \eta_{\mathcal{F},A} = \eta_{\mathcal{G},A} \text{ in } \mathbf{Set} \tag{2}$$

for the underlying map $\eta_{\mathcal{G},A}^{\mathcal{F}} : T_{\mathcal{F}}(A) \to T_{\mathcal{G}}(A)$. In such a way we obtain the required *translated internal theory* of (α, A) as the relation $(ker(ev_\alpha), (\eta_{\mathcal{G},A}^{\mathcal{F}} \times \eta_{\mathcal{G},A}^{\mathcal{F}}) \circ i_{ev_\alpha})$ in $T_{\mathcal{G}}(A)$.

Example 4. We consider for $A = \{0, 1\}$ the $\mathcal{I}$-algebra (α, A) with $\alpha(0) = 1$ and $\alpha(1) = 0$. Then we have $ev_\alpha(2n + 1, 0) = ev_\alpha(2m, 1) = 1$ and $ev_\alpha(2n, 0) = ev_\alpha(2m + 1, 1) = 0$ for all $n, m \in \mathbb{N}$, i.e., the kernel of ev_α will contain all the pairs $((2n + 1, 0), (2m, 1))$ and $((2n, 0), (2m + 1, 1))$. And $\eta_{\mathcal{D},A}^{\mathcal{I}}$ will map (n, b) to a perfect binary tree of depth n with only the constant symbol b at the leafs.

Now, it turns out that the semantical condition used in Definition 2 can be reformulated in terms of the translated theory:

Proposition 7. *For any $\mathcal{G}$-algebra (β, B) and for any map $h : A \to B$ the following conditions are equivalent:*

1. *$h : A \to B$ defines an $\mathcal{F}$-homomorphism $h : (\alpha, A) \to U_\tau(\beta, B)$ such that $h \circ ev_\alpha = h^{\mathcal{G}} \circ \eta_{\mathcal{G},A}^{\mathcal{F}}$ in $\mathbf{Alg}(\mathcal{F})$.*
2. *h is a solution of the specification $(ker(ev_\alpha), (\eta_{\mathcal{G},A}^{\mathcal{F}} \times \eta_{\mathcal{G},A}^{\mathcal{F}}) \circ i_{ev_\alpha})$ in (β, B).*

Proof.

$$
\begin{array}{ccccccc}
ker(ev_\alpha) & \xrightarrow{i_{ev_\alpha}} & T_{\mathcal{F}}(A) \times T_{\mathcal{F}}(A) & \underset{\pi_2}{\overset{\pi_1}{\rightrightarrows}} & T_{\mathcal{F}}(A) & \xrightarrow{ev_\alpha} & A \\
{\scriptstyle m}\downarrow & & \downarrow {\scriptstyle \eta_{\mathcal{G},A}^{\mathcal{F}} \times \eta_{\mathcal{G},A}^{\mathcal{F}}} & & \downarrow {\scriptstyle \eta_{\mathcal{G},A}^{\mathcal{F}}} \searrow {\scriptstyle h^{\mathcal{F}}} & & \downarrow {\scriptstyle h} \\
ker(h^{\mathcal{G}}) & \xrightarrow{i_{h^{\mathcal{G}}}} & T_{\mathcal{G}}(A) \times T_{\mathcal{G}}(A) & \underset{\pi_2}{\overset{\pi_1}{\rightrightarrows}} & T_{\mathcal{G}}(A) & \xrightarrow{h^{\mathcal{G}}} & B
\end{array}
$$

$(1) \Rightarrow (2)$ Equation (1) entails $h \circ ev_\alpha \circ \eta_{\mathcal{F},A} = h$ in **Set** thus the uniqueness of $\mathcal{F}$-extensions forces $h^{\mathcal{F}} = h \circ ev_\alpha$ in $\mathbf{Alg}(\mathcal{F})$ and thus also in **Set**. Moreover, equation (2) entails in **Set** $h^{\mathcal{G}} \circ \eta_{\mathcal{G},A}^{\mathcal{F}} \circ \eta_{\mathcal{F},A} = h^{\mathcal{G}} \circ \eta_{\mathcal{G},A} = h$ thus again the uniqueness of $\mathcal{F}$-extensions implies $h^{\mathcal{F}} = h^{\mathcal{G}} \circ \eta_{\mathcal{G},A}^{\mathcal{F}}$ in $\mathbf{Alg}(\mathcal{F})$. Since i_{ev_α} is an equalizer in **Set** we obtain in such a way the required equation

$$
\begin{aligned}
h^{\mathcal{G}} \circ \pi_1 \circ (\eta_{\mathcal{G},A}^{\mathcal{F}} \times \eta_{\mathcal{G},A}^{\mathcal{F}}) \circ i_{ev_\alpha} &= h^{\mathcal{G}} \circ \eta_{\mathcal{G},A}^{\mathcal{F}} \circ \pi_1 \circ i_{ev_\alpha} \\
&= h^{\mathcal{F}} \circ \pi_1 \circ i_{ev_\alpha} \\
&= h \circ ev_\alpha \circ \pi_1 \circ i_{ev_\alpha} \\
&= h \circ ev_\alpha \circ \pi_2 \circ i_{ev_\alpha} \\
&\quad \dots \\
&= h^{\mathcal{G}} \circ \pi_2 \circ (\eta_{\mathcal{G},A}^{\mathcal{F}} \times \eta_{\mathcal{G},A}^{\mathcal{F}}) \circ i_{ev_\alpha}
\end{aligned}
$$

$(2) \Rightarrow (1)$ Since $i_{h^{\mathcal{G}}}$ is an equalizer in **Set** the assumption ensures, in a similar way as above, $h^{\mathcal{G}} \circ \eta_{\mathcal{G},A}^{\mathcal{F}} \circ \pi_1 \circ i_{ev_\alpha} = h^{\mathcal{G}} \circ \eta_{\mathcal{G},A}^{\mathcal{F}} \circ \pi_2 \circ i_{ev_\alpha}$. But, since $ev_\alpha : T_{\mathcal{F}}(A) \to A$ is (spli) epi in **Set** there exists according to Proposition 4 a unique $h' : (\alpha, A) \to U_\tau(\beta, B)$ such that $h' \circ ev_\alpha = h^{\mathcal{G}} \circ \eta_{\mathcal{G},A}^{\mathcal{F}}$ in $\mathbf{Alg}(\mathcal{F})$ and thus in **Set**. By the equations (1), (2) and the definition of $\mathcal{G}$-extensions this provides finally $h' = h^{\mathcal{G}} \circ \eta_{\mathcal{G},A} = h$. $\qquad \square$

4   Birkhoff Construction of Free Algebras

Given any $\mathcal{F}$-algebra (α, A) we will present now a categorical analysis of the so-called BIRKHOFF-construction of free algebras [12].

Firstly, we look for an appropriate **quotient $\mathcal{G}$-termalgebra**: Since any set has up to isomorphism only a set of subsets there exists a set J (of indices), a family $((\gamma_j, C_j) \mid j \in J)$ of $\mathcal{G}$-algebras, and a family $(g_j : (\alpha, A) \to U_\tau(\gamma_j, C_j) \mid j \in J)$ of $\mathcal{F}$-homomorphisms such that for any $\mathcal{G}$-algebra (β, B) and for any $\mathcal{F}$-homomorphism $h : (\alpha, A) \to U_\tau(\beta, B)$ there is an index $j_h \in J$ such that

$$(ker(h^{\mathcal{G}}), i_{h^{\mathcal{G}}}) \cong (ker(g_{j_h}^{\mathcal{G}}), i_{g_{j_h}^{\mathcal{G}}}). \tag{3}$$

We build the product $(\prod_{j \in J}(\gamma_j, C_j), \pi_j : \prod_{j \in J}(\gamma_j, C_j) \to (\gamma_j, C_j), j \in J)$ in **Alg$(\mathcal{G})$**, where the carrier of $\prod_{j \in J}(\gamma_j, C_j)$ will be a product $\prod_{j \in J} C_j$ of the carriers in **Set** since $U_{\mathcal{G}}$ creates limits. This together with the uniqueness of $\mathcal{G}$-extensions ensures that tupling is compatible with $\mathcal{G}$-extensions

$$\langle g_j \rangle_{j \in J}^{\mathcal{G}} = \langle g_j^{\mathcal{G}} \rangle_{j \in J} : (\iota_{\mathcal{G},A}, T_{\mathcal{G}}(A)) \to \prod(\gamma_j, C_j). \tag{4}$$

Now we can construct an epi-mono factorization of $\langle g_j \rangle_{j \in J}^{\mathcal{G}}$ according to Proposition 5

Example 5. $T_\Delta(A)$ contains the trees (0), (1) of depth 0 and all binary trees that have no perfect subtree with constantly only 0 or 1 at the leafs and with depth greater than 0, because those perfect trees are mapped by $e_{\Delta,A}$ to either 0 or 1, respectively. Moreover, we will have, for instance, $\iota_{\Delta,A}((0),(0)) = (1)$, $\iota_{\Delta,A}((1),(1)) = (0)$, but $\iota_{\Delta,A}((0),(1)) = (0,1)$, $\iota_{\Delta,A}((0),(0,1)) = (0,(0,1))$, and $\iota_{\Delta,A}((0,1)),(0,1)) = (0,(0,1))$.

The claim, to be validated in the rest of this section, is that $(\iota_{\tau,A}, T_\tau(A))$ is indeed free over (α, A) w.r.t. U_τ. For this we have to find, secondly, a candidate for the **unit**:

Lemma 1. $\langle g_j \rangle_{j \in J} : A \to \prod_{j \in J} C_j$ *is a solution of the translated $\mathcal{F}$-theory of* (α, A) *in* $\prod_{j \in J}(\gamma_j, C_j)$, *and* $e_{\tau,A} \circ \eta_{\mathcal{G},A} : A \to T_\tau(A)$ *is a solution in* $(\iota_{\tau,A}, T_\tau(A))$.

Proof. According to Proposition 7 we have $(ker(ev_\alpha), (\eta_{\mathcal{G},A}^{\mathcal{F}} \times \eta_{\mathcal{G},A}^{\mathcal{F}}) \circ i_{ev_\alpha}) \subseteq$ $(ker(g_j^{\mathcal{G}}), i_{g_j^{\mathcal{G}}})$ for all $j \in J$. Moreover, $e_{\tau,A} = (e_{\tau,A} \circ \eta_{\mathcal{G},A})^{\mathcal{G}}$ is ensured by the uniqueness of $\mathcal{G}$-extensions thus we obtain by the definition of intersection, Proposition 2, Equation (4), and Proposition 3: $(ker(ev_\alpha), (\eta_{\mathcal{G},A}^{\mathcal{F}} \times \eta_{\mathcal{G},A}^{\mathcal{F}}) \circ$ $i_{ev_\alpha}) \subseteq (\bigcap ker(g_j^{\mathcal{G}}), i_\cap) \cong (ker(\langle g_j^{\mathcal{G}} \rangle_{j \in J}), i_{\langle g_j^{\mathcal{G}} \rangle_{j \in J}}) \cong (ker(\langle g_j \rangle_{j \in J}^{\mathcal{G}}), i_{\langle g_j \rangle_{j \in J}^{\mathcal{G}}}) \cong$ $(ker(m_{\tau,A} \circ e_{\tau,A}), i_{m_{\tau,A} \circ e_{\tau,A}}) \cong (ker(e_{\tau,A}), i_{e_{\tau,A}}) \cong (ker((e_{\tau,A} \circ \eta_{\mathcal{G},A})^{\mathcal{G}}), i_{e_{\tau,A}})$.   □

Lemma 1 and Proposition 7 ensure now that the map $e_{\tau,A} \circ \eta_{\mathcal{G},A} : A \to T_\tau(A)$ defines an $\mathcal{F}$-homomorphism $\eta_\alpha := e_{\tau,A} \circ \eta_{\mathcal{G},A} : (\alpha, A) \to U_\tau(\iota_{\tau,A}, T_\tau(A))$ that will be called the τ-unit for (α, A).

Example 6. The algebraic structure map in $U_\Delta(\iota_{\Delta,A}, T_\Delta(A))$ will map $(0) \mapsto (1), (1) \mapsto (0), (0, 1) \mapsto ((0, 1), (0, 1)), \dots$ And, $e_{\Delta,A} \circ \eta_{\mathcal{D},A}$ maps $0 \mapsto (0), 1 \mapsto (1)$ thus η_α becomes indeed an $\mathcal{I}$-homomorphism.

To validate that the τ-unit owns the required **universal property**, we have to show, thirdly, the **existence of mediating morphisms**: For any $\mathcal{G}$-algebra (β, B) and for any $\mathcal{F}$-homomorphism $h : (\alpha, A) \to U_\tau(\beta, B)$ we obtain according to our construction and in the same way as above

$$(ker(h^{\mathcal{G}}), i_{h^{\mathcal{G}}}) \cong (ker(g_{j_h}^{\mathcal{G}}), i_{g_{j_h}^{\mathcal{G}}}) \supseteq (\bigcap ker(g_j^{\mathcal{G}}), i_\cap) \cong (ker(e_{\tau,A}), i_{e_{\tau,A}})$$

such that the homomorphism theorem 1 ensures the existence of a unique $\mathcal{G}$-homomorphism $h^\tau : (\iota_{\tau,A}, T_\tau(A)) \to (\beta, B)$ such that

$$h^\tau \circ e_{\tau,A} = h^{\mathcal{G}} \text{ in } \mathbf{Alg}(\mathcal{F}). \tag{5}$$

Further we obtain due to the definition of $\mathcal{G}$-extensions, equation (5), and the definition of the τ-unit the required equation

$$h = h^{\mathcal{G}} \circ \eta_{\mathcal{G},A} = h^\tau \circ e_{\tau,A} \circ \eta_{\mathcal{G},A} = h^\tau \circ \eta_\alpha. \tag{6}$$

Finally, we show the **uniqueness of mediating morphisms**: Let be given any $g : (\iota_{\tau,A}, T_\tau(A)) \to (\beta, B)$ with $h = g \circ \eta_\alpha$ in $\mathbf{Alg}(\mathcal{F})$. For the underlying maps in **Set** we obtain due to the definition of η_α $h = g \circ \eta_\alpha = g \circ e_{\tau,A} \circ \eta_{\mathcal{G},A}$ thus the uniqueness of $\mathcal{G}$-extensions and equation 5 provides $g \circ e_{\tau,A} = h^{\mathcal{G}} = h^\tau \circ e_{\tau,A}$ in $\mathbf{Alg}(\mathcal{G})$. But this means $g = h^\tau$ since $e_{\tau,A}$ is epi.

5   Partitions, Coalgebras, Processcoalgebras, and Coequations

Kernels (or equivalences and congruences, respectively) are one of the most important concepts in Algebraic Specifications thus we could assume that cokernels deserve a similar important rôle in the dual setting of coalgebras. Focussing on cokernels allows for a perfect categorical dualization of all the concepts, constructions, and results from Algebraic Specifications, once we have reformulated them fully categorically, to the area of coalgebras, i.e., of Systems Specifications.

[14] presents the first steps of a firm realization of this program. Due to the limitation on space we will not "dual-copy" all the last three sections. We will only present the essential dualizations and will put more emphasis on the interpretation of the dual concepts, constructions, and results[3].

A *partition* (s, P) of a set B is a set P together with an epi (surjective map) $s : B \to P$. In usual set-theoretic terms a (canonical) partition is a set $P \subseteq \mathcal{P}(B)$

---

[3] We will use sans serif for pointing at interpretations in usual set-theoretic terms.

such that $\bigcup P = B$, $p \neq \emptyset$ for all $p \in P$, and $p_1 \cap p_2 = \emptyset$ for all $p_1 \neq p_2 \in P$. These conditions ensure that there is for each $b \in B$ exactly one $p_b \in P$ with $b \in p_b$, thus the assignment $b \mapsto p_b$ defines a surjective map $s_P : B \to P$.

We write $(s_1, P_1) \sqsubseteq_B (s_2, P_2)$, if there is a map $e : P_1 \to P_2$ such that $s_2 = e \circ s_1$, and we write $(s_1, P_1) \cong_B (s_2, P_2)$ in case $(s_1, P_1) \sqsubseteq_B (s_2, P_2)$ and $(s_2, P_2) \sqsubseteq_B (s_1, P_1)$. $P_1 \sqsubseteq_B P_2$ means, for canonical partitions, that each element in P_2 is the union of some elements in P_1, i.e., that P_2 makes more elements of B indistinguishable than P_1.

We can construct for any family $\mathcal{P} = ((s_j, P_j) \mid j \in J)$ of partitions of B the multiple pushout $(P_j \xrightarrow{e_j} \bigvee \mathcal{P} \mid j \in J)$. The e_j are epi since pushouts preserve epi's. We obtain a partition $(s_\vee, \bigvee \mathcal{P})$ of B with $s_\vee = e_j \circ s_j$, i.e., $(s_j, P_j) \sqsubseteq_B (s_\vee, \bigvee \mathcal{P})$, for all $j \in J$, called the *gluing* of $\mathcal{P}$. To compute $P_1 \vee P_2$ for two canonical partitions P_1 and P_2 we have to join $p_1 \in P_1$ with $p_2 \in P_2$ if $p_1 \cap p_2 \neq \emptyset$. By carrying out the "reflexive, symmetric, and transitive closure" of these gluings we obtain the elements of $P_1 \vee P_2$.

Using gluing we can define for each map $f : A \to B$ the *coimage of B w.r.t. f*: We build the gluing of all partitions (s, P) of A with $f = l \circ s$ for some map $l : P \to B$ and obtain a partition $(e^f, f^c(B))$ of A and a map $m^f : f^c(B) \to B$ with $f = m^f \circ e^f$. In **Set** holds also the *dual axiom of choice*: Every mono $m : A \to B$ in **Set** with $A \neq \emptyset$ is a *split mono*, i.e., there exists at least one $r : B \to A$ such that $r \circ m = id_A$. This axiom ensures that the coimage construction provides an epi-mono factorization that is equivalent to the epi-mono factorization in Proposition 1, i.e., there exists an iso $i_f : f(A) \to f^c(B)$ such that $i_f \circ e_f = e^f$ and $i_f \circ m_f = m^f$.

A *corelation* between two sets A and B is a partition (s_Q, Q) of the coproduct $A + B$, where we take here for $A + B$ the set $\{(1, a) \mid a \in A\} \cup \{(2, b) \mid b \in B\}$.

The *cokernel* $(s_f, cok(f))$ of a map $f : A \to B$ is the coequalizer of the parallel pair $(\kappa_1 \circ f, \kappa_2 \circ f)$ of maps. $cok(f)$ is a corelation in B since coequalizer are always epi. In **Set** $cok(f)$ codes the image $f(A)$: The elements of $cok(f)$ are either singleton sets $\{(i, b)\}$ with $i \in \{1, 2\}$ or two element sets $\{(1, b), (2, b)\}$, where $\{(1, b), (2, b)\} \in cok(f)$ iff $b \in f(A)$.

$$A \xrightarrow{f} B \overset{\kappa_1}{\underset{\kappa_2}{\rightrightarrows}} B + B \xrightarrow{s_f} cok(f)$$

Colimit reasoning shows that gluing of cokernels is related to cotupling of maps as expressed in

Proposition 8. *For each family $(f_j : A_j \to B \mid j \in J)$ of maps we have*
$$(s_{[f_j]}, cok([f_j]_{j \in J})) \cong_{B+B} (s_\vee, \bigvee_{j \in J} cok(f_j)).$$

Proposition 8 codes the equation $[f_j]_{j \in J}(\coprod_{j \in J} A_j) = \bigcup_{j \in J} f_j(A_j)$, i.e., $\bigvee cok(f_j)$ just collects all identifications of 'associated' elements $(1, b)$, $(2, b)$ made by the single cokernels $cok(f_j)$. Split mono's are regular mono's in any category thus the dual axiom of choice provides

Proposition 9. *Each (split) mono $m : A \to B$ is the equalizer of the pair $s_m \circ \kappa_1, s_m \circ \kappa_2 : B \to cok(m)$ of maps.*

Given a functor $\mathcal{F} : \mathbf{Set} \to \mathbf{Set}$ an $\mathcal{F}$-*coalgebra* (A, α) consists of a set A and a map $\alpha : A \to \mathcal{F}(A)$. An $\mathcal{F}^c$-*homomorphism* $f : (A, \alpha) \to (B, \beta)$ between $\mathcal{F}$-coalgebras is a map $f : A \to B$ such that $\beta \circ f = \mathcal{F}(f) \circ \alpha$.

An $\mathcal{F}$-*partition* of an $\mathcal{F}$-coalgebra (B, β) is a partition (s, P) of B with an $\mathcal{F}$-coalgebraic structure (P, ϕ) such that $s : B \to P$ defines an $\mathcal{F}^c$-homomorphism $s : (B, \beta) \to (P, \phi)$.

By $\mathbf{Alg}^c(T)$ we denote the category of all $\mathcal{F}$-coalgebras and all $\mathcal{F}^c$-homomorphisms. The assignments $(A, \alpha) \mapsto A$ and $(f : (A, \alpha) \to (B, \beta)) \mapsto (f : A \to B)$ extend to a functor $U_{\mathcal{F}}^c : \mathbf{Alg}^c(T) \to \mathbf{Set}$. $U_{\mathcal{F}}^c$ creates colimits [10]. The coproduct of two $U_{\mathcal{F}}^c$-coalgebras (A, α) and (B, β), for instance, is given by $(A + B, \alpha \oplus \beta)$ where $\alpha \oplus \beta = [\mathcal{F}(\kappa_1), \mathcal{F}(\kappa_2)] \circ (\alpha + \beta)$. Since cokernels are coequalizer, we obtain for any $\mathcal{F}^c$-homomorphism $f : (A, \alpha) \to (B, \beta)$ an $\mathcal{F}$-partition $(cok(f), \beta^f)$ of $(A + B, \alpha \oplus \beta)$.

Example 7. An $\mathcal{I}$-coalgebra (S, t) can be interpreted as a 'transition system' with a set S of 'states' and a 'state transition' map $t : S \to S$. $\mathcal{D}$-coalgebras $(S, n : S \to S \times S)$ can be understood, in this way, as a 'branching transition system'.

Finally we dualize the three crucial technical results from Algebraic Specifications. Firstly, we have also an epi-mono-factorization for $\mathcal{F}^c$-homomorphisms:

Proposition 10. *Any $\mathcal{F}^c$-homomorphism $f : (A, \alpha) \to (B, \beta)$ can be factorized as $f = m^f \circ e^f$ with $m^f : (f^c(B), \hat{\alpha}) \to (B, \beta)$ a mono in $\mathbf{Alg}^c(\mathcal{F})$ (and $m^f : f^c(B) \to B$ a (split) mono in $\mathbf{Set}$) and with $e^f : (A, \alpha) \to (f^c(B), \hat{\alpha})$ an epi in $\mathbf{Alg}(\mathcal{F})$ (and $e^f : A \to f^c(B)$ an epi in $\mathbf{Set}$).*

The dual of the homomorphism theorem could be called the *subcoalgebra theorem* since it states in terms of cokernels when a homomorphism factors through a given subcoalgebra:.

Theorem 2. *For any $\mathcal{F}^c$-homomorphisms $m : (A, \alpha) \to (B, \beta)$, $g : (C, \gamma) \to (B, \beta)$ with $m : A \to B$ (split) mono in $\mathbf{Set}$ there exists a unique $\mathcal{F}^c$-homomorphism $h : (C, \gamma) \to (A, \alpha)$ with $g = m \circ h$ iff $(s_g, cok(g)) \sqsubseteq_{B+B} (s_m, cok(m))$.*

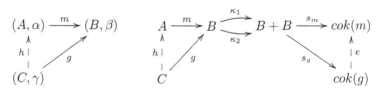

Fortunately, many functors important in applications of the theory of coalgebras do have cofree coalgebras. This holds especially for all functors we can build from constant functors, the identical functor, the product functor, the coproduct functor, the function space functor $(\_)^A : \mathbf{Set} \to \mathbf{Set}$, and the finite powerset functor $\mathcal{P}_f : \mathbf{Set} \to \mathbf{Set}$ [6, 10].

Let X be a set (of '*colors*'). An $\mathcal{F}$-coalgebra $(P_{\mathcal{F}}(X), \phi_{\mathcal{F}, X})$ together with a '*coloring*' $\varepsilon_{\mathcal{F}, X} : P_{\mathcal{F}}(X) \to X$ is *cofree over X w.r.t. $U_{\mathcal{F}}^c$* if for every $\mathcal{F}$-coalgebra (A, α) and for every coloring $c : A \to X$ there exists a unique in $\mathcal{F}^c$-homomorphism $c^{\mathcal{F}} : (A, \alpha) \to (P_{\mathcal{F}}(X), \phi_{\mathcal{F}, X})$ such that $\varepsilon_{\mathcal{F}, X} \circ c^{\mathcal{F}} = c$.

Example 8. For $X = \{0,1\}$ we can identify $P_\mathcal{I}(X)$ with the set X^ω of all (infinite) bit streams. $\phi_{\mathcal{I},X} : X^\omega \rightarrow X^\omega$ is the 'tail'-operation and $\varepsilon_{\mathcal{I},X} : X^\omega \rightarrow X$ the 'head'-operation. For a coloring $c : S \rightarrow X$ of the 'states' of a transition system (S,t) $c^\mathcal{I} : S \rightarrow X^\omega$ assigns to each state $s \in S$ the bit stream we can observe if we visit the states in (S,t) via t starting in s. $P_\mathcal{D}(X)$ consists of all infinite binary trees with bits at the branching nodes. $\varepsilon_{\mathcal{D},X} : P_\mathcal{D}(X) \rightarrow X$ provides the bit at the root and $\phi_{\mathcal{D},X}$ the both subtrees of the root.

In general, the 'elements' of $P_\mathcal{F}(X)$ could be interpreted as the 'observable behaviours' of systems of typ $\mathcal{F}$ [6, 7, 10] or as '$\mathcal{F}$-processes' [15]. Therefore, we will call $(P_\mathcal{F}(X), \phi_{\mathcal{F},X})$ the *$\mathcal{F}$-processcoalgebra over X*. Moreover, $\varepsilon_{\mathcal{F},X} : P_\mathcal{F}(X) \rightarrow X$ will be called the *$\mathcal{F}^c$-counit for X* and $c^\mathcal{F}$ the *(unique) $\mathcal{F}^c$-extension of c*.

A *coequational $\mathcal{F}$-specification over X* is a corelation $(cosp, Q)$ in $P_\mathcal{F}(X)$, where the 'cuts' $P_\mathcal{F}(X) + P_\mathcal{F}(X) \overset{ce}{\rightarrow} \mathbf{2}$ with $\mathbf{2} = \{1,2\}$ are called *coequations*[14].

A coloring $c : A \rightarrow X$ is a *solution* of $(cosp, Q)$ in an $\mathcal{F}$-coalgebra (A, α), if $(s_{c^\mathcal{F}}, cok(c^\mathcal{F})) \sqsubseteq (cosp, Q)$.

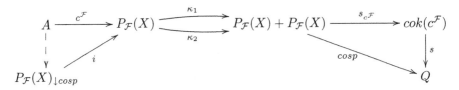

$cosp$ determines, by equalizing $cosp \circ \kappa_1$ and $cosp \circ \kappa_2$, a subset $P_\mathcal{F}(X)_{\downarrow cosp}$ of $P_\mathcal{F}(X)$ thus $c : A \rightarrow X$ is a solution of $(cosp, Q)$ in (A, α) if we can observe in the c-colored coalgebra (A, α) only the "right" processes in $P_\mathcal{F}(X)_{\downarrow cosp}$.

Remark 2. For any $e \in P_\mathcal{F}(X)$ we can define a cut $P_\mathcal{F}(X) + P_\mathcal{F}(X) \overset{c_e}{\longrightarrow} \mathbf{2}$ with $c_e(1, e) = 1$ and $c_e(i, x) = 2$ in all other cases, thus $P_\mathcal{F}(X)_{\downarrow c_e} = P_\mathcal{F}(X) - \{e\}$. This shows that the concept of coequation in [5] is covered by our definition.

Remark 3. It is well-known, that any equational $\mathcal{F}$-specification $(E, spec)$ over X gives rise to a corresponding $\mathcal{F}$-quotientalgebra $(\iota_{spec}, T_{spec}(X))$ of the $\mathcal{F}$-termalgebra $(\iota_{\mathcal{F},X}, T_\mathcal{F}(X))$. Dually, we can describe for any coequational $\mathcal{F}$-specification $(cosp, Q)$ a corresponding $\mathcal{F}$-subcoalgebra $(P_{cosp}(X), \phi_{cosp})$ of the $\mathcal{F}$-processcoalgebra $(P_\mathcal{F}(X), \phi_{\mathcal{F},X})$, where $(P_{cosp}(X), \phi_{cosp})$ will be the greatest $\mathcal{F}$-subcoalgebra (P, ϕ) of $(P_\mathcal{F}(X), \phi_{\mathcal{F},X})$ such that $P \subseteq P_\mathcal{F}(X)_{\downarrow cosp}$.

Therefore "subcoalgebras" have been proposed as a synonym for "coequational specifications" [7, 10]. The observations that Modal Logic is related to coalgebras of a certain kind, that a set of modal logical formulas determines "subcoalgebras", and the duality between "subcoalgebras" and "quotient algebras" have been developed in a row of papers under the slogan "Modal Logic is dual to Equational Logic" [7]. In contrast, our cokernel-based concept of coequational specifications is perfectly dual to equational specifications, and seems to have appropriate structural properties, as it has been indicated, e.g., in [1], where results concerning the reflection of bisimilarity along signature morphisms could be (only) proved by using corelations. Moreover, it can be easily shown that our coequational specifications "are invariant under behavioural equivalences", i.e., provide an abstract Modal Logic in the sense of [7].

6   Cofree Coalgebras and the Dual Birkhoff Construction

An unsorted signature morphism provides also for coalgebras a *coforgetful* functor but in the other direction:

Definition 3. *Any natural transformation* $\tau : \mathcal{F} \Rightarrow \mathcal{G} : \mathbf{Set} \to \mathbf{Set}$ *gives rise to a functor* $U_\tau^c : \mathbf{Alg}^c(\mathcal{F}) \to \mathbf{Alg}^c(\mathcal{G})$ *defined for any* $\mathcal{F}$-*coalgebra* (A, α) *and any* $\mathcal{F}^c$-*homomorphism* $f : (A, \alpha) \to (B, \beta)$ *as* $U_\tau^c(A, \alpha) := (A, \tau_A \circ \alpha)$ *and* $U_\tau^c(f) := f$.

$$
\begin{array}{ccccc}
A & \xrightarrow{\alpha} & \mathcal{F}(A) & \xrightarrow{\tau_A} & \mathcal{G}(A) \\
{\scriptstyle f}\downarrow & & \downarrow{\scriptstyle \mathcal{F}(f)} & & \downarrow{\scriptstyle \mathcal{G}(f)} \\
B & \xrightarrow{\beta} & \mathcal{F}(B) & \xrightarrow{\tau_B} & \mathcal{G}(B)
\end{array}
$$

Example 9. Given an $\mathcal{I}$-coalgebra (S, t) the $\mathcal{D}$-coalgebra $U_\Delta^c(S, t) = (S, \Delta_S \circ t) = (S, \langle t, t \rangle)$ consists of two identical 'transition branches'.

Now, we define cofree coalgebras for signature morphisms:

Definition 4. *Let be given a* $\mathcal{G}$-*coalgebra* (A, α). *An* $\mathcal{F}$-*coalgebra* $(P_\tau(A), \phi_{\tau,\alpha})$ *together with a* $\mathcal{G}^c$-*homomorphism* $\varepsilon_\alpha : U_\tau^c(P_\tau(A), \phi_{\tau,\alpha}) \to (A, \alpha)$ *is* cofree over (A, α) *w.r.t.* U_τ^c *if for any* $\mathcal{F}$-*coalgebra* (B, β) *and for any* $\mathcal{G}^c$-*homomorphism* $h : U_\tau^c(B, \beta) \to (A, \alpha)$ *there exists a unique* $\mathcal{F}^c$-*homomorphism* $h^\tau : (B, \beta) \to (P_\tau(A), \phi_{\tau,\alpha})$ *such that* $\varepsilon_\alpha \circ U_\tau^c(h^\tau) = h$.

$$
\begin{array}{ccc}
(P_\tau(A), \phi_{\tau,A}) & \qquad\qquad & U_\tau^c(P_\tau(A), \phi_{\tau,\alpha}) \xrightarrow{\varepsilon_\alpha} (A, \alpha) \\
\wedge & & \wedge \\
{\scriptstyle h^\tau}\,| & & {\scriptstyle U_\tau^c(h^\tau) = h^\tau}\,| \qquad\nearrow {\scriptstyle h} \\
| & & | \\
(B, \beta) & & U_\tau^c(B, \beta)
\end{array}
$$

By duality, we can construct the cofree coalgebra $(P_\tau(A), \phi_{\tau,\alpha})$ as an appropriate subcoalgebra of the $\mathcal{F}$-processcoalgebra over the carrier of (A, α). To do this, we have to "co-syntactify" the $\mathcal{G}$-coalgebra (A, α), i.e., we have to code (A, α) in terms of 'processes': We consider the $\mathcal{G}$-processcoalgebra over A. Then, the $\mathcal{G}^c$-extension of the trivial coloring $id_A : A \to A$ gives us an *unfolding* $un_\alpha := id_A^{\mathcal{G}} : (A, \alpha) \to (P_\mathcal{G}(A), \phi_{\mathcal{G},A})$, that assigns to each 'state' $a \in A$ the 'process' that can be observed in (A, α) starting in this state, thus we have

$$
\varepsilon_{\mathcal{G},A} \circ un_\alpha = id_A \text{ in } \mathbf{Set}. \tag{7}
$$

The *internal $\mathcal{G}^c$-theory of* (A, α) is given now by the cokernel $(s_{un_\alpha}, cok(un_\alpha))$. As mentioned above, the internal $\mathcal{G}^c$-theory codes, in such a way, the image $un_\alpha(A)$, i.e., all the processes that can be observed actually in (A, α). To translate the $\mathcal{G}^c$-theory we consider the $\mathcal{F}^c$-unit for A $\varepsilon_{\mathcal{G},A} : P_{\mathcal{F}}(A) \to A$ and take the corresponding $\mathcal{G}^c$-extension $\varepsilon^{\mathcal{G}}_{\mathcal{F},A} : U^c_\tau(P_{\mathcal{F}}(A), \phi_{\mathcal{F},A}) \to (P_{\mathcal{G}}(A), \phi_{\mathcal{G},A})$ with

$$\varepsilon_{\mathcal{G},A} \circ \varepsilon^{\mathcal{G}}_{\mathcal{F},A} = \varepsilon_{\mathcal{F},A} \text{ in } \mathbf{Set} \tag{8}$$

for the underlying map $\varepsilon^{\mathcal{G}}_{\mathcal{F},A} : P_{\mathcal{F}}(A) \to P_{\mathcal{G}}(A)$. In such a way we obtain the required translated internal theory of (A, α) as the corelation $(s_{un_\alpha} \circ (\varepsilon^{\mathcal{G}}_{\mathcal{F},A} + \varepsilon^{\mathcal{G}}_{\mathcal{F},A}), cok(un_\alpha))$ in $P_{\mathcal{F}}(A)$.

Example 10. We consider for $A = \{0,1\}$ the $\mathcal{D}$-coalgebra (A, α) with $\alpha(0) = (0,0)$ and $\alpha(1) = (0,1)$. Then $un_\alpha(0)$ will produce an infinite tree with only 0's at all the nodes and $un_\alpha(1)$ produces an infinite tree with only 0's except at the most right nodes, where we will have 1's. $\varepsilon^{\mathcal{D}}_{\mathcal{I},A}$ will map a bit stream into an infinite tree with only 0's (or 1's) at all nodes with the same depth $n \in \mathbb{N}$ if 0 (or 1) is the n'th bit in the stream. This means, that $un_\alpha(1)$ will be not in the image of $\varepsilon^{\mathcal{D}}_{\mathcal{I},A}$, thus the translated theory of (A, α) codes essentially only the constant bit stream 0^ω.

Also for coalgebras we have an equivalent formulation of the semantical condition in Definition 4 in terms of the translated theory:

Proposition 11. *For any $\mathcal{F}$-coalgebra (B, β) and for any map $h : B \to A$ the following conditions are equivalent:*

1. *$h : B \to A$ defines an $\mathcal{G}^c$-homomorphism $h : U^c_\tau(B, \beta) \to (A, \alpha)$ such that $un_\alpha \circ h = \varepsilon^{\mathcal{G}}_{\mathcal{F},A} \circ h^{\mathcal{F}}$ in $\mathbf{Alg}^c(\mathcal{G})$.*
2. *h is a solution of the specification $(s_{un_\alpha} \circ (\varepsilon^{\mathcal{G}}_{\mathcal{F},A} + \varepsilon^{\mathcal{G}}_{\mathcal{F},A}), cok(un_\alpha))$ in (B, β).*

Finally, we present the **dual Birkhoff construction**: Firstly, we construct an appropriate subcoalgebra of the $\mathcal{F}$-processcoalgebra over A. Since any set has up to isomorphism only a set of partitions there exists a set J (of indices), a family $((C_j, \gamma_j) \mid j \in J)$ of $\mathcal{F}$-coalgebras, and a family $(g_j : U^c_\tau(C_j, \gamma_j) \to (A, \alpha) \mid j \in J)$ of $\mathcal{G}^c$-homomorphisms such that for any $\mathcal{F}$-coalgebra (B, β) and for any $\mathcal{G}^c$-homomorphism $h : U^c_\tau(B, \beta) \to (A, \alpha)$ there is an index $j_h \in J$ such that

$$(s_{h^{\mathcal{F}}}, cok(h^{\mathcal{F}})) \cong (s_{g^{\mathcal{F}}_{j_h}}, cok(g^{\mathcal{F}}_{j_h})). \tag{9}$$

We build the coproduct $(\coprod_{j \in J}(C_j, \gamma_j), \kappa_j : (C_j, \gamma_j) \to \coprod_{j \in J}(C_j, \gamma_j))$ in $\mathbf{Alg}^c(\mathcal{F})$, where the carrier of $\coprod_{j \in J}(C_j, \gamma_j)$ will be a coproduct $\coprod_{j \in J} C_j$ of the carriers in **Set** since $U_{\mathcal{F}}^c$ creates colimits. This together with the uniqueness of $\mathcal{F}^c$-extensions ensures that cotupling is compatible with $\mathcal{F}^c$-extensions

$$[g_j]_{j\in J}^{\mathcal{F}} = [g_j^{\mathcal{F}}]_{j\in J} : \coprod (C_j, \gamma_j) \to (P_{\mathcal{F}}(A), \phi_{\mathcal{F},A}). \tag{10}$$

Now we can construct an epi-mono factorization of $[g_j]_{j\in J}^{\mathcal{G}}$ according to Proposition 10

$$\coprod (C_j, \gamma_j) \xrightarrow{\quad e^{\tau,A} \quad} (P_\tau(A), \phi_{\tau,\alpha}) \xrightarrow{\quad m^{\tau,A} \quad} (P_{\mathcal{F}}(A), \phi_{\mathcal{F},A})$$

with upper arc labeled $[g_j]_{j\in J}^{\mathcal{F}}$.

Dualizing the argumentations in section 4 we can prove that $(P_\tau(A), \phi_{\tau,\alpha})$ is indeed cofree over (A, α) w.r.t. U_τ^c. Especially, Proposition 11 ensures that the map $\varepsilon_{\mathcal{F},A} \circ m^{\tau,A} : P_\tau(A) \to A$ defines a $\mathcal{G}^c$-homomorphism $\varepsilon_\alpha := \varepsilon_{\mathcal{F},A} \circ m^{\tau,A} : U_\tau^c(P_\tau(A), \phi_{\tau,\alpha}) \to (A, \alpha)$, i.e., the τ-**counit** for (A, α).

Example 11. Taken as a $\mathcal{I}$-subcoalgebra of $(P_{\mathcal{I}}(A), \phi_{\mathcal{I},A})$ $(P_\Delta(A), \phi_{\Delta,\alpha})$ will have $P_\Delta(A) = \{0^\omega\}$, $\phi_{\Delta,\alpha}(0^\omega) = 0^\omega$, and $\varepsilon_\alpha(0^\omega) = 0$ (since $\varepsilon_{\mathcal{I},A}^\Delta(0^\omega) = un_\alpha(0)$). That is, $(P_\Delta(A), \phi_{\Delta,\alpha})$ provides a "minimal" stream-based representation of the biggest subsystem of (A, α) that obeys in infinitary, symmetric behaviour.

7   Conclusion and Future Work

We have been able to give a fully categorical account of the traditional Birkhoff construction of free algebras w.r.t. unsorted signature morphisms. Than, by dualization, we obtained new results concerning the existence and construction of cofree coalgebras. The analysis of a sample cofree construction supports the intuition that cofree coalgebras should play an important rôle in structuring and modularization of System Specifications:

   For a given unsorted signature morphism $\tau : \mathcal{F} \Rightarrow \mathcal{G}$ the forgetful functor $U_\tau : \mathbf{Alg}(\mathcal{G}) \to \mathbf{Alg}(\mathcal{F})$ forgets some 'algebraic structure' thus the corresponding free functor $T_\tau : \mathbf{Alg}(\mathcal{F}) \to \mathbf{Alg}(\mathcal{G})$ builds new 'algebraic structure' by "identifying constructions". Coalgebraically, or in terms of system behaviour, the co-forgetful functor $U_\tau^c : \mathbf{Alg}^c(\mathcal{F}) \to \mathbf{Alg}^c(\mathcal{G})$ interprets a simple, restricted behaviour as a special case of a more sophisticated, free behaviour. Therefore, the corresponding cofree functor $P_\tau : \mathbf{Alg}^c(\mathcal{G}) \to \mathbf{Alg}^c(\mathcal{F})$ reduces a system by "selecting" the part of the system obeying the restricted pattern of behaviour.

   Many results in the paper are based on the fact that $\mathbf{Alg}(\mathcal{F})$ and $\mathbf{Alg}^c(\mathcal{F})$ are concrete categories over **Set**, i.e., U_τ and U_τ^c are faithful functors. In such a way, we could formulate some results even more categorically abstract. This observation may help in generalizing the results to many-sorted signature morphisms. Having in mind the situation in Algebraic Specifications this generalization should be possible and relatively straightforward.

   A generalization to morphisms between (co)equational specifications (or even conditional (co)equational specifications) will be notationally very tedious, but should not cause principal problems, because the (dual) Birkhoff construction is based on products (coproducts) and subalgebras (partitions) and the corresponding classes of (co)algebras are closed under these constructions [5, 8, 14] (compare also the Birkhoff construction of initial algebras in [14]).

Probably the next steps should focus on the analysis of more comprehensive examples as, for instance, CSP [15], to reach a better intuition and understanding of cofree constructions and their potential rôle.

References

1. F. Bartels, A. Sokolova, and E. de Vink. A hierarchy of probabilistic system types. *TCS*, 2004. Submitted.
2. H. Ehrig, M. Große–Rhode, and U. Wolter. Applications of category theory to the area of algebraic specification in computer science. *Applied Categorical Structures*, 6(1):1–35, 1998.
3. H. Ehrig and B. Mahr. *Fundamentals of Algebraic Specification 1: Equations and Initial Semantics*, volume 6 of *EATCS Monographs on Theoretical Computer Science*. Springer, Berlin, 1985.
4. H. Ehrig and B. Mahr. *Fundamentals of Algebraic Specification 2: Module Specifications and Constraints*, volume 21 of *EATCS Monographs on Theoretical Computer Science*. Springer, Berlin, 1990.
5. H. P. Gumm. Equational and implicational classes of coalgebras. *TCS*, 260:57–69, 2001.
6. B. Jacobs and J. Rutten. A tutorial on (Co)Algebras and (Co)Induction. *Bulletin of the EATCS*, 62:222–259, June 1997.
7. A. Kurz. Coalgebras and modal Logic. Technical report, CWI Amsterdam, 2001. Lecture notes.
8. M. Löwe, H. König, and Ch. Schulz. Algebraic Properties of Interfaces. In H.J. Kreowski and G. Taentzer, editors, *tHE BOOK on Formal Methods in Software and System Modelling (This volume)*. Springer LNCS, 2004.
9. H. Reichel. *Initial Computability, Algebraic Specifications, and Partial Algebras*. Oxford University Press, Oxford, 1987.
10. J.J.M.M. Rutten. Universal coalgebra: A theory of systems. *TCS*, 249:3–80, 2000. First appeared as Report CS-R9652, CWI, Amsterdam 1996.
11. M.B. Smyth and G.D. Plotkin. The category theoretic solution of recursive domain equations. *SIAM Journ. Comput.*, 11:761–783, 1982.
12. W. Wechler. *Universal Algebra for Computer Scientists*, volume 25 of *EATCS Monographs on Theoretical Computer Science*. Springer, Berlin, 1992.
13. U. Wolter. An Algebraic Approach to Deduction in Equational Partial Horn Theories. *J. Inf. Process. Cybern. EIK*, 27(2):85–128, 1990.
14. U. Wolter. On Corelations, Cokernels, and Coequations. In H. Reichel, editor, *Third Workshop on Coalgebraic Methods in Computer Science (CMCS'2000), Berlin, Germany, Proceedings*, volume 13 of ENTCS, pages 347–366. Elsevier Science, 2000.
15. U. Wolter. CSP, Partial Automata, and Coalgebras. *TCS*, 280:3–34, 2002.

Part III

Formal and Visual Modeling

Nested Constraints and Application Conditions for High-Level Structures

Annegret Habel and Karl-Heinz Pennemann

Carl v. Ossietzky University of Oldenburg, Germany
{habel,k.h.pennemann}@informatik.uni-oldenburg.de

Abstract. Constraints and application conditions are most important for transformation systems in a large variety of application areas. In this paper, we extend the notion of constraints and application conditions to nested ones and show that nested constraints can be successively transformed into nested right and left application conditions.

1 Introduction

Constraints and application conditions are most important for transformation systems in a large variety of application areas, especially in the area of safety-critical systems e.g. the specification of railroad control systems and access control policies [10]. Application conditions for rules were investigated e.g. in [2, 6, 8, 10, 1]. They define classes of morphisms and thus restrict the applicability of their rules. Constraints, also called consistency constraints, were studied e.g. in [8, 10, 1]. They are properties on objects which have to be satisfied. In the graph case, simple constraints like the existence or uniqueness of certain nodes and edges can be expressed.

In this paper, we extend the existing theory on constraints and application conditions [1] to nested constraints and application conditions as proposed in [15]. We show that nested constraints can be transformed into nested right application conditions, and nested right application conditions can be transformed into nested left application conditions.

The transformation results are proved for high-level structures and are applied exemplarily for the case of graphs. The concepts are illustrated by a simple railroad system. The specification of a railroad system is given in terms of rail net graphs, constraints, rules for moving the trains, and application conditions. We study the integration of general rail net constraints into rail net application conditions for the movement of trains.

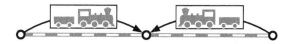

The paper is organized as follows. In section 2, we give a short introduction of adhesive HLR categories together with their basic properties. In section 3 we generalize constraints and application conditions to nested constraints and

H.-J. Kreowski et al. (Eds.): Formal Methods (Ehrig Festschrift), LNCS 3393, pp. 293–308, 2005.

application conditions in the framework of adhesive HLR categories. In sections 4, 5 and 6, we prove the main transformation results and give some applications of the results. A conclusion including further work is given in section 7.

2   Preliminaries

The main idea of high-level replacement systems is to generalize the concepts of graph replacement from graphs to all kinds of structures which are of interest in Computer Science and Mathematics. In the following sections, we will consider constraints and application conditions in adhesive HLR-categories (see [3]) and prove our transformation results on this general level. This has the advantage, that our results apply not only for graphs, but also for other high-level structures, e.g. typed attributed graphs, Petri-nets or algebraic specifications. We consider adhesive HLR-categories to benefit from the combined advantages of HLR and of adhesive categories [11]: the sound theory of the HLR framework and the much smoother requirements for adhesive categories in comparison with the variety of HLR preconditions.

Definition 1 (adhesive HLR-category). A category $\mathcal{C}$ with a morphism class M is called *adhesive HLR category*, if 1) M is a class of monomorphisms closed under compositions and decompositions ($g \circ f \in M$, $g \in M$ implies $f \in M$), 2) $\mathcal{C}$ has pushouts and pullbacks along M-morphisms, i.e. pushouts and pullbacks, where at least one of the given morphisms is in M, and M-morphisms are closed under pushouts and pullbacks, i.e. whenever a given morphism is in M, then the opposite morphism is in M, as well, and 3) pushouts in $\mathcal{C}$ along M-morphisms are VK-squares (see e.g. [3]).

Examples for adhesive HLR categories for the class M of all monomorphism include *Sets*, *Graphs*, *PT-Nets* and several other variants of graphs and nets, like typed, labeled and attributed graphs, hypergraphs and high-level nets. Further examples can be found, e.g. in [3].

Example 1 (rail net graph). The railroad system (similar to [12]) models the movement of one or more trains on a net of railroad tracks. The basic items are simple tracks from which the net is synthesized. The net together with trains on it forms the static part of the system, while movement of trains constitute the dynamic part. The static part is given by a labeled directed rail net graph:

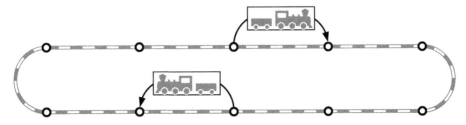

Fig. 1. A simple railway model.

tracks are modeled by undirected (resp. a pair of directed) edges, and trains are modeled by edges. Source and target nodes of a train edge encode the train's position on the track. An example of a rail net graph with two trains is given in figure 1.

Fact 1. Given an adhesive HLR-category $\langle C, M \rangle$, the following HLR conditions are satisfied.

1. Pushouts along M-morphisms are pullbacks.
2. Pushout-pullback decomposition: If the diagram $(1)+(2)$ is a pushout, (2) a pullback, and $l, w \in M$, then (1) and (2) are pushouts and also pullbacks.

$$
\begin{array}{ccccc}
A & \xrightarrow{\ b\ } & B & \xrightarrow{\ r\ } & E \\
{\scriptstyle l}\downarrow & (1) & {\scriptstyle s}\downarrow\ (2) & & \downarrow{\scriptstyle v} \\
C & \xrightarrow[\ u\ ]{} & D & \xrightarrow[\ w\ ]{} & F
\end{array}
$$

3. Uniqueness of pushout complements for M-morphisms: Given $b: A \to B$ in M and $s: B \to D$ then there is up to isomorphism at most one C with $l: A \to C$ and $u: C \to D$ such that diagram (1) is a pushout.

General Assumption. In the following, we assume that $\langle C, M \rangle$ is an adhesive HLR category with binary coproducts and epi-M-factorizations, that is, for every morphism there is an epi-mono-factorization with monomorphism in M.

3   Constraints and Application Conditions

In the following, we will consider structural constraints and application conditions. Structural constraints, short constraints, correspond to graph constraints, but not necessarily to logical constraints defined by predicate logic.

Definition 2 (constraint). *Constraints* over an object P are defined inductively as follows: For an arbitrary morphism $x: P \to C$, $\exists x$ is a *(basic)* constraint over P. For an arbitrary morphism $x: P \to C$ and a constraint c over C, $\forall(x, c)$ and $\exists(x, c)$ are *(conditional)* constraints over P. For constraints c, c_i $(i \in I)$ [over P], true, false, $\neg c$, $\wedge_{i \in I} c_i$ and $\vee_{i \in I} c_i$ are *(Boolean)* constraints [over P].

A morphism $p: P \to G$ *satisfies* a basic constraint $\exists x$ if there exists a morphism $q: C \to G$ in M with $q \circ x = p$. A morphism $p: P \to G$ *satisfies* a conditional constraint $\exists(x, c)$ $[\forall(x, c)]$ if some [all] morphisms $q: C \to G$ in M with $q \circ x = p$ satisfy c. Every morphism satisfies true, and no morphism satisfies false. A morphism p *satisfies* a Boolean constraint $\neg c$ if p does not satisfy c; p *satisfies* $\wedge_{i \in I} c_i$ $[\vee_{i \in I} c_i]$ if p satisfies all [some] c_i with $i \in I$. We write $p \models c$ to denote that p satisfies c.

An object G *satisfies* a constraint c of the form $\exists x$, $\exists(x, d)$ $[\forall(x, d)]$ if all [some] morphisms $p: P \to G$ in M satisfy c. Every object satisfies true, and no object satisfies false. An object G *satisfies* $\neg c$ if G does not satisfy c and $\wedge_{i \in I} c_i$

$[\bigvee_{i \in I} c_i]$ if it satisfies all [some] c_i with $i \in I$. We write $G \models c$ to denote that G satisfies c. Two constraints c and c' are *equivalent*, denoted by $c \equiv c'$, if, for all objects G, $G \models c$ if and only if $G \models c'$.

Constraints of the form $\exists x$ and $\neg \exists x$ with empty morphism $x: \emptyset \rightarrow C$ are denoted by $\exists C$ and $\not\exists C$, respectively. Constraints of the form $\neg \exists x$ are abbreviated by $\not\exists x$. Constraints of the form $\exists x$, $\exists(x, c)$ are said to be *existential*; constraints of the form $\forall(x, c)$ are *universal*.

In [1], negative atomic constraints are considered. A morphism $p: P \rightarrow G$ satisfies the negative atomic constraint $NC(x)$ with morphism $x: P \rightarrow C$ if there does not exist a morphism $q: C \rightarrow G$ in M with $q \circ x = p$. Negative atomic constraints are equivalent to positive ones (with negation).

Fact 2. For $x: P \rightarrow C$ in M, $NC(x) \equiv \neg \exists C$; otherwise, $NC(x) \equiv$ true.

Proof. For $x: P \rightarrow C$ in M, $G \models NC(x)$ iff, for all $p: P \rightarrow G$ in M, there does not exist a $q: C \rightarrow G$ in M such that $q \circ x = p$ iff there does not exists a $q: C \rightarrow G$ in M iff $G \models \neg \exists C$. In [1], it is shown that, for x not in M, $NC(x) \equiv$ true: Assume, there exists a G such that $G \not\models NC(x)$. Then there exist $p: P \rightarrow G$ in M and $q: C \rightarrow G$ in M with $q \circ x = p$. Then p, q in M implies x in M (contradiction).

In [1], constraints with not-M-morphisms are allowed. This does not give more expressive power. For every constraint with arbitrary morphisms, there is an equivalent constraint with morphisms in M.

Fact 3. For a morphism $x: P \rightarrow C$ not in M, the following constraints are equivalent: $\exists(x, c) \equiv \exists x \equiv \not\exists P$ and $\forall(x, c) \equiv \exists P$.

Proof. $G \models \exists x \; [\exists(x, c)]$ iff every $p: P \rightarrow G$ in M implies the existence of a $q: C \rightarrow G$ in M with $q \circ x = p$ [and $q \models c$]. Assume there is such a p. Then p, q in M implies x in M (contradiction). Thus there is no $p: P \rightarrow G$ in M and we have $G \models \not\exists P$. Vice versa, $G \models \not\exists P$ iff there is no $p: P \rightarrow G$ in M. Then all $p: P \rightarrow G$ in M can imply the existence of a $q: C \rightarrow G$ in M with $q \circ x = p$ [and $q \models c$] (because there is no p) and we have $G \models \exists x \; [\exists(x, c)]$. Furthermore, $G \models \forall(x, c)$ iff there exists $p: P \rightarrow G$ in M such that for all $q: C \rightarrow G$ in M with $q \circ x = p$ holds $q \models c$. Assume there is such a q. Then p, q in M implies x in M (contradiction). Thus there is no $q: C \rightarrow G$ in M and we have $G \models \exists P$. Vice versa, $G \models \exists P$ iff there exists an $p: P \rightarrow G$ in M. If there is no $q: C \rightarrow G$ in M with $q \circ x = p$, then all q can imply $q \models c$ and we have $G \models \forall(x, c)$.

Constraints without Boolean symbols, i.e. true, false, $\neg$, $\wedge$ or $\vee$, may have alternating quantifiers. For every such constraint with consecutive quantifiers Q, there is an equivalent constraint with single quantifier Q. In this way, equal consecutive quantifiers can be eliminated.

Fact 4. $Q(x, Q(y, c)) \equiv Q(y \circ x, c))$ for $Q \in \{\forall, \exists\}$.

Proof. Let $x: P \rightarrow C$, $y: C \rightarrow C'$ and $m: P \rightarrow G$ in M. Then $m \models \exists(x, \exists y, c))$ iff there exists $q: C \rightarrow G$ in M such that $q \circ x = m$ and there exists $q': C' \rightarrow G$

in M such that $q' \circ y = q$ and $q' \models c$ iff there exists $q': C' \to G$ in M such that $q' \circ y \circ x = m$ and $q' \models c$ iff $m \models \exists(y \circ x, c)$. Moreover, $m \models \forall(x, \forall y, c))$ iff for all $q: C \to G$ in M with $q \circ x = m$ and for all $q': C' \to G$ in M with $q' \circ y = q$ and $q' \models c$ iff for all $q': C' \to G$ in M with $q' \circ y \circ x = m$ and $q' \models c$ iff $m \models \forall(y \circ x, c)$.

Remark. The definition of constraints generalizes the ones in [8, 10, 1], because we allow arbitrary nested constraints. E.g., we allow to express constraints like "For all nodes, there exists an outgoing edge such that, for all edges outgoing from the target, the target has a loop."

Fact 5. (Counting constraints) Counting of elements is possible. The following properties of graphs can be expressed as graph constraints:

For a given $n \in \mathbb{N}_0$, all nodes have exactly n outgoing edges:

$$\exists(\bigcirc \to \bigcirc, c_{out=n})$$

There exists a node with an even number of incoming edges:

$$\forall(\bigcirc \to \bigcirc, \bigvee_{n \in \mathbb{N}_0} c_{in=2n})$$

There exists a node with same number of outgoing and incoming edges:

$$\forall(\bigcirc \to \bigcirc, \bigvee_{n \in \mathbb{N}_0} c_{out=n} \wedge c_{in=n})$$

where, given $n \in \mathbb{N}$, $c_{out=n}$ and $c_{in=n}$ are subconstraints that are satisfied, if and only if the number of outgoing and incoming edges equals n, respectively. The constraints $c_{out=n}$ and $c_{in=n}$ are defined with the help of a constraint $c(x)$ which is used to count edges. For a morphism $x: P \to C$, let $c(x) = \bigvee_{e \in \mathcal{E}} \exists(e \circ x)$, where the set $\mathcal{E}$ consists of all epimorphisms that do not identify edges. Define $c_{out=n} = c(\bigcirc \to S_n) \wedge \neg c(\bigcirc \to S_{n+1})$ and $c_{in=n} = c(\bigcirc \to T_n) \wedge \neg c(\bigcirc \to T_{n+1})$ where S_n [T_n] denotes the star with outgoing [incoming] edges as depicted below. E.g., $c(\bigcirc \to S_n)$ is satisfied, iff there exists at least n outgoing edges, and $\neg c(\bigcirc \to S_{n+1})$ is satisfied, iff there exists at most n edges.

In general, constraints are a collection of constraints combined by conjunction and disjunction. Several subconstraints may be similar. For complexity aspects, we will reduce the number of subconstraints, before the constraint is transformed into a right application condition. In the following, we sketch some equivalence-preserving rules for modifying and condensing constraints.

Fact 6. For constraints, we have the following equivalences:

(1) $\forall(o, c) \equiv c$ if c is existential.
(2) $\exists(o, c) \equiv c$ if c is universal.

where $o: \emptyset \to P$ is the empty morphism.

A constraint c *implies* a constraint c', denoted by $c \Rightarrow c'$, if, for all objects G, $G \models c$ implies $G \models c'$. A constraint c *m-implies* c', denoted by $c \dot{\Rightarrow} c'$, if, for all morphisms $p \colon P \to G$, $p \models c$ implies $p \models c'$.

Fact 7. For constraints and application conditions, we have the following implications, where $\Rrightarrow \in \{\Rightarrow, \dot{\Rightarrow}\}$.

(1) $\exists x \Rrightarrow \exists x'$ if $i \circ x' = x$ for some i in M.
(2) $\exists(x, c) \Rrightarrow \exists(x', c')$ if $c \dot{\Rightarrow} c'$ and $i \circ x' = x$ for some i in M.
(3) $\forall(x, c) \Rrightarrow \forall(x', c')$ if $c \dot{\Rightarrow} c'$ and $i \circ x = x'$ for some i in M.

We distinguish between inner and outer conjunctions and disjunctions: A conjunction [disjunction] symbol in a constraint is said to be *inner* if it occurs in a subconstraint of the form $Q(x, c)$ with $Q \in \{\forall, \exists\}$.

Lemma 1 (elimination of constraints). *Every constraint can be transformed into an equivalent constraint according to the following rules: (1) Replace subconstraints by equivalent ones and (2) condense outer [inner] conjunctions and disjunctions: (a) Eliminate c_l from $\wedge_{i \in I} c_i$ provided $c_k \Rightarrow c_l$ [$c_k \dot{\Rightarrow} c_l$] for some $k \neq l$ and (b) eliminate c_k from $\vee_{i \in I} c_i$ provided $c_k \Rightarrow c_l$ [$c_k \dot{\Rightarrow} c_l$] for some $k \neq l$.*

Example 2 (rail net constraints). Consider the railroad system in example 1. For security aspects, we formalize some rail net constraints for rail net graphs. E.g. we want to be sure that every train is on a track, that two trains do not occupy the same piece of track, and that two trains do not occupy neighboring pieces of track, except if the trains have a different direction.

(c_1) Every train occupies one piece of track:

(c_2) Different trains occupy different pieces of track:

(c_3) Two adjacent trains head into opposite directions:

In the following, we will consider application conditions for rules. Application conditions for rules were first introduced in [2]. In a subsequent paper [6], a special kind of application conditions were considered which can be represented in a graphical way. In the graph case, contextual conditions like the existence or non-existence of certain nodes and edges or certain subgraphs in the given graph can be expressed. In [8, 1] a simple form of nested application conditions are considered.

Definition 3 (rule). A *rule* $p = \langle L \leftarrow K \rightarrow R \rangle$ consists of two morphisms in M with a common domain K. Given a rule p and a morphism $K \rightarrow D$, a *direct derivation* consists of two pushouts (1) and (2). We write $G \Rightarrow_{p,m,m^*} H$ and say that $m: L \rightarrow G$ is the match and $m^*: R \rightarrow H$ is the comatch of p in H.

$$
\begin{array}{ccccc}
L & \longleftarrow & K & \longrightarrow & R \\
m \downarrow & (1) & \downarrow & (2) & \downarrow m^* \\
G & \longleftarrow & D & \longrightarrow & H
\end{array}
$$

Definition 4 (application condition for a rule). An *application condition* $a = (a_L, a_R)$ for a rule $p = \langle L \leftarrow K \rightarrow R \rangle$ consists of a constraint a_L over L and a constraint a_R over R, called *left and right application condition*, respectively. A *direct derivation* $G \Rightarrow_{p,m,m^*} H$ satisfies an application condition $a = (a_L, a_R)$, if $m \models a_L$ and $m^* \models a_R$.

Remark. The definition of application conditions generalizes the ones in [8, 1], because we allow arbitrary nested application conditions. In [1], simple nested application conditions of the form $\forall(x, \vee_{i \in I} \exists x_i)$ and $\forall(x, \wedge_{i \in I} \neg \exists(x_i))$ with morphisms x_i are considered.

Example 3 (application condition for the rule Move*).* The dynamic part of the railroad system in example 1 consists of a rule Move for the movement of trains, which is depicted in figure 2.

Move:

Fig. 2. The rule Move for movement of trains.

Application of the rule Move means to find an occurrence of the left-hand side in the rail net graph and to replace the occurrence of the left-hand side by the right-hand side of the rule. In this context, it is adequate to restrict on injective matches of the left-hand side in the rail net graph. For security aspects, we formalize application conditions for the rule Move. E.g. every train should move only on a free piece of track and, after movement, two trains should not occupy neighboring pieces of track (except if the trains have a different direction).

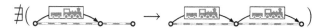

Fig. 3. A left application condition for Move.

For every derivation step $G \Rightarrow_{\text{Move},g} H$ with injective match g, $G \models c_1$ implies $H \models c_1$. Unfortunately, this does not hold for the constraints c_2 and c_3. To ensure that Move can only be applied if H satisfies the constraints c_2 and c_3, we will transform these constraints first into right application conditions and then into left application conditions for Move.

Remark. The definitions of constraints and application conditions are the same. But constraints and application conditions are used in a different way. Constraints express properties on objects which have to be satisfied. A constraint restricts the set of objects to the set of objects that satisfy the constraint. Application conditions restrict the set of matches/comatches and, thus, the applicability of the rule.

4   Transformation of Constraints

In the following, we will show that arbitrary constraints can be transformed into right application conditions.

Theorem 1. (transformation of constraints into right application conditions) *Given a rule with right-hand side R and a constraint c. Then there is a right application condition $T(c)$ such that, for all comatches $m^*: R \to H$,*

$$m^* \models T(c) \Leftrightarrow H \models c.$$

The construction is an extended version of the one for basic constraints in [1]. It is done with help of a right application condition $T_p(c)$ of c according to a morphism $p: P \to S$ in M.

Construction. Given a constraint c over P and a morphism $p: P \to S$ in M, we construct $T_p(c)$ over S according to p as follows: For a basic constraint $c = \exists x$, we construct the pushout (1) in figure 4(a) of p and x leading to $t: S \to T$ and $q: C \to T$ in M and all epimorphisms $e: T \to U$ such that $u = e \circ t$ and $r = e \circ q$ are in M. Let $\mathcal{E}$ denote the set of all these epimorphisms and, for $e \in \mathcal{E}$, $u = e \circ t$.

$$T_p(c) = \vee_{e \in \mathcal{E}} \exists u$$

For a conditional constraint $c = \forall(x, d) \, [\exists(x, d)]$, we construct $T_p(\exists x) = \vee_{e \in \mathcal{E}} \exists u$ over S according to p. The choice of an epimorphism $e \in \mathcal{E}$ determines morphisms $u = e \circ t$ and $r = e \circ q$. For the constraint d and the morphism r, we construct $T_r(d)$ of d according to r.

$$T_p(c) = \wedge_{e \in \mathcal{E}} \forall(u, T_r(d)) \quad [\vee_{e \in \mathcal{E}} \exists(u, T_r(d))]$$

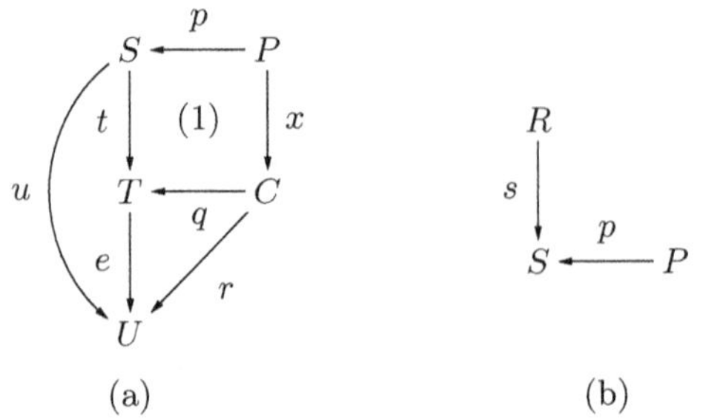

(a)   (b)

Fig. 4. Construction of $T_p(c)$ and $T(c)$.

T_p is *compatible* with Boolean operations, i.e. $T_p(\text{true}) = \text{true}$, $T_p(\text{false}) = \text{false}$, $T_p(\neg d) = \neg T_p(d)$, $T_p(\wedge_{i \in I} c_i) = \wedge_{i \in I} T_p(c_i)$, and $T_p(\vee_{i \in I} c_i) = \vee_{i \in I} T_p(c_i)$. For a universal [existential] constraint c over P and an object R, we construct $T(c)$ with help of the $T_p(c)$. Let A denote the set of all triples $a = \langle S, s, p \rangle$ with arbitrary $s : R \to S$ and $p : P \to S$ in M such that the pair $\langle s, p \rangle$ is jointly epimorphic (see figure 4(b)).

$$T(c) = \vee_{a \in A} \exists(s, T_p(c)) \quad [\wedge_{a \in A} \forall(s, T_p(c))]$$

T is compatible with Boolean operations.

Remark. For the double-pushout approach with matches in M (see [7]), the construction of $T(c)$ can be simplified: Since M is closed under decompositions, $p'' \circ s = m^*$ and $p'' : S \to H$ in M implies s in M. Therefore, it suffices to consider the subset $A' \subseteq A$ of all triples $a = \langle S, s, p \rangle$ with both $s : R \to S$ and $p : P \to S$ in M such that the pair $\langle s, p \rangle$ is jointly epimorphic.

Proof. By structural induction, we show:

(∗) For arbitrary constraints c over P and morphisms $p : P \to S$ in M, we have: For arbitrary morphisms $p'' : S \to H$ in M,

$$p'' \models T_p(c) \text{ if and only if } p' = p'' \circ p \models c.$$

We will use the following statements: (1) If $q'' : U \to H$ is a morphism in M with $q'' \circ u = p''$, then there exists a morphism $q' = q'' \circ r : C \to H$ in M with $q' \circ x = p'$. (2) If $q' : C \to H$ is a morphism in M with $q' \circ x = p'$, then there exist morphisms $e \in \mathcal{E}$ and $q'' : U \to H$ in M with $q'' \circ r = q'$ and $q'' \circ e \circ t = p''$ (see figure 5). The second statement may be seen as follows: The universal property of pushouts implies the existence of a unique morphism $h : T \to H$ with $h \circ t = p''$ and $h \circ q = q'$. Now let $q'' \circ e$ be a epi-mono factorization of h with epimorphism e and monomorphism q'' in M. Then $q'' \circ e \circ t = h \circ t = p''$ in M implies $e \circ t$ in M and $q'' \circ e \circ q = h \circ q = q'$ in M implies $e \circ q$ in M (M closed under decompositions). Thus, there exists a morphism $q'' : U \to H$ in M with $q'' \circ r = q'$ and $q'' \circ e \circ t = p''$ for some $e \in \mathcal{E}$. Now we will prove (∗) for arbitrary constraints.

For basic constraints, (∗) follows from the definitions and (1) and (2): $p'' \models T_p(\exists x)$ iff for some epimorphisms $e \in \mathcal{E}$, $p'' \models \exists u$ iff for some epimorphisms $e \in \mathcal{E}$, there exists a morphism $q'' : U \to H$ in M with $q'' \circ u = p''$ if(2) and only if(1) there exists a morphism $q' : C \to H$ in M with $q' \circ x = p'$ iff $p' \models \exists x$.

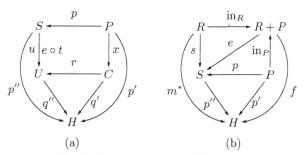

(a)   (b)

Fig. 5. Correspondence of $T(c)$ and c.

For conditional constraints, $(*)$ follows from the definitions, (1) and (2), and the inductive hypothesis: $p'' \models T_p(\forall(x, d))$ iff for all epimorphisms $e \in \mathcal{E}$, $p'' \models \forall(u, T_r(d))$ iff for all epimorphisms $e \in \mathcal{E}$ and all morphisms $q'': U \to H$ in M with $q'' \circ u = p''$, $q'' \models T_r(d)$ if$^{(1)(*)}$ and only if$^{(2)(*)}$ for all morphisms $q': C \to H$ in M with $q' \circ x = p'$, $q' \models d$ iff $p' \models \forall(x, d)$. $p'' \models T_p(\exists(x, d))$ iff for some epimorphisms $e \in \mathcal{E}$, $p'' \models \exists(u, T_r(d))$ iff for some epimorphisms $e \in \mathcal{E}$, there exists a morphism $q'': U \to H$ in M with $q'' \circ u = p''$ such that $q'' \models T_r(d)$ if$^{(2)(*)}$ and only if$^{(1)(*)}$ there exists a morphism $q': C \to H$ in M with $q' \circ x = p'$ such that $q' \models d$ iff $p' \models \exists(x, d)$.

For Boolean constraints, $(*)$ follows directly from the definitions and the inductive hypothesis.

Consequently, $(*)$ holds for all constraints.

It remains to prove the main statement: For all morphisms $m^*: R \to H$, $m^* \models T(c) \Leftrightarrow H \models c$. This is done by structural induction. We will use the following statements: (3) Given a triple $a = \langle S, s, p \rangle$ in A and a morphism p'' in M as above we define $p' = p'' \circ p: P \to H$. Then p' is in M, because p and p'' are in M. (4) Given a morphism $p': P \to H$ in M and a comatch $m^*: R \to H$, we construct the coproduct $R + P$ with injections in_R and in_P in figure 5. By the universal property of coproducts, there is a unique morphism $f: R + P \to H$ with $f \circ in_R = m^*$ and $f \circ in_P = p'$. Now let $f = p'' \circ e$ be an epi-mono factorization of f with epimorphism e and monomorphism p'' in M, and define $s = e \circ in_R$ and $p = e \circ in_P$. Then the pair $\langle s, p \rangle$ is jointly epimorphic, because e is an epimorphism, and p is in M, because $p'' \circ p = p'' \circ e \circ in_P = f \circ in_P = p'$ is in M. Hence $a = \langle S, s, p \rangle$ belongs to the set A. Moreover we have $p'' \circ s = p'' \circ e \circ in_R = f \circ in_R = m^*$ with monomorphism p'' in M.

For universal constraints we have: $m^* \models T(c)$ iff for all tuples $a = \langle S, s, p \rangle$ in A and all morphisms $p'': S \to H$ in M holds $p'' \models T_p(c)$ if$^{(3)(*)}$ and only if$^{(4)(*)}$ for all morphisms $p': P \to H$ in M holds $p' \models c$ iff $H \models c$. For existential constraints we have: $m^* \models T(c)$ iff for some tuples $a = \langle S, s, p \rangle$ in A and for some morphisms $p'': S \to H$ in M holds $p'' \models T_p(c)$ if$^{(4)(*)}$ and only if$^{(3)(*)}$ for some morphisms $p': P \to H$ in M, holds $p' \models c$ iff $H \models c$. For Boolean constraints, the statement follows directly from the definitions and the inductive hypothesis. This completes the proof.

Example 4 (transformation of constraints into right application conditions). Consider the rule Move and the constraint NoTwo $= \neg\exists(\emptyset \to \text{Two}) \equiv \not\exists\text{Two}$ where Two denotes the graph with a track edge and two parallel train edges (see the first subconstraint of c_2 in example 2), saying that two trains are not allowed to occupy the same piece of track in the same direction. Then the transformation of constraint NoTwo yields the following right application condition of Move:

$$T(\text{NoTwo}) = \neg \wedge_{a \in A} \forall(s, \vee_{e \in \mathcal{E}} \exists u).$$

The set A of all "gluings" of the right-hand side R of Move and the empty graph consists of all triples $a = \langle S, s, p \rangle$ where S is a surjective image of R. In the rail road example, we restrict to injective matches. Thus, it suffices to consider injective morphisms $s: R \to S$, i.e. the triple $a = \langle R, \text{id}, \emptyset \to R \rangle$.

R

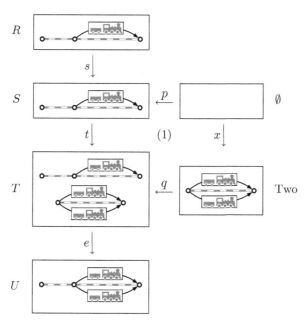

$s\downarrow$

S $\xleftarrow{p}$ (1) $x\downarrow$ $\emptyset$

$t\downarrow$

T $\xleftarrow{q}$ Two

$e\downarrow$

U

Fig. 6. Transformation of NoTwo into a right application condition of Move.

The pushout of $p\colon \emptyset \to S$ and $x\colon \emptyset \to$ Two is the disjoint union $T = S +$ Two with the injections t and q. Let $\mathcal{E}$ denote the set of all surjective morphisms $e\colon T \to U$ with injective $u = e \circ t$ and $r = e \circ q$. By the equivalence $\forall(\mathrm{id}, c) \doteq c$, we have

$$T(\mathrm{NoTwo}) \doteq \neg \vee_{e \in \mathcal{E}} \exists u \doteq \wedge_{e \in \mathcal{E}} \neg \exists u.$$

We get a conjunction of application conditions from which one is depicted in figure 6. The depicted one says that the moved train is not on a piece of track which is occupied by another train moving in the same direction.

Remark. Given a rule with left-hand side L and a constraint c, there is a left application condition $T_L(c)$ such that, for all matches $m\colon L \to G$, $m \models T_L(c) \Leftrightarrow G \models c$: Let $T_L(c)$ be the right application condition of the inverse rule $p^{-1} = \langle L \leftarrow K \to R \rangle$ and c. Then $T_L(c)$ is a left application condition of p with the wanted property.

Remark. In general, there is no transformation from application conditions into constraints: Let a be a *non-trivial* application condition, that is, there exist two morphisms m_1, m_2 with same codomain G, such that $m_1 \models a$ and $m_2 \not\models a$. Assume there exists a constraint $c(a)$ such that, for all matches $m\colon L \to G$ in M, $m \models a \Leftrightarrow G \models c(a)$. Then $G \models c(a)$ and $G \not\models c(a)$. Contradiction.

5   Transformation of Application Conditions

In the following, we will show that arbitrary right application conditions can be transformed into left application conditions.

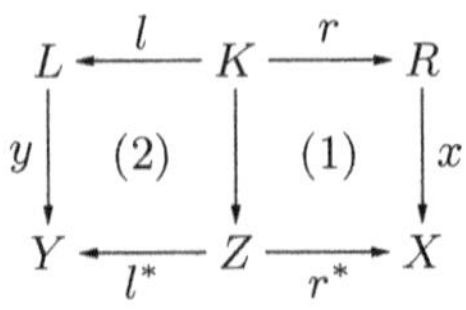

$$L \xleftarrow{\ l\ } K \xrightarrow{\ r\ } R$$

$$y \downarrow \quad (2) \quad | \quad (1) \quad \downarrow x$$

$$Y \xleftarrow{\ l^*\ } Z \xrightarrow{\ r^*\ } X$$

Fig. 7. Transformation of application conditions.

Theorem 2. (transformation from right to left application conditions)
Let a be a right application condition for p. Then there is a left application condition $T_p(a)$ such that, for all direct derivations $G \Rightarrow_{p,m,m^*} H$ we have:

$$m \models T_p(a) \Leftrightarrow m^* \models a.$$

The construction of $T_p(a)$ is an extended version of the corresponding construction for basic application conditions in [1].

Construction. Let a be a right application condition for the rule $p = \langle L \leftarrow K \rightarrow R \rangle$. Construct the left application condition $T_p(a)$ according to p as follows: for a basic right application condition $\exists x$ with morphism $x : R \rightarrow X$, define $y : L \rightarrow Y$ by two pushouts (1) and (2) in figure 7 if the pair $\langle r, x \rangle$ has a pushout complement. Let $T_p(\exists x) = \exists y$ if $\langle r, x \rangle$ has a pushout complement and false otherwise. For a conditional right application condition $Q(x, a)$ with $Q \in \{\forall, \exists\}$, construct the morphism y as above, if the pair $\langle r, x \rangle$ has a pushout complement, and the left application condition $T_{p^*}(a) = b$ according to the "derived" rule $p^* = \langle Y \leftarrow Z \rightarrow X \rangle$. Let $T_p(Q(x, a)) = Q(y, b)$ if $\langle r, x \rangle$ has a pushout complement. Otherwise, let $T_p(\exists(x, a)) = $ false and $T_p(\forall(x, a)) = $ true. T_p is compatible with Boolean operations.

Proof. Let $G \Rightarrow_{p,m,m^*} H$ be any direct derivation. We show $m \models T_p(a) \Leftrightarrow m^* \models a$ for every application condition a. The proof is done by induction on the structure of application conditions. For basic right application conditions, the statement follows immediately from the statement in [1]. For conditional right application conditions of the form $Q(x, a)$ with $Q \in \{\forall, \exists\}$, two cases may occur:

Case 1. The pair $\langle r, x \rangle$ has no pushout complement. Then $T_p(\exists(x, a)) = $ false and $T_p(\forall(x, a)) = $ true. To show is $m \models$ false $\Leftarrow m^* \models \exists(x, a)$ and $m \models$ true $\Rightarrow m^* \models \forall(x, a)$, respectively. As no morphism satisfies false and every morphism satisfies true, it suffices to show $m^* \not\models \exists(x, a)$ and $m^* \models \forall(x, a)$. Both statements hold, because there is no $q : X \rightarrow H$ with $q \in M$ and $q \circ x = m^*$. Otherwise, since the pair $\langle r, m^* \rangle$ has a pushout complement, the pair $\langle r, x \rangle$ would have a pushout complement in contradiction to case 1.

Case 2. The pair $\langle r, x \rangle$ has a pushout complement. Then the left application condition is of the form $T_p(a) = Q(y, b)$. It remains to show that $m \models Q(y, b) \Leftrightarrow m^* \models Q(x, a)$. This is done by structural induction. We will use the following statements: [α] Given a morphism $q' : Y \rightarrow G$ in M with $q' \circ y = m$ we can construct pushouts (1), (2), (5), (6) as above, where this time we first construct (6) as pullback leading in the right-hand side to a morphism $q : X \rightarrow H$ in M

with $q \circ x = m^*$. [β] Given a morphism $q \colon X \to H$ in M with $q \circ x = m^*$. From the double pushout for $G \Rightarrow_{p,m,m^*} H$ and $q \circ x = m^*$ we obtain the following decomposition in pushouts (1), (2), (5), (6): First (5) is constructed as pullback of q and d_1 leading to pushouts (1) and (5), with same square (1) as in the construction because of uniqueness of pushout complements for M-morphisms. Then (2) is constructed as pushout and we have $q' \colon Y \to G$ with $q' \circ y = m$ and pushout (6) induced by the pushouts (2) and (2)+(6). Since q is in M, z and q' are in M. [γ] Given application conditions as above and a "derived" rule $p^* = \langle Y \leftarrow Z \to X \rangle$ with morphisms $q' \colon Y \to G$ and $q \colon X \to H$ we apply the inductive hypothesis to conclude $q' \models b$ iff $q \models a$. For an universal application condition we have: $m \models \forall(y, b)$ iff for all morphisms $q' \colon Y \to G$ in M with $q' \circ y = m$ holds $q' \models b$ if$^{[\alpha],[\gamma]}$ and only if$^{[\beta],[\gamma]}$ for all morphisms $q \colon X \to H$ in M with $q \circ x = m^*$ holds $q \models a$ iff $m^* \models \forall(x, a)$. For an existential application condition we have: $m \models \exists(y, b)$ iff for some morphisms $q' \colon Y \to G$ in M with $q' \circ y = m$ holds $q' \models b$ if$^{[\beta],[\gamma]}$ and only if$^{[\alpha],[\gamma]}$ for some morphisms $q \colon X \to H$ in M with $q \circ x = m^*$ holds $q \models a$ iff $m^* \models \forall(x, a)$. For Boolean right application conditions, the statement follows directly from the definition and the inductive hypothesis.

Example 5 (transformation of right into left application conditions). Consider the rule Move in example 3 and the right application condition $T(\mathrm{NoTwo}) = \wedge_{e \in \mathcal{E}} \neg \exists u$ of example 4. Then the transformation of the right application condition $T(\mathrm{NoTwo})$ according to Move yields the left application condition

$$T_{\mathsf{Move}}(T(\mathrm{NoTwo})) = \wedge_{e \in \mathcal{E}} \neg \exists v.$$

Given a morphism $u \colon R \to U$, we have to check whether the pair $\langle r, u \rangle$ has a pushout complement, and if so, to apply the inverse rule of Move according to $u \colon R \to U$ yielding a match $v \colon L \to V$. The result of the transformation of a subcondition of $T(\mathrm{NoTwo})$ is presented in figure 9. The left application condition is obtained from the transformation of the right subconditions. It says more or less that "the next piece of track is not allowed to be occupied by a train moving in the same direction".

Remark. Given a rule p and a left application condition a_L. Then there is a right application condition a_R such that, for all direct derivations $G \Rightarrow_{p,m,m^*} H$,

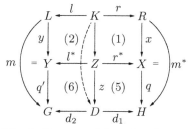

Fig. 8. Decomposition of pushouts.

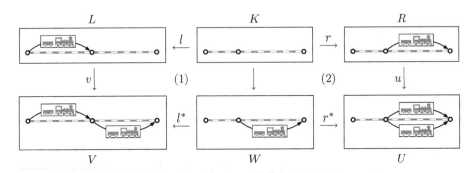

Fig. 9. Transformation of $T(\text{NoTwo})$ into a left application condition of Move.

$m^* \models a_R \Leftrightarrow m \models a_L$: Consider the inverse rule p^{-1} of p with right application condition a_L. Let $a_R = T_{p^{-1}}(a_L)$ be the left application condition of p^{-1}. Then a_R is a right application condition of p with the wanted property.

6   Application

We will use the transformation results for integrating constraints into left application conditions such that every direct derivation satisfying the application condition is constraint-guaranteeing resp. constraint-preserving.

Definition 5. Given a constraint c, a direct derivation $G \Rightarrow_{p,m,m^*} H$ is said to be *c-guaranteeing* if $H \models c$ and *c-preserving* if $G \models c$ implies $H \models c$.

Theorem 3 (guarantee and preservation of constraints). *Given a rule p and a constraint c, we can effectively construct a left application condition $a(c)$ [a'(c)] such that every direct derivation $G \Rightarrow_{p,m,m^*} H$ satisfying $a(c)$ [a'(c)] is c-guaranteeing [c-preserving].*

Proof. Let $T_L(c)$ [$T_R(c)$] be the left [right] application condition of the constraint c and $T_p(T_R(c))$ the left application condition of $T_R(c)$. Let $a(c) = T_p(T_R(c))$ and $a'(c) = T_L(c) \Rightarrow T_p(T_R(c))$ where $A \Rightarrow B$ denotes $\neg A \vee B$. Then the left application conditions have the wanted properties.

7   Conclusion

This paper is mainly based on [1]. It generalizes the notion of constraints and application conditions to nested ones and shows that nested constraints can be transformed into nested right and left application conditions. Furthermore, it presents an equivalence-preserving transformation that allows to eliminate superfluous subconstraints and subconditions, respectively. Further topics are:

- The extension of the underlying first-order logic by adding counting quantifiers as proposed in [9, 13],

- the transformation of specific application conditions into constraints and from constraints into constraints,
- a systematic study of the complexity of the transformation and a constructive, equivalence-preserving, simplifying transformation of constraints and application conditions because the integration of constraints into application conditions may yield a large conjunction/disjunction of application conditions.
- an application to typed attributed graph transformation [4] and graph-based specification of access control policies [10],
- an implementation of the transformation of nested constraints and application conditions (e.g. in the AGG tool [5]).

References

1. Hartmut Ehrig, Karsten Ehrig, Annegret Habel, and Karl-Heinz Pennemann. Constraints and application conditions: From graphs to high-level structures. In *Graph Transformations (ICGT'04)*, volume 3256 of *Lecture Notes in Computer Science*, pages 287–303. Springer-Verlag, 2004.
2. Hartmut Ehrig and Annegret Habel. Graph grammars with application conditions. In G. Rozenberg and A. Salomaa, editors, *The Book of L*, pages 87–100. Springer-Verlag, Berlin, 1986.
3. Hartmut Ehrig, Annegret Habel, Julia Padberg, and Ulrike Prange. Adhesive high-level replacement categories and systems. In *Graph Transformations (ICGT'04)*, volume 3256 of *Lecture Notes in Computer Science*, pages 144–160. Springer-Verlag, 2004.
4. Hartmut Ehrig, Ulrike Prange, and Gabriele Taentzer. Fundamental theory of typed attributed graph transformation. In *Graph Transformations (ICGT'04)*, volume 3256 of *Lecture Notes in Computer Science*, pages 161–177. Springer-Verlag, 2004.
5. Claudia Ermel, Michael Rudolf, and Gabriele Taentzer. The AGG approach: Language and environment. In *Handbook of Graph Grammars and Computing by Graph Transformation*, volume 2, pages 551–603. World Scientific, 1999.
6. Annegret Habel, Reiko Heckel, and Gabriele Taentzer. Graph grammars with negative application conditions. *Fundamenta Informaticae*, 26:287–313, 1996.
7. Annegret Habel, Jürgen Müller, and Detlef Plump. Double-pushout graph transformation revisited. *Mathematical Structures in Computer Science*, 11(5):637–688, 2001.
8. Reiko Heckel and Annika Wagner. Ensuring consistency of conditional graph grammars – a constructive approach. In *SEGRAGRA 95*, volume 2 of *Electronic Notes in Theoretical Computer Science*, pages 95–104, 1995.
9. Neil Immerman. Relational queries computable in polynomial time. *Information and Control*, 68(1-3):86–104, 1986.
10. Manuel Koch and Francesco Parisi-Presicce. Describing policies with graph constraints and rules. In *Graph Transformation (ICGT 2002)*, volume 2505 of *Lecture Notes in Computer Science*, pages 223–238. Springer-Verlag, 2002.
11. Stephen Lack and Paweł Sobociński. Adhesive categories. In *Proc. of Foundations of Software Science and Computation Structures (FOSSACS'04)*, volume 2987 of *Lecture Notes in Computer Science*, pages 273–288. Springer-Verlag, 2004.

12. Bernd Mahr and Anne Wilharm. Graph grammars as a tool for description in computer processed control: A case study. In *Graph-Theoretic Concepts in Computer Science*, pages 165–176. Hanser Verlag, München, 1982.
13. Mohamed Mosbah and Rodrigue Ossamy. A programming language for local computations in graphs: Logical basis. Technical report, University of Bordeaux, 2003.
14. Karl-Heinz Pennemann. Generalized constraints and application conditions for graph transformation systems. Master's thesis, University of Oldenburg, 2004.
15. Arend Rensink. Representing first-order logic by graphs. In *Graph Transformations (ICGT'04)*, volume 3256 of *Lecture Notes in Computer Science*, pages 319–335. Springer-Verlag, 2004.

Synthesis Revisited: Generating Statechart Models from Scenario-Based Requirements[*]

David Harel, Hillel Kugler, and Amir Pnueli

Department of Computer Science and Applied Mathematics,
The Weizmann Institute of Science, Rehovot, Israel
{dharel,kugler,amir}@wisdom.weizmann.ac.il

Abstract. Constructing a program from a specification is a long-known general and fundamental problem. Besides its theoretical interest, this question also has practical implications, since finding good synthesis algorithms could bring about a major improvement in the reliable development of complex systems. In this paper we describe a methodology for synthesizing statechart models from scenario-based requirements. The requirements are given in the language of live sequence charts (LSCs), and may be played in directly from the GUI, and the resulting statecharts are of the object-oriented variant, as adopted in the UML. We have implemented our algorithms as part of the Play-Engine tool and the generated statechart model can then be executed using existing UML case tools.

1 Introduction

Constructing a program from a specification is a long-known general and fundamental problem. Besides its theoretical interest, this question also has practical implications, since finding good synthesis algorithms could bring about a major improvement in the reliable development of complex systems.

Scenario-based inter-object specifications (e.g., via live sequence charts) and state-based intra-object specifications (e.g., via statecharts) are two complementary ways to specify behavioral requirements. In our synthesis approach we aim to relate these different styles for specifying requirements. In [10] the first two coauthors of this paper suggested a synthesis approach using the scenario-based language of live sequence charts (LSCs) [7] as requirements, and synthesizing a state-based object system composed of a collection of finite state machines or statecharts. The main motivation for suggesting the use of LSCs as a requirement language in [10] is its enhanced expressive power. LSCs are an extension of message sequence charts (MSCs; or their UML variant, sequence diagrams) for rich inter-object specification. One of the main additions in LSCs is the notion of universal charts and hot, mandatory behavior, which, among other things, enables one to specify forbidden scenarios. Synthesis is considerably harder for

---

[*] This research was supported in part by the John von Neumann Minerva Center for the Verification of Reactive Systems, by the European Commission project OMEGA (IST-2001-33522) and by the Israel Science Foundation (grant No. 287/02-1).

H.-J. Kreowski et al. (Eds.): Formal Methods (Ehrig Festschrift), LNCS 3393, pp. 309–324, 2005.

LSCs than for MSCs, and is tackled in [10] by defining consistency, showing that an entire LSC specification is consistent iff it is satisfiable by a state-based object system. A satisfying system is then synthesized.

There are several issues that have prevented the approach described in [10] from becoming a practical approach for developing complex reactive systems. A major obstacle is the high computational complexity of the synthesis algorithms, that does not allow scaling of the approach to large systems. Additional problems are more methodological, related to the level of detail required in the scenarios to allow meaningful synthesis, the problem of ensuring that the LSC requirements are exactly what the user intended, and a lack of tool support and integration with existing development approaches.

In this paper we revisit the idea of synthesizing statecharts from LSCs, with an aim of addressing the limitations of [10] mentioned above. Our approach benefits from the advances in research made since the publication of [10] – mainly the play-in/play-out approach [13], which supplies convenient ways to capture scenarios and execute them directly, and our previous work on smart play-out [11], which allows direct execution and analysis of LSCs using powerful verification techniques. We suggest a synthesis methodology that is not fully automatic but rather relies on user interaction and expertise to allow more efficient synthesis algorithms. One of the main principles we apply is that the specifier of the requirements provide enough detail and knowledge of the design to make the job easier for the synthesis algorithm. The algorithm tries to prove, using verification methods, that a certain synthesized model satisfies all requirements; if it manages to do so, it can safely synthesize the model. We have developed a prototype statechart synthesis environment, that receives as input LSCs from the Play-Engine tool [13] and generates a statechart model that can then be executed by RHAPSODY [15], and in principle also by other UML tools, see e.g., [24, 27].

The paper is organized as follows. Section 2 describes the main challenges in synthesizing statecharts from scenarios and the main principles we adopt to address them. Section 3 shows how to relate the object model of LSCs as supported by the Play-Engine tool with standard UML object models, and describes how this is supported by our prototype tool. Section 4 addresses the notion of consistency of LSCs and introduces a game view for synthesizing reactive systems. Section 5 describes our approach to statechart synthesis, while Section 6 explains the actual statechart synthesis using an example of a cellular phone system. We conclude with a discussion of related work in Section 7.

2   Main Challenges in Synthesis

In this section we discuss some of the main challenges that need to be addressed in order to make a method for synthesizing statechart models from scenarios successful. The challenges are of different nature, varying from finding a scenario-based language that is powerful and easy for engineers to learn, to dealing with the inherent computational complexity of synthesis algorithms that must handle large complex systems.

2.1   Appropriate Scenario-Based Language

An important usage of scenario notations is for communicating ideas and for documentation. For such purposes sketching an inter-object scenario on a blackboard or diagram editor can be very helpful. When our goal is synthesizing a statechart model and eventually production code from the scenarios, we need a powerful and expressive inter-object scenario-based language with rigorously defined semantics. The language should still retain the simplicity and intuitive feel that made scenario-based languages popular among engineers. In our approach we use the language of live sequence charts (LSCs) introduced in [7]. LSCs extends classical message sequence charts, which have very limited expressive power. Among other things, LSCs distinguish between behaviors that may happen in the system (existential) from those that must happen (universal). An example of a universal chart appears in Fig. 1. A universal chart contains a *prechart* (dashed hexagon), which specifies the scenario which, if successfully executed, forces the system to satisfy the scenario given in the actual chart body. For more details on LSCs see [7, 13, 14].

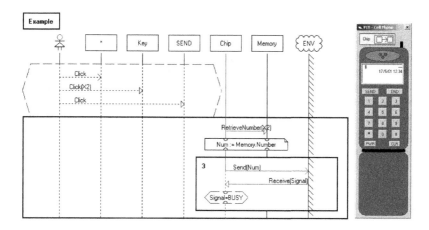

Fig. 1. Example of a universal LSC.

2.2   Sufficiently Detailed Scenario-Based Specification

Specifying requirements of a system is a very difficult task, which must be carried out in a careful and accurate manner. For this reason, it may be claimed that requirements in general, and scenario-based ones in particular, will only be partial and will focus on certain important properties and concepts of the system. According to this argument it is not possible to beneficially apply a synthesis approach for deriving a system implementation since the requirement model provides insufficient details.

We attempt to overcome this challenge by using the play-in/play-out approach introduced in [13, 14]. In play-in the user starts with a graphical representation of the system and specifies various scenarios by interacting with the GUI and demonstrating the required behavior. As this is being done, the Play-Engine tool constructs the LSC that captures what was played in. Play-in enables non-technical stake holders to participate in the requirement elucidation phase, and to contribute to building a detailed scenario model. Our experience in several projects [12] shows that the play-in/play-out approach enhanced to a large extent the efficiency of this process and allowed building rich and detailed scenario-based requirements, which can serve as a solid starting point for synthesis algorithms.

2.3   Correct Scenarios

As mentioned earlier, specifying requirements is a difficult job, and the user must be sure that the property specified is exactly what is intended. In the context of formal verification, many times when verifying a system with respect to a specified property the result shows that the system does not satisfy the property, and then the user realizes that the property specified was not exactly the intended one and refines it. In a synthesis approach the requirements themselves must be accurate, otherwise even if the synthesis algorithms work perfectly the obtained system will not be what was actually intended.

We try to address this challenge in several complementary ways. First, the requirement language of LSCs, being an extension of classical MSCs, has intuitive semantics, and allows users who are not very technical to express complex behavioral requirements, while other formalisms, e.g., temporal logic, may prove to be trickier even for advanced users. Second, play-out, the complementary process to play-in, allows one to execute the LSCs directly, giving a feeling of working with an executable system. This makes it possible to debug the requirements specification and gain more confidence that what is specified is exactly what is required.

2.4   The Complexity of Synthesis Algorithms

Solving the problem of synthesis for open reactive systems is an inherently difficult problem. In various settings the problem is undecidable, and even in more restricted settings when it is becomes decidable, the time and space requirements of the synthesis algorithm may be too large to be practical for large systems.

One way we attack this problem is by applying methods from formal verification, in ways that will be discussed below. We have in mind mainly model-checking algorithms, which in recent years – due to intensive research efforts and tool development – have scaled nicely in terms of the size of the models they can handle. Nevertheless, the models that can be treated even using state of the art technologies are still limited in size and much more work is needed here to make synthesis a practical approach.

In our current work, one of the main principles we apply is that the specifier of the requirements provide enough detail and knowledge of the design to make the job easier for the synthesis algorithm. The algorithm tries to prove that a certain synthesized model satisfies all requirements; if it manages to do that it can safely synthesize the model. This approach is not complete, since some other model may be correct and the synthesis algorithm will fail to find it. However, our hope is that for many interesting cases the synthesis will succeed.

2.5   Integration with Existing Code and System Modification

In order to make a new system development approach practical, an important requirement is that it should fit in nicely with other existing approaches. In our context of designing complex embedded software, the synthesized statechart-based model may need to interact with other software that was developed in other diverse ways. By synthesizing into a UML-based framework, we attempt to address this issue and thus to take advantage of the integration capabilities of existing commercial UML tools.

Related to this issue is our recent work on InterPlay [3]. InterPlay is a simulation engine coordinator that supports cooperation and interaction of multiple simulation and execution tools. It makes it possible to connect several Play-Engines to each other, and also to connect a statechart-based executable model in RHAPSODY to the Play-Engine. A model synthesized using algorithms described in this paper can thus be linked to the Play-Engine, allowing the scenarios to be monitored as they occur. It also supports an environment in which some subsystems run a statechart or code-based model and others execute LSCs directly, say, by play-out.

3   Transferring the Structure

Scenario-based inter-object specifications (via LSCs) and state-based intra-object specifications (via statecharts) are two complementary ways for specifying behavioral requirements. In our synthesis approach we aim to relate these different styles for specifying requirements.

According to the play-in/play-out approach the user specifies behavioral requirements by playing on a GUI representation of the system, as this is being done the Play-Engine automatically constructs corresponding requirements in LSCs.

3.1   The Play-Engine Object Model

We now introduce the object model used by the Play-Engine, which is the basis for the LSC specifications. We later explain how this object model is related to standard UML models, allowing our prototype tool to connect to models in existing UML tools, and allowing to synthesize statechart-based UML models. For a detailed explanation of the Play-Engine framework and object model see [13, 14]. An object system Sys is defined as

$$Sys = \langle \mathcal{D}, \mathcal{C}, \mathcal{O}, \mathcal{F} \rangle$$

where $\mathcal{D}$ is the set of application types (domains), $\mathcal{C}$ is the set of classes, $\mathcal{O}$ is the set of objects, $\mathcal{F}$ is the set of externally implemented functions. We refer to the user of the system as $User$ and to the external environment as Env.

A type $D \in \mathcal{D}$ is simply a (finite) set of values. The basic types supported are range, enumeration and string.

A class C is defined as:

$$C = \langle Name, \mathcal{P}, \mathcal{M} \rangle$$

where $Name$ is the class name, $\mathcal{P}$ is the set of class properties and $\mathcal{M}$ is the set of class methods.

An object O is defined as:

$$O = \langle Name, C, \mathcal{PV}, External \rangle$$

where $Name$ is the object's name, C is its class, $\mathcal{PV} : C.\mathcal{P} \to \bigcup_i D_i$ is a function assigning a value to each of the object's properties and $External$ indicates whether the object is an external object. We define the function $class : \mathcal{O} \to \mathcal{C}$ to map each object to the class it is an instance of. We also use $Value(O.P) = O.\mathcal{PV}(O.C.P)$ to denote the current value of property P in object O.

An object property P is defined as

$$P = \langle Name, D, InOnly, ExtChg, Affects, Sync \rangle$$

where $Name$ is the property name and D is the type it is based on. $InOnly \in \{True, False\}$ indicates whether the property can be changed only by the user, $ExtChg \in \{True, False\}$ indicates whether the property can be changed by the external environment, $Affects \in \{User, Env, Self\}$ indicates the instance to which the message arrow is directed when the property is changed by the system, and $Sync \in \{True, False\}$ indicates whether the property is synchronous.

An object method M is defined as:

$$M = \langle Name(D_1, D_2, \ldots, D_n), Sync \rangle$$

where Name is the method name, $D_i \in \mathcal{D}$ is the type of its ith formal parameter and $Sync \in \{true, false\}$ indicates whether calling this method is a synchronous operation.

An implemented function is defined as:

$$Func = Name : D_1 \times D_2 \times \ldots, \times D_n \to D_F$$

where $Name$ is the function name, $D_i \in \mathcal{D}$ is the type of its ith formal parameter and $D_F \in \mathcal{D}$ is the type of its returned value.

3.2 Importing a UML Model into the Play-Engine

The usual work-flow in the play-in/play-out approach as supported by the Play-Engine is that the user starts by building a GUI representation and the corresponding object model. As part of our current work we support an alternative starting point, in which a UML model is imported into the Play-Engine, (say, from RHAPSODY), and can then be used while specifying the behavior using

LSCs and the play-in process. This shows the relation between the Play-Engine object model and a standard UML model, and also from the more practical point of view it provides an easy link to models developed in existing UML tools and a good starting point for applying our synthesis approach.

The import procedure is quite straightforward, we describe here only its general principles. Types in the UML model are converted to Play-Engine types, as defined in the previous section. Currently the Play-Engine supports only simple type definitions – range, enumeration and string. A type that cannot be defined in terms of these basic type definitions is declared as `EngineVariant`, the default Play-Engine type. The Play-Engine currently does not support packages, the UML construct for grouping classes, so that when importing UML classes they all appear in a flat structure. UML attributes are mapped to Play-Engine properties, preserving their corresponding type. For each UML class, the operations are imported as Play-Engine methods, with the arguments preserving their corresponding types.

Instances in the UML model are defined as internal objects, preserving their base class. In the Play-Engine, internal objects are visualized using something resembling class diagrams, and play-in is supported by clicking and manipulating this kind of diagram in a convenient way. This allows rapid development of requirements without a need to construct a GUI. Building a GUI has many benefits in terms of visualizing the behavior, but as a first approximation importing the model and playing-in using internal objects works fine.

3.3   Synthesizing a Skeleton UML Model

Complementary to the UML to Play-Engine import described in the previous subsection, we also support the synthesis of a skeleton UML model from the Play-Engine; that is, a UML model containing the object model definitions, but without taking the LSC specifications into account and without synthesizing any statecharts. This skeleton synthesis can be useful if we have a complex Play-Engine model we have developed, and now want to go ahead and build a corresponding UML model. We can apply the synthesis of the skeleton model, thus automating the straightforward part, and then proceed to do the interesting and creative part, regarding dynamic behavior, by defining the UML statecharts manually. We may want to use this approach when we have special motivation to create the statechart model manually (see, e.g., [9] for an example), or when the automatic synthesis algorithms do not work properly. Using the InterPlay approach [3] mentioned earlier, we can then execute the statechart-based UML model linked to the Play-Engine, allowing the scenarios to be monitored as they occur.

4   Consistency of LSCs

Before being able to synthesize a statechart based model we must ensure that the LSCs are consistent. Consider the two charts `OpenAntGrad1` and `OpenAntGrad2` in Fig. 2. When the user opens the Antenna both charts are activated. However,

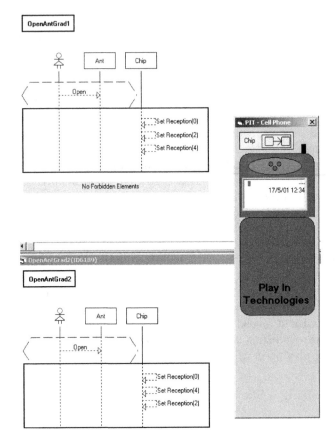

Fig. 2. Inconsistent LSCs.

there is no way to satisfy them both since after changing the reception level of the Chip to 0 (as required by both charts), the first chart requires that the reception level change to 2 and only later to 4, while the second one requires that the reception level change first to 4 and only later to 2. These are clearly contradictory. While this is a very simple example, such contradictions can be a lot more subtle, arising as a result of the interaction between several charts. In large specifications this phenomena can be very hard to analyze manually. Our tool can automatically detect some of these inconsistencies and provide information to the user. After the relevant LSCs are fixed the synthesis algorithm can again be applied. This can lead to an iterative development process at the end of which a consistent LSC specification is obtained, and a statechart model can be synthesized.

4.1 A Game View

In the study of synthesis of reactive systems a common view is that of a game between two players [6, 22]. One of the players is the environment and the other is the system. The players alternate turns each one making a move in his turn, and the requirements define the winning condition. If there exists a strategy for the system under which for any moves the environment makes the system always wins, we say the specification is realizable (consistent) and we can attempt to synthesize a system implementation.

In the Play-Engine tool while using LSCs as the requirement language, the environment can be a **User** object, as in the prechart of Fig. 1, or a more explicit environment object, as represented by the **ENV** object appearing in the main chart of Fig. 1, or an external object, an object that is designated as being implemented outside the Play-Engine specification. In principle, the clock object, which represents global time, should also be considered external, but the treatment of time is beyond the scope of this paper. All other objects are assumed to be part of the system.

The game is played as follows: the environment makes a move, consisting of performing a method call or modifying the value of an externally changeable property. The system responds by performing a superstep, a finite sequence of system events, and then it is again the environment's turn. The system is the winner of the game if all LSC requirements are satisfied, otherwise the environment is the winner.

For the finite state case, when the number of objects is finite, all types are of finite domain, and the number of different simultaneously active copies of a chart is bounded, the game can be solved using model-checking methods. An implementation of the game problem is now part of the Weizmann Institute model-checker TLV [23]. The computation complexity of the algorithms is still a major limitation in applying this game approach.

In our current work, one of the main principles we apply is that the specifier of the LSCs provide enough detail and knowledge of the design, to make the job easier for the synthesis algorithm. LSCs as a declarative, inter-object behavior language, enables formulating high level requirements in pieces (e.g., scenario fragments), leaving open details that may depend on the implementation. The partial order semantics among events in each chart and the ability to separate scenarios in different charts without having to say explicitly how they should be composed are very useful in early requirement stages, but can cause under-specification and nondeterminism when one attempts to execute them.

In play-out, if faced with nondeterminism an arbitrary choice is made. From our experience in several projects, by providing a detailed enough LSC requirement play-out can get very close to solving the game problem, and sometimes can even solve it directly. Assuming the user provided enough knowledge for the synthesis algorithm, the algorithm tries to prove that a certain synthesized model will satisfy all requirements, and if it manages to do this it can safely synthesize the model. This approach is not complete, thus a different synthesized model may be correct and the synthesis algorithm may fail to find it, but our hope is that for many interesting cases the synthesis will succeed. In a situation where for a synthesized model we have not managed to prove it correct or to find some problem with it, synthesizing a state-based model opens possibilities to try to prove its correctness using other tools and techniques, e.g., [25, 2].

5   The Synthesis Approach

In order to apply the synthesis approach we encode play-out in the form of a transition system and then apply model-checking techniques. We construct a transition system which has one process for each actual object. A state in this system indicates the currently active charts and the location of each object in these charts. The transition relation restricts the transitions of each process only to moves that are allowed by all currently active charts. We now provide some more of the details on how to translate LSCs to a transition system. The encoding of the transition relation was developed as part of our work on smart play-out [11].

An LSC specification LS consists of a set of charts M, where each chart $m \in M$ is existential or universal. We denote by $pch(m)$ the prechart of chart m. Assume the set of universal charts in M is $M^U = \{m_1, m_2, ..., m_t\}$, and the objects participating in the specification are $\mathcal{O} = \{O_1, ..., O_n\}$.

We define a system with the following variables:

act_{m_i} determines if universal chart m_i is active. It gets value 1 when m_i is active and 0 otherwise.

$msg^s_{O_j \to O_k}$ denoting the sending of message msg from object O_j to object O_k. The value is set to 1 at the occurrence of the send and is changed to 0 at the next state.

$msg^r_{O_j \to O_k}$ denoting the receipt by object O_k of message msg sent by object O_j. Similarly, the value is 1 at the occurrence of the receive and 0 otherwise.

l_{m_i, O_j} denoting the location of object O_j in chart m_i, ranging over $0 \cdots l^{max}$ where l^{max} is the last location of O_j in m_i.

$l_{pch(m_i), O_j}$ denoting the location of object O_j in the prechart of m_i, ranging over $0 \cdots l^{max}$ where l^{max} is the last location of O_j in $pch(m_i)$.

We use the asynchronous mode, in which a send and a receive are separate events, but we support the synchronous mode too. The details of encoding the transition relation are rather technical, for more information see [11].

Given this encoding we claim that play-out is correct if the following property holds.

$$\neg(EF(AG(\bigvee_{m_i \in M^U} (act_{m_i} = 1))))$$

The property specified above is a temporal logic property [8]. The operators E, A are the existential and universal path quantifiers respectively, while F and G are the eventually and always temporal logic operators. Intuitively, this formula claims that it is not the case that eventually play-out may get stuck, not being able to satisfy the requirements successfully.

We now apply the model-checker to prove this property, and if it is indeed correct we can go on and synthesize the system. The basic synthesis scheme generates a statechart for each of the participating objects, using orthogonal states for implementing different scenarios and making use of additional events to guarantee synchronization of the distributed objects along each behavioral scenario. More details are given in the next section. If the property does not hold we can apply model-checking to a variation of this property and can sometimes obtain more information on how the LSCs can be fixed so that play-out will be correct.

6   An Example of Statechart Synthesis

We use an example of a cellular phone system to illustrate our synthesis algorithms. A GUI representation of the system appears on the right-hand side of Fig. 2. The system is composed of several objects, including the **Cover**, **Display**, **Antenna** and **Speaker**. We consider a specification consisting of several universal charts.

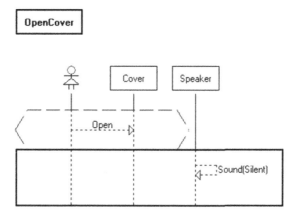

Fig. 3. Open Cover.

The chart **OpenCover**, appearing in Fig. 3, requires that whenever the user opens the **Cover**, as specified in the prechart, the **Speaker** must turn silent.

The charts **OpenAnt**, **CloseAnt**, appearing in Fig. 4, specify that whenever the user opens the **Antenna** the **Display** shows that the reception level is changed to 4, and whenever the user closes the **Antenna** the **Display** shows that the reception level is changed to 1.

The resulting statecharts for the **Antenna** and the **Display** obtained by applying the synthesis algorithms appear in Fig. 5 and Fig. 6 respectively. Consider the **Antenna** statechart of Fig. 5. The AND-**state** named Top contains two orthogonal states $OpenAnt$ and $CloseAnt$, corresponding to the scenarios of opening and closing of the **Antenna**.

The orthogonal state $OpenAnt$ has three substates, $P0, P1$ and $S0$, where $P0$ is the initial state entered, as designated by the default transition into $P0$.

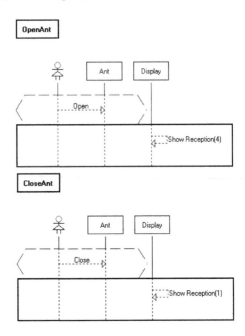

Fig. 4. Opening and Closing the Antenna.

The states $P0, P1$ and $S0$ correspond to progress of the `Antenna` object along the *OpenAnt* scenario, where we use the convention that P states correspond to prechart locations while S states correspond to main chart locations. If the `Antenna` object is in state $P0$ of the *OpenAnt* orthogonal component, and it receives the event `Open`, it takes a transition to state $P1$ and performs the action written in the label of the transition. The action has the effect of telling the other objects that the scenario of opening of the `Antenna` has been activated. This is done by sending the event `activeOpenAnt` to the other objects, i.e., the command `getItsDisplay_C()->GEN(activeOpenAnt)` generates an event `activeOpenAnt` and sends it to the `Display` object. In a similar way the event `activeOpenAnt` is generated and sent to the `Cover` and `Speaker` objects. The `Antenna` object, which is now in the sub-state $P1$ of the *OpenAnt* component, takes the null transition to state $S0$. Null transitions are transitions with no trigger event, and are taken spontaneously.

The `Display` object is originally in state $P0$ of the *OpenAnt* orthogonal state. It receives the event `activeOpenAnt` (sent by the `Antenna`), causing the transition to state $S0$ to be taken, meaning that now the object has progressed to the main chart of the scenario. From state $S0$ a null transition to state $S1$ is taken, and the reception level of the `Display` is set to volume level 4. This is done by performing the method `setReception(V_4)`, which sets the value of the attribute `reception` to `V_4`. As part of the action of the transition from state $S0$ to state $S1$ the other objects are notified that the scenario of opening of the `Antenna` is over, this is done by performing the command `getItsAnt_C()->GEN(overOpenAnt)`, and similarly for other objects. The `Display` object then takes the null transition

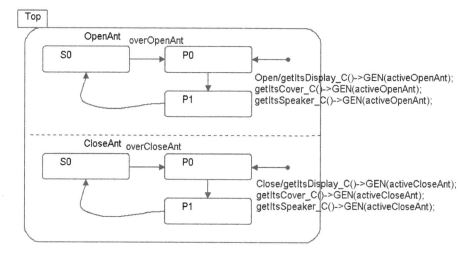

Fig. 5. Synthesized Antenna statechart.

back to state $P0$. The `Antenna` object on receiving the event `overOpenAnt` takes the transition from state $S0$ back to state $P0$. At this point the scenario of opening the `Antenna` has completed successfully. The statechart synthesis algorithm implements the scenario of closing the `Antenna` in a similar way, as reflected by the *CloseAnt* components of the `Antenna` and `Display` objects.

There are several points were the synthesis algorithm can be optimized to produce more efficient and readable models, and indeed we have a first version of such an improved algorithm. When sending an event to all other objects to notify them of some occurrence (for example when taking the transition from state $P0$ to state $P1$ in orthogonal component *OpenAnt* of the `Antenna`) it is enough in our case to send the event only to the `Display` object, since the objects `Cover` and `Speaker` do not participate and are not affected by the opening `Antenna` scenario.

A related issue is the architecture of the synthesized model: In this example, we allow each object to communicate directly with each of the other objects in the system, and we synthesize the relations in the UML model to allow this. Thus, for example, the `Antenna` object can relate to the `Speaker` by the `getItsSpeaker_C()` command. Using an optimized algorithm, if this communication is not used the corresponding relations will not be synthesized. For improved readability, if an action contains several commands of similar nature, e.g., sending an event to various objects, an optimized synthesis algorithm will define a method performing these related commands, and the label of the transition will include a call to this method, thus resulting in more readable and elegant statecharts than those of Fig. 5 and 6.

As mentioned earlier, an important part of the synthesis is to apply the play-out consistency check, as described in Section 5, which guarantees the correctness of the synthesis algorithm.

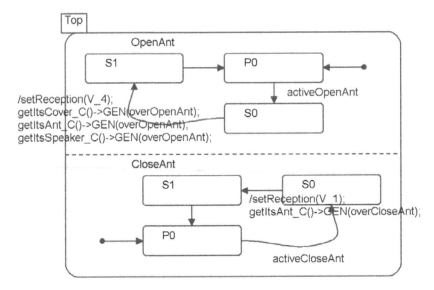

Fig. 6. Synthesized Display statechart.

7 Related Work

The idea of deriving state-based implementations automatically from scenario-based requirements has been the subject of intensive research efforts in recent years; see, e.g., [17, 18, 20, 19, 29]. Scenario-based specifications are very useful in early stages of development, they are used widely by engineers, and a lot of experience has been gained from their being integrated into the MSC ITU standard [21] and the UML [28]. The latest versions of the UML recognized the importance of scenario-based requirements, and UML 2.0 sequence diagrams have been significantly enhanced in expressive capabilities, inspired by the LSCs of [7].

There is also relevant research on statechart synthesis. As far as the case of classical message sequence charts goes, work on synthesis includes the SCED method [17] and synthesis in the framework of ROOM charts [20]. Other relevant work appears in [4, 26, 1, 19, 29]. In addition, there is the work described in [16], which deals with LSCs, but synthesizes from a single chart only: an LSC is translated into a timed Büchi automaton (from which code can be derived).

While the work in [10, 5] addressed the synthesis problem of LSCs from a theoretical viewpoint, the current paper applies new verification-based techniques and also reports on a prototype implementation. Other aspects special to our approach were described in Section 2 above. In addition to synthesis work directly from sequence diagrams of one kind or another, one should realize that constructing a program from a specification is a long-known general and fundamental problem. For example, there has been much research on constructing a program from a specification given in temporal logic (e.g., [22]).

References

1. R. Alur and M. Yannakakis. Model checking of message sequence charts. In *10th International Conference on Concurrency Theory (CONCUR99)*, volume 1664 of *Lect. Notes in Comp. Sci.*, pages 114–129. Springer-Verlag, 1999.
2. T. Arons, J. Hooman, H. Kugler, A. Pnueli, and M. van der Zwaag. Deductive Verification of UML Models in TLPVS. In *Proc. 7th International Conference on UML Modeling Languages and Applications (UML 2004)*, Lect. Notes in Comp. Sci., pages 335–349. Springer-Verlag, October 2004.
3. D. Barak, D. Harel, and R. Marelly. InterPlay: Horizontal Scale-Up and Transition to Design in Scenario-Based Programming. In *Lectures on Concurrency and Petri Nets*, volume 3098 of *Lect. Notes in Comp. Sci.*, pages 66–86. Springer-Verlag, 2004.
4. A.W. Biermann and R. Krishnaswamy. Constructing programs from example computations. *IEEE Trans. Softw. Eng.*, SE-2:141–153, 1976.
5. Y. Bontemps and P.Y. Schobbens. Synthesizing open reactive systems from scenario-based specifications. In *Proc. of the 3rd Int. Conf. on Application of Concurrency to System Design (ACSD'03)*. IEEE Computer Science Press, 2003.
6. J.R. Buchi. State-strategies for games in $F_{\sigma\delta} \cap G_{\delta\sigma}$. *J. Symb. Logic*, 48:1171–1198, 1983.
7. W. Damm and D. Harel. LSCs: Breathing life into message sequence charts. *Formal Methods in System Design*, 19(1):45–80, 2001. Preliminary version appeared in Proc. 3rd IFIP Int. Conf. on Formal Methods for Open Object-Based Distributed Systems (FMOODS'99).
8. E.A. Emerson. Temporal and modal logics. In J. van Leeuwen, editor, *Handbook of theoretical computer science*, volume B, pages 995–1072. Elsevier, 1990.
9. J. Fisher, D. Harel, E.J.A. Hubbard, N. Piterman, M.J. Stern, and N. Swerdlin. Combining state-based and scenario-based approaches in modeling biological systems. In *Proc. 2nd Int. Workshop on Computational Methods in Systems Biology (CMSB 2004)*, Lect. Notes in Comp. Sci. Springer-Verlag, 2004.
10. D. Harel and H. Kugler. Synthesizing state-based object systems from LSC specifications. *Int. J. of Foundations of Computer Science (IJFCS).*, 13(1):5–51, Febuary 2002. (Also,*Proc. Fifth Int. Conf. on Implementation and Application of Automata (CIAA 2000)*, July 2000, Lecture Notes in Computer Science, Springer-Verlag, 2000.).
11. D. Harel, H. Kugler, R. Marelly, and A. Pnueli. Smart play-out of behavioral requirements. In *Proc. 4th Intl. Conference on Formal Methods in Computer-Aided Design (FMCAD'02), Portland, Oregon*, volume 2517 of *Lect. Notes in Comp. Sci.*, pages 378–398, 2002. Also available as Tech. Report MCS02-08, The Weizmann Institute of Science.
12. D. Harel, H. Kugler, and G. Weiss. Some Methodological Observations Resulting from Experience Using LSCs and the Play-In/Play-Out Approach. Tech. Report MCS04-06, The Weizmann Institute of Science, 2004.
13. D. Harel and R. Marelly. *Come, Let's Play: Scenario-Based Programming Using LSCs and the Play-Engine*. Springer-Verlag, 2003.
14. D. Harel and R. Marelly. Specifying and Executing Behavioral Requirements: The Play In/Play-Out Approach. *Software and System Modeling (SoSyM)*, 2(2):82–107, 2003.
15. Rhapsody. I-Logix, Inc., products web page. http://www.ilogix.com/products/.

16. J. Klose and H. Wittke. An automata based interpretation of live sequence chart. In *Proc. 7th Intl. Conference on Tools and Algorithms for the Construction and Analysis of Systems (TACAS'01), volume 2031 of Lect. Notes in Comp. Sci., Springer-Verlag*, 2001.

17. K. Koskimies and E. Makinen. Automatic synthesis of state machines from trace diagrams. *Software – Practice and Experience*, 24(7):643–658, 1994.

18. K. Koskimies, T. Mannisto, T. Systa, and J. Tuomi. SCED: A Tool for Dynamic Modeling of Object Systems. Tech. Report A-1996-4, University of Tampere, July 1996.

19. I. Krüger, R. Grosu, P. Scholz, and M. Broy. From MSCs to Statecharts. In *Proc. Int. Workshop on Distributed and Parallel Embedded Systems (DIPES'98)*, pages 61–71. Kluwer Academic Publishers, 1999.

20. S. Leue, L. Mehrmann, and M. Rezai. Synthesizing ROOM models from message sequence chart specifications. Tech. Report 98-06, University of Waterloo, April 1998.

21. ITU-TS Recommendation Z.120 (11/99): MSC 2000. ITU-TS, Geneva, 1999.

22. A. Pnueli and R. Rosner. On the synthesis of a reactive module. In *Proc. 16th ACM Symp. Princ. of Prog. Lang.*, pages 179–190, 1989.

23. A. Pnueli and E. Shahar. A platform for combining deductive with algorithmic verification. In R. Alur and T. Henzinger, editors, *R. Alur and T. Henzinger, editors, Proc. 8th Intl. Conference on Computer Aided Verification (CAV'96), volume 1102 of Lect. Notes in Comp. Sci., Springer-Verlag*, pages 184–195, 1996.

24. Rational Rose Technical Developer. Rational, Inc., web page. http://www-306.ibm.com/software/awdtools/developer/technical/.

25. I. Schinz, T. Toben, and B. Westphal. The Rhapsody UML Verification Environment. In *2nd Int. Conf. on Software Engineering and Formal Methods*. IEEE Computer Society Press, 2004.

26. R. Schlor and W. Damm. Specification and verification of system-level hardware designs using timing diagram. In *European Conference on Design Automation*, pages 518–524, Paris, France, 1993. IEEE Computer Society Press.

27. Telelogic TAU. Telelogic, Inc., web page. http://www.telelogic.com/products/tau/.

28. UML. Documentation of the unified modeling language (UML). Available from the Object Management Group (OMG), http://www.omg.org.

29. J. Whittle and J. Schumann. Generating statechart designs from scenarios. In *22nd International Conference on Software Engineering (ICSE 2000)*, pages 314–323. ACM Press, 2000.

Main Concepts of Networks
of Transformation Units
with Interlinking Semantics*

Dirk Janssens[1], Hans-Jörg Kreowski[2], and Grzegorz Rozenberg[3]

[1] University of Antwerp,
Department of Mathematics and Computer Science,
Antwerp, Belgium
Dirk.Janssens@ua.ac.be
[2] University of Bremen,
Department of Mathematics and Computer Science,
Bremen, Germany
kreo@tzi.de
[3] Leiden University,
Leiden Institute of Advanced Computer Science,
Leiden, The Netherlands
rozenber@liacs.nl

Abstract. The aim of this paper is to introduce a modelling concept and structuring principle for rule-based systems the semantics of which is not restricted to a sequential behavior, but can be applied to various types of parallelism and concurrency. The central syntactic notion is that of a transformation unit that encapsulates a set of rules, imports other transformation units, and regulates the use and interaction of both by means of a control condition. The semantics is given by interlinking the applications of rules with the semantics of the imported units using a given collection of semantic operations. As the main result, the interlinking semantics turns out to be the least fixed point of the interlinking operator. The interlinking semantics generalizes the earlier introduced interleaving semantics of rule-based transformation units, which is obtained by the sequential composition of binary relations as only semantic operation.

1 Introduction

In this paper, we introduce networks of transformation units with interlinking semantics as a modelling concept and structuring principle for rule-based systems the semantics of which may be non-sequential. The key concept is a transformation unit encapsulating a set of local rules and importing other transformation units. Moreover, each transformation unit has a control condition that regulates

---

* Research partially supported by the EC Research Training Network SegraVis (Syntactic and Semantic Integration of Visual Modeling Techniques) and by the German Research Foundation (DFG) as part of the Collaborative Research Centre 637 *Autonomous Cooperating Logistic Processes – A Paradigm Shift and its Limitations*.

H.-J. Kreowski et al. (Eds.): Formal Methods (Ehrig Festschrift), LNCS 3393, pp. 325–342, 2005.

the application of the local rules and the interaction with the imported components. If a set of transformation units is closed under import, its import structure forms a network. In this way, large sets of rules can be organized and structured in such a way that each local unit may contain only a small set of rules while the effects of other units can be used by importing them.

In [4–7] transformation units have been introduced for graphs as well as for more general configurations as underlying data structures and provided with a purely sequential semantics. It is obtained by interleaving rule applications and the semantics of the imported components in such a way that the control condition is obeyed. In this paper, we generalize the framework of transformation units such that also non-sequential systems can be specified. For this purpose, we replace the underlying domain of binary relations on configurations by a domain of more general semantic entities and the sequential composition of binary relations by a set of arbitrary operations on semantic entities. But to be able to use set-theoretic operations and their properties, we assume that the domain of semantic entities is the power set of a set of semantics items.

The operations on semantic entities may be chosen as sequential, parallel or concurrent compositions or as any other operations one wants to use to model the type of semantics one is interested in. We show that the new framework covers nicely elementary net systems with their non-sequential processes as well as rule-based systems with sequential and parallel derivations, covering Chomsky grammars and various types of graph grammars in particular. This means that not only these approaches can be seen in a unified framework, but are also provided with a common structuring principle as a novel feature.

The paper is organized in the following way. In the next section, the basic notions and notations of transformation units with interlinking semantics are introduced. Networks of transformation units and their iterated interlinking semantics are studied in Section 3. Finally, the main result of this paper is formulated in Section 4. It states that the iterated interlinking semantics is the least fixed point of the interlinking operator if the used semantic operations and the control conditions are continuous. As running examples, we discuss elementary net systems and binary relations as semantic entities of grammatical systems of various kinds. Because of lack of space, the proofs are omitted.

2   Transformation Units with Interlinking Semantics

In this section, we introduce the notion of transformation units with interlinking semantics, which generalizes the formerly defined interleaving semantics.

The basis is the notion of a semantic domain (2.1) consisting of a set of semantic items together with operations on semantic entities being sets of semantic items. Typical semantic items are derivations, computations, and processes; typical operations are sequential and parallel compositions of derivations, computations, and processes or their embedding into larger context. To be able to deal with semantic entities, a semantic domain is first equipped with rules yielding a rule base (2.2) where a rule is some abstract syntactic feature that specifies a semantic entity. In many examples, rules rewrite some kind of configurations defining direct derivations and computation steps or rules are actions and events that describe elementary processes. Therefore, a set of rules provides a set of semantic entities the union of which may be closed under the operations of the semantic domain. For example, if one applies the sequential composition to direct derivations and computation steps, one gets all derivations and computations resp. Or if one applies certain kinds of parallel composition to elementary processes, one obtains all parallel processes of a set of actions or events. Often this is not enough to describe the behavior of a system. In addition, one may like to choose certain initial and terminal configurations or to regulate the rule applications by imposing a certain order or in some other way. For this purpose, a rule base is additionally equipped with control conditions (2.3) that allow one to restrict semantic effects. Formally, a control condition specifies a semantic entity depending on some environment which associates semantic entities to a given set of identifiers. The idea of this is the following. A control condition as a syntactic feature may use the identifiers to demand or forbid the applications of certain operations to the semantic entities associated to the identifiers and may restrict the free operational closure of these entities in this way.

2.1   Semantic Domains

While interleaving semantics is based on the sequential composition of binary relations, the generalization employs an arbitrary set of operations on arbitrary semantic entities. But to keep the technicalities simple, we assume that the semantic entities are the subsets of a set of semantic items such that we have union, intersection, inclusion and all other set-theoretic operations and all their properties for free.

A *semantic domain* $\mathcal{D} = (X, OP)$ consists of a set X of *semantic items* and a set OP of (partial) *operations* on the power set 2^X of X.

The arities of the operations can vary. The set of operations with arity $k \in \mathbb{N}$ is denoted by OP_k. The elements of 2^X are called *semantic entities*.

Such a semantic domain provides the operational closure for every set of semantic entities, which can be defined in the usual recursive way.

Let $M \subseteq 2^X$. Then the *operational closure* of M, $OP^*(M) \subseteq 2^X$, is recursively defined by

(i) $M \cup OP_0 \subseteq OP^*(M)$, and
(ii) $op(t_1, \ldots, t_k) \in OP^*(M)$ for $op \in OP_k$ and $t_1, \ldots, t_k \in OP^*(M)$.

Starting from the nullary operations and the given semantic entities, the operations are applied repeatedly to all semantic entities that are obtained in this way ad infinitum. The operational closure yields a set of subsets. If one wants to consider the union of them, this may be denoted by $\bigcup OP^*(M)$.

Examples

As running examples, we discuss elementary net systems with processes as semantic items (see, e.g., [3, 8, 13]) and binary relations on configurations like words and graphs as the semantic entities of grammatical rules (see, e.g., [12, 9]).

Elementary Net Systems. Let $\overline{N} = (\overline{B}, \overline{E}, \overline{F})$ be some contact-free elementary net where $\overline{B}$ is a set of *conditions*, $\overline{E}$ is a set of *events*, and $\overline{F} \subseteq (\overline{B} \times \overline{E}) \cup (\overline{E} \times \overline{B})$ is a *flow relation*. Then one may consider all processes on $\overline{N}$ as semantic items. More formally, $PROC(\overline{N})$ is the set of all pairs $proc = (N, p)$ where $N = (B, E, F)$ is an occurrence net, i.e. an acyclic and conflict-free net, and $p : N \to \overline{N}$ is a net morphism which is injective on cuts.

In particular, each acyclic and conflict-free subnet N of $\overline{N}$ together with the inclusion $incl_N : N \to \overline{N}$ yields a process. As a case $C \subseteq \overline{B}$ can be seen as a subnet $sub(C) = (C, \emptyset, \emptyset)$ with the empty set of events and the empty flow relation, C induces a particular process $proc(C) = (sub(C), incl_{sub(C)})$. In this way, the set of cases can be seen as a subset of the set of processes. Moreover, an event $e \in \overline{E}$ (together with its pre- and post conditions) induces a subnet $sub(e) = ({}^{\bullet}e \cup e^{\bullet}, \{e\}, ({}^{\bullet}e \times \{e\}) \cup (\{e\} \times e^{\bullet}))$ which is acyclic due to the contact-freeness of $\overline{N}$ and conflict-free by definition. Hence each event provides an elementary process $proc(e) = (sub(e), incl_{sub(e)})$.

There are two natural binary operations on processes: parallel and sequential compositions. Given two processes $proc = (N, p)$ and $proc' = (N', p')$ with $p(N) \cap p(N') = \emptyset$, then the parallel process is given by $proc + proc' = (N + N', <p, p'>)$ where $N + N'$ is the disjoint union of N and N' and $<p, p'>$ is the induced net morphism defined as p on N and as p' on N'.

To define the sequential composition, we need the notion of input and output conditions of an occurrence net N. The set of conditions with indegree 0 is denoted by $in(N)$ and the set of conditions of outdegree 0 by $out(N)$. Then the sequential composition of two processes $proc = (N, p)$ and $proc' = (N', p')$ requires that $p(out(N)) = p'(in(N'))$ and that there is no further overlap between $p(N)$ and $p'(N')$. The result is given $proc \circ proc' = (N + N'/out(N) = (in(N'), <p, p'>)$ where the occurrence net is the disjoint union of N and N' which is merged in each condition c of N and c' of N' with $p(c) = p'(c')$. The net morphism $<p, p'>$ is defined as in the parallel case by p on elements of N and by p' on elements on N'. It is a mapping as p and p' coincide on the merged conditions. Note that the sequential composition of processes $proc$ and $proc'$ is only partially defined by $proc \circ proc'$ if this is a process again.

Based on these preliminaries, we can consider sets of processes on $\overline{N}$ as semantic entities and extend the binary operations elementwise to such sets. Moreover, each condition $c \in \overline{B}$ provides a nullary operation $\hat{c} = \{(proc(\{c\})\} = \{(sub(\{c\}), incl_{sub(\{c\})})\}$ containing as only semantic item the process induced by $\{c\}$. Altogether we get the semantic domain $\mathcal{D}(\overline{N}) = (PROC(\overline{N}), OP(\overline{N}))$ with $OP(\overline{N}) = \{+, \circ\} \cup \{\hat{c} \mid c \in \overline{B}\}$, which is associated to the given elementary net $\overline{N}$.

It should be noted that the set of processes corresponding to cases $C \subseteq \overline{B}$ is just the closure of the nullary operations under parallel compositions. Moreover, this set is trivially closed under sequential composition because we have obviously $in(sub(C)) = C = out(sub(C))$ such that the only defined sequential composition is $proc(C) \circ proc(C)$ and yields $proc(C)$. Another significant operational closure is considered in the next subsection.

Binary Relations. Let K be a set of configurations like strings, trees, or graphs. Then the subsets of $K \times K$ can be considered as semantic entities describing, for example, input/output relations.

There is always the sequential composition $R \circ R'$ of relations $R, R' \subseteq K \times K$ given by $R \circ R' = \{(x, z) \mid (x, y) \in R, (y, z) \in R'$ for some $y \in K\}$.

If K has got some binary operation $\cdot : K \times K \to K$, this gives rise to a parallel composition $R \parallel R'$ given by $R \parallel R' = \{(x \cdot x', z \cdot z') \mid (x, z) \in R, (x', z') \in R'\}$.

A typical example is the concatenation of strings if K is the set A^* of all strings over an alphabet A. In this case, we also get an interesting unary operation *context* that embeds a given relation R into all possible contexts, i.e. $context(R) = \{(xuy, xvy) \mid (u, v) \in R, x, y \in A^*\}$.

2.2   Rule Bases

A rule base equips a semantic domain with rules as a first syntactic feature. A rule provides a semantic entity describing basic computations.

A *rule base* $\mathcal{DR} = (X, OP; \mathcal{R}, \Longrightarrow)$ consists of a semantic domain (X, OP), a class of *rules* $\mathcal{R}$, and a *rule application operator* $\Longrightarrow$ being a mapping $\Longrightarrow$: $\mathcal{R} \to 2^X$ which assigns a semantic entity $\underset{r}{\Longrightarrow} \in 2^X$ to each $r \in \mathcal{R}$.

As a rule specifies a semantic entity, a set of rules, $P \subseteq \mathcal{R}$, provides a set of semantic entities, $\{\underset{r}{\Longrightarrow} \mid r \in P\}$, which can be closed under the operations of the semantic domain. Accordingly, we denote $OP^*(\{\underset{r}{\Longrightarrow} \mid r \in P\})$ by $OP^*(P)$ for short. In this way, a set of rules P specifies a semantic entity $\bigcup OP^*(P)$, which contains all semantic items that are obtained by the operational closure of all applications of rules in P.

Examples

Elementary Net Systems. The events of $\overline{N}$ may be considered as rules. Each event $e \in \overline{E}$ induces a basic process $proc(e)$ such that the singleton set $\{proc(e)\}$ is a suitable semantic interpretation of an event as a rule. In other words, there is a rule base $\mathcal{DR}(\overline{N}) = (PROC(\overline{N}), OP(\overline{N}); \overline{E}, Proc : \overline{E} \to 2^{PROC(\overline{N})})$ with $Proc(e) = \{proc(e)\}$ for all $e \in \overline{E}$.

As $proc(e) \in PROC(\overline{N})$ for all $e \in \overline{E}$ and as the processes on $\overline{N}$ are closed under the operations in $OP(\overline{N})$, we get

$$\bigcup OP(\overline{N})^*(Proc(\overline{E})) \subseteq PROC(\overline{N})$$

for $Proc(\overline{E}) = \{Proc(e) \mid e \in \overline{E}\}$.

Conversely, let $proc = (N, p)$ with $N = (B, E, F)$ be a process on $\overline{N}$. If $E = \emptyset$, then $proc$ equals $proc(B)$ which is the parallel composition of all $\hat{c}$ for $c \in p(B)$.

For $E \neq \emptyset$, we show by induction on the number of elements in E that $proc \in \bigcup OP(\overline{N})^*(Proc(\overline{E}))$.

If $E = \{e\}$, then $proc$ is the parallel composition of $proc(p(e))$ with all $\hat{c}$ for $c \in p(B) - in(sub(p(e)))$. This case can be used as induction base.

If E has more than one element, then it is well-known that $proc$ is the sequential composition of two subnet processes $proc_i = (N_i, p_i)$ with $N_i = (B_i, E_i, F_i)$ for $i = 1, 2$ and $E_1 \neq \emptyset \neq E_2$. In particular, E_1 and E_2 are smaller sets than E such that we may assume by induction that $proc_1$ and $proc_2$ are in the operational closure of $Proc(\overline{E})$. Because of $proc = proc_1 \circ proc_2$, $proc$ is also in the closure.

Altogether, we have proved

$$\bigcup OP(\overline{N})^*(Proc(\overline{E})) = PROC(\overline{N}).$$

Binary Relations. Grammatical rules and all rules like these can be applied to some kind of configurations and derive configurations from them. Such a rule provides one with a binary relation of configurations the elements of which are often called direct derivations or computation steps.

A well-known explicit example of this type is the rule of a semi-Thue system or Chomsky grammar $p = (u, v)$ for $u, v \in A^*$ and some alphabet A. This rule specifies a binary relation $\underset{p}{\longrightarrow} \subseteq A^* \times A^*$ which is defined in infix notation by

$$xuy \underset{p}{\longrightarrow} xvy \quad \text{for all } x, y \in A^*.$$

Similarly, all kinds of graph transformation rules define a binary relation on the proper kinds of graphs by means of direct derivations.

If a rule r is composed of a pair (L, R) of configurations as in the case of the rules (u, v) with $u, v \in A^*$, then there is a simple alternative to the relation of direct derivations. This is the singleton set $simple(r) = \{(L, R)\}$.

Let us first consider the rule bases $\mathcal{DR}_1(A) = (A^* \times A^*, OP_1; A^* \times A^*, \longrightarrow)$ with $OP_1 = \{\circ\}$ and $\mathcal{DR}_i(A) = (A^* \times A^*, OP_i; A^* \times A^*, simple)$ for $i = 2, 3$ with $OP_2 = OP_1 \cup \{context\}$ and $OP_3 = OP_2 \cup \{\|\}$.

Then the following holds for a set of rules, $P \subseteq A^* \times A^*$:

$$\bigcup OP_1^*(P) = \bigcup OP_2^*(P) = \bigcup OP_3^*(P).$$

Note that the first operational closure is done for the direct derivations of P while the other two start from the simple relations $simple(p)$ for $p \in P$. The first equality follows from the obvious fact that $context(simple(p)) = \underset{p}{\longrightarrow}$ for all $p = (u, v) \in P$. The second follows from the well-known fact that the parallel composition can be expressed by context embeddings and sequential composition, i.e.

$$(xx', zz') = (xx', zx') \circ (zx', zz'.)$$

Given a set P of rules, the derivability relation $\underset{p}{\overset{*}{\longrightarrow}}$ is the sequential closure of all applications of rules in P, i.e. $\{\circ\}^*(\underset{P}{\longrightarrow})$ with $\underset{P}{\longrightarrow} = \bigcup_{p \in P} \underset{p}{\longrightarrow}$ which equals $\bigcup OP_1^*(P)$. In other words, all three rules bases $\mathcal{DR}_i(A)$ for $i = 1, 2, 3$ describe sequential derivability through sets of rules.

2.3   Rule Bases with Control Conditions

A rule base may be additionally equipped with control conditions that allow to regulate computations. For this purpose, its semantics depends on the environment given by semantic entities for a set of identifiers.

A *rule base with control conditions* $\mathcal{DRC} = (X, OP; \mathcal{R}, \Longrightarrow; ID, \mathcal{C}, SEM)$ consists of a rule base $(X, OP; \mathcal{R}, \Longrightarrow)$, a set ID of *identifiers*, a class of *control conditions* $\mathcal{C}$, and a semantic interpretation SEM which associates each condition $C \in \mathcal{C}$ and each semantic mapping $Env : ID \to 2^X$, called *environment*, with a semantic entity $SEM(C, Env) \in 2^X$.

Depending on the environment Env, a control condition C can be used to restrict the operational closure of a set M of semantic entities by means of the intersection $\bigcup OP^*(M) \cap SEM(C, Env)$.

Examples

Elementary Net Systems. An elementary net becomes an elementary net system if an initial case is added. The idea of an initial case is that semantically only processes starting in this case are considered. Initial cases are typical examples of control conditions. Let $C_{in} \subseteq \overline{B}$ be some initial case. Then its semantics $SEM(C_{in})$ consists of all processes (N, p) with $p(in(N)) = C_{in}$. Therefore, the process semantics $PROC((\overline{B}, \overline{E}, \overline{F}, C_{in}))$ of the elementary net system $(\overline{B}, \overline{E}, \overline{F}, C_{in})$ coincides with the intersection of the operational closure $\bigcup OP(\overline{N})^*(Proc(\overline{E}))$ and $SEM(C_{in})$, i.e.

$$PROC((\overline{B}, \overline{E}, \overline{F}, C_{in})) = (\bigcup OP(\overline{N})^*(Proc(\overline{E}))) \cap SEM(C_{in}).$$

Let ID be a set of identifiers and $Env : ID \to 2^X$ some semantic mapping. Then the semantics of an initial case can be extended to the environment Env in a trivial way as a case does not refer to any identifier:

$$SEM(C_{in}, Env) = SEM(C_{in}).$$

Each elementary net $\overline{N} = (\overline{B}, \overline{E}, \overline{F})$ induces a rule base with cases as control conditions $\mathcal{DRC}(\overline{N}) = (PROC(\overline{N}), OP(\overline{N}); \overline{E}, Proc; ID, 2^{\overline{B}}, SEM)$ where SEM is defined as above. We have shown that this rule base describes elementary net systems with their processes starting in the initial case as semantics by using the events as rules and the initial cases as control conditions. In 2.4 the notion of a basic transformation unit is introduced as a syntactic modelling concept that allows one the use of rules and control conditions explicitly.

Binary Relations. In grammatical systems, the most frequently used kind of control condition is the choice of start symbols or some other configurations as axioms to begin derivations and the choice of terminal alphabets to describe the configurations at which derivations may end. For example, given an alphabet A, each pair (S, T) with $T \subseteq A$ and $S \in A \setminus T$ specifies a binary relation $SEM((S,T)) = \{S\} \times T^* \subseteq A^* \times A^*$. The intersection of this relation with the derivability relation of a set of rules, $P \subseteq A^* \times A^*$, contains pairs (S, w) where S derives w and w is terminal. In other words, the intersection $(\bigcup OP_1^*(P)) \cap SEM((S,T)) = \xrightarrow[P]{*} \cap (\{S\} \times T^*)$ represents the generated language of the Chomsky grammars $G = (N, T, P, S)$ with $N = A \setminus T$ in a unique way. This type of control condition is independent of any environment in the same way as initial cases above:

$$SEM((S,T), Env) = SEM((S,T))$$

for all $Env : ID \rightarrow 2^{A^* \times A^*}$ where ID is some set of identifiers. The same remains true if S is replaced by an arbitrary start word or axiom $z \in A^*$.

In other words, we may extend the rule bases $\mathcal{DR}_i(A)$ for $i = 1, 2, 3$ into rule bases with control conditions:

$$\mathcal{DRC}_i(A) = (\mathcal{DR}_i(A); ID, A^* \times 2^A, SEM).$$

As shown above, they allow one to describe Chomsky grammars and their generated languages within our framework.

Using a more sophisticated type of control conditions, one can also specify Lindenmayer systems (see, e.g., [10, 11]) as grammatical systems with a massively parallel mode of rewriting. We introduce this mode in a general way to demonstrate the role of identifiers and environments.

Let $M \subseteq 2^{A^* \times A^*}$ be a set of binary relations on A^*, which may be some kind of basic computations. Then the sequential closure of the parallel closure of M, $\{\circ\}^*(\{\|\}^*(M))$, describes massive parallelism on the semantic level as each step of a sequence consists just of parallel computations. To express this on the syntactic level of control conditions, one needs access to the members of M for which we offer two ways. The first possibility is given by a set $P \subseteq \mathcal{R}$ of rules and the second one by a set $U \subseteq ID$ of identifiers together with an environment. Formally, we introduce the control condition $mp(P, U)$ with

$$SEM(mp(P, U), Env) = \{\circ\}^*(\{\|\}^*(\{simple(p) \mid p \in P\} \cup \{Env(id) \mid id \in U\})),$$

where mp refers to the term *massive parallelism*. If P is a set of context-free rules, i.e. $P \subseteq A \times A^*$, and U is empty, $mp(P, \emptyset)$ describes the derivation mode of 0L systems.

In order to combine massive parallelism explicitly with the rule based features for binary relations, one may consider the rule bases with control conditions $\mathcal{DRC}_i(A, mp)$ for $i = 1, 2, 3$ which are obtained from $\mathcal{DRC}_i(A)$ by adding the control conditions $\{mp(P, U) \mid P \subseteq \mathcal{R}, U \subseteq ID\}$ with $SEM(mp(P, U), Env)$ as defined above and the combined control conditions $A^* \times 2^A \times \{mp(P, U) \mid P \subseteq \mathcal{R}, U \subseteq ID\}$ with $SEM((z, T, mp(P, U)), Env) = SEM(z, T) \cap SEM(mp(P, U), Env)$.

Instead of massive parallelism, there are other derivation modes that allow one to control the application of grammatical rules. Further typical examples are $\leq k \ (= k, \geq k)$ for some $k \in \mathbb{N}$ requiring that the number of rule applications for a given set of rules is not greater than k (equals k, is not less than k) and t (for terminating) requiring that the given rules are applied as long as possible (see, e.g., [1] for more details).

2.4   Transformation Units

A rule base provides the computational framework in which rule-based specifications can be defined. The most elementary kind of such a specification in our framework is a transformation unit that comprises a local set of rules, a set of identifiers, and a control condition. The identifiers refer to used or imported components. The control condition regulates the interaction of rules and imported components.

Let $\mathcal{DRC} = (X, OP; \mathcal{R}, \Longrightarrow; ID, \mathcal{C}, SEM)$ be an arbitrary, but fixed rule base with control conditions. Then a *transformation unit* (over $\mathcal{DRC}$) is a system $tu = (P, U, C)$ where $P \subseteq \mathcal{R}$ is a finite set of rules, $U \subseteq ID$ is a finite set of identifiers, which is called the *use* or *import interface*, and $C \in \mathcal{C}$ is a control condition.

The unit is called *basic* if U is empty.

Examples are discussed together with the interlinking semantics at the end of the next subsection.

2.5   Interlinking Semantics of Transformation Units

Given a semantic entity for each import identifier, i.e. a mapping $Imp : U \to 2^X$, the transformation unit tu specifies a semantic entity which is constructed as the operational closure of the semantic entities given by the local rules and the import as far as it meets the control condition. Because the rules and the import are interlinked with each other through the semantic operations, the resulting semantic entity is called *interlinking semantics* which is formally defined as follows:

$$INTER_{Imp}(tu) = (\bigcup OP^*(P, Imp)) \cap SEM(C, Imp_+)$$

where $OP^*(P, Imp)$ is the operational closure of the semantic entities given by the rules and the import mapping, i.e.

$$OP^*(P, Imp) = OP^*(\{\underset{r}{\Longrightarrow} \mid r \in P\} \cup \{Imp(id) \mid id \in U\}),$$

and where the environment $Imp_+ : ID \to 2^X$ is the trivial extension of Imp to ID, i.e. $Imp_+(id) = Imp(id)$ for $id \in U$ and $Imp_+(id) = \emptyset$ otherwise.

Altogether, the interlinking semantics interlinks the semantic effects of the local rules of the transformation unit with the imported semantic entities according to the control condition. It should be noted that the notion of interlinking semantics of transformation units generalizes the interleaving semantics introduced in [5, 6]. The interleaving semantics concerns binary relations on graphs or configurations resp. as semantic entities and the sequential composition of binary relations as only semantic operator.

Examples

In this subsection, only examples of basic transformation units are presented. Examples of units with import can be found in 3.2 and 3.3.

Elementary Net Systems. Consider the rule base with control conditions $\mathcal{DRC}(\overline{N}) = (PROC(\overline{N}), OP(\overline{N}); \overline{E}, Proc; ID, 2^{\overline{B}}, SEM)$ for a given elementary net $\overline{N} = (\overline{B}, \overline{E}, \overline{F})$. Then an elementary net system (N, C_{in}) with a subnet $N = (B, E, F)$ of $\overline{N}$ and an initial case C_{in} can be interpreted as a basic transformation unit $tu(N, C_{in}) = (E, \emptyset, C_{in})$ such that the process semantics of (N, C_{in}) coincides with the interlinking semantics of $tu(N, C_{in})$ using the empty mapping $Empty : \emptyset \to 2^X$ as the only import mapping.

$$PROC(N, C_{in}) = INTER_{Empty}(tu(N, C_{in})).$$

This follows directly from the definition of the interlinking semantics and the considerations in 2.3.

Conversely, a basic transformation unit $tu = (E, \emptyset, C_{in})$ induces an elementary net system $N(tu) = (\overline{B}, E, F(tu), C_{in})$ with $F(tu) = \overline{F} \cap ((\overline{B} \times E) \cup (E \times \overline{B}))$ where the underlying net is the subnet of $\overline{N}$ induced by E.

Binary Relations. Consider the rule base with control condition $\mathcal{DRC}_1(A)$ as given in 2.3. It is shown there that the language $L(G) = \{w \in T^* \mid S \xrightarrow{*}_P w\}$ generated by the Chomsky grammar $G = (N, T, P, S)$ corresponds one-to-one to the binary relation $(\bigcup OP_1^*(P)) \cap SEM((S,T)) = \xrightarrow{*}_P \cap (\{S\} \times T^*)$. In other words, the grammar G gives rise to the basic transformation unit $tu(G) = (P, \emptyset, (S, T))$ over $\mathcal{DRC}_1(A)$ such that its interlinking semantics coincides with the generated language $L(G)$ up to representation. According to 2.2, this remains true if $\mathcal{DRC}_1(A)$ is replaced by $\mathcal{DRC}_2(A)$ or $\mathcal{DRC}_3(A)$.

Similarly, many other kinds of grammars, like for example tree and graph grammars, can be seen as basic transformation units such that the generated languages correspond to the interlinking semantics if one chooses the set of configurations, the set of rules, the rule application operator, single configurations as axioms and terminal configurations properly.

As a Chomsky grammar, an 0L system $G'=(A, P, z)$ with $P \subseteq A \times A^*$ and $z \in A^*$ can be modelled as the basic transformation unit $tu(G') = (P, \emptyset, (z, A, mp(P, \emptyset)))$ over one of the rule bases with control conditions $\mathcal{DRC}_i(A, mp)$ for $i = 1, 2, 3$ such that the generated language $L(G')$ corresponds again to the interlinking semantics of $tu(G')$.

2.6   Monotony of the Interlinking Semantics

The interlinking semantics depends on the imported semantic entities. If they are replaced by larger sets, the environment of a transformation unit increases automatically. The interlinking semantics and the operational closure are also increasing if the control condition and the operations are monotone. This helps to show in the following sections that the interlinking semantics is a fixed-point semantics.

An operation $op \in OP_k$ for some k is *monotone* if $op(t_1, \ldots, t_k) \subseteq op(t'_1, \ldots, t'_k)$ for all $t_i, t'_i \in 2^X$ with $t_i \subseteq t'_i$ and $i = 1, \ldots, k$. Accordingly, a set of operations is *monotone* if each of its elements is monotone. A control condition $C \in \mathcal{C}$ is *monotone* if $SEM(C, Env) \subseteq SEM(C, Env')$ for all environments $Env, Env' : ID \rightarrow 2^X$ with $Env(id) \subseteq Env'(id)$ for all $id \in ID$.

Observation 1 *Let* $tu = (P, U, C)$ *be a transformation unit over a rule base with control conditions* $\mathcal{DRC} = (X, OP; \mathcal{R}, \Longrightarrow; ID, \mathcal{C}, SEM)$, *and let* $Imp, Imp' : U \rightarrow 2^X$ *be import mappings with* $Imp \subseteq Imp'$. *Then the following hold:*

(1) $Imp_+ \subseteq Imp'_+$.
(2) $\bigcup OP^*(P, Imp) \subseteq \bigcup OP^*(P, Imp')$ *provided that* OP *is monotone.*
(3) $INTER_{Imp}(tu) \subseteq INTER_{Imp'}(tu)$ *provided that* C *is monotone in addition.*

Examples

If an operation on the powerset of a set X is the natural elementwise extension of an operation on the underlying set X, then the extension is obviously monotone. All operations considered for elementary net systems and binary relations are of this kind. Moreover, all control conditions considered in the examples are monotone because they control the composition of semantic entities independent of their size such that the results get larger if the arguments are replaced by larger sets.

3   Networks of Transformation Units with Iterated Interlinking Semantics

A transformation unit is a rule-based system with a generic import. An import identifier represents a semantic entity, but it is not fixed how it is specified. A simple way to specify the import is to assume that the identifiers name again transformation units. In this case the import structure forms a directed graph leading to the notion of a network of transformation units. If the network is finite and acyclic or if the network has no infinite path, the interlinking semantics can be defined for all transformation units in the network. If the network has a cycle or an infinite path, one may start with the empty semantic entity for each transformation unit and then iterate the interlinking semantics ad infinitum.

3.1   Networks of Transformation Units

A *network of transformation units* over a rule base with control conditions $\mathcal{DRC} = (X, OP; \mathcal{R}, \Longrightarrow; ID, \mathcal{C}, SEM)$, is a system $N = (V, \tau)$ where V is a set of nodes and τ is a mapping assigning a transformation unit $\tau(v) = (P(v), U(v), C(v))$ to each node $v \in V$ with $U(v) \subseteq V$.

A network of transformation units can be seen as a directed graph where the elements of V are the nodes and the ordered pairs of nodes (v, v') with

$v' \in U(v)$ the edges. A network of transformation units is *well-founded* if it does not contain an infinite path.

Note that a finite network is well-founded if and only if it is acyclic. In case of a well-founded network, the set of nodes can be divided into pairwise disjoint *levels* V_k for $k \in \mathbb{N}$ which are inductively defined by

(i) V_0 containing all nodes with basic units and

(ii) V_{k+1} containing all nodes $u \in V \setminus \bigcup_{i=0}^{k} V_i$ with $U(v) \subseteq \bigcup_{i=0}^{k} V_i$.

3.2   Interlinking Semantics of Well-Founded Networks

The interlinking semantics of transformation units is easily extended to well-founded networks because it can be defined inductively level by level.

Let $N = (V, \tau)$ be a well-founded network of transformation units. Then the interlinking semantics $INTER : V \rightarrow 2^X$ is inductively defined in the following way:

(1) for $v \in V_0$, we have $U(v) = \emptyset$ such that the empty mapping $Empty$ is the only choice for the semantics of the import part. Therefore, the interlinking semantics of $tu(v)$ with $Empty$ is defined yielding

$$INTER(v) = INTER_{Empty}(tu(v)).$$

(2) Let us assume that $INTER(v)$ is defined for all $v' \in \bigcup_{i=0}^{k} V_k$ for some $k \in \mathbb{N}$.

(3) Consider $v \in V_{k+1}$. Then we have $U(v) \subseteq \bigcup_{i=0}^{k} V_k$ such that $Imp_k(v') = INTER(v')$ is defined for all $v' \in U(v)$. Therefore, the interlinking semantics of $tu(v)$ with Imp_k is defined yielding

$$INTER(v) = INTER_{Imp_k}(tu(v)).$$

Examples

The first examples of networks of transformation units are well-founded and have all the same simple structure with n nodes $v_1, \ldots, v_n$ at level 0 and one node v_0 at level 1, i.e. there is a main unit importing the other units. The examples differ only in the choices of transformation units for the nodes.

Elementary Net Systems. In 2.5, an elementary net system (N, C_{in}) with $N = (B, E, F)$ is transformed into the basic transformation unit $tu(N, C_{in})$. But as each event gives rise to a basic transformation unit separately with the event as the only rule, the elementary net system can be seen as a transformation unit that imports its event units, i.e. $\tau(N, C_{in})(v_0) = (\emptyset, \{v_1, \ldots, v_n\}, C_{in})$ and $\tau(N, C_{in})(v_i) = tu(e_i) = (\{e_i\}, \emptyset, all)$ for $i = 1, \ldots, n$ and $E = \{e_1, \ldots, e_n\}$. The control condition all is a void condition the semantics of which is always the set of all processes so that the intersection with any other semantic entity has no effect. Hence, the interlinking semantics of $tu(e_i)$ is the closure of the process $proc(e_i)$ and the processes $proc(c)$ for each $c \in \overline{B}$ under parallel and sequential composition. Because the sequential composition with $proc(C)$ for $C \subseteq \overline{B}$ has no effect and the event e_i is not enabled directly after its occurrence, the sequential compositions can be ignored, and one gets as interlinking semantics of $tu(e_i)$ all cases and all single occurrences of e_i, i.e.

$$INTER(v_i) = INTER_{Empty}(tu(e_i)) = 2^{\overline{B}} \cup \{proc(e_i) + C \mid C \subseteq \overline{B} \setminus ({}^{\bullet}e_i \cup e_i^{\bullet})\}.$$

This provides also the import mapping Imp_0 for the interlinking semantics of the level-1 node v_0 consisting of all sequential and parallel compositions of the imported processes that start with C_{in} yielding the process semantics of the elementary net system (N, C_{in}), i.e.

$$INTER(v_0) = INTER_{Imp_0}(\tau(N, C_{in})(v_0)) = PROC(N, C_{in}).$$

Binary Relations. Grammar systems (see, e.g., [1]) are typical examples of the same form. The system becomes the main unit, and its components are imported. More formally, a *cooperating distributed grammar system* $\Gamma = (N, T, S, P_1, \ldots, P_n)$ consists of a set of *nonterminals* N, a set of *terminals* T, a *start symbol* $S \in N$, and a collection of finite sets of rules $P_1, \ldots, P_n$ with $P_i \subseteq (N \cup T)^* \times (N \cup T)^*$ for $i = 1, \ldots, n$. Choosing a derivation mode f (according to examples in 2.3), Γ generates the language $L_f(\Gamma)$ which contains all terminal words w that are derived from the start symbol S by a sequence of derivations in the mode f of the form

$$S \xrightarrow[P_{i_1}]{f} w_1 \xrightarrow[P_{i_2}]{f} \cdots \xrightarrow[P_{i_m}]{f} w_m = w$$

with $m \geq 1$ and $1 \leq i_j \leq n$ for $j = 1, \ldots, m$. The corresponding transformation units are defined by $\tau(\Gamma)(v_0) = (\emptyset, \{v_1, \ldots, v_n\}, (S, T))$ and $\tau(\Gamma)(v_i) = (P_i, \emptyset, f)$ for $i = 1, \ldots, n$. The interlinking semantics of the latter units coincides obviously with the derivation relations with respect to the derivation mode f. And the interlinking semantics of $\tau(\Gamma)(v_0)$ imports these, constructs the closure under the operations including sequential composition, and intersects the result with $\{S\} \times T^*$ due to the control condition. Consequently, a word w is in interlinking relation to S if and only if $w \in L_f(\Gamma)$.

Similarly, a T0L system $G'' = (A, P_1 \ldots, P_n, z)$ with an *alphabet* A, a *start word* $z \in A^*$ and a collection of finite sets of context-free rules $P_i \subseteq A \times A^*$ gives rise to a network of transformation unit. The network structure is the same as in the last two examples, and the transformation units of the nodes are defined by $\tau(G'')(v_0) = (\emptyset, \{v_1, \ldots, v_n\}, (z, A))$ and $\tau(G'')(v_i) = (P_i, \emptyset, mp(P_i, \emptyset))$ for $i = 1, \ldots, n$. Accordingly, the language generated by G'' corresponds to the interlinking semantics of the root node v_0.

3.3 Iterated Interlinking Semantics of Arbitrary Networks

The problem of networks which are not well-founded is that the semantics of the import parts of some units at the network may not be known at the moment when one wants to apply the interlinking semantics. But if one assumes to have at least a preliminary semantics for all nodes, i.e. some semantic mapping Sem : $V \to 2^X$, then the interleaving semantics is defined for the unit of each node wrt to Sem restricted to the import part, i.e. $Sem'(v) = INTER_{Sem(U(v))}(tu(v))$ for $v \in V$ with $Sem(U(v))(v') = Sem(v')$ for all $v' \in U(v)$.

The resulting semantic mapping Sem' is denoted by $INTER(Sem)$, and the operator $INTER$, that yields semantic mappings from semantic mappings by interlinking them with the semantic entities of the respective rules, is called *interlinking operator*.

In this way, one gets another semantic mapping, which may be used as a next preliminary semantics such that this process can be iterated ad infinitum whenever one starts from some semantic mapping. An obvious candidate to start is the mapping that assigns the empty set to each node of the network. Therefore, the *iterated interlinking semantics* $ITERATE : V \to 2^X$ of a network of transformation units $N = (V, \tau)$ may be defined for all $v \in V$ as follows:

$$ITERATE(v) = \bigcup_{i \in \mathbb{N}} ITERATE_i(v)$$

with $ITERATE_0(v) = \emptyset$ and $ITERATE_{i+1}(v) = INTER_{ITERATE_i(U(v))}(tu(v))$.

Examples

The concept of networks of transformation units beyond the examples in 3.2 provide new possibilities of cooperation and distribution in the framework of elementary net systems and of grammar systems. The examples of 3.2 may be reconsidered. Instead of a main unit which imports all others, the main unit imports only one of the other units which import each other. While the main unit takes care of the global control condition only, the other units do the computational work interactively.

Elementary Net Systems. Let (N, C_{in}) be an elementary net system, let $E = \{e_1, \ldots, e_n\}$ be its set of events, and let $V = \{v_1, \ldots, v_n\}$. Then the second network $\hat{\tau}(N, C_{in})$ associated to (N, C_{in}) is given by $\hat{\tau}(N, C_{in})(v_0) = (\emptyset, \{v_1\}, C_{in})$ and $\hat{\tau}(N, C_{in})(v_i) = (\{e_i\}, V, all)$ for $i = 1, \ldots, n$. Starting with the empty set of processes at each node, the first application of the interlinking operator yields the singleton set $\{C_{in}\}$ as semantics of v_0 and all cases and all single occurrences of the event e_i as semantics of v_i for $i = 1, \ldots, n$. As the import is empty at the first step, the unit at v_i behaves as $tu(e_i)$. The second application of the interlinking operator yields the set of all processes as semantics at v_i for $i = 1, \ldots, n$ as all single occurrences of all events are imported and closed under sequential and parallel composition. The semantics at v_0 contains, besides C_{in}, the sequential compositions of C_{in} with single occurrences of e_1, i.e. the single occurrence of e_1 under C_{in} if $^\bullet e_1 \subseteq C_{in}$. The third application of the interlinking operator yields the processes of N that start in C_{in} at the node v_0. The other semantic entities are kept. Further iteration is not changing the semantics.

Binary Relations. Analogously, grammar systems and T0L systems can be reconstructed as networks of transformation units of the given form.

The same remains true if the subnetwork induced by $v_1, \ldots, v_n$ is not complete, but there is a path from each node to v_1. In this case, the interlinking operator must be iterated $m + 2$ times if m is the length of the longest shortest path of a node v_i to v_1 for $i = 1, \ldots, n$.

3.4 Iterated Interlinking Semantics of Well-Founded Networks

If one applies the iterated interlinking semantics to well-founded networks, the result coincides with the ordinary interlinking semantics. This is a first indication that the interlinking semantics is meaningful.

Observation 2 *Let $N = (V, \tau)$ be a well-founded network of transformation units. Then we have*

$$INTER = ITERATE.$$

3.5 Monotony and Continuity of the Interlinking Operator

Given a semantic entity for each node of a network of transformation units, the interlinking semantics is defined for each node yielding another semantic entity. This is the basic operator which is iterated in the iterated interlinking semantics. This operator turns out to be monotone and even continuous if the used operations and control conditions are monotone resp. continuous.

An operation $op \in OP_k$ for some k is *continuous* if

$$\bigcup_{i \in \mathbb{N}} op(t_1, \ldots, t_{j-1}, t_{j_i}, t_{j+1}, \ldots, t_k) = op(t_1, \ldots, t_{j-1}, \bigcup_{i \in \mathbb{N}} t_{j_i}, t_{j+1}, \ldots, t_k).$$

for each j with $1 \leq j \leq k$ and each increasing chains of semantic entities $t_{j_0} \subseteq t_{j_1} \subseteq \ldots \subseteq t_{j_i} \subseteq \ldots$. Accordingly, a set of operations is *continuous* if each of its elements is continuous.

A control condition C is *continuous* if

$$\bigcup_{i \in \mathbb{N}} SEM(C, Env_i) = SEM(C, \bigcup_{i \in \mathbb{N}} Env_i)$$

for each increasing chain of environments $Env_0 \subseteq Env_1 \subseteq \ldots \subseteq Env_i \subseteq \ldots$ with $Env_i : ID \rightarrow 2^X$ for $i \in \mathbb{N}$.

Examples

All operations in the examples of this paper are operations on semantic items, which are extended elementwise to semantic entities. Such operations are obviously monotone and continuous. The same applies to the control conditions as they restrict the application of semantic operators independent of the content of the semantic entities.

Observation 3 *Let $\mathcal{DRC} = (X, OP; \mathcal{R}, \Longrightarrow; ID, \mathcal{C}, SEM)$ be a rule base with control conditions and $N = (V, \tau)$ be a network of transformation units over $\mathcal{DRC}$ such that OP and $C(v)$ for all $v \in V$ are continuous. Let $INTER$ be the corresponding interlinking operator. Then the following hold:*

(1) The interlinking operator is monotone, i.e.

$$INTER(Sem) \subseteq INTER(Sem')$$

for all semantic mappings $Sem, Sem' : V \to 2^X$ with $Sem \subseteq Sem'$.
(2) The interlinking operator is continuous, i.e.

$$\bigcup_{i \in \mathbb{N}} INTER(Sem_i) = INTER(\bigcup_{i \in \mathbb{N}} Sem_i)$$

for all increasing chains of semantic mappings $Sem_0 \subseteq Sem_1 \subseteq \ldots \subseteq Sem_i \subseteq \ldots$.

4   Fixed-Point Theorem

Let $N = (V, \tau)$, be a network of transformation units $N = (V, \tau)$, and $INTER$ the corresponding interlinking operator on the semantics mappings $Sem : V \to 2^X$ defined by

$$INTER(Sem)(v) = INTER_{Sem(U(v))}(tu(v))$$

for all $v \in V$. Because their domain is a power set, it is a well-known fact that the set of semantic mappings is a complete partial order with respect to the argumentwise inclusion where the union of every increasing chain is the least upper bound. In Observation 3, the interlinking operator is shown to be monotone and continuous with respect to this complete partial order such that Kleene's fixed-point theorem applies. This means that the iterated interlinking semantics is the least fixed point of the interlinking operator.

Theorem 1. *Let $\mathcal{DRC} = (X, OP; \mathcal{R}, \Longrightarrow; ID, \mathcal{C}, SEM)$ be a rule base with control conditions, and let $N = (V, \tau)$ be a network of transformation units over $\mathcal{DRC}$ such that each operation is continuous and $C(v)$ as well for each $v \in V$. Then the iterated interlinking semantics $ITERATE : V \to 2^X$ is the least fixed point of the interlinking operator $INTER$, i.e.*

$$INTER(ITERATE) = ITERATE.$$

It should be noted that this result generalizes the fixed-point theorem in [7], which deals with binary relations on some set of graphs as semantic entities and with the sequential composition of binary relations as the only semantic operator.

5   Conclusion

In this paper, we have introduced the notion of interlinking semantics of networks of transition units generalizing the purely sequential interleaving semantics of earlier work. We have demonstrated that the new concept covers parallelism and concurrency of elementary net systems and various types of grammars. The main result is a fixed-point theorem stating that the iterated interlinking semantics is the smallest fixed-point of the interlinking operator.

Future work should shed some more light on the significance of this approach in two respects at least. On one hand, it should be investigated how the fixed-point theorem can be used to analyze rule-based systems and to prove their properties. On the other hand, further case studies would be helpful to fit in further approaches to parallelism and concurrency into our new framework. In particular, we would like to relate parallelism and concurrency in the area of graph transformation, which have been intensively investigated by Hartmut Ehrig and others in the last three decades (see, e.g., the Chapters 3 and 4 in [9] and [2] for an overview), with interlinking semantics.

Acknowledgement

We would like to thank Peter Knirsch and Gabriele Taentzer for the helpful comments on an earlier version of this paper.

References

1. Jürgen Dassow, Gheorghe Păun and Grzegorz Rozenberg. Grammar systems. In G. Rozenberg and A. Salomaa, editors. *Handbook of Formal Languages, Vol. 2.* pages 155–213, Springer, 1997.
2. Hartmut Ehrig, Hans-Jörg Kreowski, Ugo Montanari and Grzegorz Rozenberg, editors. *Handbook of Graph Grammars and Computing by Graph Transformation, Vol. 3.* World Scientific, 1999.
3. Cesar Fernandez. Non-sequential processes. In W. Brauer, W. Reisig and G. Rozenberg, editors. *Petri nets: Central models and their properties, Advances in Petri nets, Part I. Lecture Notes in Computer Science*, vol. 254, pages 95–115, Springer, 1986.
4. Hans-Jörg Kreowski and Sabine Kuske. On the interleaving semantics of transformation units – A step into GRACE. In Janice E. Cuny, Hartmut Ehrig, Gregor Engels and Grzegorz Rozenberg, editors. *Proc. Graph Grammars and Their Application to Computer Science. Lecture Notes in Computer Science*, vol. 1073, pages 89–108, Springer, 1996.
5. Hans-Jörg Kreowski and Sabine Kuske. Graph transformation units with interleaving semantics. *Formal Aspects of Computing*, vol. 11, no. 6, pages 690–723, 1999.
6. Hans-Jörg Kreowski and Sabine Kuske. Approach-independent structuring concepts for rule-based systems. In Martin Wirsing, Dirk Pattison, Rolf Hennicker, editors. *Proc. 16th Int. Workshop on Algebraic Development Techniques (WADT 2002). Lecture Notes in Computer Science*, vol. 2755, pages 299–311, Springer, 2003.

7. Hans-Jörg Kreowski, Sabine Kuske and Andy Schürr. Nested graph transformation units, *International Journal on Software Engineering and Knowledge Engineering*, vol. 7, no. 4, pages 479–502, 1997.

8. Grzegorz Rozenberg. Behaviour of elementary net systems. In W. Brauer, W. Reisig and G. Rozenberg, editors. *Petri nets: Central models and their properties, Advances in Petri nets, Part I. Lecture Notes in Computer Science*, vol. 254, pages 60–94, Springer, 1986.

9. Grzegorz Rozenberg, editor. *Handbook of Graph Grammars and Computing by Graph Transformation, Vol. 1.* World Scientific, 1997.

10. Grzegorz Rozenberg and Arto Salomaa, editors. *The Book of L.* Springer, 1986.

11. Grzegorz Rozenberg and Arto Salomaa, editors. *Lindenmayer Systems.* Springer, 1992.

12. Grzegorz Rozenberg and Arto Salomaa, editors. *Handbook of Formal Languages, Vol. 1–3.* Springer, 1997.

13. R.S. Thiagarajan. Elementary net systems. In W. Brauer, W. Reisig and G. Rozenberg, editors. *Petri nets: Central models and their properties, Advances in Petri nets, Part I. Lecture Notes in Computer Science*, vol. 254, pages 26–59, Springer, 1986.

Embeddings and Contexts for Link Graphs

Robin Milner

The Computer Laboratory, University of Cambridge,
Cambridge, UK
`Robin.Milner@cl.cam.ac.uk`

1 Introduction

Graph-rewriting has been a growing discipline for over three decades. It grew out of the study of graph grammars, in which – analogously to string and tree grammars – a principal interest was to describe the families of graphs that could be generated from a given set of productions. A fundamental contribution was, of course, the *double-pushout* construction of Ehrig and his colleagues [4]; it made precise how the left-hand side of a production, or rewriting rule, could be found to *occur* in a host graph, and how it should then be replaced by the right-hand side. This break-through led to many theoretical developments and many applications. It relies firmly upon the treatment of graphs as *objects* in a category whose arrows are embedding maps.

A simultaneous development was Petri nets [13], with a quite different motivation; it was the first substantial mathematical model of concurrent processes, and gives strong emphasis to the causality relation among events. Although Petri nets are graphical, their study has been largely independent of graph-rewriting; after all, a Petri net does not change its shape – only the tokens placed upon the net actually move.

A little later came the development algebraic calculi such as CSP [1] and CCS [12] to represent interactive concurrent processes. The key concept distinguishing them from (Petri) net theory was the emphasis upon modularity. Initially at least, net theory focussed upon complete systems, developing powerful techniques such as linear algebra to analyse them. In contrast, process calculi focussed upon assembling larger systems from smaller ones using a variety of combinators, and upon defining the behaviour of the whole in terms of abstract entities that can be constructed from the behaviours of the parts by algebraic operations corresponding to the combinators. This approach was inspired by the modularity present in all good programming languages, and by the categorical formulation of algebraic theories by Lawvere [9]; in contrast with graph-rewriting methodology, here the graphs are the *arrows* in a category whose objects are interfaces.

A recent development in process calculi by Leifer and Milner [10] is the demonstration that labelled transition systems can be derived uniformly for a wide variety of calculi, using the notion of *relative pushout* (RPO), in a category where the arrows are processes. In the particular case of *graphical* process calculi such as mobile ambients [2] or bigraphs [8], where graph-rewriting is used to

H.-J. Kreowski et al. (Eds.): Formal Methods (Ehrig Festschrift), LNCS 3393, pp. 343–351, 2005.

model various kinds of mobility among processes, this naturally leads to the
need for a rapprochement between *graphs-as-objects* and *graphs-as-arrows*.

Connections between these two approaches have recently been surveyed by
Ehrig [5]. In one of these connections, previously explored for example by Gadducci et al [7], graphs-as-arrows are obtained as cospans $I \to G \leftarrow J$ of graphs-
as-objects, where the interfaces I and J are graphs of some simple form (e.g.
discrete). A second connection goes the other way; graphs-are-objects arise in a
coslice category of a category, or s-category, of graphs-as-arrows. This connection was first proposed by Cattani, and was exploited technically in the theory
of action calculi [3].

The purpose of the present paper is to examine this latter connection more
closely, in the context of *link graphs* [11], which are a constituent of bigraphs. It
is shown that, for link graphs, the coslice category is isomorphic to the natural
category of embeddings (as arrows) between so-called *ground* link graphs. The
connection almost certainly extends to full bigraphs. More generally, it is an open
challenge to characterise the classes of graphs (or other entities) and interfaces
for which this elegant isomorphism exists. I suggest that the existence of the
isomorphism is a valuable test of probity for any proposed class of graphs.

Preliminaries. Id_X will denote the identity function on a set X, and $\emptyset_X$ the
empty function from $\emptyset$ to X. We shall use $X \uplus Y$ for union of sets X and
Y known or assumed to be disjoint, and $f \uplus g$ for union of functions whose
domains are known or assumed to be disjoint. This use of $\uplus$ on sets should not
be confused with the disjoint sum '+', which disjoins sets *before* taking their
union. We assume a fixed representation of disjoint sums; for example, $X + Y$
means $(\{0\} \times X) \cup (\{1\} \times Y)$, and $\sum_{v \in V} P_v$ means $\bigcup_{v \in V}(\{v\} \times P_v)$.

If $f\colon X \to Y$ is an arrow in a category or s-category $\mathbf{C}$, we denote its *domain*
X and *codomain* Y by $\mathsf{dom}(f)$ and $\mathsf{cod}(f)$. The set of arrows from X to Y, called
a *homset*, will be denoted by $\mathbf{C}(X, Y)$.

An *s-category* is like category except that every arrow f has an associated a
finite set $|f|$, its *support*; the composition $gf\colon X \to Z$ of $f\colon X \to Y$ and $g\colon Y \to Z$
exists iff $|f| \cap |g| = \emptyset$, and then $|gf| = |f| \uplus |g|$. Furthermore, for $f\colon X \to Y$ and
an injection ρ whose domain includes $|f|$, there is an arrow $\rho \cdot f\colon X \to Y$ called
a *support translation* of f, with support $\rho(|f|)$. Support translation preserves
all structure. A general treatment of s-categories can be found in Leifer and
Milner [10], but we shall only be concerned with the special case of link graphs
where the details are obvious.

2   Link Graphs

A link graph is essentially an ordinary graph, but it carries a little more information and each edge may link any number of nodes. A family of link graphs is
determined by the kinds of nodes it has, and these are specified as follows:

Definition 1. (signature)   A *signature* $\mathcal{K}$ provides a set whose elements are
called *controls*. For each control K the signature also provides a finite ordinal
$ar(K)$, its *arity*. We write $K\colon n$ for a control K with arity n.   ∎

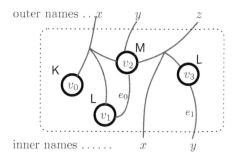

outer names ... x y z

inner names ...... x y

Fig. 1. A link graph $G\colon \{x, y\} \to \{x, y, z\}$.

We now proceed to define link graphs over a signature $\mathcal{K}$. Every node in a link graph has an associated control $K\colon n$, and has n ports. Informally, the graph consists essentially of an arbitrary linking of these ports, together with *inner* and *outer names*, which provide access to some of the links. Figure 1 shows a simple link graph, whose nodes have controls K, L and M with arities 1, 2 and 4 respectively. Assuming an infinite vocabulary $\mathcal{X}$ of *names*, we now give formal definitions.

Definition 2. (interface) An *interface* $X, Y, \ldots$ is a finite set of names drawn from $\mathcal{X}$. We call the empty interface as the *origin*, and denote it by ϵ. ∎

Definition 3. (link graph) A *link graph*

$$A = (V, E, ctrl, link)\colon X \to Y$$

has interfaces X and Y, called its *inner* and *outer names*, and disjoint finite sets V of *nodes* and E of *edges*. It also has a *control map* and a *link map*, respectively $ctrl\colon V \to \mathcal{K}$ and $link\colon X \uplus P \to E \uplus Y$, where $P \stackrel{\mathrm{def}}{=} \sum_{v \in V} ar(ctrl(v))$ is the set of *ports* of A.

A *ground* link graph is one with no inner names, i.e. with inner interface $\emptyset$.∎

We shall call the inner names X and ports P the *points* of A, and the edges E and outer names Y its *links*. Thus the link map sends points to links.

The support of a link graph consists of its nodes and edges; in terms of the definition, $|A| = V \uplus E$. If ρ is an injective map on $|A|$, the *support translation* $\rho \cdot A$ is obtained by replacing each $v \in V$ by $\rho(v)$ and each $e \in E$ by $\rho(e)$ everywhere in A.

The link graph in Figure 1 has nodes $V = \{v_0, \ldots, v_3\}$ and edges $E = \{e_0, e_1\}$; note that a link can either be an edge like e_0 or and outer name like z.

Definition 4. (s-category of link graphs) The s-category $\mathrm{LIG}(\mathcal{K})^1$ over a signature $\mathcal{K}$ has name sets as objects and link graphs as arrows. The composition

---

[1] The s-category $\mathrm{LIG}(\mathcal{K})$ involves *concrete* link graphs, whose nodes and edges have identity. Elsewhere we have denoted this s-category by $\mathrm{`LIG}$, reserving the notation LIG for the category of *abstract* link graphs in which this identity is factored out. Here we drop the accent, since we are not concerned with abstract link graphs.

$A_1 A_0 \colon X_0 \to X_2$ of two link graphs $A_i = (V_i, E_i, ctrl_i, link_i) \colon X_i \to X_{i+1}$ ($i = 0, 1$) is defined when their supports are disjoint; then their composite is

$$A_1 A_0 \stackrel{\text{def}}{=} (V_0 \uplus V_1, E_0 \uplus E_1, ctrl, link) \colon X_0 \to X_2$$

where $ctrl = ctrl_0 \uplus ctrl_1$ and $link = (\mathsf{Id}_{E_0} \uplus link_1) \circ (link_0 \uplus \mathsf{Id}_{P_1})$.

The identity link graph at X is $\mathrm{id}_X \stackrel{\text{def}}{=} (\emptyset, \emptyset, \emptyset_{\mathcal{K}}, \mathsf{Id}_X) \colon X \to X$.   ∎

To clarify composition, here is another way to define the link map of $A_1 A_0$, considering all possible arguments $p \in X_0 \uplus P_0 \uplus P_1$:

$$link(p) = \begin{cases} link_0(p) \text{ if } p \in X_0 \uplus P_0 \text{ and } link_0(p) \in E_0 \\ link_1(x) \text{ if } p \in X_0 \uplus P_0 \text{ and } link_0(p) = x \in X_1 \\ link_1(p) \text{ if } p \in P_1 \ . \end{cases}$$

We often denote the link map of A simply by A.

Note that the link map treats inner and outer names differently. Two inner names may be points of the same link, but each outer name constitutes (the target of) a distinct link. The effect is that we do not allow 'aliases', i.e. synonymous outer names.

Figure 2 shows an example of composing two link graphs; controls are not shown. The shape of nodes has no formal significance. It can be seen how the notion of s-category allows composition to preserve the identity of nodes and edges; we can tell exactly which nodes and edges in the composite came from each component. It is this feature that ensures the existence of RPOs, which is essential for the dynamic theory.

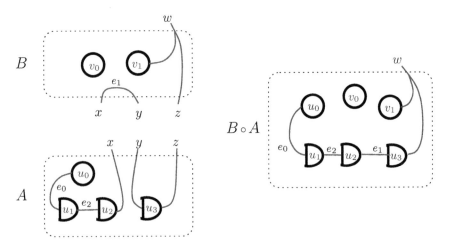

Fig. 2. Composing two link graphs.

3   Inclusions

In what follows we pay particular attention to ground link graphs $G: \epsilon \to X$, for which we write simply $G: X$. We shall denote the components of an arbitrary pair of ground link graphs $G_i: X_i$ ($i = 0, 1$) by $V_i, E_i, ctrl_i$ and $link_i$.

Definition 5. (inclusion)   Let $G_i: X_i$ be two ground link graphs with $|G_0| \subseteq |G_1|$. Then an *inclusion* $\eta: G_0 \to G_1$ is an injective map $\eta: |G_0| \uplus X_0 \to |G_1| \uplus X_1$, where $\eta = \eta^{\mathsf{v}} \uplus \eta^{\mathsf{e}} \uplus \eta^{\mathsf{n}}$ satisfies the following conditions:

(1) $\eta^{\mathsf{v}}: V_0 \hookrightarrow V_1$ is an inclusion map
(2) $\eta^{\mathsf{e}}: E_0 \hookrightarrow E_1$ is an inclusion map
(3) $\eta^{\mathsf{n}}: X_0 \to E_1 \uplus X_1$ is an arbitrary map
(4) $ctrl_1(v) = ctrl_0(v)$ ($v \in V_0$)
(5) $link_1(p) = \eta(link_0(p))$ ($p \in P_0$) .   ∎

Thus the structure of G_0 is preserved by the inclusion, except that named links may be coalesced since η^{n} need not be injective. Later we shall generalise inclusions to *embeddings*, where the two inclusions maps are replaced by injections.

Definition 6. (inclusion category)   The category INC has as objects the ground link graphs. Its arrows are inclusions, composed as functions, with identities $\mathsf{Id}_{|G|}$.   ∎

The question immediately arises: how is INC related to LIG? after all, both categories insert link graphs into bigger link graphs; in LIG this is done by composing them as arrows, and in INC by maps between them as objects. Following the idea of Cattani, we use the standard notion of a *coslice* for turning arrows into objects.

Definition 7. (coslice)   Let ϵ be any object in an s-category $\mathbf{C}$. The *coslice category* $\epsilon/\mathbf{C}$ has as objects the arrows of $\mathbf{C}$ with domain ϵ. Each of its arrows $B: G_0 \to G_1$, for $G_i: \epsilon \to X_i$ ($i = 0, 1$), consists of an arrow $B: X_0 \to X_1$ in $\mathbf{C}$ such that $BG_0 = G_1$. Composition and identities are defined by those of $\mathbf{C}$.

We define $\text{CXT}_\mathsf{I} \overset{\text{def}}{=} \epsilon/\text{LIG}$, whose objects are the ground link graphs.   ∎

Note that CXT_I is not only an s-category but a category. Also its objects are those of INC. We shall now prove that these two categories are isomorphic.

Construction 8 (link graph from inclusion)   For $G_i: X_i$ ($i = 0, 1$), we define a function
$$\mathcal{C}: \text{INC}(G_0, G_1) \to \text{LIG}(X_0, X_1) .$$

For this purpose, given an inclusion $\eta: G_0 \to G_1$ we define the components of the link graph $\mathcal{C}(\eta) = B: X_0 \to X_1$ as follows.

Set $V_B \overset{\text{def}}{=} V_1 \setminus V_0$, $E_B \overset{\text{def}}{=} E_1 \setminus E_0$ and $ctrl_B \overset{\text{def}}{=} ctrl_1 \upharpoonright V_B$. Then define the link map $link_B: X_0 \uplus P_B \to E_B \uplus X_1$, where $P_B = \sum_{v \in V_B} ar(ctrl_B(v))$:

(1) $link_B(x) \overset{\text{def}}{=} \eta(x)$ for $x \in X_0$
(2) $link_B(p) \overset{\text{def}}{=} link_1(p)$ for $p \in P_B$.   ∎

Proposition 1. (context functor) *Construction 8 yields a functor* $\mathcal{C} \colon \textsc{Inc} \to$ $\textsc{Cxt}_\textsc{I}$.

Proof (outline) Let $\eta \colon G_0 \to G_1$ and $B = \mathcal{C}(\eta)$ be as in the construction. We first establish B as an arrow in $\textsc{Cxt}_\textsc{I}$, i.e. to verify that $BG_0 = G_1$ in $\textsc{Lig}$; this is simple case analysis, using the definitions. Then by a routine case analysis we verify the functorial properties, i.e. that $\mathcal{C}(\mathsf{Id}_{|G|}) = \mathsf{id}_X$ for $G \colon X$ and that $\mathcal{C}(\eta_1 \eta_0) = \mathcal{C}(\eta_1)\mathcal{C}(\eta_0)$. ∎

We now proceed to show that the categories $\textsc{Inc}$ and $\textsc{Cxt}_\textsc{I}$ are isomorphic. For this purpose it is enough to show that a functor between them – in this case the functor $\mathcal{C}$ – is both full and faithful, i.e. bijective on each homset. To achieve this we need only show that the function $\mathcal{C}$ on each homset has an inverse.

Definition 9. (inclusion from context)   For $G_i \colon X_i$ $(i = 0, 1)$, we define a function
$$\mathcal{I} \colon \textsc{Cxt}_\textsc{I}(G_0, G_1) \to \textsc{Inc}(G_0, G_1) \ .$$

Foir this purpose, given a link graph $B \colon X_0 \to X_1$ such that $BG_0 = G_1$, we define an inclusion $\eta = \mathcal{I}(B) \colon G_0 \to G_1$ as follows:

(1) $\eta(v) \stackrel{\text{def}}{=} v$ \qquad for $v \in V_0$
(2) $\eta(e) \stackrel{\text{def}}{=} e$ \qquad for $e \in E_0$
(3) $\eta(x) \stackrel{\text{def}}{=} link_B(x)$ for $x \in X_0$.

It is a routine matter to show, using $BG_0 = G_1$, that η satisfies the five conditions of Definition 5. Thus our function $\mathcal{I}$ is well-defined. ∎

We are now in a position to prove the main theorem:

Theorem 1. (inclusions are contexts) $\textsc{Inc}$ *and* $\textsc{Cxt}_\textsc{I}$ *are isomorphic categories.*

Proof (outline) It only remains to prove that, as functions between the homsets $\textsc{Inc}(G_0, G_1)$ and $\textsc{Cxt}_\textsc{I}(G_0, G_1)$, the two functions $\mathcal{I}$ are $\mathcal{C}$ are inverse; that is, $\mathcal{I} \circ \mathcal{C} = \mathsf{Id}$ and $\mathcal{C} \circ \mathcal{I} = \mathsf{Id}$. Again, the argument is a routine case analysis. ∎

The above definitions and proofs are so natural that we may expect to find this close tie between contexts and inclusions to hold for any natural species of graph. We defer discussion of this to the concluding section.

An earlier instance of the close tie was found in shallow action graphs [3] where – due to Cattani's insight – we proved an analogous result. There, the definitions were harder, and the proof correspondingly less obvious. The theorem also served useful purpose: We needed to prove the existence of RPOs for a contextual category of shallow action graphs, and we observed that an RPO in the contextual category (or s-category) corresponds exactly to a coproduct in the coslice category. Therefore, via the isomorphism of categories, it corresponds to a coproduct in the category of inclusions; so we conducted the proof in the latter category.

4   Embeddings

The category INC of inclusions is rather thin, in the sense that, given G_0 and G_1, there is often at most one way to include G_0 in G_1. This is not surprising, because there is at most one inclusion map between two given sets. Indeed, for $G_0 : X_0$, it can be shown that if the link map of G_0 is surjective on X_0 (in link graph terminology, G_0 has no idle names) then $\text{INC}(G_0, G_1)$ has at most one member.

Let us now consider graph embeddings. In normal graph theory, an embedding of G_0 in G_1 is naturally understood to be a defined as a pair of maps, mapping respectively the nodes and the edges of G_0 into those of G_1, with the proviso that they respect the graph structure. Here, unless otherwise stated, we consider embeddings to be injective.

Injection maps are more numerous than inclusion maps. We can reflect this for link graphs by deriving embeddings from inclusions, as follows:

Definition 10. (embeddings of link graphs)   An *embedding* $\phi : G_0 \to G_1$ is a pair $\phi = (\eta, \rho)$ of a bijection ρ on $|G_0|$ and an inclusion $\eta : \rho \cdot G_0 \to G_1$.

The category EMB of embeddings has ground link graphs as objects and embeddings as arrows. The identity embedding on G is $(\text{Id}_{|G|}, \text{Id}_{|G|})$, and the composition $(\eta_1, \rho_1)(\eta_0, \rho_0)$ is the unique pair (η, ρ) of an inclusion and a bijection such that $\eta \rho = \eta_1 \rho_1 \eta_0 \rho_0$.   ∎

Can we extend our correspondence between contexts and inclusions to a correspondence between contexts and embeddings? Clearly we need a more generous notion of context, one that permits support translation of a ground link graph G before composition with a context B. This is captured by defining a thicker form of coslice:

Definition 11. (thick coslice)   The *thick coslice category* $\epsilon /\!/ \text{LIG}$ has as objects the ground link graphs. Each arrow takes the form $(B, \rho) : G_0 \to G_1$, where ρ is a bijection on $|G_0|$ and $B(\rho \cdot G_0) = G_1$ in LIG. Composition is given by

$$(C, \sigma)(B, \rho) \stackrel{\text{def}}{=} (C(\sigma \cdot B), \sigma\rho)$$

and identities take the form $(\text{id}_G, \text{Id}_{|G|})$. We shall denote $\epsilon /\!/ \text{LIG}$ by CXT_E.   ∎

Corollary 1. (embeddings are contexts)   EMB *and* CXT_E *are isomorphic categories.*

Proof   We easily find that $(B, \rho) : G_0 \to G_1$ is an arrow in CXT_E iff $B : \rho \cdot G_0 \to G_1$ is an arrow in CXT_I. The rest follows from Theorem 1.   ∎

This is the most important consequence of our main theorem. It shows that a natural class of injective embeddings can be 'defined' from an s-category of graphs-as-arrows.

As mentioned earlier, the notion of support in s-categories keeps track of the identity of nodes and edges through composition, thus ensuring the existence of RPOs for link graphs, as required by the dynamic theory. Sassone and

Sobocinski [14] have proposed G-categories (G for *groupoid*) as an alternative way to keep track of this identity. The above programme can also be followed in G-categories; the natural approach is as follows (a familiarity with 2-categories is helpful here).

A 2-category $\mathbf{C}$ is an enriched category where each homset $\mathbf{C}(X, Y)$ becomes a category itself; that is, there are second-order arrows called *2-cells* between the ordinary first-order arrows. The two compositions obey a simple structural law. Sassone and Sobocinski specialise 2-categories to *G-categories*, those in which every 2-cell is an isomorphism; they have also verified the existence of generalised RPOs, called groupoid RPOs, in suitable cases.

This leads to treating link graphs as first-order arrows in a G-category LIG_2; composition is fully defined (unlike in an s-category) via suitable renaming of nodes and edges, and the 2-cell isomorphisms keep track of the identity of nodes and edges.

2-categories also possess a generalised form of coslice (yielding a category, not a 2-category). So the previous programme can be followed, coslicing at the origin to form ϵ/LIG_2. It then appears that the programme can be completed by proving this category isomorphic to a category of embeddings between ground link graphs.

5   Conclusion

The coslice correspondence between graphs-as-arrows and graph-as-objects is rather straightforward in the case of link graphs, and this suggests that for other graphical structures it may be equally straighforward. As mentioned in the introduction, we would like to find conditions on a graphical structure that will guarantee an isomorphism between the coslice category and the embedding category.

A related question arises from the observation that the coslice construction goes from graphs-as-arrows to graphs-as-objects, while the cospan construction goes the other way. Could these constructions be inverse to one another?

It would be natural to look at this question first for link graphs; here, the cospans $I \to G \leftarrow J$ are simple because the interfaces are just discrete graphs. Thereafter the question can be investigated for more complex graphical structures with more complex interfaces.

The rationale of this paper is that, as graphical models of computing become more important – as they must do with the increasing emphasis on spatial structure and mobility in real systems – so it becomes more important to knit together different graphical formulations into a single theory.

References

1. Brookes, S.D., Hoare, C.A.R. and Roscoe, A.W. (1984), A theory of communicating sequential processes. J. ACM 31, pp560–599.

2. Cardelli, L. and Gordon, A.D. (2000), Mobile ambients. Foundations of System Specification and Computational Structures, LNCS 1378, pp140–155.
3. Cattani, G.L., Leifer, J.J. and Milner, R. (2000), Contexts and Embeddings for closed shallow action graphs. University of Cambridge Computer Laboratory, Technical Report 496. [Submitted for publication.] Available at `http://pauillac.inria.fr/~leifer`.
4. Ehrig, H. (1979) Introduction to the theory of graph grammars. Graph Grammars and their Application to Computer Science and Biology, LNCS 73, Springer Verlag, pp1–69.
5. Ehrig, H. (2002) Bigraphs meet double pushouts. EATCS Bulletin 78, October 2002, pp72–85.
6. Ehrig, H. and König, B. (2004), Deriving bisimulation congruences in the DPO approach to graph-rewriting. Proc. FOSSACS 2004, LNCS 2987, pp151–156.
7. Gadducci, F., Heckel, R. and Llabrés Segura, M. (1999), A bi-categorical axiomatisation of concurrent graph rewriting. Proc. 8th Conference on Category Theory in Computer Science (CTCS'99), Vol 29 of Electronic Notes in TCS, Elsevier Science.
8. Jensen, O.H. and Milner, R. (2004), Bigraphs and mobile processes (revised). Technical Report 580, University of Cambridge Computer Laboratory. Available from `http://www.cl.cam.ac.uk/users/rm135`.
9. Lawvere, F.W. (1963), Functorial semantics of algebraic theories. Dissertation, Columbia University. Announcement in Proc. Nat. Acad. Sci. 50, 1963, pp869–873.
10. Leifer, J.J. and Milner, R. (2000), Deriving bisimulation congruences for reactive systems. Proc. CONCUR 2000, 11th International Conference on Concurrency theory, pp243–258. Available at `http://pauillac.inria.fr/~leifer`.
11. Leifer, J.J. and Milner, R. (2004), Transition systems, link graphs and Petri nets. Forthcoming Technical Report, University of Cambridge Computer Laboratory.
12. Milner, R. (1980) *A Calculus of Communicating Systems*. LNCS 92, Springer Verlag.
13. Petri, C.A. (1962), Kommunicaten mit automaten. Schriften des Institutes für Instrumentelle Mathematik, Bonn.
14. Sassone, V. and Sobocinski, P. (2002), Deriving bisimulation congruences: a 2-categorical approach. Electronic Notes in Theoretical Computer Science, Vol 68 (2).

Towards Architectural Connectors for UML

Fernando Orejas[1] and Sonia Pérez[1,2]

[1] Departament LSI,
Universitat Politècnica de Catalunya,
Barcelona, Spain
[2] Instituto Superior Politécnico José Antonio Echevarría,
Havana, Cuba
{orejas,sperezl}@lsi.upc.es

Abstract. The notion of architectural connector was developed by Allen and Garland [1] as an important concept for the design of software architectures. In this paper, based on previous work introducing a generic approach for the definition of component-based concepts, we study how architectural connectors and components can be defined for class and sequence diagrams as a first step for applying this approach to full UML. In particular, the case of sequence diagrams is studied with some detail. A case-study of a lift system is used to illustrate these ideas.

1 Introduction

The development of component-based systems is nowadays an important area in software engineering. In this context, a lot of work has been dedicated to different issues related to approach, such as the development of methodologies and the implementation of middleware and other related tools. However, much less work has been dedicated to the modelling and specification phase of this kind of systems. In this sense, an approach that we consider very interesting is based on the use of *architectural connectors* [1,8]. In this approach architectures are built in terms of two kinds of units: components and connectors. Components are not connected directly, but through connectors. Components offer some functionality and connectors describe policies of interaction of the connected components. Originally work, the language used for the specification and modelling of components and connectors was CSP [6]. The work using this approach was followed by Fiadeiro (e.g. see [5]), who generalized in some sense the approach by putting it into a categorical context in the framework of the coordination language COMMUNITY.

In [3], we developed a very generic approach for the definition of components whose aim was to allow the definition of component concepts associated to arbitrary formal or semiformal specification methods. The idea was that one could instantiate this generic approach to any arbitrary method, as long as one could prove that it satisfied certain properties. In particular, different instantiations were sketched in terms of Petri Nets, graph transformation systems or algebra transformation systems. We also studied how the approach in [3] could be used

H.-J. Kreowski et al. (Eds.): Formal Methods (Ehrig Festschrift), LNCS 3393, pp. 352–369, 2005.

to define generic architectural connectors. In particular, preliminary results, including an instantiation to Petri Nets, were presented in [4].

In the case of UML [7], a de facto standard for many industrial applications, the notion of component [2] has a very limited nature, being essentially a kind of syntactic package. In this paper, we present some basic ideas for the instantiation of our approach to the case of UML. In particular, we study how we can define architectural connectors and components in the case of class and sequence diagrams.

In particular, with very little detail in the case of class diagrams and more detail in the case of sequence diagrams, we study how the notions of embedding and transformation (which are key notions in our approach) can be defined. And we show that these notions satisfy the required properties. Especially, the so-called extension property.

The paper is organized as follows. In the next section we review our generic approach to components and architectural connectors. First we describe the approach presented in [3] and then, in the second subsection, we see how this approach can be adapted to define architectural connectors. The third section is the main one. First, we sketch the instantiation of our approach to the case of class diagrams and, then, we study sequence diagrams in some detail. In section 4, we present a small case study, a lift system, to show how these concepts can be applied in practice. Finally, in the last section, we draw some conclusions.

2   A Generic Framework for Architectural Components and Connectors

In this section we will describe the generic component framework introduced in [3], including some variations and extensions needed in this paper. More precisely, in the first subsection we will introduce our generic approach to architectural components and, in the second subsection, we will show how to represent architectural connectors in this framework.

2.1   Generic Components

Components are self-contained units, where some details are hidden to the external user. This is achieved by providing a clear separation between the interfaces and the body of the component. There are two kinds of interfaces: import and export interfaces. The export interfaces describe external views of the behavior of the component including the services that a component offers to the outside world. On the other hand, the import interfaces describe what the component assumes about the environment, including the services used inside the component that are assumed to be provided by other components. The interfaces are used to interconnect components. In particular, we can connect or compose two components by matching an import interface of one component with an export interface of the other component [9]. In [3] we assumed that each component had just one import and one export interface. On the contrary, in this paper we will consider components including more than one import or export interface. In this way one component can be interconnected to several other components.

Our approach is generic in several senses. The first one is that we do not establish, a priori, the kind of specification or modelling technique that has to be used to describe the interfaces or the body of a component. On the contrary, the idea is that the same concepts can be used in connection with different techniques or formalisms.

Obviously, the import and export interfaces should be connected to the body in some well-defined way. However, we are also generic here. We leave open the kind of connectors used to relate the interfaces and body. Intuitively, we assume that the import connections (binding the import interfaces to the body) are some kind of inclusion or embedding, in the sense that the functionality defined in the body is built upon the import interfaces. We also assume that the export connections (binding the export interfaces to the body) are some kind of transformation describing a *refinement* of each export interface.

Now, to define an adequate semantics, ensuring the compositionality of the interconnection operations, we must impose some requirements on the kinds of inclusions and transformations considered for the given specification formalism (for more details see [3]): we assume that a transformation framework T consists of a class of transformations, which is also closed under composition and includes identical transformations and a class of embeddings, which is a subclass of the class of transformations and that is closed under composition and includes the identical embeddings, and such that the following *extension property* is satisfied: For each transformation $trafo\colon SPEC_1 \Rightarrow SPEC_2$, and each inclusion $i_1\colon SPEC_1 \subseteq SPEC_1'$ there is a selected transformation $trafo'\colon SPEC_1' \Rightarrow SPEC_2'$, with inclusion $i_2\colon SPEC_2 \subseteq SPEC_2'$, called the *extension* of $trafo$ with respect to i_1, leading to the following extension diagram:

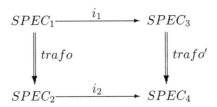

It must be pointed out that, in a given framework T, given $trafo$ and i_1 as above, there may be several $trafo'$ and i_2, that could satisfy this extension property. Our assumption means that only one such $trafo'$ and i_2 are chosen, in some well-defined way, as the extension of $trafo$ with respect to i_1.

Essentially, this extension property means that if one can apply a transformation on a certain specification, then it should be possible to apply the "same" transformation on a larger specification.

As said above, we consider that the import interfaces of a component must be, in some sense, independent. More precisely, we characterize independence in terms of three conditions. The first one states that if two (or more) specifications are embedded into a given one $SPEC$, then we can transform or refine in parallel each of these specifications and obtain a new specification $SPEC'$ which includes all these refinements and is a transformation of $SPEC$. The second condition states that extensions are a special kind of parallel extensions. Finally, the last condition states that we can iterate this kind of parallel transformations. In particular, the last two conditions imply that all these refinements may be done not in parallel, but sequentially, leading to the same result.

Definition 1. *A family of embeddings* $i_j : SPEC_j \to SPEC, (j = 1..n)$ *is independent if the following properties are satisfied:*

- *For every family of transformations* $t_j \in Trafo(SPEC_j), (j = 1..n)$, *there exists a selected transformation* $t \in Trafo(SPEC)$ *and selected independent embeddings* $i'_j, (j = 1..n)$ *such that the diagram in figure 1 commutes.*
  t is called the parallel extension of $\{t_j\}_{j=1..n}$ *with respect to* $\{i_j\}_{j=1..n}$ *and is denoted as* $PE_{\{i_j\}_{j=1..n}}(\{t_j\}_{j=1..n})$
- *For any* $SPEC_j, 1 \le j \le n$, *given the extension diagram of figure 2 and any* $SPEC_k, 1 \le k \le n$, *we have that the diagram in figure 3 is a parallel extension diagram, where* i''_k *is the composition of* i_k *and* t''. *Note that, in this case, we are asking that the composition of the embedding* i_k *and the transformation* t'' *should be an embedding*
- *Parallel extension diagrams can be composed vertically.*

In this context, we can define our generic notion of component:

Definition 2. *A component consists of a body specification with a list of import specifications together with the corresponding embeddings, which are pairwise independent, and a list of export specifications together with the corresponding transformations into the body specifications. Thus, a component will have this general form*

$$(B, \langle b_1 : I_1 \to B, ..., b_n : I_n \to B \rangle, \langle e_1 : E_1 \Longrightarrow B, ..., e_n : E_n \Longrightarrow B \rangle)$$

A possible graphical representation is given in figure 4.

2.2 Architectural Connectors

The notion of architectural connector was developed by Allen and Garland [1] as an important concept for the design of component systems. This approach was then used in a more categorical context by Fiadeiro and Lopes [5] as the

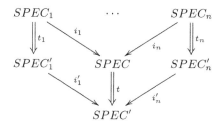

Fig. 1. Parallel Extension.   **Fig. 2.** Extension Diagram.

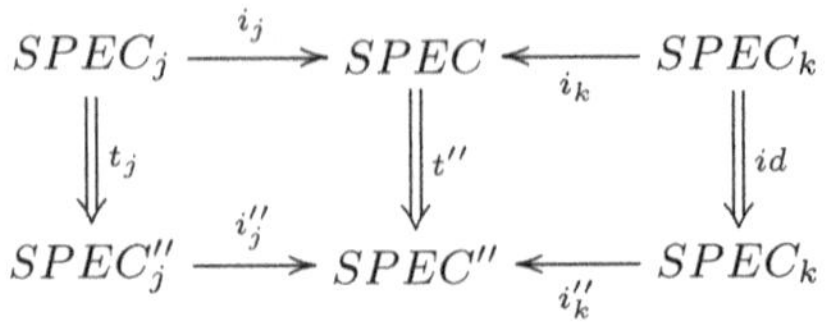

$$SPEC_j \xrightarrow{i_j} SPEC \xleftarrow{i_k} SPEC_k$$

$$\Big\Vert t_j \qquad \Big\Vert t'' \qquad \Big\Vert id$$

$$SPEC_j'' \xrightarrow{i_j''} SPEC'' \xleftarrow{i_k''} SPEC_k$$

Fig. 3. Extension as parallel extension.

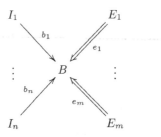

Fig. 4. Diagram of a component.

basis for their COMMUNITY approach. In this approach architectures are built out of two kinds of units: components and connectors. Components are not connected directly, but through connectors. Components offer some functionality and connectors describe policies of interaction of the connected components. For instance, in the example used as a case study in this paper (a system to control a lift, or a set of lifts, of a building) a component describes the functionality of a lift, another component describes the set of buttons that we have in the floors and another component describes the set of buttons that we have inside the lift, In this context, a connector used to interconnect these three components may describe the policy to satisfy the existing calls at a given moment.

In our generic approach, Allen and Garland's components can be seen as components having just export interfaces, called ports in [1], while connectors can be seen as components having just import interfaces, called roles in [1]. More precisely, this means that, in this context, a component $COMP = (B, \langle e_1 : E_1 \Longrightarrow B, ..., e_n : E_n \Longrightarrow B \rangle)$ for $n \geq 0$ is given by the body B and a family of export interfaces E_i with export transformations $e_i : E_i \Longrightarrow B$ for $i \in \{1, ..., n\}$. A connector $CON = (B, \langle b_1 : I_1 \to B, ..., b_n : I_n \to B \rangle)$ for $n \geq 2$ is given by the body B and a family of import interfaces I_i with body embeddings $b_i : I_i \to B$ for $i \in \{1, ..., n\}$.

Now we can define formally how a connector connects different components. Given a connector $CON = (B, b_1, ..., b_n)$ of arity n, and n components $COMP_i = (B_i, e_{i_1}, ..., e_{i_{m_i}})$ of arity m_i with connector transformations $con_i : I_i \Longrightarrow E_{i_k}$ with $1 \leq k \leq m_i$ for $i \in \{1, ..., n\}$ then we obtain the **connector diagram** in Figure 5:

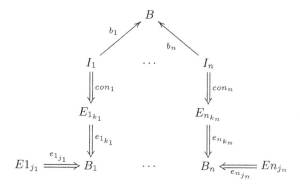

Fig. 5. Connector Diagram.

Now we are able to define the composition of components.

The **composition of n components by a connector of arity** n is defined as follows: Given the corresponding connector diagram (see Figure 5, where, for every i, $t_i = e_{i_{k_i}} \circ con_i$) we construct the corresponding parallel extension diagram (1) in Figure 6. The result of the composition of the components $COMP_1, ..., COMP_n$ by the connector CON with the connection transformations $con_1, ..., con_n$ is again a component.

Based on this kind of connector diagrams, we could define a notion of connectors architecture as a diagram involving several components and connectors interconnecting them in a non-circular manner. In [4] we proved that such architectures denote, after the evaluation of all the composition operations involved, a component which is independent on the order of evaluation of these operations.

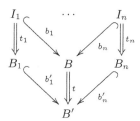

Fig. 6. Composition.

3   Architectural Connectors for UML Diagrams

An UML specification is a set of diagrams of different kinds. Therefore, an obvious approach for defining an instantiation of our generic notion of connectors and components for UML is to instantiate the corresponding concepts for each kind of diagram. In this paper we will only consider the cases of class (with not much detail) and sequence diagrams, but similar ideas can be used in the definition of components and connectors for activity, state and collaboration diagrams. One may consider this approach in some sense insufficient. In particular, we may think that these concepts (e.g. the notion of refinement) should consider all kinds of diagrams at once and not separately to avoid inconsistencies between the various refinements involved. This is certainly true. However, the inconsistency between the different parts of an UML specification is a general problem of UML specifications that we think that should be studied independently. Nevertheless, at some points below, when defining the notion of refinement for sequence diagrams we may mention some additional conditions related to class diagrams.

Now, defining these instantiations for a given class of diagrams means to define notions of embedding and transformation or refinement and showing that they satisfy the required properties (e.g. the extension property). In the following subsections we will define embeddings and transformations for each kind of diagram considered. However, we will study these diagrams without full detail. We think that considering all kinds of details and variations associated to a given class of diagrams will make the paper too boring and would not add any interesting insights to the problem.

3.1 Class Diagrams

We will consider that a class diagram C_1 is embedded in a class diagram C_2 if C_1 is a subdiagram of C_2 up to the renaming of some of the labels (e.g. the names of a class, a method, etc.). This means that, obviously, the requirements for embeddings are satisfied. In particular, embeddings are closed under composition and the identity is an embedding.

Now, in order to define a notion of transformation between class diagrams we must first provide some intuition. We consider that a class diagram is, essentially, the definition of a signature. In particular, in a class diagram we define (we give name to) classes, methods, attributes, etc. This is essentially syntax. The semantics of these elements is given by means of other kinds of diagrams or by an associated OCL specification (not considered in this paper). Only some of the existing relations between classes of a given diagram provide some semantics. In this context, we consider that a refinement of a class diagram is just another class diagram where the latter involves some new elements (classes, methods, relations, etc.) that may be used to refine some elements of the former or that they are just considered hidden when going into a higher level of abstraction. But we think that no additional "semantic" relation is needed to define a refinement. In this sense, we consider that a transformation of class diagram is also just an embedding.

As a consequence, all the required conditions to instantiate our generic concepts to the case of class diagrams are trivially satisfied. In particular, extensions would just be defined in terms of pushouts (unions with some possible renamings to avoid name clashes).

3.2 Sequence Diagrams

In the case of sequence diagrams components would typically consist of several diagrams and not just of a single one. As a consequence, we will first define embeddings between single diagrams and, then, extend this notion to sets of diagrams. In particular, we will consider that a sequence diagram is embedded into another one if the latter describe the same set of interactions (up to renaming) in the same partial order, perhaps intertwined with some other additional interactions. To be more precise:

Definition 3. *A sequence diagram S over a set of messages M is a triple (L, Loc_L, I), where L is the set of lifelines corresponding to the objects that are shown in the diagram; for each $l \in L$ Loc_l is the totally ordered set of locations corresponding to the lifeline l; and I is a set of interactions, where an interaction ε is a triple (loc_1, loc_2, m) where loc_1 and loc_2 are locations associated to some lifelines in L and m is a message in M and such that two interactions never occur simultaneously, i.e. if $(loc_1, loc_2, m), (loc_3, loc_4, m') \in I$, with $(loc_1, loc_2, m) \neq (loc_3, loc_4, m')$ and $loc_i, loc_j \in Loc_l$, $i \in [1, 2]$, $j \in [3, 4]$ then $i \neq j$.*

Moreover, we assume that sequence diagrams must satisfy that the precedence relation defined over the set of interactions of the diagram, $prec_I$, is a partial order, where $prec_S$ is the reflexive and transitive closure of the least relation satisfying that given $(loc_1, loc_2, m), (loc'_1, loc'_2, m') \in I$ if loc_i, loc'_j $(i, j \in [1, 2])$ are in the same lifeline and $loc_i < loc'_j$ then $(loc_1, loc_2, m), (loc'_1, loc'_2, m') \in prec_I$.

A sequence diagram may involve any number of "useless" locations, i.e. locations that do not take part in any interaction.

Definition 4. *A sequence diagram $S = (L, Loc_L, I)$ is minimal if for every location $loc \in Loc_L$ there is an interaction $(loc_1, loc_2, m) \in I$ such that $loc = loc_1$ or, $loc = loc_2$.*

In what follows, we will assume that all diagrams are minimal.

Definition 5. *A message renaming $h : M \to M'$ is an injective mapping on sets of messages.*

Let S be a sequence diagram over M, the renaming of $S = (L, Loc_L, I)$ through $h : M \to M'$, denoted $h(S)$ is the sequence diagram (L, Loc_L, I'), where I' is the set of interactions:

$$\{(loc_1, loc_2, h(m))/(loc_1, loc_2, m) \in I\}$$

Let $S = (L, Loc_L, I)$ be a sequence diagram over M, $S' = (L', Loc'_{L'}, I')$ a sequence diagram over M' and $h : M \to M'$ a message renaming such that $L \subseteq L'$. An h-based embedding $i : S \Rightarrow S'$ is an L-indexed family of injective mappings $t = \{i_l : Loc_l \to Loc'_{L'}\}_{l \in L}$ preserving the order relations (i.e., they must be poset monomorphisms), such that:

1. *For every $(loc_1, loc_2, m) \in I$, with $loc_1 \in Loc_{l_1}$ and $loc_2 \in Loc_{l_2}$, we have that $(i_{l_1}(loc_1), i_{l_2}(loc_2), h(m)) \in I'$.*
2. *If $m' \in h(M)$ then for every $(loc'_1, loc'_2, m') \in I'$ there is $(loc_1, loc_2, m) \in I$ such that $(i_{l_1}(loc_1), i_{l_2}(loc_2), h(m)) = (loc'_1, loc'_2, m') \in I$.*

Note that, according to this definition, all the interactions in the diagram S must be present in the diagram S' exactly in the same order (up to message renaming, but S' may include additional interactions. However, we consider that these additional interactions should be new, i.e. should not be in $h(M)$. The intuition is that the embeddings should preserve in some sense the behavior. In particular, let us suppose that a diagram describes that, after sending the message a, an object sends a message b and then a message c. We think that this behavior would not be preserved by a diagram describing that the same object sends first the message a, then c, then b and then c. However, we could consider this behavior preserved if in between the sequence a,b,c, another message d is sent which could be considered not visible in S.

Now, we can extend this definition to deal with sets of sequence diagrams. The most obvious extension would be to define embeddings of sets of diagrams in a pointwise manner, i.e. as a family of embeddings between the two sets where each diagram in the first set is embedded into another diagram in the second set. However, we think that this may be insufficient. It must be possible to embed a diagram in terms of a set of diagrams, or rather, in terms of the composition of a set of diagrams. Probably, several kinds of composition of diagrams may be sensible, however we have chosen just a simple kind which we feel that should be enough for most purposes. In particular:

Definition 6. *Given two diagrams $S = (L, Loc_L, I)$ and $S' = (L', Loc'_{L'}, I')$ over a set of messages M, we define the* composition $S + S'$ *as the diagram over M $(L \cup L', Loc''_{L \cup L'}, I \cup I')$ where for each $l \in L \cup L', Loc''_l$ is the poset $Loc_l + Loc'_l$,[1] where $+$ denotes disjoint union and all the elements in Loc_l are considered smaller than all the elements in Loc'_l.*

It may be noted that this composition operation is associative, but not commutative. Now, we can define embeddings on sets of diagrams:

Definition 7. *Let $\mathcal{S}$ be a set of sequence diagrams over M, $\mathcal{S}'$ be a set of sequence diagrams over M' and $h : M \to M'$ a message renaming. An h-based embedding $i : \mathcal{S} \Rightarrow \mathcal{S}'$ consists of an h-based diagram embedding for each diagram $S \in \mathcal{S}$: $i = \{i_S : S \Rightarrow S'/S'$ is the composition of a set of diagrams in $\mathcal{S}'\}_{S \in \mathcal{S}}$.*

Obviously, if we consider full UML specifications, including (at least) class diagrams and sets of sequence diagrams, then the renamings corresponding to the embeddings defined for each class of diagrams must be consistent.

It should be obvious that the requirements for embeddings are satisfied. In particular, embeddings are closed under composition and the identity is an embedding.

Now, as we did for the embeddings, we will define transformations for sets of sequence diagrams, extending a definition of diagram transformation. In our opinion, when considering refinement relations between sequence diagrams we may consider two different kinds of intuitions. A straightforward one is to consider a refinement relation as an implementation relation, i.e. a sequence diagram is refined by another one if the latter can be seen as an implementation of the former. In particular, if we consider that single interactions (sending messages)

---

[1] If l is not in L we assume Loc_l to be the empty set and, similarly, if l is not in L' we assume $Loc'_{l'}$ to be empty.

are refined or implemented by other sequence diagrams, then we could define that a sequence diagram D_1 is refined by another sequence diagram D_2 if D_2 is the composition of diagrams implementing the interactions in D_1. However, this is not the only intuition in our context. In particular, if D_2 is a refinement of D_1, the latter diagram may be just an abstraction of D_2, in the sense that some of the interactions described in the body are hidden in D_1 because they are considered irrelevant detail. In particular, this means just that D_1 is embedded in D_2. Putting these two intuitions together, we have that a transformation is a combination of an "implementation" and an embedding:

Definition 8. *Let L and L' be sets of lifelines such that $L \subseteq L'$, M and M' sets of messages and S a set of sequence diagrams whose sets of lifelines are included in L'. Let T be a set of L-typed interactions over M, i.e. a subset of $L \times L \times M$. An implementation $\mathcal{I}$ of T by (L', S) is a pair of mappings $(\mathcal{I}_{Lines} : L \to 2^{L'}, \mathcal{I}_{Mess} : T \to S)$, such that:*

1. *For every lifeline l in L, $l \in \mathcal{I}_{Lines}(l)$*
2. *If $l_0 \neq l_1$ then $\mathcal{I}_{Lines}(l_0) \cap \mathcal{I}_{Lines}(l_1) = \emptyset$*
3. *If $\mathcal{I}_{Mess}(l_0, l_1, m) = (L_1, Loc1_L, I_1)$ then $L1 = \mathcal{I}_{Lines}(l_0) \cup \mathcal{I}_{Lines}(l_1)$*

If $\mathcal{I}$ is an implementation and $\varepsilon = (loc_0, loc_1, m)$ is an interaction in the diagram $S = (L, Loc_L, I)$, we define $\mathcal{I}(\varepsilon)$ as follows: if $loc_0 \in l_0$ and $loc_1 \in l_1$ and $\mathcal{I}_{Mess}(l_0, l_1, m) = S'$ then $\mathcal{I}(\varepsilon) = S'$; otherwise, $\mathcal{I}(\varepsilon)$ is the diagram consisting only of the interaction ε.

Let $S = (L, Loc_L, I)$ be a diagram over M and $\mathcal{I}$ an implementation by (L', S) of a set T of L''- typed interactions over M, where $L \subseteq L''$ then the application of $\mathcal{I}$ to S, denoted $\mathcal{I}(S)$ is defined as follows. Let $\langle \varepsilon_1, \ldots, \varepsilon_n \rangle$ be a sequence of interactions, with $I = \{\varepsilon_1, \ldots, \varepsilon_n\}$, such that $(\varepsilon_j, \varepsilon_k) \in prec_I$ then $j < k$; then $\mathcal{I}(S) = \mathcal{I}(\varepsilon_1) + \ldots + \mathcal{I}(\varepsilon_n)$.

Finally, if S is a set of diagrams over M and $\mathcal{I}$ an implementation by (L', S') of T, then the application of $\mathcal{I}$ to S, denoted $\mathcal{I}(S)$, is the set of all diagrams $\mathcal{I}(S)$ such that $S \in S$.

The intuition of implementations is, on one hand, that in the refinement of each lifeline other lifelines may be involved, which are considered hidden at a higher abstraction level. In this sense, the first condition states each lifeline is part of its own refinement. The second condition states that a given lifeline cannot be involved in the implementation of two different lifelines. On the other hand, the third condition, states that if a given (typed) interaction is implemented by a certain diagram, then this diagram includes only the lifelines which implement the lifelines occurring in the interaction. Then, applying an implementation to a diagram means replacing all the interactions by the corresponding diagrams defined by the implementation. Note that we allow to apply an implementation to diagrams whose sets of lifelines do not coincide, but are included, in the set of lifelines implemented by $\mathcal{I}$.

Note also that, given a message renaming $h : M \to M'$, the translation of a sequence diagram D over M and an h-based refinement associated can be seen as a special cases of implementations.

It may be proved that the definition of $\mathcal{I}(S)$ is independent of the specific sequence of interactions chosen. In particular if ε and ε' are independent interactions then $\mathcal{I}(\varepsilon) + \mathcal{I}(\varepsilon') = \mathcal{I}(\varepsilon') + \mathcal{I}(\varepsilon)$. The reason is that, if the two interactions are independent then the lifelines involved in the interactions are disjoint and, as a consequence, the sets of lifelines involved in the diagrams $\mathcal{I}(\varepsilon)$ and $\mathcal{I}(\varepsilon')$ are also disjoint. In this context, we can define a notion of transformation or refinement over sequence diagrams.

Definition 9. *Let $S = (L, Loc_L, I)$ be a sequence diagram over M and $S' = (L', Loc'_{L'}, I')$ a sequence diagram over M'. A transformation $t : S \Rightarrow S'$ is a pair $(\mathcal{I}, i)$, where $\mathcal{I}$ is an implementation by (L', S') of a set T of L-typed interactions over M and i is an embedding of $\mathcal{I}(S)$ into S'.*

Now, we can extend this definition to deal with sets of sequence diagrams in a similar way as we did with embeddings, i.e. by allowing a diagram to be refined in terms of the composition of several other diagrams. However, to be consistent in the refinement of the different diagrams we will assume that the implementation used is always the same one:

Definition 10. *Let $\mathcal{S}$ be a set of sequence diagrams over M and $\mathcal{S}'$ a set of sequence diagrams over M'. A transformation $t : \mathcal{S} \Rightarrow \mathcal{S}'$ is a pair $(\mathcal{I}, i)$, where $\mathcal{I}$ is an implementation by $(L'', \mathcal{S}'')$ of a set T of L''-typed interactions over M and where L'' includes all the sets of lifelines of the diagrams in $\mathcal{S}$, and i is an embedding of $\mathcal{I}(\mathcal{S})$ into $\mathcal{S}'$.*

It should be clear that this notion of transformation satisfies that is closed under composition and that the identity is a transformation. Therefore, we just have to prove that the the two notions of embedding and transformation satisfy the extension property. We will do this in four steps. First, we will show that if $i_1 : S_0 \to S_1$ and $i_2 : S_0 \to S_2$ are embeddings then we can define a diagram S_3 that embeds S_1 and S_2. Actually, the construction is a pushout in a category of embeddings, although we will not prove it. The second step will be to show that if $i_1 : S_0 \to S_1$ is an embedding and $\mathcal{I}_2$ is an implementation such that $\mathcal{I}_2(S_0) = S_2$ then we can define an implementation $\mathcal{I}_1$ such that $\mathcal{I}_1(S_1)$ embeds S_2. From these two properties, we can easily conclude the extension property for single diagrams. Finally, we will prove the extension property for sets of diagrams.

Proposition 1. *Let $S_j = (L_j, Loc^j_{L_j}, I_j)$, be sequence diagrams over M_j, for $j = 0, 1, 2$, respectively. Let $h_1 : M_0 \to M_1$ and $h_2 : M_0 \to M_2$ be message renamings, $i_1 : S_0 \to S_1$ and $i_2 : S_0 \to S_2$ are h_1 and h_2-based embeddings, respectively. Let M_3 be the pushout (on the category of sets) defined on figure 7.*

Let L_3 be $L_2 + (L1 \setminus L0)$ (i.e. the pushout in the category of sets of L_2 and $L1$ sharing $L0$) and let $Loc3_l$, for each $l \in L_3$ be the pushout (on the category of posets) defined on figure 8.

Finally, let S_3 be the diagram $(L_3, Loc3_{L_3}, I_3)$ over M_3, where the set of interactions

Fig. 7. Definition of M_3. **Fig. 8.** Definition of $Loc3_l$.

$$I_3 = \{(i'_1(loc), i'_1(loc'), h'_1(m)/(loc, loc', m) \in I_1\}$$
$$\cup \{i'_2(loc), i'_2(loc'), h'_2(m)/(loc, loc', m) \in I_2\}$$

Then, $i'_1 : S_1 \rightarrow S_3$ and $i'_2 : S_2 \rightarrow S_3$ are h'_1 and h'_2-based embeddings, respectively.

By construction, the proof of this proposition should be obvious.

Definition 11. *Let $h_1 : M_0 \rightarrow M_1$ be a message renaming, L_j sets of lifelines, for $j = 0, 1$, such that $L_0 \subseteq L_1$, T a set of L_0-typed interactions over M_0, and $\mathcal{I}$ an implementation of T by $(L, \mathcal{S})$ for given sets L of lifelines and $\mathcal{S}$ of diagrams over a set of messages M. Then, we define T' and $\mathcal{I}'$, called the extensions of T and $\mathcal{I}$ with respect to h_1 and L_1, as follows: T' is the set of L_1-typed interactions over M_1:*

$$T' = \{(l_1, l'_1, h_1(m))/(l_1, l'_1, m) \in T_0$$

and $\mathcal{I}'$ is the implementation of T' by $(L' + (L_1 \setminus L_0), \mathcal{S})$ be defined as follows:

- *For every $l \in L_0$, $\mathcal{I}'_{Lines}(l) = \mathcal{I}_{Lines}(l)$ and for every $l \in (L_1 \setminus L_0)$, $\mathcal{I}'_{Lines}(l) = \{l\}$*
- *For every $(l_1, l'_1, h_1(m)) \in T'$, $\mathcal{I}'_{Mess}(l_1, l'_1, h_1(m)) = \mathcal{I}_{Mess}(l_1, l'_1, m)$*

Proposition 2. *Let $S_j = (L_j, Loc^j_{L_j}, I_j)$ be sequence diagrams over M_j, for $j = 0, 1$, respectively. Let $h_1 : M_0 \rightarrow M_1$ be a message renaming, $i_1 : S_0 \rightarrow S_1$ an h_1-based embedding, $\mathcal{I}$ an implementation of a set T of L_0-typed interactions over M_0 by $(L, \mathcal{S})$, and let T' and $\mathcal{I}'$ be the extensions of T and $\mathcal{I}$ with respect to h_1. Then, the diagram $\mathcal{I}(S_0)$ is embedded into $\mathcal{I}'(S_1)$.*

Proof. According to the definition of $\mathcal{I}'$, we have that if $\langle \varepsilon'_1, \ldots, \varepsilon'_n \rangle$ is an ordered sequence of the interactions in I_1 then:

$$\mathcal{I}'(S_1) = \mathcal{I}'(\varepsilon_1) + \ldots + \mathcal{I}'(\varepsilon_n)$$

On the other hand, by definition, we know that for every interaction $(l_0, l'_0, m_0) \in S_0$, $\mathcal{I}(l_0, l'_0, m_0) = \mathcal{I}'(i_1(l_0), i_1(l'_0), h_1(m_0))$. This means that $\mathcal{I}(S_0)$ is the sum

$$\mathcal{I}'(S_1) = \mathcal{I}'(\varepsilon_{j_1}) + \ldots + \mathcal{I}'(\varepsilon_{j_m})$$

of a subset of $\{\mathcal{I}'(\varepsilon_1), \ldots, \mathcal{I}'(\varepsilon_n)\}$. This directly implies the embedding.

Proposition 3. *Let $S_j = (L_j, Loc^j_{L_j}, I_j)$, be sequence diagrams over M_j, for $j = 0, 1, 2$, respectively. Let $h_1 : M_0 \rightarrow M_1$ be a message renaming, $i_1 : S_0 \rightarrow S_1$ an h_1-based embedding, and $t_2 : S_0 \rightarrow S_2$, $t_2 = (\mathcal{I}_2, i_2)$ a transformation.*

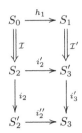

Fig. 9. Extension of diagram transformations.

Let S_3' be $\mathcal{I}'(S_1)$ where $\mathcal{I}'$ is the extension of $\mathcal{I}$ with respect to h_1 and let i_2' be the embedding whose existence was proved in the previous proposition.

Let S_3 be the the diagram associated to the embeddings i_2 and i_2', defined according to proposition 1, which embeds S_2' via i_2'' and S_3' via i_3' (for a graphical explanation, see figure 9).

Then, S_3 embeds S_2 via $i_2'' \circ i_2'$ and refines S_1 via $t_1 = (\mathcal{I}', i_3')$

Proof. The proposition is a direct consequence of the previous two propositions.

Theorem 1. *Let $\mathcal{S}_j$ be sets of sequence diagrams over M_j, for $j = 0, 1, 2$, respectively. Let $h_1 : M_0 \rightarrow M_1$ be a message renaming, $i_1 : \mathcal{S} \Rightarrow \mathcal{S}'$ an h_1-based embedding and $t_2 : \mathcal{S}_0 \Rightarrow \mathcal{S}_2$ a transformation, $(\mathcal{I}, i_2)$. Then, there is a set of diagrams $\mathcal{S}_3$, such that $\mathcal{S}_2$ is embedded into $\mathcal{S}_3$ and $\mathcal{S}_3$ refines $\mathcal{S}_1$.*

Proof. Let $\mathcal{S}_3$ be $h'2(\mathcal{S}_2) \cup \mathcal{I}'(\mathcal{S}_1')$, where $\mathcal{S}_1' = \mathcal{S}_1) \cup \{i_1(S_0)/S_0 \in \mathcal{S}_0$ and $\mathcal{I}'$ is the extension of $\mathcal{I}$ with respect to h_1.

Now, by construction, $\mathcal{S}_3$ obviously embeds $\mathcal{S}_2$ since $\mathcal{S}_3$ includes a renaming of the diagrams in $\mathcal{S}_2$. On the other hand, $\mathcal{S}_3$ refines $\mathcal{S}_1$ since $\mathcal{I}'(\mathcal{S}_1)$ is included in $\mathcal{S}_3$.

4   An Example

In this section we will present a small example of the use of this kind of component system. For brevity we will only use sequence diagrams, which means that the corresponding class diagrams will remain implicit. The example describes a lift system including just one lift. However, a system including several lifts would not be difficult to describe using the same components, but a more complex connector.

We consider that a lift system can be built (at a certain level of abstraction) out of three kinds of components: the elevators themselves, including the doors and the engines to move the lift; the buttons that are located inside the lift; and the set of buttons which are located in each floor (for simplicity we will consider that there is only one button per floor and not two, as it usually happens). Also for simplicity, will only describe the normal scenario describing the system, i.e. we will not consider abnormal situations. Now, let us model these components.

The body of the elevator can be described by the diagram in figure 10.

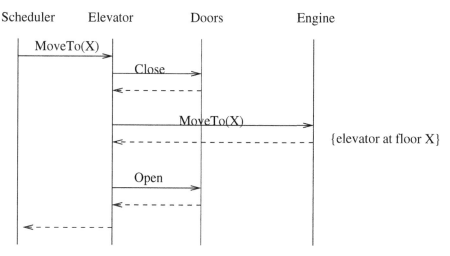

Fig. 10. Elevator Body.

This diagram describes the following scenario. Someone, which we have called the scheduler, tells the elevator to move to floor X. This causes the doors to close and when they are closed (an ack is received), the elevator sends a message to the engine to move to floor X. When the elevator is at floor X, the doors open and the scheduler is acknowledged that the operation has been completed. Now for the interface there are details that can be abstracted from this diagram. In particular, for the use (as a component) of the elevator, we do not need to know about how doors are opened or how the engine works. So, the elevator interface is just the diagram in figure 11.

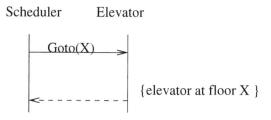

Fig. 11. Elevator Interface.

Obviously, this interface is refined by the body of the component (actually the transformation is just an embedding). Now, the body of the component describing the buttons inside the lift is presented in figure 12.

In particular, when a user presses the button to go to floor X, the light associated to that button is switched on and a message is sent to the elevator to move to floor X. When the elevator is at that floor the light will be switched off. We have considered that it is the elevator who sends the message to switch off the light. Instead, we may have considered that when elevator is at floor X,

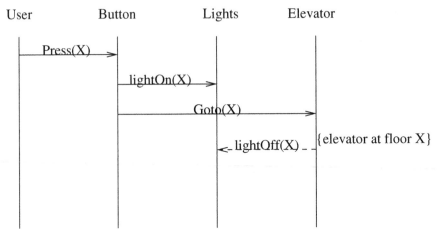

Fig. 12. Buttons Body.

it will send an acknowledgement to the button object who, then, will switc
off the light. Now, according to this body diagram, the interface describing th
connection to the interacting components can be seen in figure 12.

Again, the refinement between this interface and the body of the componen
is just an embedding. The component associated to the set of buttons which ar
located in each floor could be described exactly in the same way as the previou
one. Note that this would not have been true if there would be two buttons pe
floor.

Now, if we want to build a lift system, including just one lift, we need t
connect these three components. The body of this connector would consist c
three diagrams, where two of them would be almost identical. In particular th

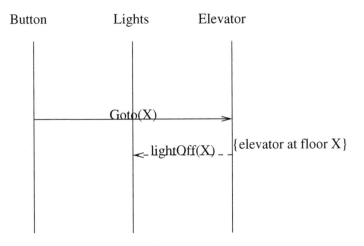

Fig. 13. Buttons Interface.

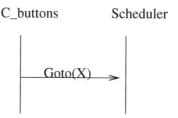

Fig. 14. Connector Body 1.

first diagram (see figure 14) describe that, when a request is received from some set of buttons (for instance the cabin buttons located inside the elevator, C-buttons), this request is received by a scheduler (which will probably store the request in some queue). A similar diagram would be needed to describe the situation when the request is received from the buttons located in the floors. We have not shown this diagram. The third diagram (see figure 15) describes that, when the first request to serve refers to floor X, the scheduler sends a message to the elevator to go to that floor. When the elevator acknowledges that the elevator is at floor X, then the scheduler asks the two sets of buttons to switch off the lights corresponding to that floor.

Now, the connector would have three interfaces, the first two which are again almost identical would consist of two diagrams. The first one would coincide with the first body diagram (figure 14). The second one, see figure 16, describes the interaction for switching off the lights upon arrival at a given floor. The third interface describes the interaction with the elevator and would be identical to the elevator interface (figure 11).

The composition of the connector with the three components would provide the expected global specification of the lift system. The connection of the eleva-

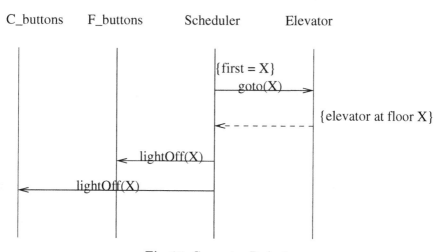

Fig. 15. Connector Body 2.

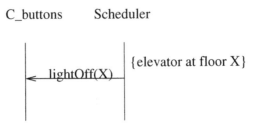

Fig. 16. Connector Interface 2.

tor interface with the corresponding connector's interface is trivial, since both interfaces are equal. In the case of the buttons, the connection must be made via a transformation. In particular, we would need to say that the C-buttons life-line in the connector's interface is implemented in terms of the lifelines Button and Lights from the button component; and, similarly, the lightOff(X) message is implemented by a diagram that includes only one interaction, consisting of sending the message lightOff(X) from the Scheduler to the Lights lifeline.

5   Conclusion

In this paper, we have presented some basic ideas for the definition, in the context of UML, of connectors and components for the architectural design of software systems, following the approach presented in [1]. This was done by adapting and instantiating the generic approach presented in [3]. In particular, we studied how we can define architectural connectors and components in the case of class and sequence diagrams, defining and studying the notions of embedding and transformation, which are needed for the application of the approach presented in [3].

It must be said that, yet, there is much work to be done for the complete definition, in the context of UML, of the framework presented in [1]. On one hand, other kinds of diagrams have to be considered. Although, we think that the ideas presented in this paper could be useful when dealing with these other diagrams. On the other hand, we just studied the foundations for the definition of connectors and components for the kinds of diagrams considered, but we did not want to present a specific constructions for this aim. Finally, the development of new tools (or the customization of existing ones) for providing support for this kind of architectural design would also be needed.

Acknowledgements

This work is partially supported by the Spanish project MAVERISH (TIC2001-2476-C03-01) and by the CIRIT Grup de Recerca Consolidat 2001SGR 00254, and by the European RTN project SegraVis (RTN2-2001-00346). The stay of Sonia Pérez in Barcelona was supported by the European Alfa Net CORDIAL (AML/B7-311-97/0666/II-0021-FA).

References

1. R. Allen, D. Garlan. A Formal Basis for Architectural Connection. In *ACM TOSEM '97*, pp. 213–249.
2. J. Cheesman, J. Daniels. *UML Components*. Addison-Wesley, 2001.
3. H. Ehrig, F. Orejas, B. Braatz, M. Klein, M. Piirainen. A Generic Component Framework for System Modeling. In *Proc. FASE 2002*, Springer LNCS 2306 (2002), pp. 33–48.
4. H. Ehrig, J.Padberg, B. Braatz, M. Klein, M. Piirainen, F. Orejas, S. Perez, E. Pino. A Generic Framework for Connector Architectures based on Components and Transformations. Proc. FESCA 2004, Barcelona.
5. J.L. Fiadero, A. Lopes Semantics of Architectural Connectors. *Proc TAPSOFT '97*, Springer LNCS 1214 (1997), pp. 505–519.
6. C. A. R. Hoare: Communicating Sequential Processes Prentice-Hall 1985
7. J. Rumbaugh, I. Jacobson, G. Booch. The Unified Modeling Language Reference Manual. Addison Wesley (1999).
8. Mary Shaw and David Garlan Software Architecture: Perspectives on an Emerging Discipline Prentice Hall, 1996
9. A.M. Zaremski, J.M. Wing. Specification Matching of Software Components. In *ACM TOSEM '97*, pp. 333–369.

Loose Semantics of Petri Nets[*]

Julia Padberg[1] and Hans-Jörg Kreowski[2]

[1] Technische Universität Berlin,
Fakultät IV, Informatik und Elektrotechnik,
Berlin, Germany
padberg@cs.tu-berlin.de
[2] Universität Bremen,
Fachbereich für Mathematik und Informatik,
Bremen, Germany
kreo@tzi.de

Abstract. In this paper, we propose a new, loose semantics for place/transition nets based on transition systems and generalizing the reachability graph semantics. The loose semantics of a place/transition net reflects all its possible refinements and is given as a category of transition systems with alternative sequences of events over the net. The main result states that each plain morphism between two place/transitions nets induces a free construction between the corresponding semantic categories.

1 Introduction

Petri nets are one of the most thoroughly investigated approaches with a multitude of extensions and variants. They are one of the most prominent specification techniques for modeling concurrency and have a wide range of application areas in practice. In this paper, we introduce a new semantics for Petri nets which is based on transition systems. The semantics of a net is given by a class of models corresponding to all possible refinements of a net with respect to transition refinement. In this sense, it is a loose semantics as known and well accepted in the area of data type specification (see, e.g. [24]).

The semantics we define here is developed in view of system specification. It is suitable for relating different stages of refinement. This is obviously important for the vertical structuring, but as well for horizontal structuring with abstraction mechanisms like parameterization and modularization.

The reachability graph is a standard model of a place/transition net describing all possible sequences of firings of transitions starting from an initial marking. Our new semantics generalizes this net semantics in such a way that a firing of a transition can be refined by sequences of events. Moreover, we allow alternative possibilities for each such refinement. Typical examples of alternative sequences are the interleavings of independent events. Altogether, the loose semantics of a place/transition net consists of the class of transition systems with alternative sequences of events including the reachability graph. This class forms a category

---

[*] Research partially supported by the EC Research Training Network SegraVis.

H.-J. Kreowski et al. (Eds.): Formal Methods (Ehrig Festschrift), LNCS 3393, pp. 370–384, 2005.
© Springer-Verlag Berlin Heidelberg 2005

in a natural way. As the main result of the paper, we show that each plain morphism between two place/transition nets induces a free construction between the corresponding semantic categories. This is the key result that allows one to consider Petri nets as building blocks of parameterization and modularization.

We continue this paper by introducing state transition systems that capture our idea of alternatives and refinement. In Section 3, we show that transitions systems over Petri nets can be considered as a loose semantics of the corresponding Petri net. Next we show that we obtain a free functorial construction of the place/transition net semantics, based on contravariant forgetful functor. Subsequently in Section 5 we treat the relation to other approaches in some detail and discuss at last the impact of a loose semantics for Petri nets in Section 6.

2   Transition Systems with Alternative Sequences

In this section, we recall the notion of state transition systems and add a new feature to them: a relation of alternative sequences of events. Transition systems with alternative sequences will be combined with place/transition nets in the next section.

A state transition system $STS = (S, E, TS, \widehat{s})$ is given by a set of states S, a set of events E, the set of transitions $TS \subseteq S \times E \times S$, and the initial state $\widehat{s} \in S$.

If one reads the events along the paths in state transition systems, one gets sequences of events. More formally, we write $s_0 \xrightarrow{w} s_n$ if there is some sequence of transitions $(s_{i-1}, e_i, s_i) \in TS$ for $i = 1, .., n$ and $w = e_1 e_2 ... e_n \in E^*$ or if $w = \lambda$ and $s_0 = s_n$.

Next we want to consider some of these sequences of events as alternatives to each other. To make this precise, let $AS \subseteq E^* \times E^*$ be some relation on E^* and AS^{Con} denote its congruence closure, i.e. the closure of AS that is reflexive, symmetric, transitive, and congruent with respect to concatenation. Moreover, let $E^\diamond$ denote the quotient factoring E^* through AS^{Con} and $[\_] : E^* \to E^\diamond$ the canonical function with $[\_](w) = [w]$ for all $w \in E^*$ where $[w] = \{w' | (w, w') \in AS^{Con}\}$ is the congruence class of $w \in E^*$.

Definition 1 (Transition Systems with Alternative Sequences). *A transition system with alternative sequences* $TSA = (S, E, TS, \widehat{s}, AS)$ *is given by a state transition system* $(S, E, TS, \widehat{s})$ *and the relation of alternative sequences* $AS \subseteq E^* \times E^*$ *subject to the following consistency condition:*

$$\forall w' \in [w] : s \xrightarrow{w} s' \iff s \xrightarrow{w'} s'$$

The consistency condition ensures that alternatives are alternatives at all states. So, they are global alternatives in the following sense: Whenever there is a state where the sequence w occurs the alternative sequence $w' \in [w]$ has to occur as well.

Next we examine morphisms between transition systems with alternative sequences. We allow mapping one event $e_1 \in E_1$ to a congruence class of sequences of events by a morphism $f_E : E_1 \to E_2^\diamond$ with $f_E(e_1) = [w]$. This denotes the refinement of one event by alternative sequences of events. The morphism $f_E : E_1 \rightarrow E_2^\diamond$ can be extended uniquely by $f_E^\diamond : E_1^\diamond \rightarrow E_2^\diamond$ defined for $w = e_1 \cdot \ldots \cdot e_n \in E_1^*$ by $f^\diamond([w]) = f(e_1) \cdot \ldots \cdot f(e_n)$, where the concatenation of congruence classes is defined by the congruence class of the concatenation, i.e. $[u] \cdot [v] = [uv]$.

Definition 2 (TSA-Morphisms). *Given transition systems with alternative sequences $TSA_i = (S_i, E_i, TS_i, \widehat{s_i}, AS_i)$ for $i = 1, 2$, then a TSA-morphism is given by $f : TSA_1 \rightarrow TSA_2$ with $f = (f_S, f_E)$ and $f_S : S_1 \rightarrow S_2$ and $f_E : E_1 \rightarrow E_2^\diamond$ such that the following conditions hold:*

1. *Existence of a path: For all $(s_1, e_1, s_1') \in TS_1$ and for all $e_2^1 \cdot e_2^2 \cdot \ldots \cdot e_2^n \in f_E(e_1)$*

   there is a path $f_s(s_1) \xrightarrow{e_2^1} s_2^1 \xrightarrow{e_2^2} s_2^2 \xrightarrow{*} s_2^n \xrightarrow{e_2^n} f_s(s_1')$.
2. *Reachability of initial state: We have $\widehat{s_2} \xrightarrow{*} f_S(\widehat{s_1})$.*
3. *Preservation of alternatives: Given $(w, w') \in AS_1$ then we have $f_E^\diamond([w]) = f_E^\diamond([w'])$ for the unique extension $f_E^\diamond : E_1^\diamond \rightarrow E_2^\diamond$ and $w \in E_1^*$.*

Then we obtain:

- *Composition $g \circ f : TSA_1 \rightarrow TSA_3$ of the morphisms $f : TSA_1 \rightarrow TSA_2$ and $g : TSA_2 \rightarrow TSA_3$ is given by the composition of its components with $(g \circ f)_S = g_S \circ f_S$ and $(g \circ f)_E : g_E^\diamond \circ f_E$, where $g_E^\diamond : E_2^\diamond \rightarrow E_3^\diamond$ is the unique extension.*
- *Identity $id_{TSA} : TSA \rightarrow TSA$ is given by $id_{TSA} = (id_S, [\_]_E)$.*

Hence, we have the category **TSA** *of transition systems with alternative sequences.*

Note, condition 3 obviously implies $f_E^\diamond([w]) = f_E^\diamond([w'])$ for any $w' \in [w] \in E_1^\diamond$ (see [21]) and the composition is well-defined as we have congruence with respect to concatenation.

Example 1 (Transition Systems with Alternative Sequences). Here we give a short example of some transition systems with alternative sequences, where we concentrate on the events and depict the states merely as $\bullet$, and the initial state by $\rightarrow \bullet$. The numbers adjacent to the states are merely used to illustrate morphisms later on. First, we investigate the examples in Fig. 1 to illustrate our notion of morphisms and subsequently we give an interpretation of the example.

All states are mapped injectively. TSA_1 is mapped to TSA_2 by f, where $f_E(s) = [s] = \{s\}$ and $f_E(d) = [d] = \{d\}$. Preservation of alternatives is satisfied as $AS_1 = \emptyset$. The TSA morphism $g : TSA_2 \rightarrow TSA_3$ is defined for the events by $g_E(s) = [t] = \{t, uv\}$, $g_E(s') = [s'] = \{s'\}$, $g_E(d) = [d] = \{d\}$, and $g_E(d) = [d] = \{d\}$. Preservation of alternatives is satisfied since we have $g_E^\diamond([sd]) = [td] = \{td, uvd, s'd'\} = [s'd'] = g_E^\diamond([s'd'])$. The composition $g \circ f$ is for the events obviously given by $g_E^\diamond \circ f_E(s) = g_E^\diamond([s]) = [t] = \{t, uv\}$ and $g_E^\diamond \circ f_E(d) = g_E^\diamond([d]) = [d] = \{d\}$. Again preservation of alternatives is satisfied since $(g_E \circ f_E)^\diamond([sd]) = g_E^\diamond \circ f_E^\diamond([sd]) = [td] = \{td, uvd, s'd'\} = [s'd'] = (g_E \circ f_E)^\diamond([s'd'])$. The interpretation of this example is that the transition system TSA_1 describes a simple system with the following events s for start, d for distribute, r for

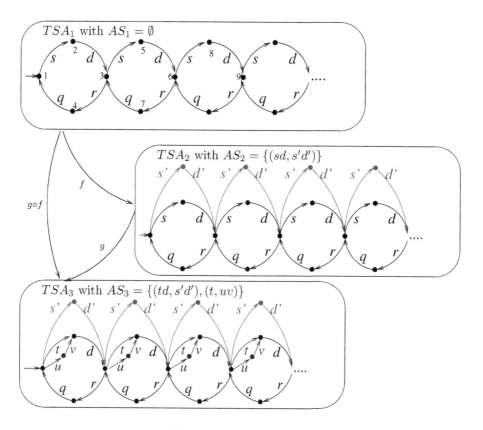

Fig. 1. Transition systems.

receive, and q for quit. These events follow each other as depicted in Fig. 1. The transition system TSA_2 states that the sequences of events sd and $s'd'$ are alternatives. Whenever one of both occurs at a certain state the other does so too. This describes independently of the syntactic specification that a system has different, but equally desired sequences of events. As in the case of our example, they need not be induced by single events. In TSA_2 two alternatives for starting and distributing namely sd or $s'd'$ result in the same state, and they do that in any case. Nevertheless neither s and s' nor d and d' are alternatives. By the morphism $g : TSA_2 \rightarrow TSA_3$ we refine the event s by $[t] = \{t, uv\}$, that is s can be expressed either by the event t or the sequence uv.

3 Transition Systems with Alternative Sequences over Net Systems

In this section, we associate place/transition systems with transition systems with alternative sequences. We use place/transition nets in the usual way with weighted arcs so that the pre- and post-domains of transitions as well as the markings are multisets over the sets of places (as the algebraic notation in [14]). Given a set P, let the set of finite multisets over P be denoted by $P^\oplus$.

Then a place/transition net is given by $N = (P, T, pre, post, \widehat{m})$ where P is the set of places, T the set of transitions, $pre, post : T \to P^\oplus$ are mappings associating a pre- and a post-domain to each transition, and $\widehat{m} \in P^\oplus$ is the initial marking.

The set of finite multisets over P is the free commutative monoid over P. An element $w \in P^\oplus$ can be presented either by the natural function $w : P \to \mathbb{N}$ or as a linear sum $w = \sum_{p \in P} \lambda_p \cdot p$, and we can extend the usual operations and relations on $\mathbb{N}$ as $\oplus, \ominus, \leq$, and so on to $P^\oplus$. Moreover, we need to state how often is a basic element with in an element of the free commutative monoid given. We define this for an element $p \in P$ and a linear sum $w = \sum_{p \in P} \lambda_p \cdot p \in P^\oplus$ with $w_{|p} = \lambda_p$ for $p \in P^\oplus$ and $w_{|Q} = \sum_{p \in Q} \lambda_p \cdot p$ for a subset $Q \subseteq P$.

The pre-set $\bullet x$ and the post-set $x \bullet$ are defined as usual, and so is the set of reachable markings $[\widehat{m} >$.

The set of enabled transitions is $[T > = \{t \in T | m[t > m'$ for some $m \in [\widehat{m} >\}$.

We now define transitions systems that can be viewed as models of a net, where a refinement of the enabled transitions and the representation of the states relate the net to the transition system. In particular, we allow refinements of transitions to be equivalence classes of alternative sequences.

Definition 3 (Transition Systems with Alternative Sequences over Place/Transition Nets). *A transition system with alternative sequences TSA over a place/transition net $N = (P, T, pre, post, \widehat{m})$ consists of a transition system with alternative sequences $TSA = (S, E, TS, \widehat{s}, AS, rep, ref)$ and two functions $rep : [\widehat{m} > \to S$ and $ref : [T > \to E^\diamond$ subject to the following conditions:*

1. Representation of markings: *The function $rep : [\widehat{m} > \to S$ represents the reachable markings.*
2. Refinement of transitions: *The function $ref : [T > \to E^\diamond$ refines the transitions.*
3. Reachability of initial marking: *We have $\widehat{s} \xrightarrow{*} rep(\widehat{m})$.*
4. Existence of a path: *For all $m[t > m'$ with $m \in [\widehat{m} >$ we have a path for all $w \in ref(t)$ so that $rep(m) \xrightarrow{w} rep(m')$.*

Note that we may have $ref(t) = [\lambda]$ only for transitions where for all $m, m' \in [\widehat{m} >$ with $m[t > m'$ we have $rep(m) = rep(m')$. Next we establish the category of transition systems with alternative sequences over N. Hence, this category is the loose semantics of a net N.

Definition 4 (Category TSA(N) of Transition Systems with Alternative Sequences over N). *The category $\mathbf{TSA(N)}$ of transition systems with alternative sequences over the place/transition net $N = (P, T, pre, post, \widehat{m})$ is given by the class of transition systems $TSA = (S, E, TS, \widehat{s}, AS, rep, ref)$ over N, and by TSA-morphisms $f : TSA_1 \to TSA_2$ with $rep_1 : [\widehat{m_1} > \to S_1$ and $ref_1 : [T_1 > \to E_1^\diamond$ (resp. $rep_2 : [\widehat{m_1} > \to S_2$ and $ref_2 : [T_1 > \to E_2^\diamond)$ satisfying the following conditions:*

1. Preservation of representation: $f_S \circ rep_1 = rep_2$.
2. Preservation of transition refinement: $f_E^\circ \circ ref_1 = ref_2$.

Now we have a class of transition systems for each net. Moreover, it is a category so we have morphisms, that denote refinements of events with alternatives. To illustrate this new type of net semantics, we now present the well-known producer-consumer net and discuss its loose semantics.

Example 2 (Producer-Consumer).
In Fig. 2 we have the well-known producer-consumer net with the transitions s for start, d for distribute, r for receive, and q for quit. This net is obviously closely related to the transition systems in

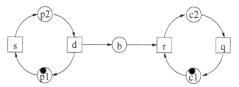

Fig. 2. Producer-Consumer net PCN.

Fig. 1. This producer-consumer net PCN denotes a part of the category **TSA(PCN)** of transition systems over this net.

The transition system TSA_1 is extended by the representation function $rep_1 : [\widehat{m} > \; \rightarrow \; S_1$ with $rep_1(p1 \oplus c1) = 1$, the follower marking is mapped by $rep_1(p2 \oplus c1) = 2$, $rep_1(p1 \oplus b \oplus c1) = 3$, and so one, where the numbers denote the states of transition system TSA_1 in Fig. 1. The transitions are refined trivially by themselves, $ref_1(s) = [s] = \{s\}$, $ref_1(d) = [d] = \{d\}$, and so forth. The transition systems TSA_2 and TSA_3 with suitable representations and refinements are clearly transition systems over PCN.

The transition system TSA_1 is isomorphic to the reachability graph of PCN and hence it is initial in this category (see Section 5).

The loose semantics for some place/transition net N comprises all possible refinements, where we use refinement in a very broad sense: A transition can be refined by various alternatives of event sequences including the empty sequence. The only requirement is that the source of these event sequences needs to be the representation of the marking before firing the transition, and the target of the sequence needs to be the representation of the marking after firing the transition. The initial object is the usual reachability graph $R(N)$ of a net N (see Section 5). So, the classical semantics of a net is a distinguished member of the loose semantics and any transition system TSA in **TSA(N)** is a refinement of $R(N)$, as there is a unique morphism from $R(N)$ to TSA.

4 Free Construction of the Loose Semantics over Plain Morphisms

Based on the algebraic notion of Petri nets [14] we use simple homomorphisms that are generated over the set of places. These morphisms map places to places and transitions to transitions. Morphisms are the basic entity in category theory; they can present the internal structure of objects and relate the objects. So they are the basis for the structural properties a category may have and can be used successfully to define various structuring techniques.

Definition 5 ((Plain) Morphisms). *A plain morphism $f : N_1 \to N_2$ is given by $f = (f_P, f_T)$ with $f_P : P_1 \to P_2$ and $f_T : T_1 \to T_2$ so that $pre_2 \circ f_T = f_P^\oplus \circ pre_1$ and post analogously.*

Moreover, for the initial marking we have for all $p \in P_1$:
$\widehat{m_1}(p) \leq \widehat{m_2}(f_P(p))$ *for the natural function associated to a multiset.*

Lemma 1 (Plain morphisms preserve firing). *Plain morphisms $f : N_1 \to N_2$ preserve firing in the following sense:*

$$m[t > m' \text{ implies } f_P^\oplus(m)[f_T(t) > f_P^\oplus(m) \text{ for } m, m' \in P_1^\oplus \text{ and } t \in T_1.$$

Then we define $\widehat{f}_P : [\widehat{m_1} >\!\!\to [\widehat{m_2} > \text{ with } \widehat{f}_P(m) = f_P^\oplus(m) \oplus m_2^R$ where we have $\widehat{m_2} = f_P^\oplus(\widehat{m_1}) \oplus m_2^R$.

Note, by induction over the length of the firing sequence we can show that $\widehat{f}_P$ is well-defined and preserves firing as well: $m[t > m'$ with $m \in [\widehat{m_1} > $ implies $\widehat{f}_P(m)[f_T(t) > \widehat{f}_P(m')$

Theorem 1 (Forgetful Functor of Transition Systems with Alternative Sequences over N). *A plain morphism $f : N_1 \to N_2$ induces the following forgetful functor (if necessary subscripted with the corresponding net morphism) $V = V_f : \mathbf{TSA(N_2)} \to \mathbf{TSA(N_1)}$. This functor $V(TSA_2) = TSA_1$ is defined by $TSA_2 = (S_2, E_2, TS_2, \widehat{s_2}, AS_2, rep_2, ref_2)$ with $rep_2 : [\widehat{m_2} >\!\!\to S_2$ and $ref_2 : [T_2 > \to E_2^\diamond$, where $TSA_1 = (S_2, E_2, TS_2, \widehat{s_2}, AS_2, rep_1, ref_1)$ and we have*

- *the following representation*
  $rep_1 := rep_2 \circ \widehat{f}_P : [\widehat{m_1} >\!\!\to S_2$, *and*
- *the following refinement*
  $ref_1 := ref_2 \circ f_T : [T_1 > \to E_2^\diamond$.

A TSA-morphism $h : TSA_2 \to TSA_2'$ is mapped by $V(h) = h$.

Proof. TSA_1 is a transition system over N_1:

1. Representation: rep_1 is well-defined.
2. Refinement: ref_1 is well-defined.
3. Reachability of of initial state: $\widehat{s_2} \xrightarrow{*} rep_2(\widehat{m_2}) = rep_2(\widehat{f}_P(\widehat{m_1})) = rep_1(\widehat{m_1})$
4. Existence of a path:
   for any $m[t > m'$ with $m \in \widehat{m_1} >$ we have $\widehat{f}_P(m)[f_T(t) > \widehat{f}_P(m')$ and hence there is the path $rep_2(\widehat{f}_P(m)) \xrightarrow{ref_2 \circ f_T(t)} rep_2(\widehat{f}_P(m'))$ that is the path $rep_1(m) \xrightarrow{ref_1(t)} rep_1(m')$.

Given TSA-morphism $h : TSA_2 \to TSA_2'$ then $VP(h) : TSA_1 \to TSA_1'$ with $V(h) = h$ is well-defined:

1. preservation of representation :
   $h_s \circ rep_1 = h_S \circ rep_2 \circ \widehat{f}_P = rep_2' \circ \widehat{f}_P = rep_1'$

2. preservation of transition refinement:
$h_E^\diamond \circ ref_1 = h_E^\diamond \circ ref_2 \circ f_T = ref_2' \circ f_T = ref_1'$ See the diagrams below:

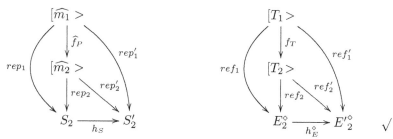

Theorem 2 (Free Functor of Transition Systems with Alternative Sequences over N). *A plain net morphism $f : N_1 \to N_2$ induces the following free functor $F = F_f : \textbf{TSA}(N_1) \to \textbf{TSA}(N_2)$. This functor $F(TSA_1) = TSA_2$ is defined by $TSA_1 = (S_1, E_1, TS_1, \widehat{s}_1, AS_1, rep_1, ref_1)$ and $TSA_2 = (S_2, E_2, TS_2, \widehat{s}_2, AS_2, rep_2, ref_2)$ as given in the proof.*

A TSA-morphism $h : TSA_1 \to TSA_1'$ is mapped by $F(h) = \overline{h}$.

Proof. 1. First we give the construction for TSA_2.
In **Set** we construct the pushout PO1
below and obtain S_2, and hence rep_2 :
$[\widehat{m_2} > \rightarrowtail S_2$. We define $\widehat{s}_2 = u_S(\widehat{s}_1)$.
The construction of E_2 is given by $E_2 = E_1 \uplus [T_2 > \setminus f_T([T_1 >)$ in **Set**.
Then we define $AS_2 = AS_1 \uplus \{(w_1, w_2)|$
$f_T(t_1) = f_T(t_2)$ for some $w_1 \in ref_1(t_1)$
and $w_2 \in ref_1(t_2)\}$.
Then we have $E_2^\diamond$, and we define $u_E := [\_] \circ inc_E : E_1 \to E_2^\diamond$ and hence
$u_E^\diamond : E_1^\diamond \to E_2^\diamond$. This is well-defined as $E_1 \subseteq E_2$ and $AS_1 \subseteq AS_2$.
We now define $ref_2 : [T_2 > \rightarrow E_2^\diamond$ by

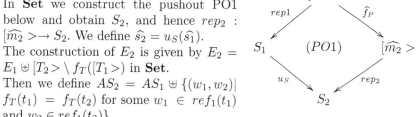

$$ref_2(t) := \begin{cases} [t] & t \notin f_T([T_1 >) \\ u_E^\diamond \circ ref_1(t') & t = f_T(t') \text{ and } t' \in [T_1 > \end{cases}.$$

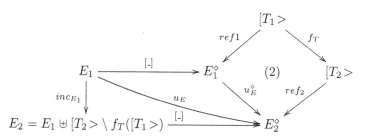

The square (2) commutes due to the quotient construction, since we have
$E_2^\diamond = E_2^*{}_{|AS_2^{Eq}}$.
We define $TS_2 \subseteq S_2 \times E_2 \times S_2$ using the transition system TS_1, all new firing
paths of N_2 and then construct all alternatives event sequences inductively:

(a) If $(s, e, s') \in TS_1$
    then $(u_S(s), e, u_S(s)) \in TS_2$.
(b) If $m[t > m'$ in N_2 and $t \notin f_t([T_1 >)$
    then $(rep_2(m), t, rep_2(m')) \in TS_2$.
(c) If $m[t > m'$ in N_2, $m \notin \widehat{f_P}(\widehat{m_1})$, $t = f_T(t_1)$, and with some $m_1[t_1 > m'_1$
    then for all $w = e_0....e_n \in ref_2(t)$
    with $rep_1(m_1) \xrightarrow{e_0} s_1...s_n \xrightarrow{e_n} rep(m'_1) \in TS_1$
    we have $(rep_2(m), e_0, u_S(s_1)) \in TS_2$ and
    $(u_S(s_n), e_n, rep_2(m')) \in TS_2$.
So, we have $TSA_2 = (S_2, E_2, TS_2, \widehat{s_2}, AS_2, rep_2, ref_2)$.
It is obviously well-defined.

2. We have a free construction:
   There is $u : TSA_1 \rightarrow V \circ F(TSA_1)$.
   Note that, $V \circ F(TSA_1) = V(TSA_2)$
   $= (S_2, E_2, TS_2, \widehat{s_2}, rep_2 \circ \widehat{f_P}, ref_2 \circ f_T)$.
   So we define $u = (u_S, u_E)$ where u_S and
   u_E are given in PO1 and (2) above.
   u is well-defined as $u_S \circ rep_1 = rep_2 \circ \widehat{f_P}$
   and $u_E^\diamond \circ ref_1 = ref_2 \circ f_T$.

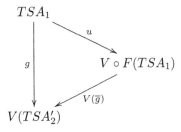

Given $g : TSA_1 \rightarrow V(TSA'_2)$ in $\mathbf{TSA(N_1)}$ defined by $g = (g_S, g_E)$ with $g_S : S_1 \rightarrow S'_2$ and $g_E : E_1 \rightarrow E'^\diamond_2$ then we have to construct $\overline{g} : TSA_2 \rightarrow TSA'_2$ in $\mathbf{TSA(N_2)}$. We have $\overline{g}_S$ induced by PO1.
And we obtain $\overline{g}_E : E_2 \rightarrow E'^\diamond_2$ due to the coproduct E_2.

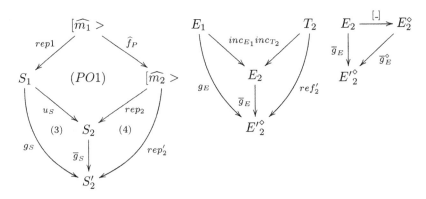

$\overline{g}$ is well-defined in $\mathbf{TSA(N_2)}$, as (4) commutes,
and we have $\overline{g}_E^\diamond \circ ref_2 = \overline{g}_E^\diamond \circ [\_] \circ inc_{T_2} = \overline{g}_E \circ inc_{T_2} = ref'_2$.
Now we prove that $g = V(\overline{g}) \circ u$:
We have $g_S = \overline{g}_S \circ u_S$ due to (3).
And we have $\overline{g}_E^\diamond \circ u_E = \overline{g}_E^\diamond \circ [\_] \circ inc_{E_1} = \overline{g}_E \circ inc_{E_1} = g_E$ $\checkmark$

Example 3 (Refining the Producer-Consumer).

In Fig. 3 we again have the producer-consumer net *PCN*. This net is refined by the morphism $f : PCN \rightarrow PCN'$ to the producer-consumer net *PCN* where we can directly feed and empty the buffer. The morphism f maps the states and transitions injectively. Now, we have the categories **TSA(PCN)** and **TSA(PCN')** that are related by the forgetful and the free functor as given in the Theorems 1 and 2. In Fig. 4 we illustrate the two functors, that form the adjunction $F \dashv V$.

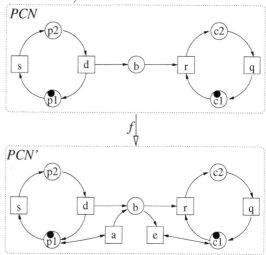

Fig. 3. *PCN* and *PCN'*.

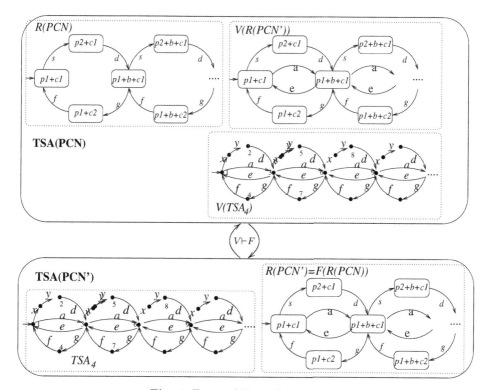

Fig. 4. Free and Forgetful Functor.

We have the transition system $R(PCN)$ in $\textbf{TSA}(\textbf{PCN})$, that is the reachability graph of the net PCN. This transition system $R(PCN)$ is mapped by the free functor $F{:}\textbf{TSA}(\textbf{PCN}){\rightarrow}\textbf{TSA}(\textbf{PCN}')$ to the transition system $F(R(PCN))$. This is isomorphic to the transition system $R(PCN')$. $F(R(SynPCN))$ is constructed by adding the reachable markings and the new transitions in the net N_2. The forgetful functor $V : \textbf{TSA}(\textbf{PCN}') \rightarrow \textbf{TSA}(\textbf{PCN})$ maps the transition system TSA_4 to the transition system $V(TSA_4)$ and $R(PCN')$ to $V(R(PCN'))$ by keeping the transition system and redirecting the representation and the refinement functions.

5   Related Work

In the course of the last 40 years there has been developed a lot of different Petri net semantics: reachability or marking graph [23, 6], event structures [17, 3, 18], trace languages [13, 10, 15], partial orders semantics [2, 12], and others more. All these semantics have in common that they relate a Petri net to one semantic object.

In this section we relate our loose semantics to the two closest Petri net semantics, namely reachability graph and trace languages, in a provisional way. Since most of the above semantics are related to each other in a significant way (see [25, 19]) the results from the discussion below can be adopted accordingly.

Reachability Graph of Place/Transition Net Systems. The reachability graph of a place/transition net is given by the reachable markings and the firing transitions in-between. Hence, a suitable definition of the reachability graph is a transition systems with the empty alternative sequence of events. So, more formally the reachability graph $R(N) = ([\widehat{m} >, T, TS, \widehat{m}, \emptyset)$ with $TS \subseteq [\widehat{m} > \times T \times [\widehat{m} >$ defined by $TS = \{(m, t, m')|m[t > m'$ for any $m \in [\widehat{m_1} >\}$ of a place/transition net N is a TSA over N, where $rep(m) = M$ and $ref(t) = [t] = \{t\}$. Moreover, the reachability graph $R(N)$ is initial in the category $\textbf{TSA}(\textbf{N})$ for the proof see [21]. This means, that every transition system over N is $\textbf{TSA}(\textbf{N})$ can be considered a refinement of the reachability graph along a unique morphism.

Trace Equivalences. The relation to trace equivalences is the following.

Local trace equivalences [10, 15] denote the set of independent events following a sequence of events. So in a sense the interleaving of independent events are alternative sequences. But our approach states alternatives globally (in Definition 1).

Let us denote with $|_-|_e : E^* \rightarrow \mathbb{N}$ the family of functions, that counts the number of times an event $e \in E$ occurs in a sequence.

So given a transition system with alternative sequences, we can compute multisets of independent events. A multiset $m \in E^{\oplus}$ consists of independent events, if for any linearization $v \in Lin(m) = \{w \in E^*|m_{|e} = |w|_e$ for all $e \in E\}$ we have $Lin(m) \subseteq [v]$.

The other way round we can consider for the set of alternatives AS the set of linearization $Lin(m)$ of each multiset $m \in M$, the set M of multisets of independent events i.e. $AS = \bigcup_{m \in M} Lin(m) \times Lin(m)$.

6   Discussion of the Impacts of a Loose Semantics

Parameterized Petri Nets
Based on the ideas of parameterization for data types, Petri nets can be parameterized by distinguishing a subnet as the parameter. Analogously to algebraic specifications, we map the formal parameter net PAR by an inclusion to the target net TAR. In Fig. 5 we have the formal parameter net, that denotes the transition s can be replaced by an actual parameter net.

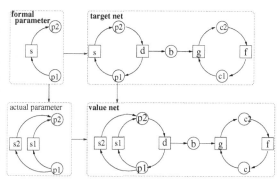

Fig. 5. Actualization.

The semantics of a parameterized Petri net (PAR, TAR) maps each transition system over the PAR to the corresponding transition net over TAR. The semantics is given by the free functor $F : \textbf{TSA}(\textbf{PAR}) \to \textbf{TSA}(\textbf{TAR})$. In Fig. 5 the net is refined by the actual parameter, where the transition s is substituted by the subnet containing the transitions $s1, s2$.

Constructor Semantics for Petri Net Modules. Most attempts to Petri net modules (among others [4, 5, 11]) do not provide Petri nets as interfaces. For a not so recent survey see [1]. There are either places or transitions, but no full Petri nets in the interface. When modeling software components these notions of Petri net modules are not powerful enough, since they do not allow specifying behavior in the interfaces. Our approach to Petri net modules [20] has been achieved by a transfer from the concept of algebraic module specifications presented in [7]. The main motivation for our approach to Petri net modules is the modeling of component-based systems. The component concept as suggested in [16, 26] for *Continuous Software Engineering (CSE)* is the underlying concept for our approach.

A Petri net module $MOD = (IMP, EXP, BOD)$ consists of three Petri nets, namely the import net IMP, the export net EXP and the body net BOD. Two Petri net morphisms $m : IMP \to BOD$ and $r : EXP \to BOD$ connect the interfaces to the body.

$$\begin{array}{c} EXP \\ \downarrow r \\ IMP \xrightarrow{m} BOD \end{array}$$

The import interface specifies resources which are used in the construction of the body, while the export interface specifies the functionality available from the Petri net module to the outside world. The body implements the functionality specified in the export interface using the imported functionality. The import morphism m is a *plain morphism* and describes how and where the resources in the import interface are used in the body. The export morphism r is a *substitution morphism* and describes how the functionality provided by the export interface is realized in the body. The class of substitution morphism is as generalization of plain morphisms, where a transition is replaced by a subnet. Nevertheless, the forgetful functor constructions can be given for substitution morphisms as well (explicitly in [21]).

In [22] a transformation-based semantics for Petri net modules has been introduced based on the transformation-based approach to generic components [8]. There the semantics is defined based on all transformations the import may undergo. The advantage of our approach is that it is constructive: The semantics of a module is based on the loose semantics presented in this paper. It gives for each possible transition system over the import net the according transition system over the export net.

So we obtain a functor mapping the category of transition systems over the import net **TSA(IMP)** to the category of transition systems over the export net **TSA(EXP)**.

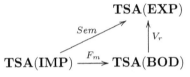

This semantic functor naturally depends on the morphisms that relate the interfaces to the body of the module. We define the functor $Sem : \mathbf{TSA(IMP)} \to \mathbf{TSA(EXP)}$ by $Sem = V_r \circ F_m$. V_r and F_m are constructed using the morphisms r and m. Now, we have the constructive semantics of a Petri net modules analogously to [7]. Each model of the import net is mapped to the corresponding model of the export. This module semantics takes some transition system over the import net and then constructs freely along the plain morphism $m : IMP \to BOD$, yielding a transition system over the body net BOD. Then it forgets along ' the forgetful functor V_r the internal details of the body and only represents the part specified by the export net EXP. Based on this notions we then obtain directly: internal and model correctness, compositional semantics with respect to module operations as union, composition (as given for Petri net modules [20]) or general module operations, based on schemes.

References

1. L. Bernadinello and F. De Cindio. A survey of basic net models and modular net classes. *Advances in Petri Nets 1992*, LNCS 609,pp: 304–351, Springer, 1992.
2. E. Best and J. Desel. Partial order behaviour and structure of Petri nets. *Formal Aspects of Computing*, pages 123–138, 1990.
3. A. Corradini, H. Ehrig, M. Löwe, U. Montanari, and F. Rossi. An event structure semantics for safe graph grammars. In *Proc. PROCOMET'94, IFIP TC2 Working Conf., San Miniato 1994*, pages 417–439. IFIP TCS, 1994.
4. S. Christensen and L. Petrucci. Modular analysis of Petri nets. *Computer Journal*, 43(3):224–242, 2000.

5. J. Desel, G. Juhás, and R. Lorenz. Petri Nets over Partial Algebras. In H. Ehrig, G. Juhás, J. Padberg, and G. Rozenberg, editors, *Advances in Petri Nets: Unifying Petri Nets*, volume 2128 of *LNCS*. Springer, 2001.

6. J. Desel and W. Reisig. Place/transition Petri nets. In W. Reisig and G. Rozenberg, editors, *Lectures on Petri Nets: Basic Models*, pages 122–173. Springer Verlag, LNCS 1491, 1998.

7. H. Ehrig and B. Mahr. *Fundamentals of Algebraic Specification 2: Module Specifications and Constraints*, volume 21 of *EATCS Monographs on Theoretical Computer Science*. Springer Verlag, Berlin, 1990.

8. H. Ehrig, F. Orejas, B. Braatz, M. Klein, and M. Piirainen. A Generic Component Concept for System Modelling. In R. Kutsche, H. Weber (Eds.)*Proc. FASE 2002: Formal Aspects of Software Engineering*, LNCS 2306, pages 33–48. Springer, 2002.

9. P. Huber, K. Jensen, and R.M. Shapiro. Hierarchies in Coloured Petri Nets. In G. Rozenberg, editor, *Advances in Petri nets 1990*, LNCS 483, pages 313–341. Springer, 1991.

10. P. W. Hoogers, H. C. M. Kleijn, and P. S. Thiagarajan. A trace semantics for Petri nets. *Information and Computation*, 117(1):98–114, 1995.

11. G. Juhás and R. Lorenz. Modelling with Petri modules. In B. Caillaud, X. Xie, and L. Darondeau, Ph.and Lavagno, editors, *Synthesis and Control of Discrete Event Systems*, pages 125–138. Kluwer Academic Publishers, 2002.

12. E. Kindler. A compositional partial order semantics for petri net components. In Azéma, P. and Balbo, G., editors, *18th Int. Conf. on Application and Theory of Petri Nets*, LNCS 1248, pages 235–252. Springer-Verlag, 1997.

13. A. Mazurkiewicz. Basic notions of trace theory. In de Bakker, J.W. et al., editors, *Linear Time, Branching Time and Partial Order in Logics and Models for Concurrency.*, LNCS 354, pages 285–363. Springer, 1989.

14. J. Meseguer and U. Montanari. Petri Nets are Monoids. *Information and Computation*, 88(2):105–155, 1990.

15. R. Morin and B. Rozoy. On the semantics of place/transition nets. In *CONCUR 99*, LNCS 1664, pages 447–462. Springer, 1999.

16. H. Müller and H. Weber, editors. *Continuous Engineering of Industrial Scale Software Systems*. IBFI, Schloß Dagstuhl, Dagstuhl Seminar Report #98092, 1998.

17. M. Nielsen, G. Plotkin, and G. Winskel. Petri Nets, Event Structures and Domains, Part 1. *Theoretical Computer Science*, 13:85–108, 1981.

18. M. Nielsen and V. Sassone. Petri nets and other models of concurrency. In W. Reisig and G. Rozenberg, editors, *Lectures on Petri Nets I: Basic Models*, LNCS 1491, pages 587–642. Springer, 1998.

19. M. Nielsen, V. Sassone, and G. Winskel. Relationships Beween Models of Concurrency . In G. Rozenberg, J.W. de Bakker, W.-P. de Roever, editors, *A Decade of Concurrency*, pages 425 – 476. LNCS 803, 1993.

20. J. Padberg. Petri net modules. *Journal on Integrated Design and Process Technology*, 6(4):105–120, 2002.

21. J. Padberg. Transition systems with alternatives: an approach to a loose semantics of place/transition nets. Technical Report 2003-18, Technical University Berlin, 2003.

22. J. Padberg and H. Ehrig. Petri net modules in the transformation-based component framework. *Journal of Logic and Algebraic Programming*, 2003. submitted.

23. W. Reisig. *Petri Nets*, volume 4 of *EATCS Monographs on Theoretical Computer Science*. Springer Verlag, 1985.

24. H. Reichel. Specification semantics. In E. Astesiano, H.-J. Kreowski, and B. Krieg–Brückner, editors, *Algebraic Foundations of System Specification*, IFIP State–of–the–Art Reports, chapter 5, pages 131–158. Springer Verlag, 1999.

25. B. Rozoy. On distributed languages and models for concurrency. In G. Rozenberg, editor, *Advances in Petri Nets*, LNCS 609, pages 267–291. Springer, 1992.

26. H. Weber. Continuous Engineering of Communication and Software Infrastructures. In J.P. Finance (ed);*Fundamental Approaches to Software Engineering (FASE'99)*, LNCS 1577, 1999, pages 22–29. Springer Verlag, Berlin, Heidelberg, New York, 1999.

A Formal Framework for the Development of Concurrent Object-Based Systems*

Leila Ribeiro[1], Fernando Luís Dotti[2], and Roswitha Bardohl[3]

[1] Instituto de Informática, Universidade Federal do Rio Grande do Sul,
Porto Alegre, Brazil
leila@inf.ufrgs.br
[2] Faculdade de Informática, Pontifícia Universidade Católica do Rio Grande do Sul,
Porto Alegre, Brazil
fldotti@inf.pucrs.br
[3] International Conference and Research Center for Computer Science
Schloss Dagstuhl, Germany
rosi@dagstuhl.de

Abstract. In this paper we present a framework for developing concurrent object-based systems. The framework is based on graph grammars and includes techniques for specification, simulation, animation and verification.

1 Introduction

The development of methods and techniques to aid the construction of correct concurrent software systems has been a challenge for computer scientists for many years. In particular with the boom of intra- and Internet systems as well as the development of highly parallel and vectorial computer architectures, the commercial importance of software applications for distributed and concurrent systems increased significantly and the need for correctness gained a new perspective. This triggered a lot of research activities in the areas of semantical models (e.g. CCS [30], event structures [39], I/O automata [28] – see [40] for an overview), formal specification languages (e.g. LOTOS [37], Estelle [24], TLA [26], Petri nets [32,33], graph grammars [20,18]), verification techniques and tools (e.g. SMV [29], SPIN [23]). Many of such activities resulted in quite powerful models that were able to express and reason about concurrent and distributed systems, and were applied successfully to protocol specifications (system states are quite simple because usually they do not have any structure, and important events are state changes triggered by signals or messages). However, software engineers often use (semi-formal) languages like the UML [6] to specify real-life systems with complex states that cannot be captured easily and naturally by existing formal techniques. The formal background of users in industry was rather

---

* This work was partially sponsored by GRAPHIT (CNPq/DLR), ForMOS (FAPERGS/CNPq), PLATUS (CNPq), IQ-Mobile II (CNPq/CNR) and DACHIA (FAPERGS/IB-BMBF) Research Projects, and partially developed in collaboration with HP Brasil.

H.-J. Kreowski et al. (Eds.): Formal Methods (Ehrig Festschrift), LNCS 3393, pp. 385–401, 2005.

poor, and they were not familiar with the required mathematical notation for using formal specification languages to describe systems and their properties. Due to the size and complexity of distributed and concurrent software applications nowadays, the costs of finding and correcting bugs in the implementation are even higher in this kind of systems. Thus, the software development process would be greatly improved with a comprehensive analysis of functional as well as performance requirements during the specification phase. For such an analysis a formal model of the application would be useful. The development of a formal model, however, is not supported by most of the UML languages used in industry.

With the aim of bridging the gap between the semi-formal languages used in practice and the formal notation required to prove correctness, an international cooperation project between Brazil and Germany, leaded by Prof. Hartmut Ehrig, started in 1994: the GRAPHIT project (DLR/CNPq) [3]. Originally, the project involved two universities (TU-Berlin and UFRGS) and two industrial partners (MSB and Nutec); later further partners joined the project (Universitĩ Stuttgart and PUCRS). The idea was to develop a formal specification language with a graphical layout, preferably following the object-oriented style used in practice, and relying on few simple and powerful concepts. This would allow users to easily understand and build specifications with this language. Graph grammars seemed to be a perfect basis due to some of its inherent characteristics: graphical description of states (even complex ones can be better understood using a suitable graph representation), changes of states can be easily specified via relationships between graphs (rules), concurrency is naturally described (implicitly, through parallel applications of rules). In the following years a lot of research activities took place in order to define a suitable specification language as well as structuring and analyzing techniques for distributed and concurrent object-based systems in the GRAPHIT and in follow-up projects. In this paper we will review the main results we obtained for specification, simulation, animation and verification of concurrent object-based systems, and present the necessary steps to be followed in order to improve the software development process.

One of the results of the GRAPHIT project was the development of the specification language Object-Based Graph Grammars (OBGG) [12]. This language has been strongly influenced by the composition operators for graph grammars presented in [34] (a Ph.D. thesis advised by Prof. Hartmut Ehrig). The main idea is that a system is constructed by composing objects belonging to different classes. Each class is specified with a graph grammar (with special characteristics giving the language an object-based style natural for non-academic users). In section 2 we present the main concepts of Object-Based Graph Grammars.

To be really useful in practice, there must be techniques (and possibly tools) for the analysis of a specification. With this aim we have built a simulation tool for OBGG (the PLATUS tool[1]). With this tool it is possible not only to execute

---

[1] The underlying concepts of the simulator as well as a prototype of the tool were developed in the PLATUS project (CNPq).

the system for validating strategies, but also to make quantitative analysis (e.g. count the number of messages exchanged in the system, the number of created objects). In section 3 we describe how OBGG specifications are simulated.

Another result of the GRAPHIT cooperation was the development of the GENGED concepts and environment [4] (a Ph.D. thesis advised by Prof. Hartmut Ehrig). GENGED is a visual environment based on graph grammars for the visual definition of visual languages and the generation of language-specific visual environments. In section 4 we show how to use GENGED to provide animation of specifications written in OBGG[2].

The possibility of analyzing specifications was one of the main aims since the beginning of GRAPHIT. After defining the specification language OBGG it was possible to define verification techniques that could make use of existing model checking tools in a suitable way (due to the restrictions imposed in OBGG, verification of many properties became feasible – see [27] for examples). We have used different approaches for verification: translation to PROMELA (the input-language of the SPIN model checker [23])[3], translation to π-calculus [31][4] and we use unfolding semantics to verify properties directly for the OBGG specification [2]. The results about verification[5] are presented in section 5.

2   Object-Based Graph Grammars

In this section we propose a formal method to describe concurrent object-based systems. The main characteristics of this formalism are: it has a graphical layout, it is based on rules, it allows for a natural description of complex states, concurrency and non-determinism are inherent to the formalism, and it supports an object-based style.

The specification of each kind of object (class) that will be part of the composition of an object-based system is done via an (object-based) graph grammar. Before detailing the description of a class, we will present the kind of graphs and rules that are used for the specification of object-based systems (Sect. 2.1). These graphs are called *object-based graphs* and were introduced in [12].

2.1   Object-Based Graphs and Rules

Each graph in an object-based graph grammar is composed of instances of the vertices and edges shown in Figure 1(a). The vertices represent object identities/classes and abstract data types, whereas messages and attributes of objects are modeled as hyperedges (edges with one destination and many source

---

[2] The use of GENGED for animation of OBGG was a result of the German-Brazilian cooperation project DACHIA (IB-BMBF/FAPERGS), involving the universities TU-Berlin, Universität Stuttgart, UFRGS and PUCRS, and the company MSB.

[3] Investigations on this topic were carried out in the scope of the ForMOS (FAPERGS/CNPq) and CASCO (in collaboration with HP-Brazil) projects.

[4] Results obtained in the IQ-Mobile (Italian-Brazilian cooperation CNPq/CNR) and ForMOS projects.

[5] Investigations on verification were obtained in the scope of the ForMOS, CASCO, IQ-Mobile and and DACHIA projects.

vertices). We defined a distinguished graphical representation for these graphs to increase the readability of specifications (see Figure 1(b)). Elements of abstract data types are allowed as attributes of classes and/or parameters of messages. Note that the graph in Figure 1 defines a scheme only, indicating which kinds of vertices and edges may occur in a specification, and does not oblige classes or messages to have attributes. For example, this graph specifies that, if a class has attributes, they must be either of type ADT or of type Class.

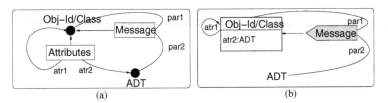

Fig. 1. (a) Object-Based Graph Scheme (b) Graphical Representation.

A rule expresses the reaction of an object to the receipt of a message. A rule of an object-based graph grammar consists of:

- a *left-hand side L*: describes the items which must be present in the current state to enable the rule application. The restrictions imposed by left-hand sides of rules are:
    - There must be exactly one message vertex, called trigger message (this is the message treated by this rule);
    - Only attributes of the class which is the target of the trigger message may appear (in the graphical representation, not all attributes of this class must appear, only the ones necessary for the treatment of this message);
    - Items of type ADT may be variables, which will be instantiated at the time of rule application. Operations defined in the ADTs may be used.
- a *right-hand side R*: describes the items which will be present after the application of the rule. It consists of:
    - Objects: all objects and attributes present in the left-hand side of the rule, as well as new objects (created by the application of the rule). The values of attributes may change but attributes cannot be deleted;
    - Messages to all objects appearing in R are allowed.
- a *condition*: that must be satisfied for the rule to be applied. This condition is an equation over the attributes of left- and right-hand sides.

Formally, we use typed attributed hypergraphs, and rules are (partial) graph homomorphisms with application conditions. See [19] for more details on this graph transformation approach.

2.2   Specification of Classes

A *class* is composed of a *type graph*, a *set of rules* and an *initial graph*.

Type Graph: a graph containing information about all attributes of this class (either ADT types or references to other classes) and messages sent/received by objects of this class. This graph can be seen as an instantiation of the object-based graph scheme described above.

Set of Rules: these rules specify how objects of this class will react to messages. For the same kind of message, there may be many rules specifying the intended behavior. Depending on the conditions imposed by these rules (conditions on attribute values and/or parameters of the message), they may be mutually exclusive or not. In the latter case, one of them will be chosen non-deterministically to be executed. Note that the behavior of an object when receiving a message is not specified as a series of steps that shall be executed, but rather as an atomic change of the values of the object attributes together with the creation of new messages to other (or the same) objects. That is, there is no control structure to govern the application of the rules specifying the behavior of an entity. Our approach is data driven, and therefore unnecessary sequentializations of computation steps are avoided because the specifier only has to care about the causal dependencies between events.

Initial Graph: this graph can be seen as a *template* for the creation of objects. It specifies initial attribute values for objects of this class, as well as messages which must be sent to these objects when they are created. These values can be either concrete or variables. If they are variables, they shall be seen as instantiation parameters (see discussion on object creation below).

There are two ways of creating objects in an object-based system: either they are created in the beginning (initial state) or they are created during the execution of the system. These are called *static* and *dynamic creation*, respectively. In both cases, the created objects must use the initial graph template of the corresponding class.

Static Object Creation: This situation occurs when the user composes the initial state of a system by creating instances of various different classes and linking them. To create an object one must instantiate the attributes which correspond to variables in the definition of the initial graph of the corresponding class, as well as assign concrete objects to the attributes which are references to other objects. All messages belonging to the definition of the initial graph of the class will be automatically sent to this object in the initial state of the system.

Dynamic Object Creation: During the execution of the system, new objects may be created. Objects of any class may request the creation of objects of other (or the same) class. The only requirement is that the the structure of the *creation message* of the class of the newly created object is known by the object that requests the creation. Each class has an associated *creation message*, which is a message that has the values needed to create the initial state of an object of that class as parameters (instantiation parameters). Moreover, there is a rule, called *creation rule*, which treats this creation message: it removes the message and creates the internal structure of the object according to definitions of the initial graph of the class and the parameters of the creation message. Only after the application of this rule the object is

executable (that is, before the application of this rule, the vertex corresponding to the object may be part of the system, but messages will be neither
treated nor sent by this object). The creation message and creation rule are
automatically obtained from the definition of the initial graph of each class,
and cannot be changed by the user.

2.3   Specification of Object-Based Systems

To build an object-based system, the user shall choose all classes that will compose his/her system, connect them in a suitable way, and create an initial state
for the system consisting of objects of these classes. More concretely, the steps
to be followed are:

Step 1: Choose the classes that will compose the system.

Step 2: Connect the classes chosen in step 1. This is accomplished by relating
the class vertices and message edges in the type graphs of the involved graph
grammars (classes). This is necessary because the names of classes/messages
in each type graph may not be syntactically the same. With this relationship
we say which items are semantically equivalent and are different, regardless
of the names used in the specification of the classes.

Step 3: Generate a grammar that corresponds to the whole system, using the
grammars chosen in step 1 and the connection defined in step 2.[6] Note that
the relationship among the type graphs defined in step 2 induces a relation
on the initial graphs and rules of the involved classes.

Step 4: Create the desired objects of each of the classes that compose the system. This corresponds to the *static creation* of objects discussed above.

2.4   Behavior

Each *state* of a computation described with a graph grammar is a graph containing objects (with their internal structure) and messages to be treated. A
rule r is *enabled* in a state S by a message m if m is the message that triggers r (the message deleted by r), each attribute which appears in the left-hand
side of r has the necessary value in S,[7] and the condition required by the rule
is true in S. The effect of this *rule application* is that all attributes of object
which are not preserved will receive new values as defined in the rule, message
m will be deleted and the new messages that appear in the right-hand side of
the rule will be created (new objects may be created as well). Nothing else in
the state is modified by this rule. Formally, this corresponds to a pushout in the
corresponding category [12].

At each execution state, several rules may be enabled (and therefore are
candidates for execution at that moment). Rule applications only have local

---

[6] Formally, this step is implemented as a composition (via a universal construction)
of the corresponding graph grammars using the morphisms induced by the relationship defined in step 2. More details about this composition operator can be found
in [34, 35].

[7] A rule may require that an attribute has a certain value, for example, rule *SymStart*
of Figure 2 can only be applied if acquire has the value true in state S.

effects on the state. However, there may be several rules competing to update the same portion of the state. To determine which set of rules will be applied at each time, we need to choose a set of rules that is consistent, that is, in which at most one rule has write access to the same resources (non-conflicting rule applications). Due to the restrictions imposed in object-based graph grammars, conflicts can only occur among rules of the same class. When such a conflict occurs, one of the rules is (non-deterministically) chosen to be applied. The semantics of the whole system is the set of all possible computations [25] (or, equivalently, a tree containing all computations [34]).

2.5   Example: Dining Philosophers

In this section we model the dining philosophers problem using OBGGs. Traditionally the dining philosophers problem is described by the following scenario: There are N philosophers sitting at a table with N forks (one fork between each philosopher). The philosophers spend some time thinking, and from time to time a philosopher gets hungry. In order to eat a philosopher must acquire exclusively its left and right forks. After eating a philosopher releases both forks and starts thinking again. Using OBGGs we modeled the problem with the two classes *Fork* and *Phil*, depicted in Figures 2 and 3[8]. Each *Fork* object is composed of a boolean attribute (*acquired*) determining if the fork is currently in use by a philosopher (*acquired* true) or not (*acquired* false). Each *Phil* object is composed of two reference attributes to *Fork* objects (modeling its left (*leftFork*) and right (*rightFork*) forks) and five boolean attributes: *acquire* (the philosopher is trying to acquire the forks), *eat* (the philosopher is eating), *release* (the philosopher is releasing its acquired forks), *asym* (indicates if the philosopher starts getting the left fork (false), or the right fork (true)), and *forks* (used to control the number of acquired forks). The rules for the *Phil* and *Fork* objects represent the behavior of these objects (creation rules are omitted). A philosopher starts executing, rules (rules *AcquireLeft* and *AcquireRight*). If the philosopher can acquire the fork (rule *Acquire*), he tries to acquire the other fork (rules *SymLeft* or *AsymRight*). If the philosopher can acquire it too (rules *SymRight* or *AsymLeft*), he starts eating (rule *Eating*). After eating the philosopher releases his forks and starts all over again.

When building a model for the dining philosophers using OBGG, after choosing the two classes above (step 1) and suitably connecting them (step 2), we can generate a grammar for the complete system (step 3), and choose an initial graph for this system (step 4). In Fig. 4 we show an initial graph for an asymmetric solution for the problem. This solution is asymmetric because the philosopher *Phil2* has its *asym* attribute set to true, meaning that he will try to acquire the right fork first, differently from the other philosophers.

---

[8] The numbers inside the circles (indexig each class of the type graph) are used to allow a clearer graphical representation of the type graph (preventing arrows which would cross the picture) as well as to indicate how objects appearing in rules and initial graphs are mapped to the type graph.

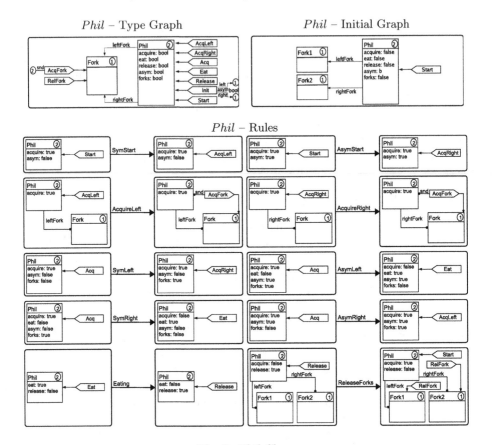

Fig. 2. *Phil* Class.

3   Simulation of OBGG Specifications

One of the main advantages of using simulation models while developing an application is the possibility to *validate* strategies as well as control algorithms even before their implementation. For instance, simulation models may be used to: check if the components of an application behave as expected; if they are independent from each other such that the replacement of a simulated component by a more sophisticated version becomes possible; check the application behavior under various environment conditions (e.g. failure simulation); obtain quantitative results such as the number of messages exchanged, number of rules applied for each object, number of objects created dynamically, among others.

The simulation of OBGGs must ensure that the formalism is faithfully represented. Analyzing the model characteristics we conclude that an OBGG model is a discrete-state system: the state of the objects changes as rules are executed in response to the messages. Thus, the use of discrete-event simulation is a natural approach to evaluate such systems.

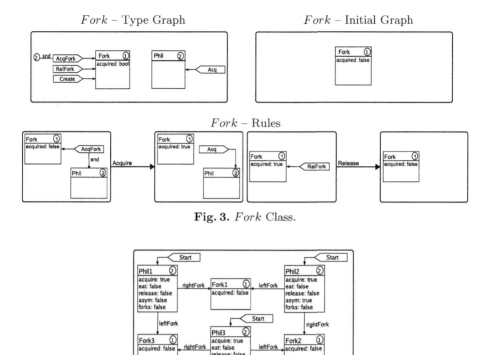

Fig. 3. *Fork* Class.

Fig. 4. Initial Graph for an Asymmetric Solution.

An OBGG specification is mapped to a simulation model in two main parts: one is the simulation kernel, the other is the representation of internal activities of objects. The kernel is responsible for passing messages among objects (i.e. dispatching events to simulation entities) and managing the global simulation time. The simulation time is needed to ensure the progress of the simulation process and to simulate the parallel execution of rule applications. OBGG messages are not timed or ordered in any way. The chosen kernel algorithm follows a conservative approach and the simulation time has a centralized control. The kernel keeps a list of messages (or events), known as event list. The simulation process consists of selecting event(s) (message(s)) according to the OBGG semantics; delivering those events (messages) to the target simulation entities (objects); awaiting each recipient entity (object) to handle the message; advancing the simulated time and starting selecting events again.

In an OBGG model we may have concurrent activities in different objects or within a same object. A multi-threaded approach is natural to model concurrency among objects. Each object will have an associated thread to independently process messages (events) from its input buffer. The internal thread of each object is responsible for processing the incoming messages and applying appropriate rules, according to the OBGG semantics. Within an object, all non-conflicting enabled rules may be applied in parallel[9]. For more details on the internal behavior of a simulation entity (representing an OBGG object) see [7, 10].

Based on the ideas discussed above, a simulation environment for OBGGs was constructed [8, 7, 10] and used in various applications. The simulation environment is composed of a library that offers: the kernel functionality; the basic internal behavior of an object; the basic structure for a rule; the basic structure for a message. More concretely, this library is offered in an object-oriented language (Java) and these are the main classes comprising the library. OBGG models are mapped to simulation models in Java as follows: i) each OBGG class maps to one class specializing class Entity from the library, which will have the corresponding attributes and an internal thread performing as discussed above; ii) each rule specifying part of the behavior of an OBGG class is mapped to a class specializing class Rule; iii) messages handled by OBGG classes map to specializations of class Message; iv) the initial graph is mapped accordingly to a initial state of the simulation environment through the creation of the appropriate entity instances, messages and attributes initialization.

With this, to simulate an OBGG specification we have the following steps:

Step 1 – Translation of the OBGG to a Simulation Model: According to the ideas discussed above, an algorithm is used to translate an OBGG specification to a Java program extending pre-defined classes and interacting with the simulation kernel;

Step 2 – Simulate: Once the initial state is built, the simulation takes place generating a log with all messages exchanged among instances and the kernel and rules applied in each instance, along with the simulation time these events occurred.

Step 3 – Analysis of Results: The user may then use the log to analyze the phenomena of interest. Simulation was shown to be very useful to find specification errors (e.g. missing behavior (missing rules), wrong left-hand side specification, etc.) as well as to estimate the communication behavior in terms of total number of exchanged messages and helping to estimate the number of messages in a round (number of causally-dependent messages generated/deleted to complete a service). Alternatively, the simulation log may be submitted for animation using the ideas described in the next section.

4 Animation

In order to have a better visualization of the states and behavior of a system, it is useful to define an animation for the specification, if possible, using symbols that are close to the application domain to ease the understanding. To animate the specification of an object-based system *OBSys* using OBGGs, there are 4 steps to be followed:

Step 1 – Construction of Animation Grammars: The specification of each class *C* of *OBSys* is extended by animation information given by attributes used during the animation and/or by messages that trigger animation events.

---

[9] The encapsulation of object-based modeling prevents conflicts among objects.

This new kind of grammar is called animation grammar for class C ($AnimC$). Note that this information shall not interfere with the behavior of the original OBGG C (just extra attributes may change and extra messages may be sent with each rule application).

Step 2 – Definition of Animation Interfaces: Now the user must define the attributes/messages that shall be visible for the one that will define the concrete animation module for the application. This is an abstraction process, resulting in a grammar for each class, called animation interface. This interface is a tool for the users of this class: they do not have to know the internal structure and behavior of objects of this class to choose the desired graphical representation (this guarantees encapsulation). In fact, the same class may have many different graphical representations, depending on the way it is used (see next step).

Step 3 – Definition/Choice of Animation Modules: An *animation module* consisting of concrete graphical representations for the items in each class must be defined (or chosen, if a library of animation modules is available). An animation module can be specified using GENGED, an approach with tool support for the visual definition of visual languages based on graph grammars and graph transformation, respectively. Using GENGED, the animation of each class is considered as visual language. Note that, in order to build the animation of a class, only the information present in the animation interface is necessary (other attributes and messages can only be seen by the developer of a class).

Step 4 – Animate: For the animation, we first generate the initial visualization of the system according to the initial state of the object-based system. This means, we automatically generate visualizations for all objects of the initial system graph (cf. step 4 in sect.2.3). Then, while simulating the OBGG corresponding to the whole system, animation messages shall be sent to the animation view generated by GENGED(these messages trigger the changes in the graphical representation).

In general, the specification of a visual language using GENGED [4, 5] is given by an alphabet and a grammar. The alphabet defines the language vocabulary, i.e. the symbols and how they are connected: symbol descriptions consist of a textual name and a corresponding graphic, connections between symbols are associated to graphical constraints. The type graph of a class can be used to generate an alphabet, i.e. it establishes a type system for all instances: vertices are modeled as attributed graph nodes where the attributes describe the graphics and arcs are modeled as graph edges. The animation using GENGED will be triggered by message passing. For this purpose, messages vertices become symbols in the alphabet. Since messages should not be visualized, message symbols have no associated graphical representation. The initial graph and the rules defined using GENGED have to fit to the corresponding components of the interface. Each rule describes how the animation shall change when receiving messages defined in the animation interface of the module. Other rules may be defined for animation purposes.

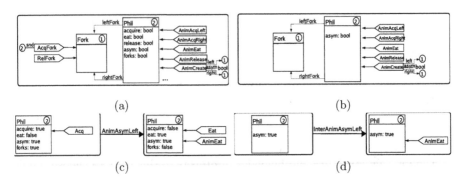

Fig. 5. (a)(c) Animation Grammar (b)(d) Animation Interface.

Figure 5 shows an example of animation components for the dining philosophers. In (a) we can see the the animation grammar type graph for the *Phil* class (this type graph consists also of all messages that appear in Figure 2, but these were omitted here). A rule of this grammar is depicted in (c) – it is an extension of the corresponding rule of the *Phil* grammar. The animation interface type graph is presented in (b), consisting only of items that shall be visible for the users of the class and animation module developers. The animation interface rule that corresponds to the concrete rule (c) is shown in (d).

5   Verification of OBGG Specifications

Although simulation is a useful analysis method, it does not allow one to make definitive assertions about the behavior of a system. Specially when building concurrent systems, it is not obvious that certain properties hold. In order to guarantee the desired behavior, verification techniques can be used. Model checking is an appealing verification method since it does not require advanced users know-how during the verification process (in contrast to theorem proving). Methods and tools to allow the model checking of OBGG specifications were developed. Instead of developing a model checking tool for graph grammars or OBGGs we focused on translating our models to formal languages that serve as input for existing model checking tools. Our approach seeks to reuse the existing implementations of model checkers, as well as take immediate advantage of their enhancements.

Starting with [22], efforts were made to translate OBGG specifications to π-Calculus. The π-calculus [31] is a well known and established formalism for description of semantics of concurrent systems. There are some automatic checkers for this formalism, for example, HAL [21] and MWB (Mobility Workbench) [38]. Objects and messages are defined as processes of the π-calculus (agents) that communicate through local channels. The source object and the parameters of each message are represented as parameters of message agents. The objects reactions when receiving a message are described by rule agents that compose the object agent. A rule describes the procedure to treat a message. Each kind of message can have several procedures for treating it, so we may have several rule agents that describe the treatment of the same kind of message. The choice of procedure to be executed is non-deterministic. This is described by composing rule agents (for the same kind of message) with the Sum operator (+) without guards. The concurrency between objects is modeled by parallel composition of object and message agents. So, each object can treat its messages in parallel. The internal concurrency is modeled by recursion of object agents. Although the semantic compatibility of the translation could be shown, two main problems were faced: i) models had to be abstracted/restricted such that OBGG objects had no internal state (there is a great overhead to model attributes in π-calculus); ii) when using existing π-Calculus model checkers, there are some limitations, specially to support the replication operator of the π-Calculus. Moreover, π-calculus tools were not very optimized, and thus only small examples could be translated.

A second translation was proposed in [11], translating OBGG to PROMELA (PROcess/PROtocol MEta LAnguage) [1], the input language of SPIN (Simple Promela INterpreter) [23]. PROMELA is based on processes and provides shared memory as well as (CSP-like) channels for interprocess communication. According to the proposed translation, OBGG objects are mapped to PROMELA processes. For verification purposes, attributes of OBGG objects are restricted to the types supported in PROMELA (boolean, char, int, array). Moreover, OBGG messages are translated to PROMELA messages, and the receipt of messages is done through an asynchronous channel (that is defined for every object), which is also used as reference of translated objects. Rules for OBGG objects are mapped to a condition structure inside the translated object, and the OBGG initial graph becomes the initial process in PROMELA. Concurrency among objects is naturally preserved by the concurrency between translated objects (processes). More details about the translation can be found in [11], including a discussion of the semantic compatibility for the generated PROMELA model and the original OBGG model.

Complementary to the model translation to the input language of a model checker, an approach for the specification of properties using a given temporal logic is needed. Up to now, we have considered properties about the history of rule applications starting from the initial state. In [11] and [36] we define how to specify properties over OBGG models using Linear Temporal Logics (LTL) – the same temporal logic used in SPIN. LTL properties are defined over events of the model, where an event is a rule application. Moreover, a graphical presentation of counter-examples in terms of OBGG abstractions, instead of the translated PROMELA model, is provided.

Considering the above discussed ideas, we have the following steps for model checking of OBGG specifications:

Step 1 – Translate the OBGG Specification to a Verification Model:
An OBGG specification is translated, following an algorithm, to a PROMELA model which serves as input to the SPIN model checker;

Step 2 – Definition of Properties in Temporal Logic: LTL formulas can be defined taking into consideration rule applications (that are the events);

Step 3 – Model Checking: The model checking tool can be initiated with the model and the properties to be verified;

Step 4 – Analyze Results: Step 3 may have different outcomes: i) the model satisfies the property; ii) the model does not satisfy the property and a counter example is generated which can be visualized in terms of OBGG abstractions (instances, messages and rule applications); iii) the verification terminates abnormally due to insufficient size of channels. To respect the OBGG semantics, channels must always have room for incoming messages since otherwise the sending process would block and this would mean blocking an OBGG instance while sending a message. In such cases the specifier may define larger channels and try again; and iv) the verification terminates due to insufficient resources, such as memory. It is known that model checking techniques can be very time and space consuming.

For the dinning philosophers example the following properties were shown:

– Liveness: the asymmetric version is shown to be deadlock free while deadlocks may occur in the symmetric version. To prove this, an LTL formula was used to specify that "it is always possible that some philosopher will eat". More concretely, it is proven that it is always possible, in the future, to apply the *Eating* rule.
– Safety: in a setting with up to three philosophers it is sufficient to prove that "two philosophers will not be eating at the same time". The approach taken is to prove that, if a philosopher starts eating (rule *Eating*) then until he releases the forks (rule *ReleaseForks*) no other philosopher will start eating (rule *Eating*).

6   Conclusions and Future Work

The development of concurrent and distributed systems is challenging. This paper presented how to use OBGG for modeling and analyzing such systems. Beyond the basic abstractions provided by the language, OBGG is attractive due to the diversity of development methods and tools supported (see [9] for a description of the tools). To validate this specification language, we carried out a series of case studies. Using simulation, we analyzed mobile code applications [12], active networks [17], and a pull-based failure detector [13]. With verification we checked the readers and writers problem [36], where it is shown that writers can starve while readers proceed; the distributed election in a ring algorithm [14], where we have shown that "eventually there will be one element of the ring elected" and "there cannot be two elements of the ring simultaneously elected"; and message ordering [15], where it could be shown that a message ordering mechanism delivers the messages in the expected order.

In order to allow the specification and reasoning about distributed systems in the presence of faults, in [16,13] an approach for the insertion of classical fault behaviors in OBGG models of distributed systems has been presented. It was shown that, due to the abstractions provided by OBGGs, it is possible to transform an OBGG model to embed the behavior of a given fault model. We have analyzed distributed systems in the presence of faults using simulation, and are currently investing efforts in analyzing such systems using the model checking approach.

The object-based nature of the language and its application to open distributed systems naturally calls for verification methods for reasoning over partial systems. Thus, efforts are being made towards methods for the verification of partial systems described using OBGGs (first results can be found in [15]).

Moreover, the OBGG language itself can be improved, for example, by adding inheritance and module concepts. This would greatly simplify an integration with UML, in the sense that OBGG can be used as a semantical model for many UML languages, integrating in a smooth way the different views provided by UML. For this, translations from these UML languages into OBGG should be defined. In this way, we could use all techniques and tools available for OBGG also for specifications using the UML languages.

Acknowledgments

Altogether the series of achievements (and cooperation projects) presented in this paper were influenced (directly or indirectly) by the ideas and stimulus of Hartmut Ehrig. Besides his own technical contributions to Computer Science and qualified technical advisoring, Hartmut is also concerned with the academic maturity of his students, encouraging them to present their results to the academia and giving them the opportunity to work in cooperation projects to learn how to combine efforts to reach more relevant contributions to Computer Science. We would like to thank him heartily for the technical discussions (that led to many results presented here), carrier advices (that helped us to continue the work started in Berlin) and friendship (encouraging us to pursue our ideas).

References

1. Promela language reference. http://spinroot.com/spin/Man/promela.html, 2003.
2. P. Baldan, A. Corradini, and B. Koenig. Verifying finite-state graph grammars: an unfolding-based approach. In *Proc. of CONCUR 2004*, LNCS. Springer, 2000.
3. B. Bardohl, R. Bardohl, P. Castro, B. Copstein, H. Ehrig, M. Korff, A. Martini, D. Nunes, L. Ribeiro, and H. Schlebbe. GRAPHIT: Graphical support and integration of formal and semi-formal methods for software specification and development. In *6th German-Brazilian Workshop on Information Technology*, 2000.
4. R. Bardohl. *Visual Definition of Visual Languages based on Algebraic Graph Transformation*. PhD thesis, Technical University of Berlin, Germany, 2000.
5. R. Bardohl, C. Ermel, and I. Weinhold. Specification and Analysis Techniques for Visual Languages with GenGED. Technical Report 2002–13, Technical University Berlin, Dept. of Computer Science, September 2002. ISSN 1436-9915.
6. G. Booch, J. Rumbaugh, and I. Jacobson. *The Unified Modeling Language user guide*. Addison-Wesley, 1998.

7. B. Copstein, M. C. Móra, and L. Ribeiro. An environment for formal modeling and simulation of control systems. In *33rd Annual Simulation Symposium*, pages 74–82, USA, 2000. IEEE Computer Society Press.

8. B. Copstein and L. Ribeiro. Specifying simulation models using graph grammars. In *10th European Simulation Symposium*, pages 60–64, UK, 1998. SCS.

9. F. L. Dotti, L. Duarte, L. Foss, L. Ribeiro, D. Russi, and O. Santos. An environment for the development of concurrent object-based applications. *Electronic Notes in Theoretical Computer Science (International Workshop on Graph-Based Tools)*, 2004.

10. F. L. Dotti, L. M. Duarte, B. Copstein, and L. Ribeiro. Simulation of mobile applications. In *2002 Communication Networks and Distributed Systems Modeling and Simulation Conference*, pages 261–267, USA, 2002. The Society for Modeling and Simulation International.

11. F. L. Dotti, L. Foss, L. Ribeiro, and O. M. Santos. Verification of object-based distributed systems. In *6th International Conference on Formal Methods for Open Object-Based Distributed Systems*, volume 2884 of *LNCS*, pages 261–275, France, 2003. Springer.

12. F. L. Dotti and L. Ribeiro. Specification of mobile code systems using graph grammars. In *4th International Conference on Formal Methods for Open Object-Based Distributed Systems*, volume 177 of *IFIP Conference Proceedings*, pages 45–64, USA, 2000. Kluwer Academic Publishers.

13. F. L. Dotti, L. Ribeiro, and O. M. Santos. Specification and analysis of fault behaviours using graph grammars. In *2nd International Workshop on Applications of Graph Transformations with Industrial Relevance*, volume 3062 of *LNCS*, pages 120–133, USA, 2003. Springer.

14. F.L. Dotti, L. Foss, L. Ribeiro, and O.M. Santos. Specification and formal verification of distributed systems (in portuguese). In *17th Brazilian Symposium on Software Engineering (SBES)*, pages 225–240, 2003.

15. F.L. Dotti, F. Pasini, and O.M. Santos. A methodology for the verification of partial systems modelled with object based graph grammars (in portuguese – accepted for publication). In *18th Brazilian Symposium on Software Engineering (SBES)*, 2004.

16. F.L. Dotti, O.M. Santos, and E.T. Rödel. On the use of formal specifications to analyse fault behaviors of distributed systems. In *First Latin-American Symposium on Dependable Computing*, volume 2847 of *LNCS*, pages 341–360. Springer, 2003.

17. L. Duarte and F.L. Dotti. Development of an active network architecture using mobile agents – a case study. Technical Report TR-043, PPGCC-FACIN-PUCRS, Brazil, 2004.

18. H. Ehrig. Introduction to the algebraic theory of graph grammars. In *1st International Workshop on Graph Grammars and Their Application to Computer Science and Biology*, volume 73 of *LNCS*, pages 1–69, Germany, 1979. Springer.

19. H. Ehrig, R. Heckel, M. Korff, M. Löwe, L. Ribeiro, A. Wagner, and A. Corradini. Algebraic Approaches to Graph Transformation II: Single Pushout Approach and Comparison with Double Pushout Approach. In G. Rozenberg, editor, *Handbook of Graph Grammars and Computing by Graph Transformation, Volume 1: Foundations*, chapter 4. World Scientific, 1997.

20. H. Ehrig, M. Pfender, and H. J. Schneider. Graph grammars: an algebraic approach. In *14th Annual IEEE Symposium on Switching and Automata Theory*, pages 167–180, 1973.

21. G. Ferrari, S. Gnesi, U. Montanari, M. Pistore, and G.Ristori. Verifying Mobile Processes in the HAL Environment. In *International Conference on Computer Aided Verification*, volume 1427 of *LNCS*, pages 511–515, Vancouver, CA, 1998. Springer.

22. L. Foss and L. Ribeiro. A translation of object-based hypergraph grammars into π-calculus. *Electronic Notes in Theoretical Computer Science*, 95:245–267, 2004.

23. G. J. Holzmann. The model checker SPIN. *IEEE Transactions on Software Engineering*, 23(5):279–295, 1997.

24. ISO. Information processing systems – Open systems interconnection – Estelle – a formal description technique based on an extended state transition model, 1989.

25. M. Korff. *Generalized graph structure grammars with applications to concurrent object-oriented systems*. PhD thesis, Technical University of Berlin, Germany, 1996.

26. Leslie Lamport. The temporal logic of actions. *ACM Transactions on Programming Languages and Systems*, 16(3):872–923, 1994.

27. A. B. Loreto, L. Ribeiro, and L. V. Toscani. Decidability and tractability of a problem in object-based graph grammars. In *17th IFIP World Computer Congress – Theoretical Computer Science*, volume 223 of *IFIP Conference Proceedings*, pages 396–408, Canada, 2002. Kluwer Academic Publishers.

28. N. Lynch and M. Tuttle. An introduction to input/output automata. *CWI-Quarterly*, 2(3):219–246, 1989.

29. K. L. McMillan. *Symbolic Model Checking*. Kluwer Academic Publishers, 1993.

30. R. Milner. *Communication and Concurrency*. International Series in Computer Science. Prentice Hall, London, 1989.

31. R. Milner. *Communicating and mobile systems: the π-calculus*. Cambridge University Press, USA, 1999.

32. C.A. Petri. *Kommunikation mit Automaten*. PhD thesis, Schriften des Institutes für Instrumentelle Mathematik, Bonn, 1962.

33. Wolfgang Reisig. *Petri Nets*, volume 4 of *EATCS Monographs on Theoretical Computer Science*. Springer, 1985.

34. L. Ribeiro. *Parallel Composition and Unfolding Semantics of Graph Grammars*. PhD thesis, Technical University of Berlin, Germany, 1996.

35. L. Ribeiro. Parallel Composition of Graph Grammars. *Applied Categorical Structures*, 7(4):405–430, 1999.

36. O. M. Santos, F. L. Dotti, and L. Ribeiro. Verifying object-based graph grammars. *Electronic Notes in Theoretical Computer Science (Proc. 2nd Graph-Transformation and Visual Modeling Techniques)*, 2004.

37. P.H.J. van Eijk, C. A. Vissers, and M. ((editors) Diaz. *The formal description technique LOTOS*. Elsevier Science Publishers, 1989.

38. B. Victor and F. Moller. The Mobility Workbench – a tool for the π-calculus. In David Dill, editor, *Proc. International Conference on Computer Aided Verification, CAV*, volume 818 of *LNCS*, pages 428–440. Springer, 1994.

39. G. Winskel. An introduction to event structures. In *Linear Time, Branching Time and Partial Order in Logics and Models for Concurrency*, pages 364–397. Springer, 1989.

40. G. Winskel and M. Nielsen. Models for concurrency. Technical Report BRICS RS-94-12, University of Aarhus, 1994.

A Formal Description of the Basic Concepts of System Theory for Transportation

Eckehard Schnieder and Jörg R. Müller

Technical University of Braunschweig,
Institute for Traffic Safety and Automation Engineering,
Braunschweig, Germany
{schnieder,mueller}@iva.ing.tu-bs.de

Abstract. In this paper some of the basic concepts of system theory are presented in a formal way. This is done with the help of the formal modeling language petri-nets. An example out of the transportation is used to illustrate the discussed concepts.

1 Introduction

There are models, principles and laws that apply to generalized systems or their subclasses, irrespective of their particular kind, the nature of their component elements, and the relations or "forces" between them. So it seems legitimate to ask for a theory not for systems of a more or less special kind, but of universal principles applying to systems in general. The aim of this paper is to give a general formalization of some of the basic concepts, such as "boundary of a system", or "emergent property" used in systems of different kinds (see [5]). The formalization of some fundamental concepts of system theory shall provide a common basis for all branches of science. In each of these branches the common (abstract but formal) definitions made here can be instanciated with domain-specific terms.

In chapter 2 we introduce some of these concepts in an informal way. Afterwards, in chapter 3 we use petri-nets as a formal modeling language to formalize the concepts presented in the second chapter. Chapter 4 focusses on an example from the transportation to clarify the introduced concepts.

2 Basic Concepts of System Theory

In this chapter we'll outline the basic concepts of system theory in an informal way (see [1]). An idea of these concepts is the basis for the following chapters.

2.1 What Is a System?

To start with the beginning: One can imagine, that in the beginning there existed only a totally homogenious and unformed *primary-matter*. In order to create something, one has to establish a *distinction*. Where these distinctions come from (whether from some kind of mystic creator or likewise an observer), does

H.-J. Kreowski et al. (Eds.): Formal Methods (Ehrig Festschrift), LNCS 3393, pp. 402–411, 2005.

not matter here. But how can these distinctions be made? The primary-matter elements are ordered in some way. Due to this ordering they can be distinguished from other elements. So, these ordered elements fulfil special relations and conditions (in contrast to the non-ordered elements) – as a consequence, the *ordered elements can be separated from the non-ordered ones.* The established relation between the ordered elements is also called *organisation* or *structure.* This organisation itself determines, how the system operates and which processes can be performed. One must not misunderstand "organisation" as a static concept; here an organisation consists of *operations* and *processes* – the organisation of a system determines, how the system operates, depending on a concrete order of system elements. The set of all conditions that are fulfilled by an element is called the element's *state.* The state of the system is defined as the state of all the system's elements. As a consequence, the organisation and the state of the system determine, how the system operates.

So, a system consists of (a set of) elements (selected from all possible elements), that are related in a special manner and therefore enable certain processes.

2.2   A System and Its Environment

That, whereof a system can be distinguished (that, what is beyond the system), is called *environment* – the system embeds its own elements, thus the environment is excluded. The effect is: Environment is environment only in relation to a special system; there is no environment without a system. As a consequence, the unformed primary-matter in the beginning is not an environment in our sense.

All the parts of a system's organisation that depend directly on its environment or that affect directly its environment form the *boundary of a system.*

2.3   Subsystems and Emergence

If one examines a system without taking its environment into account, it is apparent that the system is divided into functional parts: For every operation there are components or *subsystems.* These subsystems are achieved due to differentiations: "The establishment of new system/environment-differences as part of the initial system."([2])[1] The above mentioned elements can be regarded as subsystems, if they form a functional part.

Interdependencies between subsystems may lead to a new organisation on the system level. If this transformation of organisation is not deducible from only the subsystems but with taking into account the interdependencies between them, we call these transformations *emergence.*

3   Formal Modeling with Petri-Nets

In this chapter the basic notations of petri nets [3] are outlined. We use petri nets here, as an instrument to formalize the concepts introduced in chapter 2.

---

[1] "Etablierung neuer System/Umwelt-Differenzen innerhalb des Ursprungssystems." ([2], translation J. R. M.).

After each formal definition, we try to work out a point to point correspondence between the informal definition made in chapter 2 and the formal counterpart made here.

In chapter 4 we'll use petri-nets as a modeling language to illustrate the concepts introduced so far on a concrete transportation-oriented example.

Definition 1 (Place/Transition-Net)

A *place/transition-net (p/t-net)* is a quadruple $\mathcal{N} = (P, T, F, L)$, with:

$$P \; : \; \text{is a set of } \textit{places},$$
$$T \; : \; \text{is a set of } \textit{transitions} \text{ with } P \cap T \; = \; \emptyset,$$
$$F \subseteq (P \times T) \cup (T \times P) \text{ is a } \textit{flow relation}$$
$$L \; : \; F \longrightarrow \mathbb{N}.$$

For every arc $(a, b) \in F$ $L(a, b)$ is called the *label* of the arc (a, b); especially in the context of p/t-nets the labels are called *(arc) weights*.

Remark 1. A petri-net $\mathcal{N} = (P, T, F, L)$ models the organisation of the systems taken into account:

- the elements in the set P, model the possible (local) states of the considered systems,
- the elements in T and F model relations and *operations* and so provide the basis for *processes*,
- the elements in L are just used to refine relations and operations.

(see example in chapter 4.1)

Remark 2. In definition 1 only the organisation (or: structure) of a system is defined. Note that a system's-structure may exist without any system-elements.

Definition 2 (Marking)

M is a *marking* of a p/t-net $\mathcal{N} \; = \; (P, T, F, L)$, iff

$$M : P \longrightarrow \mathbb{N}.$$

Remark 3. $M(p)$ specifies the number of elements that are in a defined local state p. So, each element is in a defined local state and related to other system-elements due to the system's-structure.

Definition 3 (Marking of a P/T-Net)

Let $\mathcal{N} = (P, T, F, L)$ be a p/t-net; the *marking of all places of a net* $M(P)$ is specified with a column vector

$$M(P) = \begin{pmatrix} M(p_1) \\ M(p_2) \\ \vdots \\ M(p_3) \end{pmatrix}.$$

For every place $p \in P$, it is

$$M(p) = \big(M(P)\big)(p).$$

Remark 4. $M(P)$ gives for each local state $p \in P$ the number of elements in that state. So, we can call $M(P)$ the *global state* of a system.

Definition 4 (Activated, Firing, Follower Marking)
Let $\mathcal{N} = (P, T, F, L)$ be a p/t-net and M be a marking; we say $t \in T$ is *activated* or *enabled* under M, if

$$\forall p \in {}^{\bullet}t : M(p) \geq L(p, t).$$

If t is activated under M (in symbols $M[t\rangle$), t may *fire* and transfer the marking M to a *follower marking* M' (in symbols $M[t\rangle M'$), with

$$M'(p) := \begin{cases} M(p), & \text{if } p \notin {}^{\bullet}t \cup t^{\bullet} \\ M(p) - L(p, t), & \text{if } p \in {}^{\bullet}t \setminus t^{\bullet} \\ M(p) + L(t, p), & \text{if } p \in t^{\bullet} \setminus {}^{\bullet}t \\ M(p) - L(p, t) + L(t, p), & \text{if } p \in {}^{\bullet}t \cap t^{\bullet}. \end{cases}$$

Remark 5. Definition 4 defines the *firing-rule*. The firing rule defines how a system's state changes to a follower state, due to some kind of *atomic process* (the firing of a single transition). Each system process, determined by the system's organisation (see definition 1) and the system's state (see 2), consists of a sequence of atomic processes.

Definition 5 (Firing Sequence)
Let $\mathcal{N} = (P, T, F, L)$ be a p/t-net with $t_1, t_2, \ldots, t_n \in T$.
   $\sigma = t_1, t_2, \ldots, t_n$ is a *firing sequence* iff there exist markings $m_0, m_1, \ldots m_n$ such that

$$m_0[t_1\rangle m_1[t_2\rangle \ldots [t_n\rangle m_n$$

holds.
   The set $\{M \mid \exists\sigma: \; M_0[\sigma\rangle M\}$ of all reachable markings is symbolized with $[M_0\rangle$.

Remark 6. The correspondence between system-processes (as described in chapter 2) and the concept of firing sequences as introduced here, is obvious (see example in chapter 4.1).

Definition 6 (P/T-System)
Let $\mathcal{N} = (P, T, F, L)$ be a p/t-net and M_0 be a marking of $\mathcal{N}$. We call $(\mathcal{N}, M_0)$ a *p/t-system*. M_0 is called the *initial marking*.

Remark 7. So, we define a system as a pair, consisting of a structure or organisation (of ordered elements) and a global state (see example in chapter 4.1).

Definition 7 (Subnet)

We call $\mathcal{N}' = (P', T', F', L')$ a *(finite) subnet* of the p/t-Net $\mathcal{N} = (P, T, F, L)$, if

$$P' \subseteq P,$$
$$T' \subseteq T, \quad \text{und}$$
$$F' \subseteq F \cap ((P' \times T') \cup (T' \times P')),$$
$$L'(f') := L(f') \; \forall f' \in F'.$$

Additionaly we require the sets P', T' and F' to be finite.

Remark 8. With a p/t-net we model the structure of a system (see definition 1). So, a subnet specifies some kind of *substructure* or *suborganisation* – that means nothing but the organisation of a subsystem:

Definition 8 (Subsystem)

Let $(\mathcal{N}, M_0)$ be a p/t-system and $\mathcal{N}'$ a subnet of $\mathcal{N}$; let further be

$$M_0'(p') := M_0(p') \; \forall p' \in P'.$$

Then we call $(\mathcal{N}', M_0')$ a *subsystem* of $(\mathcal{N}, M_0)$.

Remark 9. So, a subsystem is a system itself. It consists of a suborganisation and a substate. In general there are (in relation to the surrounding system) both, less (local) states are selected and less relations are forced. As a consequence, a more approximate look is forced: not all elements of the surrounding system are taken into account when looking at the subsystem.

Definition 9 (Union of Systems)

Let $(\mathcal{N}_1, M_1) := ((P_1, T_1, F_1, L_1), M_1)$ and $(\mathcal{N}_2, M_2) := ((P_2, T_2, F_2, L_2), M_2)$ be p/t-systems. If $P_1 \cap P_2 = \emptyset$ and $T_1 \cap T_2 = \emptyset$, then the *union of the systems* $(\mathcal{N}_1, M_1) \cup (\mathcal{N}_2, M_2)$ is defined as follows:

$$(\mathcal{N}_1, M_1) \cup (\mathcal{N}_2, M_2) := ((P_1 \cup P_2, T_1 \cup T_2, F_1 \cup F_2, L_1 \cup L_2), M)$$

with the function M defined as follows:

$$M(p) := \begin{cases} M_1(p), & \text{if } p \in P_1 \\ M_2(p), & \text{if } p \in P_2. \end{cases}$$

Definition 10 (Reachability Graph)

Let $\mathcal{N} = (P, T, F, L)$ be a p/t-net and M_0 a marking of $\mathcal{N}$. The *reachability graph* $RG(\mathcal{N}, M_0) = (V, E)$ of $(\mathcal{N}, M_0)$ is a directed graph with

$$V = [M_0\rangle \quad \text{as set of nodes}$$
$$E = \{(M, t, M') | \; M \in [M_0\rangle \wedge t \in T \wedge M[t\rangle M'\}$$

as set of labelled arcs.

$$L(E) = \{t \in T | \exists (M, t, M') \in E\}$$

is the set of labels. Usually the labels (M, t, M') are abbreviated by t.

Remark 10. The reachability graph is used to visualize all reachable states and executable firing sequences (processes).

Definition 11 (Emergence)
Let $(\mathcal{N}, M_0) := ((P, T, F, L), M_0)$ be a p/t-system, with subsystems $(\mathcal{N}_1, M_1) := ((P_1, T_1, F_1, L_1), M_1)$ and $(\mathcal{N}_2, M_2) := ((P_2, T_2, F_2, L_2), M_2)$ such that

$$P_1 \cap P_2 = \emptyset, \quad P_1 \cup P_2 = P \quad \text{and}$$
$$T_1 \cap T_2 = \emptyset, \quad T_1 \cup T_2 = T.$$

Let further be Σ the set of all firing sequences in the system $(\mathcal{N}, M_0)$ starting at M_0 and Σ_{12} be the set of all firing sequences in the union of the systems $(\mathcal{N}_1, M_1)$ and $(\mathcal{N}_2, M_2)$ starting at M_0, too. If

$$\exists \sigma \in \Sigma_{12} \wedge \sigma \notin \Sigma$$

the set of firing sequences Σ in $(\mathcal{N}, M_0)$ differs from the set Σ_{12}. We call this difference *emergence*.

Remark 11. We speak about emergence, if the behavior of a system (that means the set of all processes) cannot be deduced from its subsystems. When a system is divided into subsystems the interdependencies between them get lost. Merging the subsystems does not retrieve them. This "new behavior" yields in a different reachability graph (see example in chapter 4.3).

Definition 12 (Preset, Postset)
Let $\mathcal{N} = (S, T, F)$ be a petri-net. For all $x \in P \cup T$ we call

$$^\bullet x := \{y | (y, x) \in F\} \quad \text{the preset of } x \text{ and}$$
$$x^\bullet := \{y | (x, y) \in F\} \quad \text{the postset of } x.$$

Definition 13 (Boundary of a System)
Let $(\mathcal{N}, M_0)$ be a system with a subsystem $(\mathcal{N}', M_0')$. The *boundary* $B((\mathcal{N}', M_0'))$ of the system $(\mathcal{N}', M_0')$ is defined as follows:

$$B((\mathcal{N}', M_0')) := (P_b, T_b), \quad \text{with}$$
$$P_b := \{p' \in P' | \exists t \in T \backslash T' \colon p' \in {}^\bullet t \vee p' \in t^\bullet\},$$
$$T_b := \{t' \in T' | \exists p \in P \backslash P' \colon p \in {}^\bullet t' \vee p \in t'^\bullet\}.$$

Remark 12. So, the boundary of a (sub-)system consists of all possible states (here P_b) and relations or functions (here T_b) that belong to that system and are directly influenced by (or dependent from) the system's environment (see example in chapter 4.2).

4   Application

In this chapter we apply the concepts introduced in chapter 2 and 3 to a transportation system.

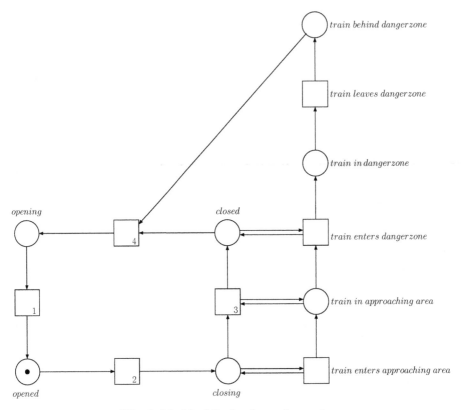

Fig. 1. Model of the level-crossing system.

Our model of a level-crossing (see figure 1) consists of a gate-model with the set of local states {*opened, closing, closed, opening*} and of a train-model with the set of states {*train in approaching area, train in dangerzone, train behind dangerzone*}. To allow a train to enter the approaching area, it's necessary that the gate is in the state "closing". A train in the approaching area enables the gate to get closed. The gate itself has to be in this state, to admit the train to enter the dangerzone. Until the train is behind the dangerzone, the gate stays closed and can then change to state "opening". In the initial state, the gate is open.

Note that the only intended purpose of this model is to apply the concepts introduced so far – the model therefore was kept very easy.

4.1   System, Subsystem, Process and Organisation

As mentioned above, our *system* consists of the *subsystems* "gate" and "train (see figure 2). In the global system, these subsystems are related (or ordered) in a special way – for example: The train only may enter the approaching area, if the precondition "gate is in state closing" is fulfilled – this relation is specified

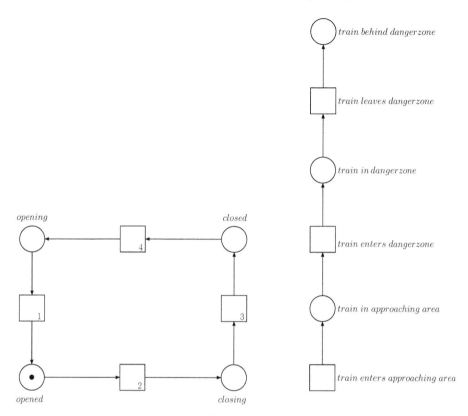

Fig. 2. Model of the gate (left) and model of the train (right).

in the model with the arrows *(closing, train enters approaching area)*. Due to this relations between the system's elements, certain *processes* are enabled: The operation "train enters approaching area" has to take place before transition t_3 fires and the gate gets close. The operations "train enters dangerzone" and "train leaves dangerzone" have to take place before transition t_4 fires and the gate can change to state "opening". So, the special *organisation* of the two subsystems lead to certain system-processes.

4.2   Environment and Boundary

The only *environment* of the subsystem "gate" taken into account is the system of the train (in our model, only the behavior of the gate is modeled, its physical parts are irrelevant here). The *boundary* B_g of the gate is defined as follows $B_g := (\{closing, closed\}, \{3, 4\})$. These are the only relations/operations and local states of the subsystem "gate" that are affected by or that affect its environment.

4.3   Emergence

Assume that a system consists of the two subsystems "gate" and "train", but this time these subsystems are completely independet (one can regard the two

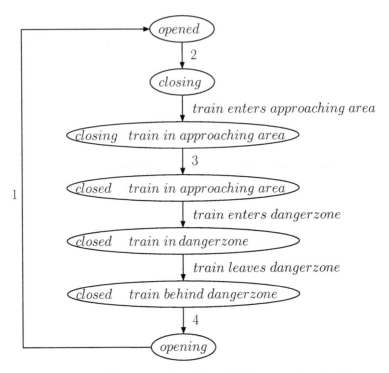

Fig. 3. Reachability Graph of the global system (see fig. 1).

systems in figure 2 as one global system). As the four local states of the gate
are independent reachable from the three states of the train, there are twelve
global states reachable (the number of states in the cross-product). As it appears
in the reachability graph of our level-crossing-system (see fig. 3) there are only
seven global states reachable. So, without looking at the processes, it's obvious,
that there exist processes in the union of the two subsystems, that are not executable in the level-crossing-system. Due to new restrictions (on account of new
relations between the two subsystems) the level-crossing system fulfils certain
properties – for example: When the train is in dangerzone, the gate is closed.
This condition can't be deduced neither from the subsystem "gate" nor from the
subsystem "train", but with the interdependencies between these two systems
and is therefore called an *emergent behavior*.

5   Conclusion

In this paper some of the basic concepts of system theory (as "system", "boundary of a system" and "emergence") were presented. After the informal introduction of these in chapter 2, they were defined in a formal way in chapter 3 with
the help of petri-nets. In doing so, we kept in mind the correspondence between
the informal and the formal specification. In the last chapter, an example from
the transportation was used to exemplify the presented concepts.

We hope that this article leads to the insight, that an abstract but formal definition of the central concepts of system theory may have a promising advantage: the abstract but unique and formal definition may serve as a kind of template to be instanciated with domain specific terms in each branch of science.

References

1. David J. Krieger. *Einführung in die allgemeine Systemtheorie, 2. Auflage.* Wilhelm Fink Verlag, UTB für Wissenschaft: Uni-Taschenbücher; 1904, München, 1996.
2. Niklas Luhmann. *Soziale Systeme. Grundriß einer allgemeinen Theorie.* Frankfurt am Main, 1984.
3. C. A. Petri. *Kommunikation mit Automaten.* Schriften des Institutes für instrumentelle Mathematik, Bonn, 1962.
4. E. Schnieder. *Prozeßinformatik - Einführung mit Petrinetzen.* Vieweg-Verlag, Braunschweig, 2. erweiterte Auflage, 1993.
5. Ludwig von Bertalanffy. *General System Theory: Foundations, Development, Applications.* George Braziller, 1976.

Author Index

Lecture Notes in Computer Science

For information about Vols. 1–3291

please contact your bookseller or Springer